Foundations of Interconnect and Microstrip Design

Foundations of Interconnect and Microstrip Design

Third Edition

T. C. Edwards
Engalco, UK

M. B. Steer
North Carolina State University, USA
University of Leeds, UK

JOHN WILEY & SONS, LTD

Chichester • New York • Weinheim • Brisbane • Singapore • Toronto

John Wiley & Sons, Ltd
Baffins Lane, Chichester,
West Sussex, PO19 1UD, England

National 01243 779777
International (+44) 1243 779777
e-mail (for orders and customer service enquiries): cs-books@wiley.co.uk
Visit our Home Page on http://www.wiley.co.uk or http://www.wiley.com

Other Wiley Editorial Offices

John Wiley & Sons, Inc., 605 Third Avenue,
New York, NY 10158-0012, USA

WILEY-VCH Verlag GmbH
Pappelallee 3, D-69469 Weinheim, Germany

Jacaranda Wiley Ltd, 33 Park Road, Milton,
Queensland 4064, Australia

John Wiley & Sons (Canada) Ltd, 22 Worcester Road
Rexdale, Ontario, M9W 1L1, Canada

John Wiley & Sons (Asia) Pte Ltd, 2 Clementi Loop #02-01,
Jin Xing Distripark, Singapore 129809

Library in Congress Cataloging-in-Publication Data

Edwards, T. C. (Terence Charles)
Foundations of interconnect and microstrip design / T.C. Edwards, M.B. Steer. —3rd ed.
p.cm.
Rev. ed. of: Foundations for microstrip circuit design. 2nd ed. c1991.
Includes bibliographical references and index.
ISBN 0-471-60701-0 (cloth : alk. paper)
1. Microwave integrated circuits. 2. Strip transmission lines. I. Steer, M. B. II. Edwards, T. C. (Terence Charles). Foundations for microstrip circuit design. III. Title.

TK7876 .E35 2000
621.381'32—dc21 00-043468

British Library Cataloguing in Publication Data

A catalogue record for this book is available from the British Library

ISBN 0-471-60701-0

Image of chip on cover was supplied by International Business Machines Corporation.

Produced from PostScript files supplied by the authors.
Printed and bound by Antony Rowe Ltd, Eastbourne
This book is printed on acid-free paper responsibly manufactured from sustainable forestry, in which at least two trees are planted for each one used for paper production.

To our wives

Patricia

and

Mary

Contents

Preface

Interconnects have achieved a prominent position in determining the performance of high-speed digital, RF and microwave circuits. In digital circuits, interconnect delay exceeds that of individual gates and is the primary determinant of clock speed. In RF and microwave circuits, interconnects and passive elements defined using them are critical circuit components.

This design text is both a sequel and an update to the original well-received first and second editions. The expanded text provides foundations for the accurate design of microstrip components and of circuits applicable to microwave, millimetre-wave and high-speed digital sub-systems.

The text is primarily intended for design engineers and research and development specialists who are active in these areas. It has been our attempt to show the commonalities in the design of interconnects in high-speed digital, RF and microwave applications. This is done by showing the common principles of signal transmission. It is also likely to prove useful to instructors and students in advanced undergraduate and graduate electronics and computer engineering courses.

The direction is strongly toward explaining the fundamentals of operation, and towards useful design formulas and approaches — a repeat coverage of well-documented analyses of microstrip structures has been considered unnecessary and out of place here, but is fully cited.

The work is partly based on research and teaching extending over two decades. Microwave and interconnect courses were presented at La Trobe University (Melbourne, Australia), the University of Bradford (Great Britain), North Carolina State University (Raleigh, North Carolina, U.S.A.), and the University of Leeds (Great Britain). The work is also based on short courses on the signal integrity of and interconnect design for high-speed digital circuits. The majority of the research forming the basis of important sections of this book was undertaken at North Carolina State University and at the Royal Military College of Science (Shrivenham, England).

The text is organized into eleven chapters, leading from the physical principles of signal transmission on interconnects, through the fundamental aspects of interconnect and microstrip design, on to circuit applications in RF, microwave, millimetre-wave and high-speed digital circuits. Additional material, including colour figures, are available at the website for this book http://www.wiley.co.uk/commstech.edwards.html or htp://www4.ncsu.edu/ mbs/foundations.html .

The design of high-speed interconnects for digital circuits and of RF and microwave transmission lines has significant common elements, but also significant differences. There are common underlying physical principles, and throughout the text this is stressed. The successful design of the highest performance digital interconnects, for example a clock distribution net, requires considerable transmission line knowledge.

Generally, in treatments in papers and chapters of relevant books 'just enough' transmission line theory and technology is presented. Not all of the options are covered. Our approach has been to provide the digital interconnect designer with a comprehensive treatment beginning with physical principles in Chapter 1, as well as more pragmatic approaches in Chapter 2, answering such questions as 'when are inductive effects important?' However, the interconnect treatment provides the digital designer with the tools for interconnect design now and in the future. The final chapter considers a number of clock distribution designs and, drawing from the material presented throughout the book, illustrates the importance of transmission line knowledge in the design of the highest performance interconnect. In contrast to how this material is often presented to digital designers, we contend that providing just enough knowledge is not enough to develop advanced and competitive interconnect designs.

This book provides a solid basis for RF, microwave and millimetre-wave design. The material enables the designer to make technology choices, and provides insight that supports the early stages of design. The many examples in the book show how these technology choices are made.

A basic review of interconnects and of TEM-mode transmission line theory is presented in Chapter 1. This is intended to provide the fundamentals for concepts and expressions used in many later chapters. Chapter 2 addresses the unique aspects of interconnects in high-speed digital interconnects. Chapter 3 considers interconnect technology, including interconnect and transmission line structures and the effect of substrate and metallization. This chapter may be used as a source of initial interconnect technology decisions.

Chapters 4 through 8 consider specific transmission line structures and interconnect discontinuities. Considerable insight is provided by using current and charge profiles of the various structures.

The text is also intended to be used in short courses and in graduate level courses. Chapter 9 considers power and current handling capability, transitions between different transmission line structures, and measurement techniques. Design studies are considered in the last two chapters of the book, with Chapter 10 looking at applications of passive interconnects to realize circuit functionality such as filters and lumped element components. The realization of lumped element components is particularly important in analog and RF integrated circuits, and Chapter 11 addresses the use of interconnect design technology in active circuits. For example, in Chapter 11 transmission line principles are used in the development of a digital clock architecture capable of supporting clocking at ten gigahertz or more.

Through these chapters the book presents a unifying foundation for the design of interconnects and microstrips. It then shows application of these lines in a variety of passive and active digital, analog, RF and microwave circuits.

T. C. Edwards and M. B. Steer
May 2000

Acknowledgements

During the preparation of all editions of this text there were a number of people who made significant contributions and thanks to whom the quality of the material was decidedly enhanced. Most of these were at La Trobe University (Melbourne, Australia), the University of Bradford (England), North Carolina State University (NCSU, Raleigh, North Carolina, U.S.A.), the Royal Military College of Science (RMCS, Shrivenham, England) and the University of Leeds (Great Britain). The authors wish to particularly thank Dr. Ashwain Rayit whose dissertation on co-planar waveguides was used as the basis for some of the material used in Chapter 6. La Trobe, Bradford, Leeds and NCSU universities were indeed fortunate with their quality of undergraduate, graduate and postgraduate students who were responsible for guiding the development of the text, and both learnt from and contributed to the eventual form of the text.

Terry wishes to thank several of his former colleagues at RMCS, particularly at what is now known as the School of Electrical Engineering and Science: past (Honorary) Professors Michael Potok, Chris Harris and Jim James for their advice and stimulation during the preparation of the original manuscript. Dr. Ann Henderson also read parts of the original text, and her suggestions proved extremely helpful. At La Trobe University, thanks are extended to the late Professor Daryl Hooper who exhibited infectious drive and enthusiasm. Also to Dr. Laurie Cahill for many deep and helpful ideas. Dr. Roger Owens was of great assistance with his critical reading of the original manuscript, and made many useful suggestions for the second edition.

Michael wishes to express gratitude to several of his former graduate students whose research led to many of the results presented here, as well as to design insight. In particular, he wishes to thank Dr. Mark Basel, Mr. Jeffrey Kasten, Mr. Steven Golberg, Mr. Steven Lipa, Mr. Alan Glasser, Mr. Bari Biswas, Mr. Meta Ozkar and Mr. Carlos Christoffersen. These individualas are now involved in leading edge design of high-speed digital circuits, RFICs, MMICs and in the computer aided engineering industry. Michael also wishes to express gratitude to his colleague and friend Professor Paul Franzon, who undertook many joint interconnect research projects, and with whom he presents well attended interconnect short courses. The feedback received from the working engineers who took part in these short courses helped refine the way in which this material is presented to people with both rudimentary knowledge in electromagnetics as well as those on the leading edge of high-speed interconnect design. Thanks are also extended to Professor Franzon for critiqueing the material related to interconnects for high-speed digital circuits. This book was written when Michael was at the University of Leeds, Great Britain, during 1999 and 2000, and he wishes to thank Professor Roger Pollard for his academic leadership during this period.

The attention of the reader is drawn to those at John Wiley & Sons. Mr. Mark Hammond, the editor, and to Ms. Sarah Hinton, Ms. Sarah Corney, Ms. Valerie Simmons and Mr. Jeremy Thompson for their strong support, backing, and patience that led to the production of this text.

Finally, we want to express our gratitude to the forbearence and support of our families during the preparation of this text. Terry thanks his wife Patricia and Michael thanks his wife Mary and his children Cormac, Fiona and Killian.

Terry Edwards and Michael Steer

1

Fundamentals of Signal Transmission on Interconnects

1.1 INTERCONNECT AS PART OF A PACKAGING HIERARCHY

Electrical connections are part of a hierarchy that connects individual active devices at the lowest level to system-level connections at the highest [1–3]. One way of defining the interconnect hierarchy is as follows:

Level 1 Interconnect: Chips
Level 2 Interconnect: MultiChip Modules (MCM, Package)
Level 3 Interconnect: Printed Circuit Board[1], PCB (Package)
Level 4 Interconnect: Backplane (Package)
Level 5 Interconnect: Rack, Connect Systems (Package)

The size of interconnects varies with interconnect level with both the smallest interconnects, both in cross-section and in length, being at the chip level. The way they are designed and modelled also depends on the physical size of the connections, whether digital or analog circuitry is being interconnected, and on the clock or operating frequency.

Interconnect delay has become an important factor in the performance of high-speed digital circuits and interconnects, or transmission lines, have always been critical circuit components in RF and microwave circuits. The lower the level in the interconnect hierarchy generally the shorter the connection and the higher the performance in terms of delay to transmit a signal from one point to another. A high performance system realizes a greater proportion of interconnections at the lower levels than does a lower cost system with less critical performance requirements. Not every system has all of the levels listed above, and even some interconnect networks, especially at microwave frequencies, need not involve active devices. This book focuses on the design and modelling of interconnects, and also on fundamental understanding of signal propagation on interconnects. Today's high performance digital systems have many interconnect issues in common with microwave (generally 1 to 30 GHz) and millimetre wave (above 30 GHz) systems and it is appropriate to consider the interconnect issues as a continuum. For example, it will be seen in Chapter 2 that

---

[1] Printed Wiring Board (PWB) is the preferred usage in North America, while printed circuit board is used elsewhere

the long haul interconnects on a chip (comparable in length to the edge dimensions of the chip) are conceived of and designed with the same criteria used to realize millimetre-wave circuits.

A generalization of the interconnect hierarchy is that the higher levels of interconnect 'breakout' the lower levels, that is, smaller and lossier interconnects at the chip level to larger interconnects (with lower loss and more controlled characteristics) at the next higher level for longer haul connections. This breakout process is repeated as long as necessary. Every aspect of the interconnect path must be modelled and designed. The complete path can involve nearly every interconnect technology, but there is considerable physics that is in common.

Before proceeding we need to define the term 'net': a net is a group of interconnects that carry the same signal, generally from one source to one or more loads or receivers.

1.2 THE PHYSICAL BASIS OF INTERCONNECTS

Electrical interconnections are generally made using metallic conductors, although conductive paths can be made using doped semiconductors and superconductors. In this book we restrict ourselves to metallic interconnects. Metallic conductors, such as copper and aluminium, the most common types used, are crystals with positively charged ions locked into position in a regular lattice. The ion here is made up of the nucleus of an atom which is positively charged and a complement of electrons local to each atom which almost, but not quite, balance the positive charge. In a metallic crystal there are some 'free' electrons shared by several ions with the overall effect that the positive and negative charges are balanced and the free electrons can wander around the lattice. The wandering electrons travel in random directions at a speed which is a substantial portion of c, the ultimate speed of light. This speed can typically be $c/3$ at room temperature in copper. These electrons are moving randomly with no overall average movement, therefore they do not transmit information. When an electric force field, E, is applied, the electrons begin to accelerate in the opposite direction to E. This gives rise to an average movement of electrons in one direction and this movement of charge carriers, crossing a fixed position per unit of time, constitutes current I. At absolute zero temperature the lattice is motionless and the electrons can, essentially, move through the lattice unimpeded. With an applied E field (the voltage V divided by the distance over which E is applied) the electrons would eventually reach a speed only limited by speed-of-light considerations. As temperature increases the lattice begins to vibrate and the electrons start colliding with it. These collisions impede the movement of the electrons and this constitutes electrical resistance R. Therefore, the ultimate average speed of the electrons is limited, and thus so is the current. The higher the temperature the higher the electrical resistance and the lower the average speed or current. This is the underlying physical basis of Ohm's law ($V = IR$). The movement of charge, and the electric and magnetic fields that this establishes, is also the basis for information transfer in interconnects. Note, however, that signals can move much faster than the electrons themselves.

The movement of charge results in a magnetic field, and hence magnetic energy storage. The ability of a structure to store magnetic energy is described by its inductance L. Similarly the rearrangement of charge to produce localized net positive or negative charge results in electric field, and thus electric energy storage with the capacitance C indicating the amount of energy that can be stored. The ratio of the energy stored in the magnetic and electric forms ($\propto \sqrt{L/C}$) and the rate at which the energy can be moved ($\propto 1/\sqrt{LC}$) determine the characteristics of an interconnect.

Our first important point is that to transfer information energy must be transferred. Indeed in the quest to develop Integrated Circuits (ICs) that consume very little power it is common to talk about the amount of energy required to transfer a bit of information, or the amount of energy required to switch a transistor. To come to grips with the fundamentals of interconnects we must become comfortable with the notion of energy transfer, how it is transferred and how the transfer of energy is disrupted (that is, signal integrity degraded) by interrupting the quantities (e.g. the electric and magnetic field or the current path) transferring energy from one point to another.

1.2.1 WHAT AN INTERCONNECT IS AND HOW INFORMATION IS TRANSMITTED

An interconnect can take various forms, either delivering power to part of an electronic circuit or being the means by which information is transmitted from one point to one or more other points in a circuit. We are concerned with the design of interconnects to ensure reliable undistorted transmission of information. In low frequency analog and digital circuits interconnects can be viewed simply as wires and, provided that the wire has sufficiently low resistance and current carrying capability, the interconnect can be largely ignored. However, if the transmission must be over a considerable distance then the interconnect must be considered as part of a circuit.

The earliest fundamental understanding of signal transmission on interconnects was for the transmission of digital signals — telegraph signals. Initially signal integrity was poor and telegraphy could only be used over short distances. It was not until Oliver Heaviside in 1887 [4] developed the first understanding of signal transmission on interconnects that telegraphy could be used over more than short distances. The key ingredient that Heaviside brought to the fundamental understanding of interconnects was to consider the behaviour of the signal in the frequency domain. This is still the best way of developing a fundamental understanding of interconnects, even if they are used for purely digital signals.

The key determinant of whether an interconnect can be considered as an invisible connection is whether the signal anywhere along the interconnect has the same value at a particular instant. If the value of the signal (say voltage) varies along the line (at an instant), then it may be necessary to consider transmission line effects. Jumping ahead a little, a typical criterion used is that if the length of the interconnect is less than one twentieth of the wavelength of the highest frequency component of a signal, then transmission line effects can be safely ignored and the circuit can be modelled as a single RLC circuit. The actual threshold used, 1/20 th, 1/10 th or 1/5 th, is based on experience. For example, an interconnect on a silicon chip clocking at 2 GHz has an appreciable frequency component at 10 GHz. An interconnect reaches the 1/10 th threshold when it is 0.9 mm long. This is less than the dimensions of most chips which can be up to 1 cm on a side. Thus it takes a finite time for the variation of a voltage at one end of an interconnect to impact the voltage at the other end. The ultimate limit is determined by the speed of light, but this is reduced by the relative permittivity of the material in which the fields exist. The best term describing this physical phenomenon is retardation. In addition to retardation other properties of the interconnect must be considered, including its resistance, current carrying capability, interaction of a signal with those on other interconnects, and other effects to be discussed in this book.

At high clocking speeds, and at RF and microwave frequencies, retardation can be significant and an interconnect cannot be considered to be an instantaneous connection. The interconnect can have an appreciable impact on the operation of the circuit. As a result, it can be used as a circuit element in microwave and milimeter-wave circuits and even in quite sophisticated circuits; e.g. filters can be constructed by configuring interconnects in a prescribed manner.

1.3 THE PHYSICS, A GUIDED WAVE

A good deal of electrical and electronic engineering is based on manipulating the energy stored in an electromagnetic field which we generally consider as separate electric and magnetic fields. While it is hard to appreciate, since they are introduced at very early stages in electro-technology education, current and voltage concepts are abstractions — albeit very convenient ones. (Current is less abstract than voltage except at microwave frequencies, which will be seen when we look at current flow on interconnects.) The fundamental understanding of what an interconnect is, and how information is transmitted, requires that we consider electromagnetic fields. The insight required to manage signal integrity is greatly enhanced by understanding these. To be rigorous, information (energy) is transferred by the movement of the electromagnetic field and not by current, voltage or charge.

1.3.1 TRANSMISSION OF A PULSE

When a voltage pulse is applied to an interconnect an electric field in the interconnect is produced. We will use the coaxial line configuration shown in Figure 1.1(a) to illustrate this discussion. The positive voltage side of the pulse is applied to the centre conductor of the coaxial line and results in an electric field which is essentially directed from the centre conductor to the outer conductor, see Figure 1.1(b). However, the electric field will also be slightly directed along the line. The direction of the electric field is the direction in which positive charges would move if released into the field. The component of the field that is directed along the shortest path from the centre conductor to the outer conductor (in what is called the transverse plane) will be denoted E_T, and the component directed along the line denoted E_L. (The subscripts T and L denote transverse and longitudinal components respectively.) Thus while $E_L \ll E_T$, it is necessary to accelerate electrons on the conductors and so give rise to current flow to support electromagnetic fields. Introducing E_L presents some difficulties as there is no analytic way of solving for its value, and only recently developed electromagnetic field simulation programs using long computer simulation times are capable of modelling it. However, it is small and so usually we restrict ourselves to talking about the transverse voltage and field, but restricting ourselves to this alone misses out on essential physics. The transverse voltage, V_T, is given by E_T multiplied by the distance between the inner and outer conductors: $V_T = E_T(a - b)$. This is a good measure provided that the transverse dimensions are sufficiently small compared to a wavelength.

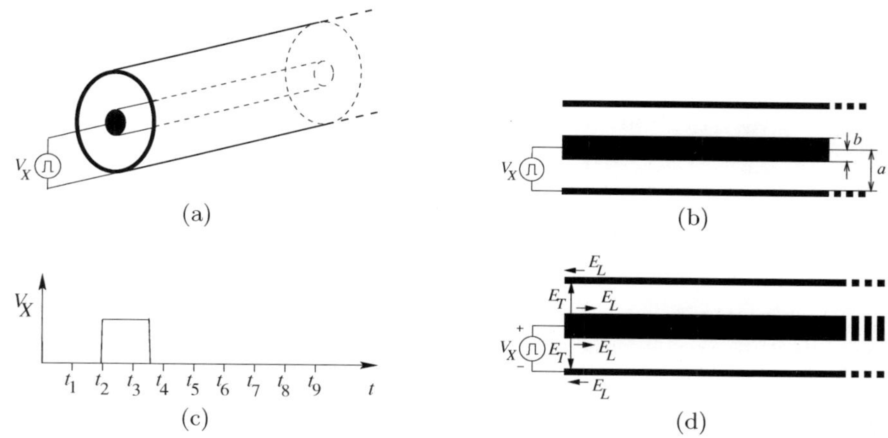

Figure 1.1 Coaxial transmission line with a voltage source applying a pulse: (a) three dimensional view; (b) its longitudinal cross-section; (c) the output of the pulse generator in time; and (d) the fields at the beginning of the line at time t_2.

Now let's get back to talking about E_L. E_L is small but is essential in producing a signal that moves along the line. The movement of a signal on the line is shown in Figure 1.2 at various times which are referred to those shown in Figure 1.1(c). At time t_1 the signal has not yet been initiated and at time t_2 the voltage pulse has just been applied to the line. The small longitudinal field, E_L, (exaggerated here) is directed as shown in Figure 1.1(d) and again in Figure 1.2. We know that it is in this direction as the voltage at the very beginning of the line is positive and the voltage along the line away from the start of the line is zero. There is only a longitudinal field very close to the beginning of the line initially as the influence of the electric field cannot travel faster than c, the speed of light in a vacuum. E_L causes electrons in the conductors to start moving in a direction opposite to that of the electric field. The electrons of course constitute current which is in the opposite direction to electron flow. The influence of the electric field moves along the line comparable to the speed of light and over time initiates current flow further down the line as shown in Figure 1.2(b). At time t_4 the voltage of the source drops to zero, as in Figure 1.2(c), and the influence of this also moves along the line at a speed near c. The electric field moving down the line under this influence is zero and so current drops to zero, see Figure 1.2(d) at time t_5. At all times displacement of the electrons 'unmasks' the positively charged ions of the lattice and so an electric field directed away from the ions towards the electrons is produced with the effect that E_L is opposed and the net longitudinal electric field is very small.

The signal that travels down the line is a packet of energy and thus of information. We can see energy in the pulse as moving along in two equivalent ways. The first, and physically correct, interpretation is to consider that the current gives rise to a magnetic field H. This field too is composed of a transverse component H_T and a very small longitudinal component H_L. The energy contained in the pulse is the integral over volume of the product of E and H (actually $\int_V E_T \times H_T.\mathrm{d}V$). The other way of calculating the energy in the pulse is to integrate the product of the current and voltage over the duration of the pulse, but this is only valid at low microwave frequencies and below in most instances where voltage and current can be easily determined.

The wave moves down the line at what is called the propagation velocity v_p, which is determined by the physical properties of the region between the conductors. The permittivity ε describes the energy storage capability associated with E, and the energy storage associated with H is described by the permeability μ. (Both ε and μ are properties of the medium — the material.) It has been determined that[2]

$$v_p = 1/\sqrt{\mu\varepsilon} \tag{1.1}$$

In a vacuum $\varepsilon = \varepsilon_0$, the 'free space' permittivity, and $\mu = \mu_0$, the 'free space' permeability. These are physical constants and have the values:

Permittivity of free space $\varepsilon_0 = 8.854 \times 10^{-12}$ F/m
Permeability of free space $\mu_0 = 4\pi \times 10^{-7}$ H/m

1.3.2 TRANSVERSE ELECTROMAGNETIC LINES (TEM-LINES)

In coaxial lines virtually all of the fields are in the plane transverse to the direction of propagation along the line, i.e. $E_T \gg E_L$ and $H_T \gg H_L$. Such lines are called Transverse Electromagnetic Lines or TEM.

Cross-sections of some common TEM-mode lines which have been fully analysed and are well understood are shown in Figure 1.3. The parallel line structure of Figure 1.3(a) is rarely used for carrying radio-frequency signals because, being unshielded, it radiates energy and is easily disturbed electrically by any nearby objects. Notice that both the parallel line and the coaxial line, shown in Figure 1.3(b), need only have dielectric supports at intervals along their lengths to maintain correct separation of the conductors. Furthermore, these supports may be filamentary, so that the 'dielectric filling' can consist almost entirely of air and the dielectric losses are low. On the other hand, the stripline, Figure 1.3(c), inherently requires a solid, low-loss, dielectric filling. The stripline is known as a planar interconnect as it can be built up using slabs of material either grown, in the case of a semiconductor, or sandwiched in the case of a PCB. The metallic connection is fabricated by etching (removing) material from a continuous sheet of metal.

Another common planar interconnect is the microstrip line shown in cross-section in Figure 1.4. This cross-section is typical of what would be found with a semiconductor or multichip module. In the deposited multichip module, or MCM-D, the metal layer is deposited primarily by sputtering or electroplating on a layer of dielectric applied in the liquid state by spinning. This is built up on a supporting substrate of ceramic or silicon. The interconnect is defined using a photolithographic process and etching. The line is finally covered by another layer of deposited dielectric to produce the conformal shape. Current flows in both the top and bottom conductor, but in the opposite direction. Electric field lines start on one of the conductors and finish on the other and are located almost entirely in the transverse plane. The magnetic field

[2] This is derived from a set of physical principles called Maxwell's equations; there is no underlying theory for these, but they have been verified by numerous experiments.

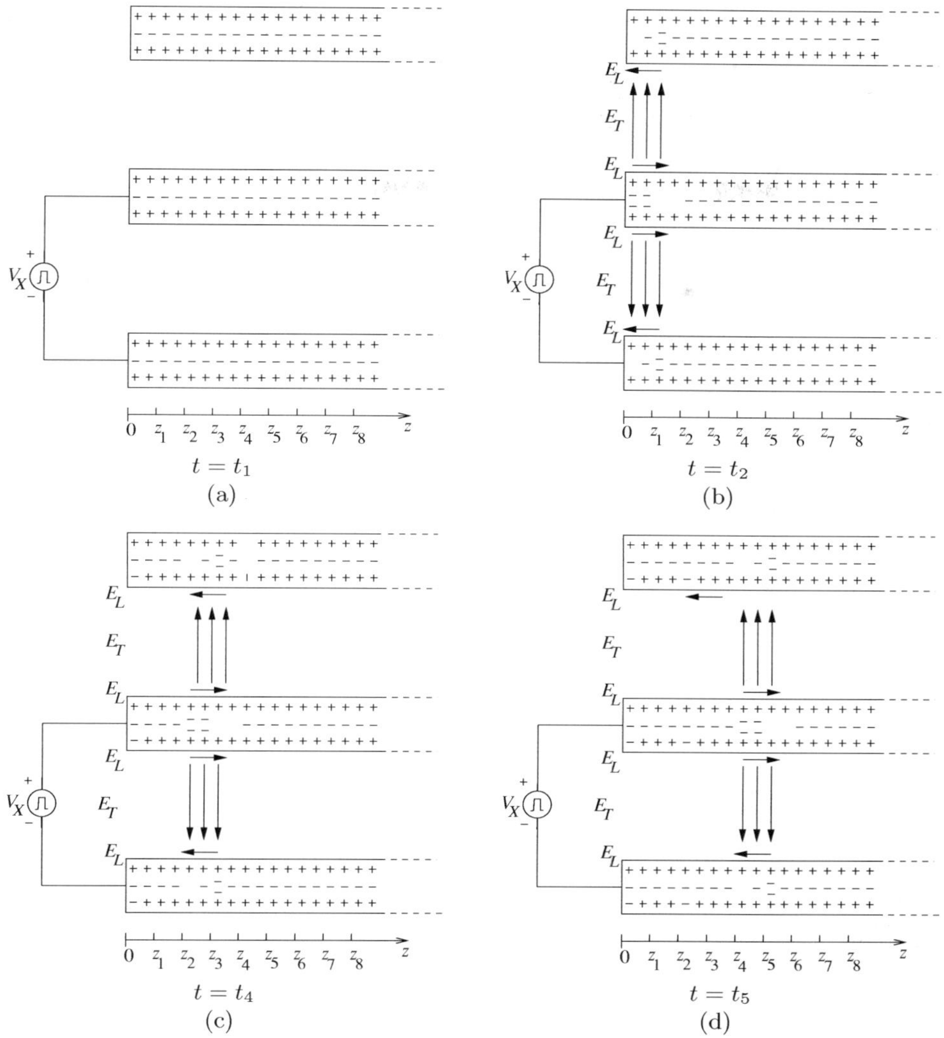

Figure 1.2 Coaxial transmission line with pulsed voltage source showing voltage at different times. The plus and minus signs indicate net charge.

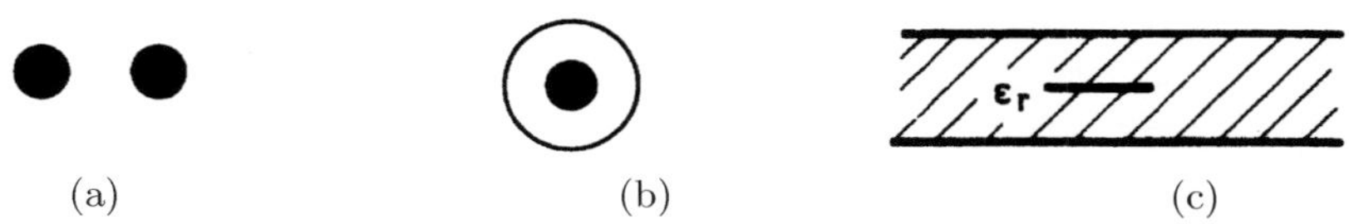

Figure 1.3 Cross-sections of some common TEM-mode transmission lines: (a) parallel two-wire transmission line; (b) coaxial line; and (c) stripline (triplate line).

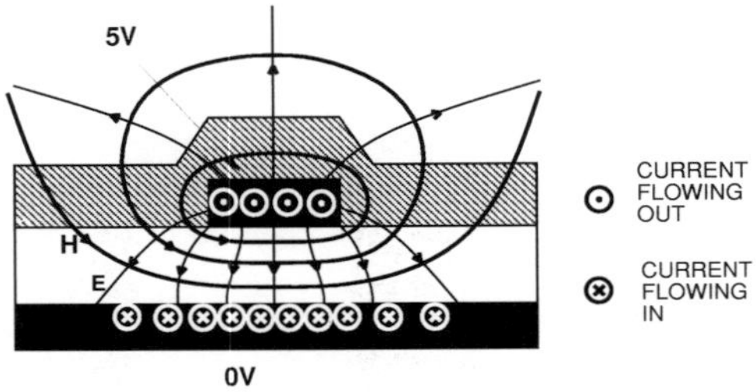

Figure 1.4 Cross-sectional view showing the electric and magnetic field lines and current flow for a microstrip interconnect. The electric and magnetic fields are in two media — the dielectric and air.

is also mostly confined to the transverse plane. This line is more commonly called a quasi-TEM line, as the longitudinal fields are not as negligible as with the coaxial line that we considered above. The relative level of the longitudinal fields increases with frequency, but below about 10 GHz, the line is still essentially TEM. Figure 1.4 illustrates an important point — current flows in the strip and a return current flows in what is normally regarded as the grounded conductor. More about this in Section 1.11 on page 25.

1.3.3 MULTIMODING

The TEM mode is one possible solution of Maxwell's equations that describe electromagnetic fields with conductors. However there are other solutions that can exist at high frequencies (when the conductors are $\lambda/2$ or more apart, or when there are more than two conductors, e.g. coupling). These situations result in two or more different field configurations being supported — each configuration being called a mode. This is usually an undesirable situation as the modes travel at different speeds. Thus the transverse dimensions of an interconnect must be kept small enough, and separations of adjacent interconnects sufficiently large, to ensure only one mode per interconnect. One of the rare exceptions is when coupling is desired as in some microwave and millimetre-wave circuits.

1.3.4 THE EFFECT OF DIELECTRIC

The presence of a material between the conductors alters the electrical characteristics of the interconnect. With a dielectric, the application of an electric field moves the centres of positive and negative charge at the atomic and molecular level. Moving the charge centres changes the amount of energy stored in the electric field — a process akin to storing energy in a stretched spring. The extra energy storage property is described by the relative permittivity ϵ_r, which is the ratio of the permittivity of the material[3] to that of free space:

$$\varepsilon = \varepsilon_r \varepsilon_0. \tag{1.2}$$

The relative permittivities of materials commonly used with interconnects range from 2.08 for Teflon™, used in high performance PCBs, 11.9 for silicon (Si), 3.8–4.2 for silicon dioxide (SiO_2), and 13 for gallium arsenide (GaAs) chips. (Values for other materials are given in Table 3.3.)

When the fields are in more than one medium (a nonhomogeneous transmission line), as for the microstrip line shown in Figure 1.4, the effective relative permittivity ε_{eff} is used. The characteristics of the nonhomogenous line are then, more or less, the same as for the same structure with a uniform dielectric of permittivity ε_{eff}. The ε_{eff} changes with frequency as the proportion of energy stored in the different regions changes. This effect is called dispersion and causes a pulse to spread out as the different frequency components of the pulse travel at different speeds. Dispersion is elaborated on more in Section 1.3.6 on page 11.

A similar effect on energy storage in the magnetic field occurs for a few materials. The magnetic properties of materials are due to the magnetic dipole moments resulting from alignment of electron spins — an intrinsic property of electrons. In most materials the electron spins occur in pairs with opposite signs with the result that there is no net magnetic moment. However, in magnetic materials some of the electron spins are not cancelled by matching pairs and there is a net magnetic moment. This net magnetic moment aligns itself with an applied H field and so provides a mechanism for additional storage of magnetic energy. The relative permeability μ_r describes this effect and

$$\mu = \mu_r \mu_0. \tag{1.3}$$

Nearly all materials used with interconnects have $\mu_r = 1$. One notable exception is nickel which has a high permeability, is a very convenient processing material, and often appears in packages.

1.3.5 FREQUENCY-DEPENDENT CHARGE DISTRIBUTION

Interconnects on chips, PCBs and backplanes are essentially planar like the stripline shown in Figure 1.3(c) and the microstrip line shown in Figure 1.4. Such lines have what at first seems an unusual frequency dependent behaviour. Consider the charge distribution for the microstrip line shown in Figure 1.5(a). The charge distribution shown here applies when there is a positive DC voltage on the strip (the top conductor). In this case there will be positive charges on the top conductor generally arranged in a fairly uniform distribution. The individual positive charges (caused by the absence of some balancing electrons exposing positively charged ions) do tend to repel each other, but this has little effect on the charge distribution for practical conductors with finite conductivity. (If the conductor has zero resistance then these charges would be confined to the surface of the conductor.) The bottom conductor is known as the ground plane and there are balancing negative charges, or surplus of electrons, so that that electric field lines begin on the positive charges and terminate on the negative charges. The negative charges on the ground plane are uniformly

---

[3] From here on we will just use the term permittivity to refer to relative permittivity.

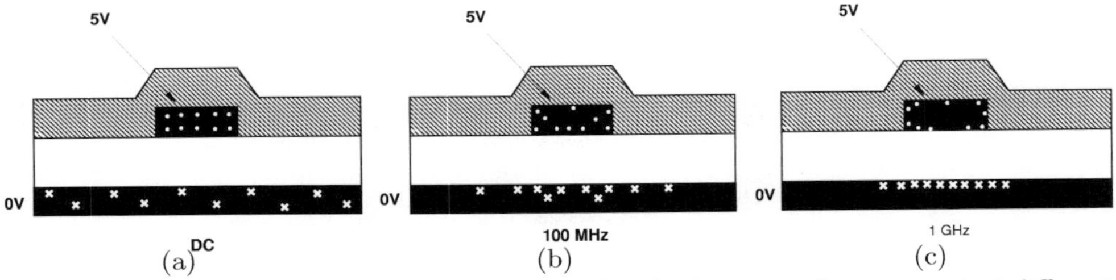

Figure 1.5 Cross-sectional view of the charge distribution on an interconnect at different frequencies. The dots and crosses indicate charge concentrations of different polarity.

distributed over the whole of the ground plane. An important point is that where there are unbalanced, or net, charges there can be current flow. So the charge distribution at DC, shown in Figure 1.5(a), indicates that for the top conductor there will be current flow uniformly distributed throughout the top conductor, and the return current in the ground plane will be distributed over the whole of the ground plane.

As the frequency of the signal increases, that is a sinusoidally varying voltage is applied between the top and bottom conductor, the charges do not have sufficent time to rearrange and the charges do not spread out uniformly. Also, the electromagnetic fields, now being time-varying, are not able to penetrate the conductors as much as the frequency increases — this effect is known as skin effect. This is shown in Figure 1.5(b) at 100 MHz. The situation is more extreme as the frequency continues to increase, e.g. to 1 GHz as in Figure 1.5(c). There are several important consequences of this. On the top conductor the negative charges are not uniformly distributed with respect to the depth of penetration into the conductor. Consequently, current flow is mostly concentrated near the surface of the conductor and the effective cross-sectional area of the conductor, as far as the current is concerned, is less. Thus, the resistance of the top conductor increases. A more dramatic situation exists for the charge distribution in the ground plane. Now the charge is not uniformly distributed over the whole of the ground plane, but instead becomes more concentrated under the strip. Thus the resistance of the ground plane increases as again the effective cross-sectional area has reduced from something that could have been quite large. The electric field lines shift as a result of the different distributions of charge, with more of the electric energy being in the dielectric.

At high frequencies the fundamental result of the field rearrangement is that the capacitance of the line increases, but this change can be quite small — typically less than 10% over the range of DC to 100 GHz. (This effect is described by the frequency dependence of the effective permittivity of the transmission line which is first defined in Section 4.5.2 on page 94.) The redistribution of the current results in a change of the inductance with frequency. This is principally because, for the same current, magnetic energy is stored inside as well as outside the conductors at low frequencies. As frequency increases the magnetic field becomes confined to the region outside the conductors and the inductance remains approximately constant. At intermediate frequencies the inductance reduces with frequency. Only above 1 GHz or so can the inductance be treated as approximately constant for the interconnect dimensions we are concerned with. The most significant frequency dependent effect is on the resistance of the line which increases indefinitely approximately as the square

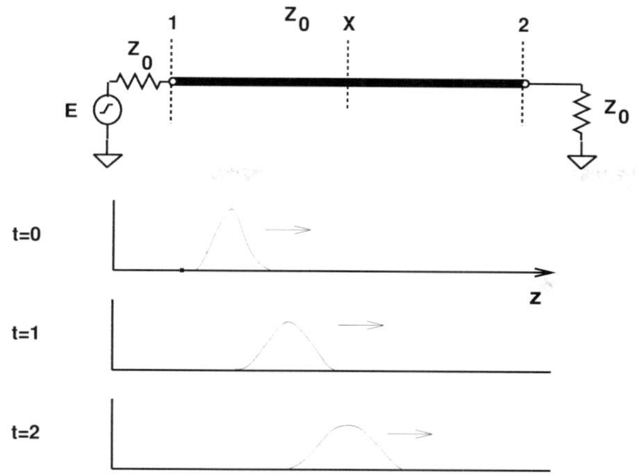

Figure 1.6 Dispersion of a pulse along an interconnect.

root of frequency.

1.3.6 DISPERSION

Dispersion is principally the result of the velocity of the different frequency components of a signal being different. To a lesser extent, dispersion is also the result of other parameters changing with frequency, such as an interconnect's resistance. The effect of dispersion can be seen in Figure 1.6 for a pulse travelling along an interconnect. The pulse spreads out as the different frequency components travel at different speeds along the line. It becomes harder to recognize the pulse, and in a digital system the level of the pulse could become less than that required to trigger a receive circuit. You can imagine a long telegraph line with this problem — successive pulses would start merging and the signal would become unintelligible.

Different interconnect technologies have different dispersion characteristics. For example, with the microstrip line of Figure 1.4 the effective permittivity changes with frequency as the proportion of the electromagnetic energy in the air-region to that in the dielectric regions changes. Dispersion is reduced if the fields are localized and cannot change orientation with frequency. This is the case with coplanar interconnects considered in Chapter 6 — in particular, coplanar waveguide (CPW) and coplanar strip (CPS) lines which have lower dispersion characteristics than does microstrip (for small geometries). The stripline of Figure 1.3(c) also has low dispersion as the fields are confined in one medium and the effective permittivity is just the permittivity of the medium. Thus there are interconnect choices that can be made that affect the integrity of a signal being transmitted.

1.4 WHEN AN INTERCONNECT SHOULD BE TREATED AS A TRANSMISSION LINE

If the level of a signal is reasonably constant along the entire length of an interconnect, then it can safely be treated as a lumped element and need not be treated as a transmission line. If the signal is a sinusoid then it is generally agreed that the signal does not change much in time over an interval equal to one twentieth of the period, T, of the signal. Consider a sinusoidally varying signal $x(t) = A\sin(\omega t)$, where $\omega = 2\pi f$ is the radian frequency of the signal and $f = 1/T$ is its frequency so that $x(t) = A\sin(2\pi t/T)$. The maximum change occurs when the signal goes from a time $t = -\alpha T/2$ to $t = +\alpha T/2$ where α is the fraction of the period. So in one twentieth of the period ($\alpha = 0.05$) the signal can change by 16% of its maximum possible change in value. So when the interconnect line length is less than 1/20 th of the wavelength, λ, of the signal, it is regarded as safe to use a circuit model of resistors and capacitors for the interconnect. Specifying a length in terms of a fraction of a wavelength is the same as using the signal duration in time as a fraction of the period. (In 1/10 th of the period it can change by 31%, and in 1/5 th of the period it can change by 59%). With digital signals the time of flight delay t_f is compared to the rise time t_r of the signal in determining whether or not a transmission line model is necessary. The rise time t_r is defined as the time required for the signal to change from 10% to 90% of its final value. The time of flight delay

$$t_f = l/\nu \tag{1.4}$$

where l is the interconnect length and ν is the propagation speed. The general guideline is that transmission line modelling is necessary when

$$t_r < 2.5t_f \tag{1.5}$$

as then the signal has changed by more than 40% of its value. It is generally regarded that the line can be modelled by capacitors and resistors when

$$t_r > 5t_f \tag{1.6}$$

The criteria are summarized in Table 1.1.

When a transmission line model is necessary, either a Spice-compatible distributed RLC model is used or else a full transmission line model is needed. Which is chosen depends on the accuracy needed and also the capability of an available circuit simulation program. Note that the RLC model cannot fully capture the retardation phenomenon completely.

For both sinusoidal and pulsed signals there is a grey area in which it is not clear which type of model should be used. Only experience and the accuracy needed for a particular application are a guide as to which type of modelling should be used. It is definitely preferable to use a lumped element model if possible. The commonality between the two criteria comes from considering the harmonic content of the rising edge of a pulse. However, the relative levels of the harmonics to the total signal swing depend on the technology of the circuit used to produce the pulse (e.g. the characteristics of a CMOS driver). Typically about five harmonics are significant in

Table 1.1 Modelling criteria for sinusoidal and digital signals.

Signal Type	Model Required	
	Transmission Line, or RLC	Lumped R, C
Sinusoid	$l > \lambda/10$	$l < \lambda/20$
Digital Pulse	$t_r < 2.5t_f$	$t_r > 5t_f$

a rising pulse and the level of the fifth harmonic is about one third that of the total signal swing. This leads to the very approximate relationship

$$t_r \approx \beta \frac{3}{10} nT \tag{1.7}$$

where n is the index of the maximum significant harmonic, β is its level relative to the total pulse swing, T is the period of the fundamental component (so that nT is the period of the n th harmonic). The criterion for modelling an interconnect as lumped elements for a sinusoid, $l < \delta\lambda$, becomes, after using $\lambda = \nu/f = \nu T$,

$$l < \nu T/20. \tag{1.8}$$

Using Equation (1.4) the criterion can be expressed as

$$t_f < T/20 \tag{1.9}$$

Now using Equation (1.7) the criterion is

$$t_f < \frac{1}{20}\frac{10}{3}\frac{1}{\beta n} t_r. \tag{1.10}$$

With $n = 5$, and $\beta = 1/6$, the criterion from sinusoidal considerations becomes

$$t_r > 10t_f \tag{1.11}$$

This differs from the guideline generally used for digital signals by a factor of 2. This is because, in general, modelling accuracy is more critical with analog signals.

The above discussion of interconnects is pertinent to off-chip interconnects and to on-chip interconnects of sufficient size (cross-section) that the resistance of the interconnect is not very large. Consequently RF and microwave chips have relatively large interconnects of reasonably low resistance but the high compression of digital circuitry means that nearly all on-chip interconnects for digital ICs have high loss. This is because when the interconnect is very narrow (e.g. around a few microns in width or less) the resistance of the line dominates and the signal dissipates before transmission line lengths are significant. (In practice buffer circuits, possibly inverters, are used to boost the signal level along a long line.) So narrow on-chip interconnects can nearly always be modelled by lumped element circuits. In addition, with digital circuits the close packing of the interconnects on-chip, and thus coupling of the interconnects to each other, requires that computer-based model extraction be used to develop a large RLC-like circuit describing the interconnect network. A program that does this is called a *parasitic extractor*. We will talk more about parasitic extraction in Chapter 2.

Table 1.2 Comparison of critical length criteria for different digital technologies. The critical length criterion is $t_r = 2.5t_f$. (For Si $\varepsilon_r = 11.9$, SiO_2 $\varepsilon_r = 3.8$, GaAs $\varepsilon_r = 13$, FR-4 $\varepsilon_r = 4.4$, Alumina $\varepsilon_r = 9.8$, and Teflon™ $\varepsilon_r = 2.08$.)

Technology (Interconnect) Medium)	On-Chip		Off-Chip		
	Rise Time ps	Critical Length cm	Rise Time ps	Substrate	Critical Length cm
CMOS (SiO_2)	50–700	0.3–4.3	1000–4000	Alumina FR-4 Teflon	3.8–15 4.7–19 8.3–33
Bipolar Si (SiO_2)	25–100	0.15–0.6	100–400	Alumina FR-4 Teflon	0.1–1.5 0.12–1.9 0.8–3.3
MESFET, PHEMT, HBT (GaAs)	5–100	0.02–0.33	50–200	Alumina FR-4 Teflon	0.2–0.8 0.25–1.0 0.4–1.7

1.5 THE CONCEPT OF RADIO FREQUENCY TRANSMISSION LINES

A review of important TEM-mode radio-frequency transmission line theory is essential here, because interconnects with uniform cross-sections along their length can be treated, at least to a first approximation, by assuming the static-TEM situation. We shall see that many significant interconnect design calculations require TEM-mode results from transmission line theory.

When transmission line structures have dimensions which are substantially smaller than the wavelengths of the signals being transmitted, then the structures may be satisfactorily analysed on the basis of line voltages and currents. As the frequency goes higher, and therefore the wavelength becomes smaller, it becomes necessary to set up a complete electromagnetic field solution in order to analyse the line structures. Here we can often assume that small line dimensions prevail, and therefore a number of useful results are obtained on a voltage and current basis. The theory thus developed is called distributed circuit theory.

1.6 PRIMARY TRANSMISSION LINE CONSTANTS

In this section we are concerned only with uniform transmission lines. Regardless of the actual structure, such lines are shown schematically as in Figure 1.7(a). Primary constants are indicated in Figure 1.7(b) and can be defined as follows:

Resistance along the line	$= R$	all specified 'per loop metre.'
Inductance along the line	$= L$	
Conductance shunting the line	$= G$	
Capacitance shunting the line	$= C$	

R, L, G and C are also referred to as resistance, inductance, conductance and capacitance per unit length.

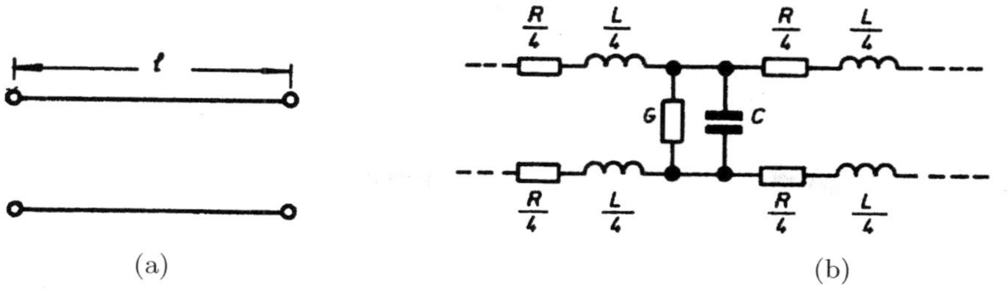

Figure 1.7 The uniform transmission line: (a) transmission line of length l; and (b) primary constants assigned to a lumped element model of a transmission line.

The 'per loop metre' specification means that each quantity is determined on a 'go and return' basis. For example, to find the total value of R per loop metre for a coaxial line we must add the resistance of one metre of the inner conductor to that of the outer conductor. The return signal path is just as important as the signal path in determining the parameters of an interconnect. In most radio-frequency transmission lines the effects due to L and C tend to dominate, because of the relatively high inductive reactance and capacitive susceptance, respectively. In such cases we refer to 'loss-free,' or lossless, lines although in practice some information about R or G may be necessary to determine actual power losses. The lossless concept is just a useful and good approximation. Note that the lossless approximation is not valid for narrow on-chip (digital) interconnections, as then resistance is very large. The propagation of a wave along the line is characterized by the (complex) propagation coefficient γ:

$$\gamma = \sqrt{(R + j\omega L)(G + j\omega C)} \tag{1.12}$$

or

$$\gamma = \alpha + j\beta \tag{1.13}$$

where

α = attenuation coefficient, in Nepers per metre
β = phase-change coefficient, in degrees, or radians, per metre

At sufficiently high radio frequencies, Equations (1.12) and (1.13) yield the important result:

$$\beta = \omega\sqrt{LC}. \tag{1.14}$$

1.7 SECONDARY CONSTANTS FOR TRANSMISSION LINES

In one complete wavelength λ_g along the line the travelling wave must experience 2π radians of phase shift. Thus

$$\beta\lambda_g = 2\pi \tag{1.15}$$

or

$$\beta = 2\pi/\lambda_g. \tag{1.16}$$

(Hence , β is also called the 'wave number.') From the relationship for the velocity $v_p = f\lambda_g$ we can also write the velocity of propagation as

$$v_p = \omega/\beta \ \text{m/s}. \tag{1.17}$$

By using Equation (1.14) we also obtain

$$v_p = 1/(\sqrt{LC}). \tag{1.18}$$

The velocity of propagation is also given in terms of the absolute permeability μ and permittivity ε of the medium through which the wave passes:

$$v_p = \frac{1}{\sqrt{\mu\varepsilon}} = \frac{c}{\sqrt{\mu_r\varepsilon_r}} \tag{1.19}$$

where $c = 2.99793 \times 10^8$ m/s, the velocity of light in free space — often approximated to $c = 3 \times 10^8$ m/s. In coaxial lines designed for millimetre-wave (above 30 GHz) applications the velocity is very close to c because there is little dielectric filling, only well-separated spacers. (This is not true for coaxial cables, whose dielectric fills the space between the conductors and reduces the velocity of a signal.)

Most lines do not possess any ferromagnetic materials, and thus have $\mu_r = 1$. For the stripline, Figure 1.3(c), the filling consists of a uniform dielectric, and Equation (1.19) then gives the velocity as

$$v_p = c/\sqrt{\varepsilon_r}. \tag{1.20}$$

We note that the wave is slowed by the dielectric medium.

By noting that $c = f\lambda_0$ and $v_p = f\lambda_g$, we can use Equation (1.20) to show that

$$\lambda_g = \lambda_0/\sqrt{\varepsilon_r}. \tag{1.21}$$

(The subscript 'g' means 'guide'; we can look upon the transmission line as a 'guiding' structure as the electromagnetic fields are guided along the interconnect.) This expression, Equation (1.21), is particularly significant. It explicitly states that the wavelength is reduced according to the square root of the relative permittivity of the material in which the line is embedded. For example, a coaxial transmission line is filled with alumina ($\varepsilon_r \approx 10$) and excited at 10 GHz. The free-space wavelength would be 3 cm, but the wavelength in the coaxial line is only $3/\sqrt{10} = 0.95$ cm. Thus, distributed components become shorter and occupy less space, if high-permittivity materials can be used. This has important implications for microstrip and all line structures.

The final secondary constant to be discussed is the characteristic impedance Z_0. This is generally given by

$$Z_0 = \sqrt{\frac{R + j\omega L}{G + j\omega C}} \tag{1.22}$$

At high radio frequencies this simplifies to

$$Z_0 = \sqrt{\frac{L}{C}} \tag{1.23}$$

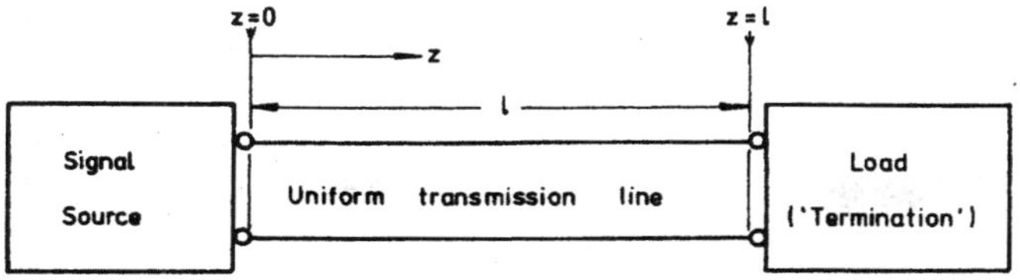

Figure 1.8 Notation applicable to a general, uniform line.

which is a purely resistive result. For coaxial lines the most common value is $Z_0 = 50$ Ω, although at some frequencies (e.g. some television cables) often $Z_0 = 75$ Ω as this corresponds to coaxial line dimensions that yield the lowest loss. The coaxial line dimensions corresponding to $Z_0 = 50$ Ω do not have much higher loss, but are easier to manufacture. On-chip it is preferable to have a higher characteristic impedance as this reduces the current drive requirements and I^2R (resistive) losses.

Changing Z_0 requires a change in the dimensions of the line. Specifically, in the case of coaxial lines, it requires changing the ratio of the radii applicable to the inner and outer conductors. Dimensional alterations are much more easily carried out with stripline or its derivatives, and therefore such structures can be fabricated for a wide range of characteristic impedances (typically 20 to 150 Ω for stripline). It is not often appreciated that Equation (1.23) can be re-expressed, with the aid of Equation (1.18), in the following alternative forms:

$$Z_0 = v_p L \tag{1.24}$$

or

$$Z_0 = 1/(v_p C). \tag{1.25}$$

Equation (1.25) is particularly useful in establishing some fundamental microstrip parameters.

1.8 TRANSMISSION LINE IMPEDANCES

The input impedance to a transmission line varies with the distance progressed along the line. Therefore, a suitable distance notation is required, which is shown in Figure 1.8. Many standard textbooks derive the expressions given as Equations (1.26) to (1.32) here.

By setting down the distance limits into expressions for the voltage and current at any point along the line we obtain, for the load impedance Z_L the following expression for the input impedance at $z = 0$:

$$Z_{\text{in}} = Z_0 \left(\frac{Z_L \cosh \gamma l + Z_0 \sinh \gamma l}{Z_o \cosh \gamma l + Z_L \sinh \gamma l} \right). \tag{1.26}$$

This expression is completely general. For lossless lines the result simplifies to

$$Z_{\text{in}} = Z_0 \left(\frac{Z_L + jZ_0 \tan \beta l}{Z_0 + jZ_L \tan \beta l} \right). \tag{1.27}$$

The input impedances of lines having short-circuit or open-circuit terminations (see Figure 1.9) follow from Equation (1.27), and these impedances can now be easily written down.

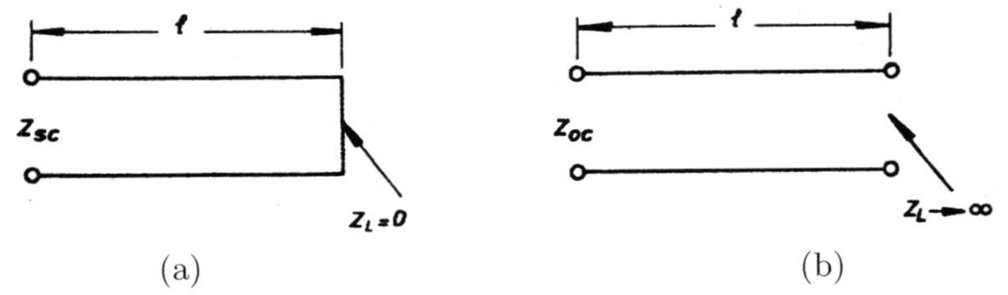

Figure 1.9 Line terminations: (a) short-circuit; and (b) open-circuit.

For the short-circuit case $Z_L = 0$ and Equation (1.27) gives

$$Z_{sc} = jZ_0 \tan \beta l. \tag{1.28}$$

For the open-circuit case $Z_L = \infty$ and Equation (1.27) gives

$$Z_{oc} = -jZ_0 \cot \beta l. \tag{1.29}$$

Short-circuit or open-circuit terminated lines are very useful structures in a wide variety of circuits.

By choosing the correct lengths (l) inductive or capacitive circuit elements are automatically realized in transmission line form, as can be seen from Equations (1.28) and (1.29). There are many applications in resonators, filters, and matching and coupling networks. When the line is terminated in a load impedance exactly equal to the characteristic impedance (Z_0) of the line itself, it is perfectly broadband matched. The denominator is equal to the numerator in the right-hand parentheses of Equation (1.26) and hence

$$Z_{\text{in}} = Z_0. \tag{1.30}$$

This completely matched condition is often used in test and measurement procedures.

1.9 REFLECTION

1.9.1 REFLECTION AND VOLTAGE STANDING-WAVE RATIO (VSWR)

In all cases, except the completely matched condition, the load termination reflects some of the energy originally sent down the line. Interference between the incident and reflected waves, travelling at the same velocity but in opposite directions, causes a 'standing-wave' field pattern to be set up.

The voltage reflection coefficient Γ (the ratio of reflected to incident voltage at the load) is given by

$$\Gamma = \frac{Z_L - Z_0}{Z_L + Z_0}. \tag{1.31}$$

The ratio of maximum to minimum amplitude of the standing wave is called the Voltage Standing-Wave Ratio (or 'VSWR'), given by

$$\text{VSWR} = \frac{1 + |\Gamma|}{1 - |\Gamma|}. \tag{1.32}$$

Notice that, in general, Γ may be complex but VSWR is necessarily always real. For the matched condition, $\Gamma = 0$ and VSWR=1. The reflection coefficients and standing-wave ratios for short-circuit and open-circuit terminated conditions are -1 and $+1$, respectively, and in both cases the VSWR is infinite.

1.9.2 FORWARD AND BACKWARD TRAVELLING PULSES

Forward and backward travelling pulses are shown in Figure 1.10, when the resistance at the end of the line is lower than the characteristic impedance of the line ($Z_L < Z_0$). The voltage source is a step voltage which is zero for times $t < 0$. At time $t = 0$ the step is applied to the line, and it begins travelling down the line as shown at time $t = 1$. At time $t = 2$ the leading edge of the step reaches the load, and as the load has lower resistance than the characteristic impedance of the line, the reflection coefficient Γ is negative, and the total voltage on the line, which is all we can directly observe, drops. A reflected, smaller and opposite step signal travels in the backward direction and adds to the forward travelling step to produce the waveform shown at $t = 3$. The impedance of the source is matched to the transmission line impedance so that the reflection at the source is zero. The signal on the line at time $t = 4$, the round trip propagation time of the line, therefore remains at the lower value. The easiest way to remember the polarity of the reflected pulse is to consider the situation with a short-circuit at the load. Then the total voltage on the line at the load end must be zero. The only way this can occur when a signal is incident is if the reflected signal is equal in magnitude but *opposite* in sign, in this case $\Gamma = -1$. So whenever $|Z_L| < |Z_0|$, the reflected pulse will tend to subtract from the incident pulse.

The opposite situation occurs when the resistance at the load end is higher than the characteristic impedance of the line, Figure 1.11. Then the reflected pulse has the same polarity as the incident signal. Again, to remember this, think of the open-circuited case. The voltage across the load does not need to be zero and indeed doubles as the reflected pulse has the same sign as well as magnitude as that of the incident signal, in this case $\Gamma = +1$.

A more illustrative situation is shown in Figure 1.12, where a signal like that of a pulse is incident on a load that has a resistance higher than that of the characteristic impedance of the line. The peaking of the voltage that results at the load is a typically the design objective in many long digital interconnects, as less overall signal energy needs to be transmitted down the line, or equivalently a lower current drive capability of the source is required to achieve first incidence switching. This is at the price of having reflected signals on the interconnects, but these can be dissipated through a combination of the interconnect and absorption of the reflected signal at the driver.

1.9.3 EFFECT ON SIGNAL INTEGRITY

Multiple reflections can affect the integrity of the signal in a digital system. A typical waveform resulting from multiple reflections is shown in Figure 1.13(a). One of the classic results of poorly defined matching of an interconnect, a driver and a receiver is the production of a metastate, as shown. The metastate often results from reflection at first incidence of a signal that has an initial level that is below that required for switching (the minimum logic 1 switching level is V_{IH}) This situation occurs when the

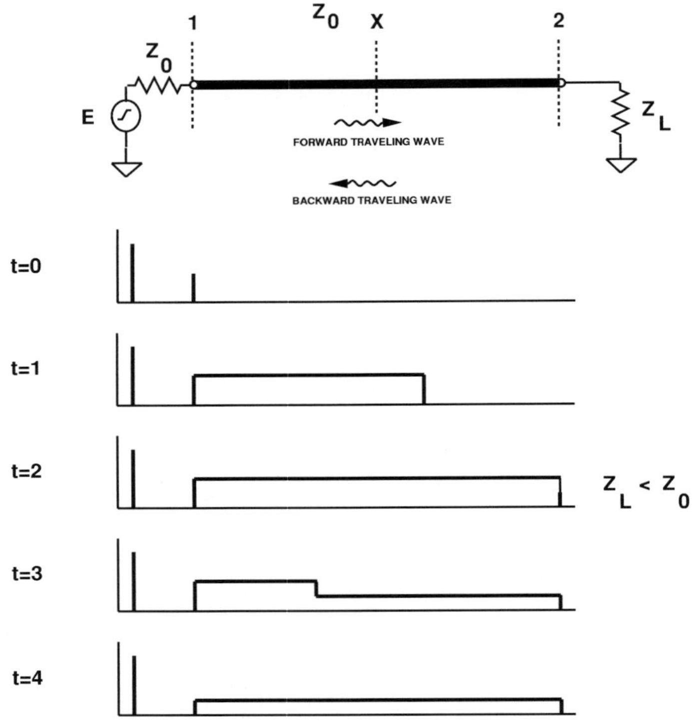

Figure 1.10 Reflection of a pulse at a load which is lower in resistance than the characteristic impedance of the line.

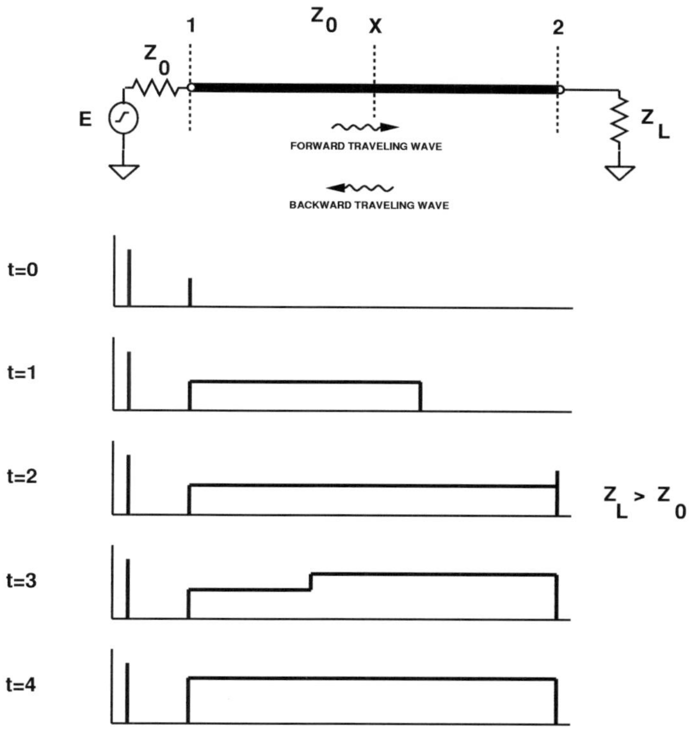

Figure 1.11 Reflection of a pulse at a load which is higher in resistance than the characteristic impedance of the line.

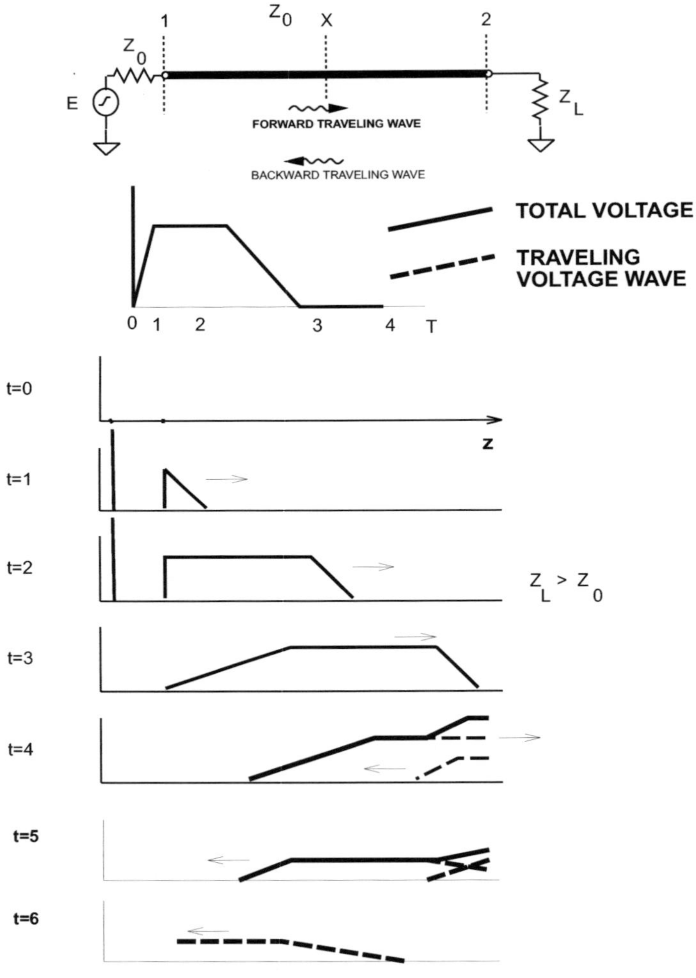

Figure 1.12 Reflection of a pulse on an interconnect showing forward and backward travelling pulses.

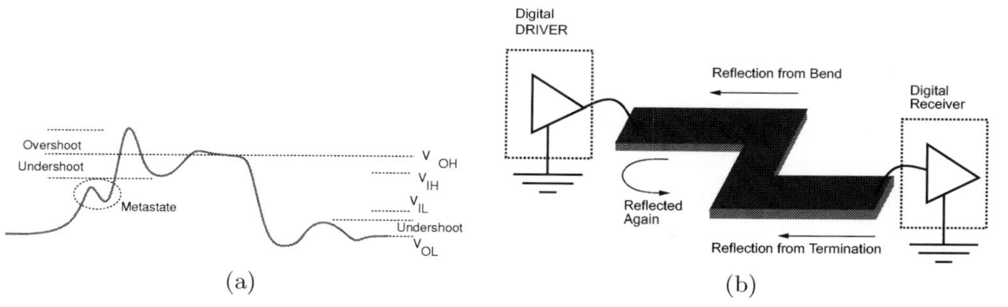

Figure 1.13 Waveform (a) resulting from multiple reflections such as those indicated in (b). (© 2000 P. Franzon, used with permission.)

current drive capability of the digital driver is lower than that required to establish the full output high voltage level V_{OH} on the line. This is a necessary compromise with most digital circuits in order to keep the transistors small. A disastrous situation can occur when the metastate crosses the switching threshold multiple times. If the circuitry is switched asynchronously (i.e. is not clocked) false switching may occur. Reflection noise can also result in ringing. Similar situations occur at the falling edge.

1.10 MULTIPLE CONDUCTORS

Any parallel-coupled pair of transmission lines, regardless of their practical realization, may be described by the four-port configuration indicated schematically in Figure 1.14. The dashed lines indicate mutual coupling; input and output connections and terminations have been omitted for simplicity.

In terms of y-parameters the characteristics of the coupled interconnects are described by

$$\begin{bmatrix} I_1 \\ I_2 \\ I_3 \\ I_4 \end{bmatrix} = \begin{bmatrix} y_{11} & y_{12} & y_{13} & y_{14} \\ y_{21} & y_{22} & y_{23} & y_{24} \\ y_{31} & y_{32} & y_{33} & y_{34} \\ y_{41} & y_{42} & y_{43} & y_{44} \end{bmatrix} \begin{bmatrix} V_1 \\ V_2 \\ V_3 \\ V_4 \end{bmatrix}. \tag{1.33}$$

Coupling is described by the terms y_{13} ($=y_{31}$), y_{14} ($=y_{41}$), y_{23} ($=y_{32}$) and y_{24} ($=y_{42}$). Coupling, however, is understood by considering the effect of overlapping electromagnetic fields produced by each of the coupled interconnects.

To examine the coupling modes which are established, we now concentrate on a pure TEM-mode type of structure. The simplest structure to consider is the dual, coaxial, coupled-line system, in which electric and magnetic fields couple through a narrow slot cut into the adjacent outer conductor along a certain length. Such a structure is shown in Figure 1.15. Notice that the consistent field patterns correspond to waves travelling in opposite directions along the respective lines, and it may be stated that any 'parallel'-coupled pair of transmission lines yield contra-directional travelling waves.

The coupled coaxial arrangement constitutes a pure TEM-mode system, assuming either a uniform dielectric filling or an air-filled structure. At any instant the relative polarities of the voltages, taken at any specific plane along the structure, will either be

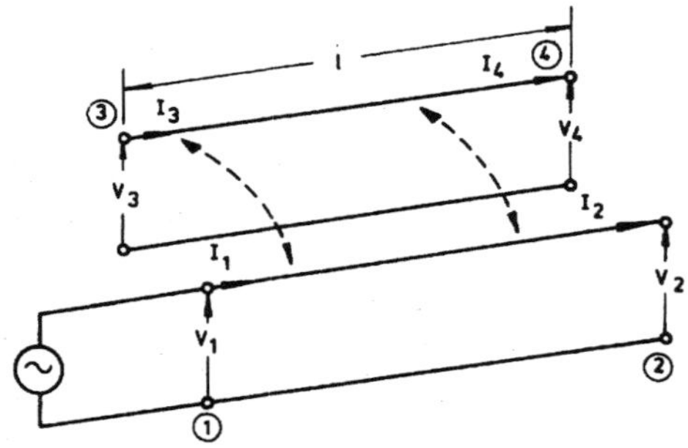

Figure 1.14 Two parallel coupled transmission lines.

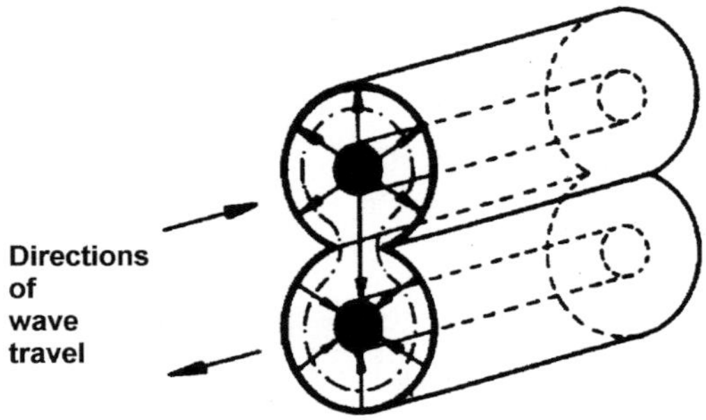

Figure 1.15 Parallel-coupled coaxial lines showing fields: — - — magnetic field; and —— electric field.

alike or opposite. We refer to different field configurations set up by such polarities as the even-mode and the odd-mode respectively. The diagram of Figure 1.16 illustrates these effects. The complete behaviour of the coupled structure can be obtained by superposition of the effects due to these two modes.

Due to the assumed uniform dielectric filling, any signal travelling in such a pure TEM-mode system will always travel at the same velocity, $v = c/\sqrt{\varepsilon_r}$. Thus the velocity associated with the even mode is identical to that associated with the odd mode here.

The equivalent primary constants for the coupled lines having even- or odd-modes, taken separately, must differ because of the different field distributions. As a result of this, two distinct characteristic impedances may be defined, one for each

Figure 1.16 Modes on a parallel-coupled coaxial lines: (a) even-mode; and (b) odd-mode. (For simplicity only the electric field is indicated.)

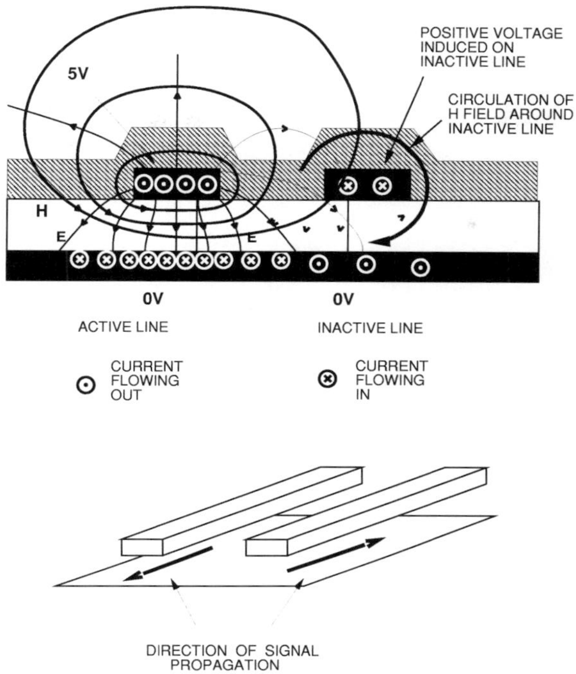

Figure 1.17 Coupled microstrip lines in perspective. (© 2000 P. Franzon, used with permission.)

mode, expressed in terms of the different primary constants by equations of the form of Equation (1.23). We term these impedances the even-mode and odd-mode characteristic impedances, denoted by Z_{0e} and Z_{0o}. This applies to any TEM-mode or quasi-TEM mode parallel-coupled structure, and the characteristic impedances are major parameters in design procedures.

In many planar interconnects the fields extend indefinitely and this is the case with the two parallel microstrip interconnects shown in Figure 1.17 where only the left line is driven. Part of the magnetic field on the driven line encompasses the quiet line on the right inducing current flow as shown. The electric field also extends from the driven interconnect and induces a voltage on the quiet (or victim) line. The result is that a reverse travelling wave is induced on the quiet line. The signal induced is called crosstalk noise, and the crosstalk resulting from many neighbouring nets can constitute a signal voltage that could result in false switching in a digital system. The type of crosstalk signal induced by a pulse travelling down a microstrip line is illustrated in Figure 1.18.

1.11 RETURN CURRENTS

We have already introduced the concept of return currents when we considered the current flowing on a microstrip line, see Figure 1.4. The physics is such that if there is a signal current there must be a return signal current which will tend to be as close to the signal current as possible to minimize stored energy. The provision of a

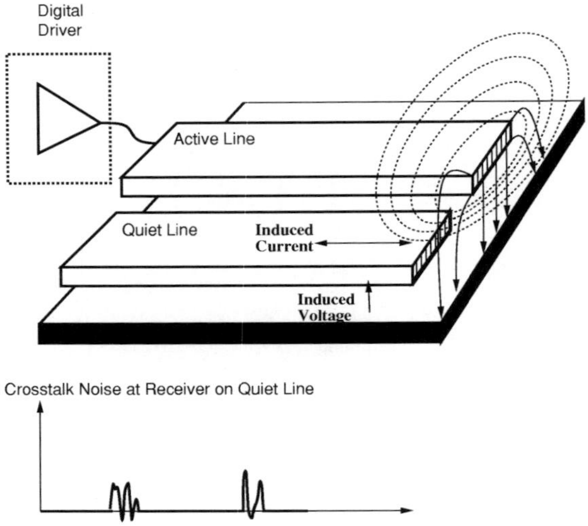

Figure 1.18 Coupled microstrip lines showing the signal induced on the quiet (or victim) line due a pulse on the driven line.

signal return path is important in maintaining the integrity (i.e. predictable signal waveform) of an interconnect. In most interconnects each interconnect is confined to its own conductor, but the return signal path is used by signals on many other conductors.

The return signal path for a complete system is shown in Figure 1.19. The most common signal integrity problem that is introduced at the design stage is not making provision for the signal return path. An example is the provision of a return (say, ground) pin remote from the signal pin at a chip-to-PWB interface. As well as slowing the signal down considerably this situation also increases the level of crosstalk. Of course it is not feasible to provide an individual signal return path for every net on a chip. However, the practice of providing return paths dedicated to two or just a few signal pins is becoming more common. Provision of individual return current paths is appearing for long haul nets on large digital chips, e.g. clock and data buses, which can be close to the chip-edge dimension in length. Commonly this is done using ground planes or using differential lines such as the coplanar strip lines considered in Chapter 6.

Silicon-based RFIC chips generally use differential signalling to overcome the limitations of MOS devices. The currents on each of the differential signal paths balance each other and so each provides the signal return path for the other. This design practice effectively eliminates RF currents that would occur on ground conductors. The off-chip RF interconnects interfacing the chip to the outside world also require differential signalling but now, because of the larger electrical dimensions, differential transmission lines are required. These are discussed in Section 6.11 on page 218.

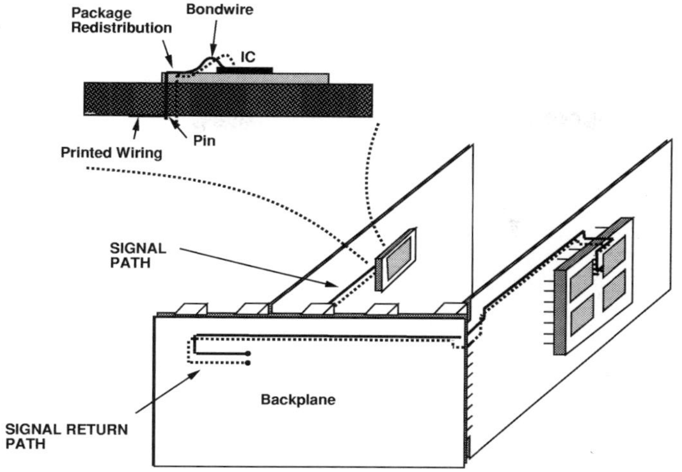

Figure 1.19 Signal paths in a system. (© 2000 P. Franzon, used with permission.)

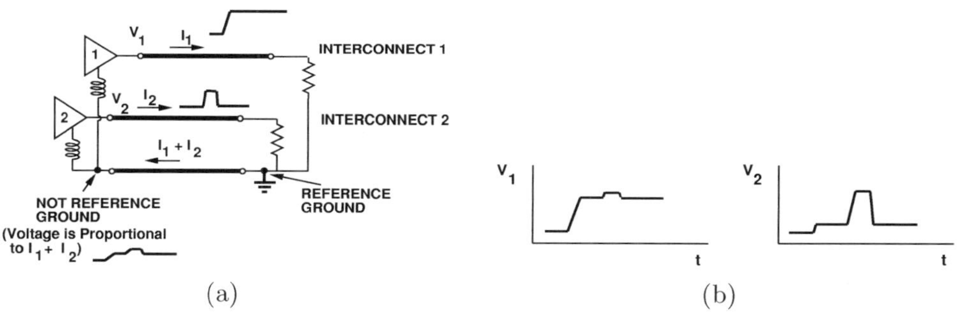

Figure 1.20 Shared current return path forming a common impedance connection (a) and the voltage waveforms resulting (b).

1.11.1 COMMON IMPEDANCE COUPLING

Sharing of a return path results in coupling of signals and crosstalk — the unintentional inducing of a signal on one net by the signal on a second net. Coupling resulting from the sharing of a signal return path is shown in Figure 1.20. This type of coupling is often called common impedance coupling as there is a circuit element common to two or more interconnects. The simplest situation is a shared impedance, rather than a shared transmission line return, so that the return current attributed to one interconnect induces a voltage across the common impedance element. This signal then appears as though it was on the victim net. This is the situation with ground bounce, a common problem that occurs when connecting a packaged chip to the next higher level in the packaging hierarchy. The problem can be illustrated by considering a ground pin, forming part of the current return path, that is utilized by several off-chip drivers. The ground pin has an effective impedance resulting from bond wire inductances which connect the die to the package and from the package pin itself. Return current spikes effectively boost the voltage of the on-chip ground

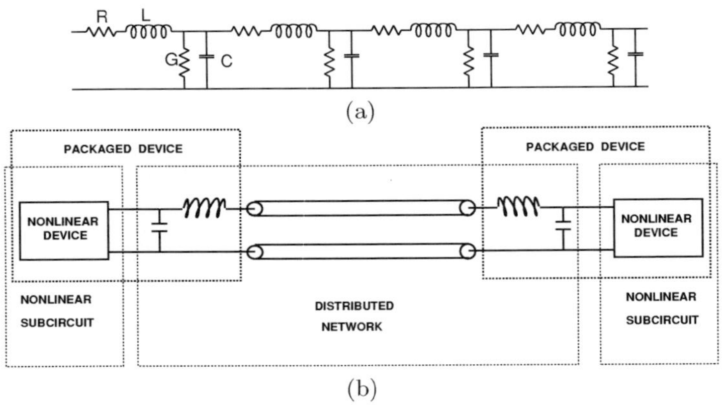

Figure 1.21 *RLC* model (a) of an interconnect and a model (b) incorporating a driver and a receiver.

above that of the off-chip ground so that the drive signal of one chip appears at the output of other drivers. This phenomenon is known as 'ground bounce.' While the false signal resulting from one driver is small, the switching of many drivers can result in an appreciable signal. Since the common impedance element is inductive, the voltage induced is proportional to the switching speed of the drivers. The only solutions are to use many ground connections so that few nets share the same current return path — now a common practice with high performance parts — and to slow down the transitions of off-chip signals — a necessary requirement in any case with current technology. Also to reduce the current levels, it is becoming necessary to use higher impedance drivers off-chip so that current levels are lower. However, given the reduced voltage levels being used in digital systems the energy contained in these off-chip nets is low, and so more susceptible to electromagnetic interference.

1.12 MODELLING OF INTERCONNECTS

A range of models are used for interconnects depending on:

- The accuracy required — nets carrying analog signals need to be modelled more accurately than do those carrying digital signals.
- The amenability of the net to modelling.
- The frequency of operation.

Short on-chip interconnects are commonly modelled as *RLC* networks where the inductor and capacitor networks are arrived at separately using static calculations of the effect of very small segments of interconnect on other small segments.

Uniform interconnects (with regular cross-section) can be modelled by determining the characteristics of the transmission line, e.g. Z_0 and γ versus frequency, or arriving at a distributed lumped element circuit as shown in Figure 1.21(a). This model can be combined with driver and receiver models to realize a circuit model for a complete system.

Entire microwave and millimetre-wave structures can be modelled using electromagnetic modelling software. This software yields the network parameters of the structure that can readily be used in microwave circuit simulation programs.

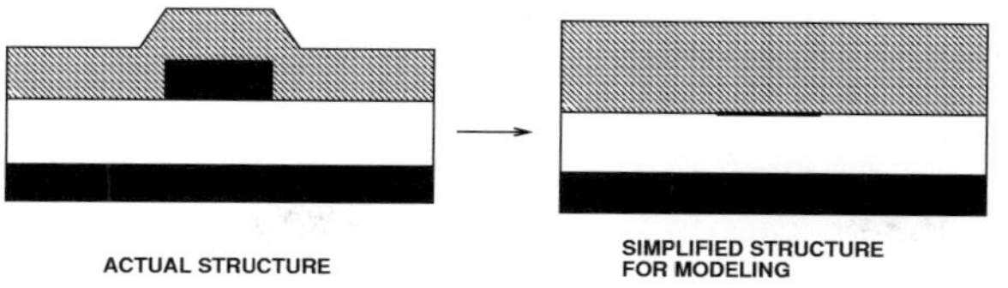

Figure 1.22 An interconnect modelled as an idealized zero thickness microstrip line (cross-sectional view.)

The most useful electromagnetic modelling software models planar interconnects as having zero thickness, as shown in Figure 1.22. This is reasonable for microwave interconnects as the thickness of a planar strip is usually much less than the width of the interconnect. Many analytic formulas have also been derived for the characteristics of uniform interconnects. These formulas are important in arriving at synthesis formulas that can be used in design, i.e. arriving at the physical dimensions of an interconnect structure from its required electrical specifications.

1.13 SUMMARY

This chapter introduced the basics of signal transmission on interconnects. Interconnects determine the limits on the performance of digital systems and the characteristics of analog circuits. Indeed at RF and microwave frequencies, interconnects are part of the circuit and must be designed accordingly.

Chapter 2 of this book focuses on on-chip interconnects and the special properties of these. The short haul interconnects on digital chips, locally connecting transistors, have small dimensions. These interconnects are unique, mainly in that the signal return path is poorly defined and losses are high. Chapter 3 looks at interconnect technologies — the different way interconnections can be fabricated. Chapters 4 to 8 inclusive describe the characteristics of different types of interconnects and present design equations and strategies that enable the physical parameters to be derived from the electrical specifications. While powerful electromagnetic tools are available to model interconnects, and to optimize interconnect networks for specific properties, it is always preferable to synthesize a design. This is becoming increasingly so with manufacturability issues becoming more important. In one sense manufacturability requires that the risk in a design must be minimized, and this requires that performance be synthesized and controlled. Chapters 9 to the end of the book are concerned with specific circuits utilizing interconnects as design elements.

2

On-Chip Interconnects for Digital Systems

2.1 OVERVIEW OF ON-CHIP INTERCONNECTS

Up to the early 1990s on-chip digital signals had components much below 1 GHz; relatively short run lengths (constrained by the dimensions of the chip); widths of several microns; and a height of a micron or more. The first two factors resulted in the electrical lengths of on-chip interconnects being much less than a wavelength of the signals present (i.e. of the highest frequency components present). Then an interconnect could be adequately modelled as a shunt lumped capacitance. A series resistance completed the model (a so-called RC model), but this was small because of the relatively large cross-sections of the interconnect.

This situation has changed because of three main developments, primarily for digital circuits, but affecting analog circuitry because of the rise of mixed signal systems. The main developments are

(a) Faster clocks, of a gigahertz and above.
(b) Longer interconnects as the lateral dimensions of large chips are around 2 cm.
(c) Fine lithography leading to interconnects having cross-sectional dimensions of less than a micron.

The consequences of these developments are that RC modelling is not always adequate. However on-chip digital interconnects are highly irregular and densely packed and often do not even approximate interconnections of uniform transmission line segments.

The typical arrangement of interconnects is shown in Figure 2.1. The first metal layer here is tungsten and provides local interconnect with short interconnect lengths of small cross-section. Higher level interconnects use lower resistivity metal, here copper, and have larger cross-sections, thus further reducing line resistance. Greater attention is made to the provision of signal return paths at the higher metal levels. The higher level metal lines also have lower coupling to the silicon which, having finite conductivity (but not very high) is a significant source of loss. Copper diffuses into silicon oxide (SiO_2) and into silicon (Si) with disastrous effects so that barriers are required. Currently titanium nitride (TiN) and tantalum (Ta) are used [5].

The provision of signal return paths for digital and mixed signal chips has only been a recent consideration and then only for a few nets. The exceptions are clock

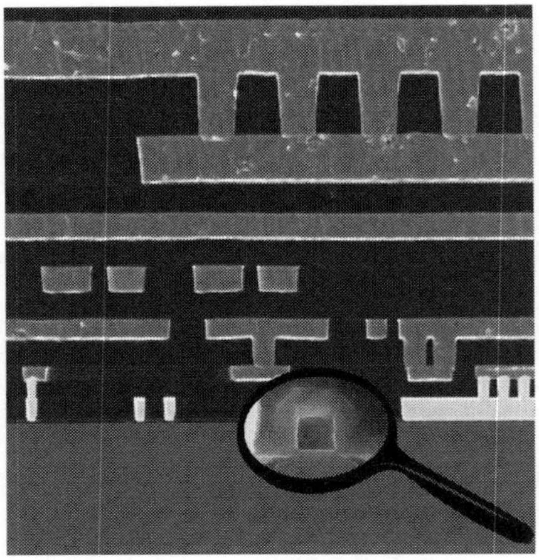

(a)

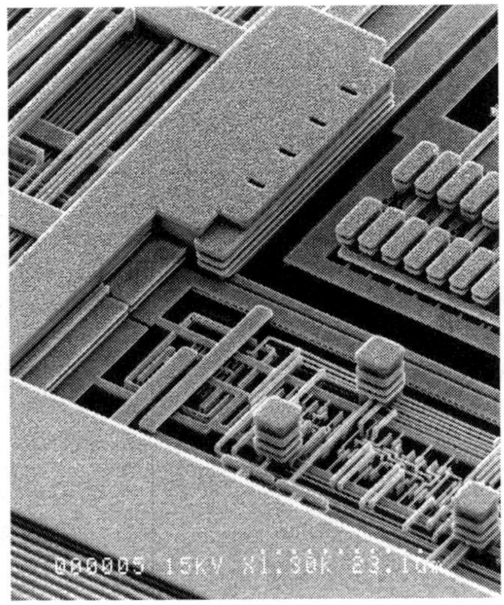

(b)

Figure 2.1 Focused ion beam-scanning electron micrographs (FIB-SEMs) of a six-level interconnect structure developed by IBM. The first (bottom) level is tungsten as are the vias. The other metal levels are copper: (a) cross-section; and (b) perspective after etching of oxide dielectric. (© IBM, from IBM web page, used with permission.

distribution nets and other long nets (such as certain data buses) where signal integrity is paramount. Another important characteristic was that little attention was given to providing signal return paths. In contrast, with RF and microwave chips it has always been considered necessary to provide current return paths so that the electromagnetic field produced by a signal on an interconnect is uniform, contributing to well defined electrical characteristics — hence the term controlled impedance interconnect. Such an interconnect could be conveniently modelled as a transmission line. With RF and microwave chips there are relatively few active devices and so the provision of interconnects large enough (to effectively minimize interconnect resistance) and of defined ground planes for good current return paths can be accommodated at reasonable cost.

The characterization of and modelling required for these interconnects are the main subjects of this chapter. The main issues to be resolved in this chapter are

(a) When is it necessary to model an interconnect as an *RLC* circuit?
(b) When is it necessary to use a transmission line model?
(c) What is the significance of the current return path?
(d) How do the characteristics of interconnects vary with frequency or clock frequency?

2.1.1 TYPES OF ON-CHIP INTERCONNECTS

On-chip interconnects have several distinguishing characteristics that can be used in determining the type type of modelling required. It needs to be emphasized that an interconnect is not just a wire connection from one point in a circuit to another. There is always a current return path and this path (and its proximity to the signal path) is equally important in determining the electrical characteristics of the interconnect. Electromagnetic fields between the two conductors making up a single interconnection between two points and anything that interferes with this field will have an effect on signals being transmitted. Two obvious physical conjectures establish that there must be a return path: without it charges would build up at some point in the circuit to very high levels; also the electric field supported by the charges (and induced by the voltage source originally) on the signal conductor of an interconnect must terminate on matching charges, and these are located on the signal return path. A signal interconnect is usually a conductor which is constrained in its transverse dimensions, but this is not the case with the signal return path. If a ground plane carries the return signal, then the return path can be as wide as the ground plane at DC, but as the frequency of the signal increases, the width of the current return path narrows as the charges on the signal return path have limited time to redistribute. The return path at high frequencies tends to be as close to the interconnect bearing the signal as possible. The proximity, and metal width available, can significantly affect the resistance, capacitance and inductance of the interconnect.

On-chip interconnects can be broadly classified by the type of signals conveyed, and geometrical and circuit characteristics as described below.

Interconnect networks (or *nets*) on digital chips can be categorized by the type of connections made:

i. *Local connections with maximum lengths of 1–3 mm*
The majority of connections on a chip are of this type. The utilization of space is a premium and so minimum line widths are used as dictated by the technology and these are in the range of 0.1 to 0.5 μm, known as deep submicron. These lines have high resistance and are driven by transistors of minimum dimensions, and hence of low current drive capability, and have high output (or drive) impedance $Z_{\rm DRV}$. It is not economical to provide metal for individual signal return paths.

ii. *Medium length connections with minimum lengths of 1–3 mm*
The maximum usable length is determined by the characteristics of the technology and of the interconnect itself (e.g. width and hence resistance). Typically, the maximum usable length is around 2–5 mm. The longer usable length is achieved principally by making the line wider, but also by paying attention to the current return path. Higher current drivers with lower $Z_{\rm DRV}$ are required.

iii. *Long connections up to a chip-edge in length*
There are three families of connections that are in this category each with different characteristics. However, what they have in common is that the drivers of these connections must have high current capability, and low $Z_{\rm DRV}$, so that the receiver (at the end of the line) switches correctly when the signal makes its first transition down the line. The families are:

— Data buses: These are often wide (meaning that there are many parallel interconnections, perhaps 128 for example) and can be about half the chip-edge in length. Their purpose is to convey data from one part of a chip to another, such as from the central processor to an on-chip data cache. Such lines switch at the same time and coupling of the lines is a considerable concern, also the delay on the lines is affected by the switching pattern as will be seen in Section 2.3.1 on page 37. Another characteristic of these lines is that they have a small load at the end as they are often driving one or a few receivers.

— Control lines: these transmit signals that must be widely distributed. A good example is a reset signal. There are relatively few of these types of lines with speed issues being less important.

— Clock signal: this must be very widely distributed with integrity of the signal being paramount. As well as ensuring first incidence switching, the clock skew, that is the difference between the arrival times of the clock edges in different parts of the chip, must be precisely controlled. The clock distribution net can consume a very large proportion of the chip power both because of the gates that are driven at the end of the clock distribution net and because of efforts required to minimize skew.

Many long nets have intermediate buffers along them to boost the signal to compensate for attenuation of the signal.

iv. *Power and ground distribution buses*

It is perhaps unusual to think of the power and ground buses as interconnects as the voltage on them is, hopefully, constant. However, these buses carry large switching currents that change at the clocking frequency. The pulsing current transmits a signal just as real as do voltage variations. With fast chips it is necessary to model and design these as high speed interconnections. Ideally these nets are of very low impedance so that the current fluctuations have little effect on voltage levels. Low impedance can be achieved using metal planes for supply and ground and by large capacitors between the supply and ground nets.

Interconnects for digital signals can be contrasted to interconnects for other types of signals. The density of analog and RF circuits does not approach that of digital circuits and in any event, controlling the characteristics of transistors (impedance levels, pole frequencies etc.) is critical to the performance of these frequency sensitive circuits.

Microwave and millimetre-wave on-chip circuits (usually called MMICs for Microwave Monolithic Integrated Circuits) usually have interconnections that are a substantial portion of a wavelength, but also have ground planes or similar structures forming a highly regular signal return path. Thus, transmission line modelling is necessary and is also convenient.

2.2 EXPERIMENTAL CHARACTERIZATION OF AN ON-CHIP INTERCONNECT

Biswas *et al.* [6, 7] experimentally characterized interconnects fabricated at SEMATECH using a deep submicron process. The cross-section of the interconnect is shown in Figure 2.2. The epi-layer is highly doped and forms a ground plane. On this a layer of dielectric is grown, and then a thin layer of titanium tungsten (TiW) is deposited before a thick layer of aluminium (Al). Dielectric is then deposited, and the surface is then planarized (the CMP-process) so that the tops of the dielectric and metal interconnect to form a plane. The process is repeated until finally a

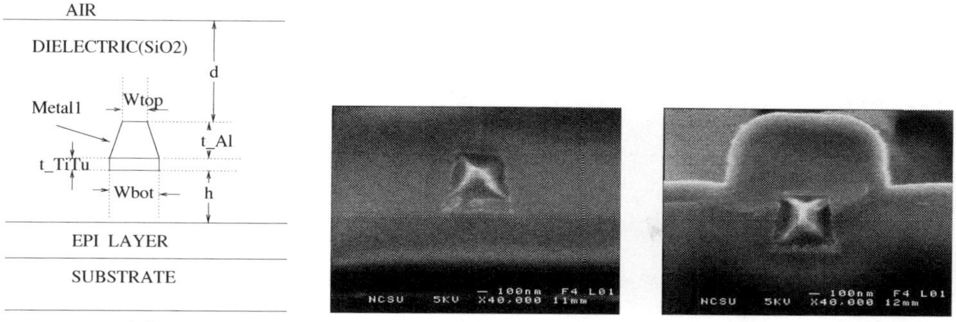

(a) (b) (c)

Figure 2.2 Cross-section of a deep-submicron interconnect: (a) line-diagram; (b) SEM cross-section of bottom metal line; and (c) SEM cross-section of a top metal line.

passivation layer is deposited. The metallization consists of multiple metal types including, in addition to the basic aluminium interconnect metallization, buffer layers of, for example, titanium and tungsten vias. Higher performance interconnects use copper as the main metallization, but this must be coated with a buffer layer to prevent reaction of the copper with the dielectric. Further complicating our ability to model this structure are changes in the density of the dielectric, and thus the dielectric constant, around the metal. Biswas *et al.* used microwave frequency domain measurement techniques. Time domain techniques enable alternative experimental characterization using signals that are much closer to the types of signals encountered on a chip, but without the dynamic range of frequency domain measurements [8–12]. The conductors were aluminium and a Chemical-Mechanical Planarization (CMP) process was used to ensure planarization — this is an important step when the aspect ratio (height-to-width) is appreciable. A particularity of the measurements is that it is not possible to extract all four frequency dependent parameters, either the four R, L, G, C parameters or the four components (i.e. the real and imaginary parts) of Z_0 and γ, from measurements. Generally, it is assumed that the capacitance per unit length of the line does not vary much with frequency and this has been extensively investigated [16]. The characteristic impedance and propagation constant of an interconnect of nominal width 0.25 μm is shown in Figure 2.3 and its R and L parameters shown in Figure 2.4. The parameters are arrived at using two lengths of line of the same cross-section. The Through Line (TL) technique was used, which involves just two calibration structures and reduces the test area otherwise required substantially [15, 17]. This is recommended as a standard technique for the frequency domain characterization of on-chip interconnects. In Figures 2.3 and 2.4 two sets of results are presented — obtained by considering three lines so that two extractions were possible. (This was done for purposes of cross-checking results. Any discrepancies that were found were always traced back to contact problems as reliable probe contact up to 20 GHz is a potential problem.) The irregular results shown at low frequencies can be ignored as the lines of necessity must have short electrical lengths here, and hence resolution is poor. However, the complete electrical properties of the interconnects are contained in Z_0 and γ and these are smooth. The per-unit-capacitance, C, was obtained using separate low frequency measurements and was

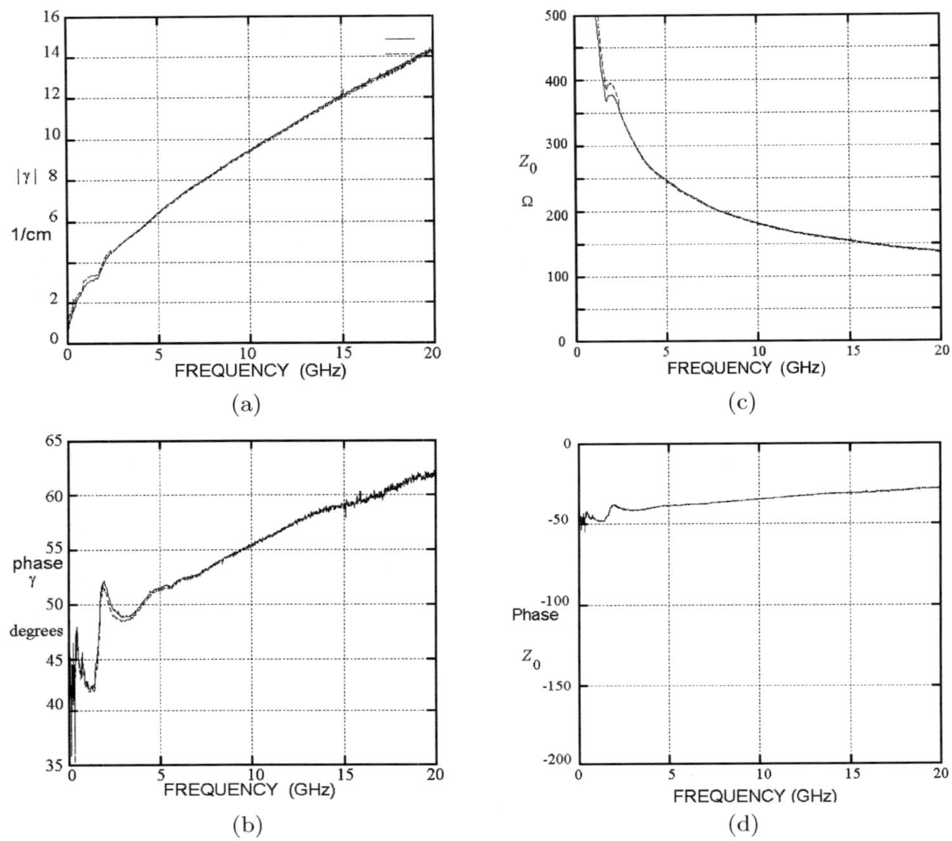

Figure 2.3 Experimentally derived characteristic propagation constant and characteristic impedance of a nominal 0.35 μm wide interconnect: (a) magnitude and (b) phase of propagation constant; and (c) magnitude and (d) phase of characteristic impedance.

827 fF/cm. G was taken as negligible. One notable feature is the high resistance of the line due to the small cross-section.

2.3 *RC* MODELLING ON-CHIP INTERCONNECTS

The reason for modelling interconnects is to verify signal integrity and provide an insight into allowing redesign if necessary. The ultimate goal of course is to take electrical specifications and to develop the physical layout requirements, but this is not as well developed as would be liked. The main signal integrity issues are:

(a) Will the chip operate at the rated speed? Is the timing correct, that is, do the signals at the receiver end of a net arrive at the required switching levels within the timing budget?
(b) Is signal integrity adequate? That is, is crosstalk within acceptable levels?
(c) Will it perform reliably? Reliability includes electromigration and fuse effects, as

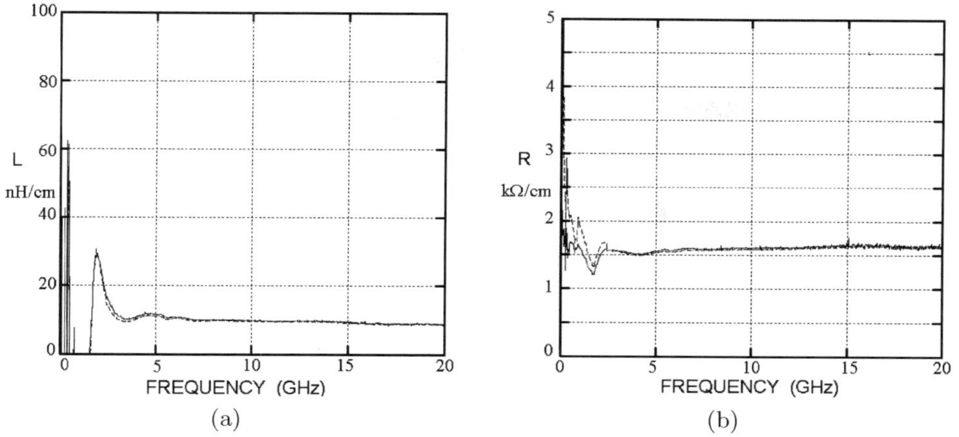

Figure 2.4 Experimentally derived inductance (a) and resistance (b) unit length parameters. The capacitance was 827 fF/cm.

well as signal integrity issues.

Modelling on-chip interconnect effects is daunting as there are millions of nets on a typical submicron IC and the ultimate clock speed of a circuit is determined by the slowest of these interconnects and the nets that have the highest crosstalk. An idea of the complexity can be seen by examining the interconnects involved in a single CMOS inverter, shown in Figure 2.5.

It is clearly not possible to incorporate the interaction of every net on every other other one. So the signal integrity verification problem is driven by the need to minimize the modelling complexity as much as possible. This is done by first identifying those nets, the critical nets, that are most likely to cause a problem, and to model these using simple modelling. These nets are not assured of actually being problems, but if there is a problem we can have a very high confidence that the problem net will be one of the critical nets. Critical nets are identified using relatively loose criteria, but are still sophisticated enough that only a small fraction of the total number of nets are so identified. The critical nets are then examined in considerable detail to determine if there is a signal integrity violation. Thus in a second pass, the *at risk* nets are identified. Generally, the *at risk* nets must be examined by a human to evaluate whether or not there is a signal integrity violation.

The simplest electrical models that will capture the effect are used, for example, in increasing complexity: a simple delay model; an *RC* model consisting of resistors and capacitors only; an *RLC* model consisting of inductors as well; and finally, a distributed transmission line model. These types of models are discussed below.

2.3.1 DELAY MODELLING

Nearly all of the drivers on a chip are sized near their minimum functional dimensions and so have their current capability limited. The consequence of this is that a driver, driving a net, does not appear like a linear driver, but instead as sourcing or sinking

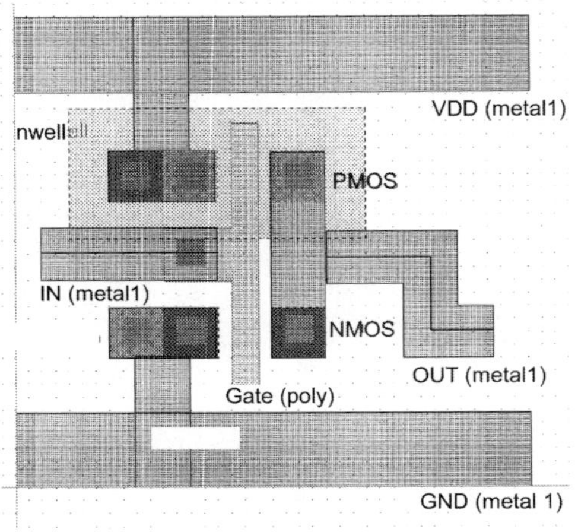

Figure 2.5 The layout of a CMOS inverter.

essentially a fixed current until the desired voltage level is obtained. Thus the voltage on a short interconnect only rises according to the rate at which charge is sourced or sunk from the line, and the time required to charge to the desired voltage level is proportional to the capacitance of the interconnect. Thus the delay on a line is directly related to the time constant of the line, which is equal to the time for a single pole circuit, an RC circuit for example, to change by $(1 - 1/e)$ $\approx$65% of the total change that will occur. Typically, this is about the voltage change required to alter the logic state of a digital circuit. The total effective time constant, τ_T, of a series cascaded circuit with K poles with time constant τ_k for the k th pole is, approximately,

$$\tau_T = \sum_{k=1}^{K} \tau_k \tag{2.1}$$

where the k th time constant

$$\tau_k = R_k C_k. \tag{2.2}$$

R_k is the resistance of the k th segment of interconnect, and C_k is the capacitance of the segment. So that an estimate for the delay t_d through such a circuit is

$$t_d = \tau_T = \sum_{k=1}^{K} \tau_k \tag{2.3}$$

Thus a good delay model of the cascaded circuit is

$$t_d = \sum_{k=1}^{K} R_k C_k. \tag{2.4}$$

Equation (2.1) only gives us an estimate of the effective overall time constant of a circuit and there are restictions on its validity. The primary restriction is that it is useful when the individual time constants τ_k are comparable. However, it provides the best simple estimate of timing.

Short lines can be modelled by considering the RC representation of the output of a driver and dividing the line into just two segments. If R_L and C_L are the total line resistance and capacitance, respectively, then the delay, called the Elmore delay model [18], is

$$t_d = R_D C_D + (R_L/2)(C_L/2) + (R_L/2)(C_L/2) \tag{2.5}$$
$$= R_D C_D + (R_L C_L)/2 \tag{2.6}$$

where R_D is the output resistance of the driver and C_D is its parasitic output capacitance. This is generalized for a net so that the following formula has traditionally been used for short lines [26]:

$$t_d = R_D C_D + (R_N C_N)/2 \tag{2.7}$$

where R_N and C_N are the net resistance and effective capacitance. Thus $R_D C_D$ can be referred to as the gate delay and $(R_N C_N)/2$ as the interconnect delay. For very short nets the resistance and capacitance of the net are negligible and the first term in Equation (2.7) dominates. In other words, gate delay dominates for very short nets. C_N in Equation (2.7) is the driving point capacitance (the effective capacitance seen by the driver) and because of capacitive coupling to neighbouring nets, C_N will be data-dependent varying according to the switching activity on neighbouring nets. The origins of this are described in greater detail in Appendix D.

If the driven net can be considered as a single transmission line, with resistance and capacitance per unit length of R and C, respectively, then $R_N = Rl$ and $C_N = Cl$ so that the simple delay model of Equation (2.7) becomes

$$t_d = R_D C_T + (RCl^2)/2 \tag{2.8}$$

indicating that the delay increases in proportion to the square of the length of the line. However, for very short nets the gate delay dominates and length-dependent delay becomes a problem for medium to long length lines.

Unless specific considerations are made at the design stage, nets are strongly coupled to neighbours. If J neighbours are regarded as being significantly coupled then the worst case situation, resulting from the worst possible data switching pattern, results in the driving point capacitance C_N of the i th net being

$$C_N = C_{ii} \sum_{\substack{j=1 \\ i \neq j}}^{J} r_j C_{ij} = C_{ii} \left(1 + \sum_{\substack{j=1 \\ i \neq j}}^{J} r_j K_{ij} \right) \tag{2.9}$$

where C_{ij} is the mutual capacitance of the i th and j th nets, C_{ii} is the self capacitance (or just the capacitance alone) of the i th net, $K_{ij} = C_{ij}/C_{ii}$ is the capacitive coupling factor, and r_j is the capacitance multiplier:

- $r_j = 0$ if the j th net switches concurrently with the i th net (that is, they are at the same voltage).
- $r_j = 1$ if the j th net stays at a constant level while the victim line switches from low to high.

$r_j = -1$ if the j th net stays at a constant level while the victim line switches from high to low.

Thus the worst case situation — resulting from the worst possible data switching pattern — with the highest effective capacitance is

$$C_N = C_{ii}\left(1 + \sum_{\substack{j=1 \\ i\neq j}}^{J} K_{ij}\right). \tag{2.10}$$

K_{ij} is strongly influenced by the proximity of the i th and j th nets and by the provision of current return paths (or ground) that can partly shield the nets and reduce K_{ij}. However, if care is not taken and many nets are significantly coupled to each other the driving point capacitance can be many times the self capacitance of the victim net (C_{ii} for the i th net). Capacitive coupling along with inductive coupling (which will be considered later) dominate the performance of long and medium length lines. This is especially so, since there are relatively few nets of these types and low impedance drivers with higher current drive capability (both resulting from larger output transistors) ensure that the gate delay is very small.

The highest performance digital circuits use specific strategies to reduce coupling by providing a current return path (or ground) as nearby medium-to-long nets as possible. Specific strategies include reserving on-chip layers from ground planes and using differential lines such as the coplanar strip (CPS) line discussed in Section 6.11 on page 218. These strategies, especially the provision of one or more ground planes, have long been the practice with printed wiring boards and with multichip modules.

To summarize, the delay model discussed here is useful for interconnects with drivers of limited current capability so that the voltage on the interconnect must rise gradually as the line is charged. Faster signal transmission is achieved using a large driver which functions as a voltage source transmitting a 'logic-level' voltage signal down the line. The delay is determined more by time-of-flight considerations.

2.3.2 RC MODELLING

The first step in developing an RC model is to develop the capacitance model of a subset of nets. As the density of nets is very large, it is not sufficient to consider a net in isolation. Neither is it feasible to consider the interaction of a net with every other net on the chip. A successful approach is to focus on the signal integrity of one net due to both the delay effects and to crosstalk from other nets. Fortunately the electromagnetic properties help us here. A voltage and current carrying interconnect generates electric and magnetic, or electromagnetic, fields which fall off with distance as $1/R^N$ where R is distance and $N \geq 1$. (N does not fall off much faster as the electric and magnetic fields are essentially confined in a plane.) This is the justification for considering a region of influence such as that shown in Figure 2.6. Here the region of influence (the box) is centred on an individual net, the victim net. It is assumed that every conductor within the region of influence will be coupled to the victim net. The capacitances describing this coupling are generally developed using a number of template models. For example, if the victim net and another net cross each other (the

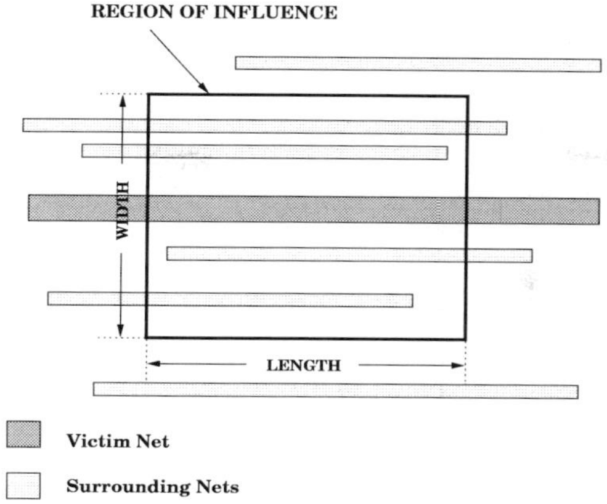

Figure 2.6 Neighbouring nets and the region of influence used in capacitance extraction.

nets having X- and Y-orientations, respectively) a parallel plate capacitor model is used for the capacitance. The templates of a large of number of simple structures that can commonly occur are generally developed in advance for a particular process. The templates could consist of precise data, each template developed for a structure of a particular geometry, and a very large number of templates are needed. Alternatively fewer templates of the polynomial type with geometry variables such as separation distance could be used and the coefficients of the polynomials developed by fitting to electromagnetic simulation of several variants of a particular template structure. With the multiple metal composition of interconnects and fabrication-dependent dielectric density variations, these templates should be calibrated with actual measurements. Pattern recognition is used to identify candidate structures (that match a template). If the two nets run parallel to each other then a standard capacitance calulation is performed — again essentially another parallel plate capacitance calculation, but using fringing capacitance corrections. These capacitance templates are typically developed using three-dimensional electromagnetic simulation codes to build up a standard library.

It is not really practical to use electromagnetic simulation to develop a capacitance matrix for a target net with its neighbours, although this can be done for the most critical nets. Modelling proceeds by meshing the interconnect structure, as shown in Figures 2.7(a) and 2.7(b) for the CMOS inverter considered previously. Here each interconnect is subdivided into a large number of segments. The coupling effect of each segment on every other segment is evaluated and the capacitance matrix extracted following a number of measurement-like simulations to extract one capacitance at a time. The procedure to extract the capacitance matrix is outlined in Appendix D. A faster electromagnetic analysis can be developed by considering zero-thickness metallization as shown in Figure 2.7(c) (resistance of the interconnect is correctly handled by treating it as a surface resistance). However this is not sufficiently accurate for today's high aspect ratio digital interconnects. The use of templates is the only practical approach for performing parasitic extraction in reasonable time.

The accurate calculation of resistance is a much simpler problem than the extraction of capacitances. This is because nets can be considered on their own and mutual

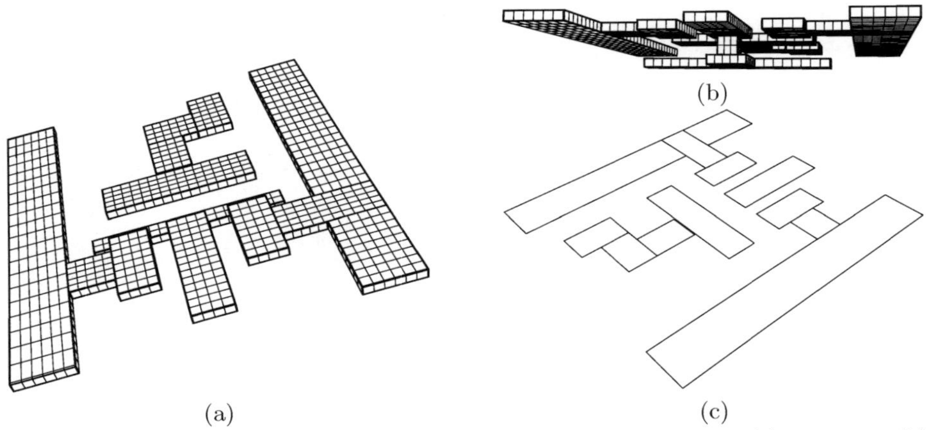

Figure 2.7 The interconnects of the CMOS inverter of Figure 2.5: (a) top view; (b) bottom view; and (c) a top view with a different perspective and the lines treated as though they had zero thickness.

resistances (through the substrate or dielectric layers) are negligible. We simply use the resistivity of the conductors and the cross-sectional dimensions of the interconnect. Note, however, that resistance in the generally shared current return paths can have a significant effect on coupling due to the common impedance effects discussed in Section 1.11.1 on page 27.

RC extraction is a database manipulation problem with issues such as

(a) Storage of geometric data and relating that to each active device so that the electrical model can be recalculated without performing another parasitic extraction.
(b) Support of hierarchical electrical model data, such as starting with a delay model, then replacing this by a capacitance only model, then a resistor-capacitor model and so on.
(c) Ability to quickly develop circuit models of selected nets or regions.
(d) Support of incremental extraction to handle only the changed regions of an IC.

2.4 MODELLING INDUCTANCE

Modelling the inductance of long high performance digital interconnects is now required. Nets of this type are relatively few, but they include the on-chip clock distribution net. For these nets *LRC* modelling (an excellent approximation of a transmision line as far as digital inteconnects are concerned) must be used. The aggressive nets are either constructed as signal lines over a ground plane (a microstrip structure) or with shield lines between neighbouring signal lines to form a coplanar waveguide-like structure. In both cases, the aim is to achieve a well controlled signal return line. It is possible to achieve very high performance from nets without a reference conductor following in tandem with each signal net, but this requires extremely good design discipline and model extraction tools. For structures with reference conductors, extracting inductances is much easier than it would be otherwise. Complicated structures without the reference conductors require full-wave three-dimensional electromagnetic field solutions [19].

2.4.1 WHEN ARE INDUCTANCE EFFECTS IMPORTANT?

The signal response on an interconnect depends on the properties of the interconnect as well as the characteristics of the driver. In nearly every situation the driver, through design or otherwise, appears as either a Thevenin equivalent voltage source with a voltage step generator and an equivalent driver impedance Z_{DRV}, or as a current source of fixed current drive which charges a line. With the latter type of driver, when the voltage on the line corresponds to the appropriate logic level, the current source turns off. The highest performance is obtained using the Thevenin equivalent driver as this has the greatest chance of resulting in first incidence switching at the retardation time. This type of driver must be relatively large so that it can provide the current levels to support a full voltage signal travelling away from the driver. The current sourcing (or sinking) capability is equal to the voltage change to be imposed on the line divided by the characteristic impedance of the line. So if the voltage change required to change the logic level is 3 V and the characteristic impedance of the line is 50 Ω, the current drive capability is 60 mA. This reduces to 20 mA if Z_0 is 100 Ω and the level change is 2 V. Thus, reducing the operating voltage and using interconnects of high characteristic impedance (Z_0) is important in limiting current drive levels to reasonable values. The required current drive levels are further increased by the resistive divider effect of the driver output impedance Z_{DRV}. Thus it is also important to keep Z_{DRV} low. For most on-chip drivers it is not feasible to provide the current drive levels required to generate a full level voltage signal on the line, so instead they operate as fixed current sources until the required voltage levels on the line are obtained. This restricts such drivers to short connections, as then the slow RC-like response is not critical to determining system performance.

In a series of significant papers [8,9,20] Deutsch and her colleagues at IBM presented definitive studies that answer the question as to when inductive effects must be considered. The developments below draw heavily on this work.

We begin by considering a lossy line of infinite extent and derive the characteristic of the line and then the response to a voltage step at the start of the line. The voltage on the line, $v(x,t)$ and the current, $i(x,t)$ on the line are conveniently represented by the phasor voltage, $\mathbf{V}(x)$ and phasor current $\mathbf{I}(x)$ when the signal on the line is sinusoidal. Thus the relationship between the voltage and current is described by two coupled first order differential equations:

$$\frac{\mathrm{d}\mathbf{V}}{\mathrm{d}x} = -Z\mathbf{I} \tag{2.11}$$

and

$$\frac{\mathrm{d}\mathbf{I}}{\mathrm{d}x} = -Y\mathbf{V} \tag{2.12}$$

where x is the distance along the line, $Z = R + j\omega L$, $Y = G + j\omega C$ and R, L, C, and G are the resistance, inductance, capacitance and conductance of the line per unit length. Weber [21]ignored the conductivity G and solved the coupled equations, Equations (2.11) and (2.12), and transformed the solution into the time domain. The result is that the time domain voltage response, to a unit voltage step at the start of the line (i.e. at $x = 0$) at a position $x = l$ along the line and at time t is

$$\begin{aligned} v(l,t) &= \left(e^{-Rl/(2Z_0)} + \frac{Rl}{2Z_0} \int_{\tau=l\sqrt{LC}}^{t} \left\{ \frac{e^{-Rl/(2Z_0)}}{\sqrt{\tau^2 - l\sqrt{LC}}} \right.\right. \\ &\left.\left. \times I_1 \left[\frac{R}{2Z_0} \sqrt{\tau^2/(LC) - l^2} \right] \mathrm{d}\tau \right\} \right) \times u\left(t - l\sqrt{LC}\right) \end{aligned} \tag{2.13}$$

where $I_1(.)$ is the modified Bessel function of the first kind. Introducing the retardation[1] time τ_0 which is the earliest time a signal can be detected at $x = l$ and using the following definitions:

β_0	$= \omega\sqrt{LC}$	the lossless propagation factor
τ_0	$= l\sqrt{LC}$	the retardation time
α_C	$= R/\sqrt{LC}$	the conductive attenuation
α'_C	$= \alpha_C = Rl/\sqrt{LC}$	the conductive attenuation factor

the waveform response of Equation (2.13) becomes (after some rearrangement)

$$v(l,t) = e^{-\alpha'_C} \left[1 + v_T(l,t)\right] u(t - \tau_0) \tag{2.14}$$

where $v_T(l,t)$ is the unattenuated transient response:

$$v_T(l,t) = \int_{\tau=\tau_0}^{t} \left\{ \frac{I_1 \left[\alpha'_C \sqrt{(\tau/\tau_0)^2 - 1} \right]}{\sqrt{(\tau/\tau_0)^2 - 1}} \right\} \mathrm{d}\tau. \tag{2.15}$$

Using the time normalized to the retardation time, $t' = t/\tau_0$ and $\tau' = \tau/\tau_0$, this equation becomes:

$$v_T(l,t) = \int_{\tau'=1}^{t'} \left\{ \frac{I_1 \left[\alpha'_C \sqrt{(\tau')^2 - 1} \right]}{\sqrt{(\tau')^2 - 1}} \right\} \mathrm{d}\tau'. \tag{2.16}$$

A wealth of information is contained in this response:

(i) The first is that there is no response before the retardation time, τ_0 as the step function $u(\ )$ is defined as:

$$u(t - \tau_0) = \left\{ \begin{array}{ll} 0 & t < \tau_0 \\ 1 & t \geq \tau_0 \end{array} \right. . \tag{2.17}$$

It is a physical necessity that there can be no response before the 'time of flight'. This can be constrasted to a circuit consisting of Rs, Ls and Cs. There will always be an instantaneous response, even if it is very small. Simplifying the model will simply make the anonomously early response worse.

---

[1] The concept of retardation was introduced on page 4, and refers to the finite time of flight of signals.

(ii) Secondly, the response is scaled by the overall attenuation:

$$e^{-\alpha'_C} = e^{-\alpha_C l} = e^{Rl/(2Z_0)}. \tag{2.18}$$

At a critical length $l_{\text{crit}} = 2Z_0/R$ the attenuation is $1/e = 37\%$. At lengths above this it is difficult to achieve first incidence switching [22]. As the loss of the line increases, Z_0 remains relatively unchanged but the attenuation increases.

(iii) The term in the integral describes the effect of the response due to the interaction of R, L and C.

(v) The time duration of the transient response is essentially related to the normalized times. Thus, approximately, the transient response at a position $x = 2l$ has twice the duration as it does at $x = l$.

(iv) The form of the transient response is determined by two terms which we will examine separately. Let

$$T_F(\tau') = \frac{1}{\sqrt{(\tau')^2 - 1}} \tag{2.19}$$

and

$$T_S(\tau') = I_1\left[\alpha'_C\sqrt{(\tau')^2 - 1^2}\right] \tag{2.20}$$

so that

$$v(l,t) = e^{-\alpha'_C}\left\{1 + \int_{\tau'=1}^{t'} [T_F T_S \mathrm{d}\tau']\right\} u(\tau' - 1). \tag{2.21}$$

T_F and T_S are different components of the transient response with T_F being independent of the resistance of the line and capturing the fast rising LC transient response of the line. In addition, T_S is dependent on the resistance of the line. It is interpreted as capturing the slow RC response of the line [8].

If the total resistance (Rl) of the line is low the transient response is initially dominated by T_F – the LC response — as T_S has little effect. Thus with low resistance or a short distance from the driver (i.e. low Rl), the initial transient has overshoot as v_T is additive to the direct retarded response. Eventually this response is damped as T_S becomes significant. With high overall resistance (low Rl) — longer lines or larger R — then T_S dominates the transient response, and the rise time of the response is degraded with the transient resembling an RC response.

2.4.2 INDUCTANCE EXTRACTION

Extracting the inductances of a group of interconnects is a much more difficult problem than that of capacitance extraction. This is because the current return paths must be known as well as the current flow through the net itself. This contrasts with the capacitance extraction problem where the possible location of charges is all that needs to be known. Capacitance derives from storage of energy in the electric field and almost all of these lines conveniently terminate on conductors. So it is possible to limit the scope of the problem using a 'region of influence' as discussed in Section 2.3.2 on page 40. The return path that must be determined in inductance extraction is not confined just to conductors, but includes displacement currents through coupling capacitors. Inductance, which describes the current storage capability of magnetic fields, requires that the majority of magnetic fields be included, and unfortunately these can extend for considerable distances. Leaving out more distant interactions does, as expected, lead to a less accurate model but, what is particularly worrisome, leads to unstable interconnect models [19].

There are two dominant approaches used in inductance extraction. The first we will consider uses three-dimensional magnetostatic numerical analysis[2] [23–25]. In these techniques the conductors are segmented into current filaments or sheets and the interaction of one filament (or sheet) with other filaments (or sheets) calculated. In the light of the previous discussion the obvious drawback is that the current return paths must be known to ensure that all of the relevent interactions are known.

The major alternative to the numerical field analysis above is the Partial Element Equivalent Circuit (PEEC) technique used to extract partial inductances [19, 26–28]. The distinguishing characteristic of this technique is that the current return paths do not need to be known.

2.5 DESIGN APPROACHES TO HANDLING INTERCONNECT EFFECTS

Architectural approaches are described in several books and publications [29–31]. These include using buffers on long interconnects and partitioning large sections of chips into smaller units in which the retardation effects are insignificant. Two design approaches are discussed in the following subsections and a novel clock architecture scheme is considered in Chapter 11. The reader is directed to the references to explore this topic further.

2.5.1 PERFORMANCE-DRIVEN ROUTING

Performance-driven routing for VLSI circuits aims at automatically optimizing the interconnect topology for reducing delay, particularly the delays of critical nets. That is, given electrical specifications and the maximum allowable delay, what are the physical dimensions and layout that satisfy this? This at first seems like a much simpler problem than the other synthesis problems considered in this book. It is complicated by the very large number of nets that must be considered. A global optimum is required rather than the optimum of one net at a time. This mainly concerns the tradeoffs in the layouts of nets. The delay of a particular net is essentially determined by interconnect resistance and capacitance as discussed in Section 2.3.1 on page 37.

2.5.2 TRANSMISSION LINE RETURN PATHS

The provision of a structure for a definite current return path is the easiest architectural approach to relieving the interconnect delay problem. Typical structures

[2] The term magnetostatic rather than electromagnetic is used as in the analysis only magnetic interactions are considered. While not as comprehensive as a full electromagnetic analysis, analysis can be performed in a reasonable time and leads directly to the desired inductance only circuit.

that provide close and well defined current return paths are the microstrip, coplanar waveguide (CPW) and differential line or coplanar strip (CPS). (These will be discussed in greater detail in Chapter 3 — Chapter 6 extensively — and the structures are shown in Figure 3.1.) Using these structures certainly makes modelling much easier. The provision of a defined current return path is not the only consideration. Modern ICs use high aspect ratio interconnects with heights that can be twice the minimum conductor width [19]. As a result the capacitive coupling of neighbouring conductors can have a very significant effect on the performance of the interconnect in light of the discussion concerning Equation (2.10). In addition to affecting the delay of an interconnect, the large mutual capacitive coupling results in high crosstalk. Only for high performance interconnects, such as for clock distribution, can the conductors be separated to reduce the mutual coupling and take advantage of reference planes. In most cases, if a defined return path is required, shield lines are often the best option resulting in the embedded CPW structures. The shield lines can also be used to distribute ground and power as well as reducing mutual capacitances and providing current return paths.

3

Interconnect Technologies

3.1 INTRODUCTORY REMARKS

Microstrip and similar planar interconnects, and circuits using them, form the principal subjects of this book. First it is important to establish the general setting with which we are concerned and also to briefly examine alternative transmission line structures that sometimes exhibit special advantages over others. In this chapter we shall therefore cover the following topics:

(a) microwave frequency and wavelength ranges; applications
(b) a selection of alternative transmission line structures
(c) choice of substrate for hybrid microwave microcircuits
(d) thin-film manufacture
(e) thick-film manufacture
(f) monolithic technology
(g) printed circuit boards
(h) multichip modules.

3.2 MICROWAVE FREQUENCIES AND APPLICATIONS

The use of the term 'microwave' is actually rather ambiguous — particularly at the lower end of the range. Sometimes signals at frequencies of only a few hundred megahertz are referred to as 'microwaves.' For the present purposes we shall assume that appropriate frequencies extend from around 1 GHz to several hundred gigahertz. Corresponding wavelengths in free space are 30 cm at 1 GHz, 3 cm at 10 GHz, and 3 mm at 100 GHz. For digital systems, where the fifth (or higher) harmonic is contained in the digital pulses, our discussions relate to circuits clocking above 200 MHz.

The bulk of microwave technology occupies the spectrum including Ultra High Frequency UHF through to Super High Frequency (SHF), i.e. from roughly 1 to 30 GHz. At lower frequencies more conventional radio techniques take over. In the Extra High Frequency (EHF) bands (above 30 GHz) the wavelengths are extremely short, being less than 10 mm, and hence signals at such wavelengths are often called 'millimetric.'

A strong and significant core of applications exists for analog signals at microwave frequencies, among them:

- radars (a very large family with a wide variety of demands and techniques)
- point-to-point and point-to-multipoint multichannel radio links ('microwave radio')
- satellite communications systems (both fixed and mobile)
- tropospheric scatter ('troposcatter') systems
- cellular (mobile) communications systems — mobiles and infrastructure
- cable television (CATV) and associated cable telecommunications
- local multimedia distribution services (LMDS)
- Microwave/Millimetre-wave Video Distribution Systems (MVDS)
- subspace communications systems (e.g. high-flying aircraft)
- microwave transceivers for optical fibre transmission
- Wireless Local Area Networks (WLANs)
- microwave instrumentation.

At least as significant as any of the applications listed above, high speed interconnects on semiconductor chips (ICs for Integrated Circuits) and multichip modules (MCMs) (and also on card) are now extending into gigabit rates which means that transmission line design issues apply. MCMs are used to provide small, high precision interconnects that can be cost effectively produced with larger areas than for an IC, have lateral dimensions that are comparable or perhaps larger than on-chip, and have higher performance than interconnects on Printed Circuit Boards (PCBs). Generally the distinguishing feature is that bare (unpackaged) die can be used with PCBs, but this is usually not acceptable with PCB assembly. MCM substrates and technologies are discussed thoughout this chapter and reviewed in Section 3.9 on page 74. Printed Circuit Boards are also referred to as Printed Wiring Boards (PWBs), but we will use the terms Printed Circuit Board or PCB here. The PWB term is more common in North America, but PCB is preferred elsewhere. PCBs are briefly considered at the end of this chapter in Section 3.8 on page 71 as many of the features of these are covered in the topics considered throughout the chapter.

The appropriate type of technology and its impact on systems costing will vary greatly depending upon the application. At one extreme we can consider a relatively simple vehicle proximity radar, based upon a milliwatt-level Doppler or clutter-reference system, attracting a unit price of only a few dollars. On the other hand, an earth-station for a satellite communications system represents a considerable capital investment costing up to hundreds of thousands of dollars.

Circuits using microstrip can be economically implemented in many low-to-medium power radars, associated electronic countermeasures, some segments of point-to-point microwave radio links, and several modules within satellite communications systems. Microstrip transmission lines are widely used in Microwave (hybrid) Integrated Circuits (MICs), Multichip Modules (MCMs), single chip packages and, increasingly, in monolithic integrated circuits for these systems.

Microstrip transmission lines require a dedicated reference ground plane for the return signal and in high performance digital integrated circuits result in a significant cost penalty because of the required addition of a metal layer. Other structures considered here provide more economical controlled impedance lines, but with different performance.

Before proceeding further, let us pause momentarily to consider the position of coaxial and rectangular waveguide[1] design. The evolution of Microwave Integrated Circuit (MIC) technologies does not mean that coaxial line and rectangular waveguide-based design are to be abandoned; on the contrary, such propagating structures assume continuing and significant roles. However, these structures are clearly inappropriate for integration although some novel use of micromachining techniques presents some interesting possibilities.

Carefully constructed coaxial lines and rectangular waveguides are used in precision instrumentation equipment — the network analyser, for example — and this situation is not likely to change radically in the foreseeable future. Waveguides also remain useful in at least three areas:

(a) high power transmission (e.g. kilowatt-to-megawatt transmitters)
(b) circuits where a very high-Q component is deemed necessary (e.g. very low loss filters)
(c) some millimetric-wavelength systems. (e.g. some millimetre-wave automotive radar use these as they conveniently form an antenna)
(d) very low loss transmission systems.

Developments in planar technology are already tending to overcome problems in areas (b) and (c), but not (a) or (d).

The dramatic evolution of transistor switching speeds and microwave semiconductor devices has added extra impetus to the development of hybrid microelectronics (i.e. Multichip Modules (MCMs) for digital applications, and MICs for microwave applications) and, more recently, RFICs (for Radio Frequency Integrated Circuits) and MMICs (for Microwave Monolithic Integrated Circuits). MCMs are considered in Section 3.9 on page 74 and this discussion draws heavily on the topics in the sections up to then.

For many years both Bipolar Junction Transistors (BJTs) and Field-Effect Transistors (FETs) have been readily available for operation as microwave transistors. More recently Heterostructure Bipolar Transistors (HBTs) and High Electron Mobility Transistors (HEMTs) have become of great interest. These devices based on silicon (BJTs and MOSFETs) are useful up to 10 GHz. Compound semiconductor, GaAs and InP, devices are also capable of providing useful power, gain and/or low-noise performance above 1 GHz, and up to 100 GHz. Also diode devices, such as Schottky, Gunn and IMPATT, are available to cost effectively detect or generate power at millimetre-wave frequencies. Furthermore, it has become apparent that hybrid microcircuit technology can readily be adapted to provide MICs. This combination of thick- and thin-film hybrid microcircuit technology and 'chip' components such as microwave transistors has formed a powerful design vehicle for modern integrated systems. The hybrid MIC approach extends to microwave and high-speed digital MultiChip Modules (MCMs).

[1] Rectangular waveguide is a transmission structure which consists of a tube of rectangular cross-section that 'guides' an electromagnetic wave. It is quite different from the interconnect structures considered in this book that consist of two or more conductors.

3.3 TRANSMISSION LINE STRUCTURES

Many planar transmission line structures have been conceived and variants are still frequently being developed; Figure 3.1 depicts a collection of such structures. Each one comprises a combination of metal, shown as dense black, and dielectric indicated by its permittivity ε. The region with permittivity ε_0 is free-space or air. In most cases the dielectric principally supports the metal pattern, acting as a substrate, and clearly influences the wave propagation, but the image guide is an exception in that propagation is actually via the dielectric slab. The actual choice of structure depends upon several factors including the type of circuit or subsystem and its operating frequency. For example, the image line is only suitable for consideration in millimetric wavelength applications, whereas microstrip is capable of propagating DC, or AC at frequencies up to some tens of gigahertz at least. Each type of line structure has potential advantages for various applications, and therefore each is briefly discussed here. First we expand on the term TEM for Transverse ElectroMagnetic mode.

As discussed in Chapter 1, the information in a signal is contained in the electromagnetic wave and when the electric and magnetic fields are in the plane perpendicular to the direction of travel the fields are said to be TEM. If they are nearly confined to the transverse plane then they are called quasi-TEM modes. With non-TEM transmission lines the electric and magnetic fields, together, are not even approximately TEM.

3.3.1 IMAGELINE (FIGURE 3.1(i))

At very high microwave frequencies, into hundreds of gigahertz, the wavelengths become comparable with practically realistic dielectric 'slab' cross-sectional dimensions. If such a dielectric 'slab', in the form of a continuous strip of dielectric, is placed on a backing sheet of conductor it forms what is known as an imageline.

This structure then behaves like a dielectric waveguide which propagates a range of TE and TM modes[2] trapped within it, due principally to the dielectric-air interface. A substantial amount of work on this and related structures has been reported in the literature [35–37]. Imageline is formed by adhering a dielectric slab (or thick strip) onto a metal ground plane, and typical dielectrics which may be used include boron nitride, alumina and high-resistivity silicon. The dielectric slab is usually machined to a regular, approximately rectangular, shape.

Although unloaded Q-factors of several thousand are common and operation above 100 GHz is best, many problems remain. There is poor compatibility with active devices, mutual coupling, radiation from discontinuities and bends. It is also apparent that the dielectric/metal adhesives are very lossy and this reduces the practical, loaded, Q-factor. It is therefore clear that much more work is necessary before imageline, or some derivative thereof, becomes a really attractive practical proposition.

Due to these difficulties, some further low-cost contenders have been considered for operation at frequencies around and exceeding 100 GHz. One interesting possibility

---

[2] TE and TM modes are a departure, but not total departure, from TEM modes. With TE modes the electric fields are confined to the transverse plane, and with TM modes the magnetic fields are.

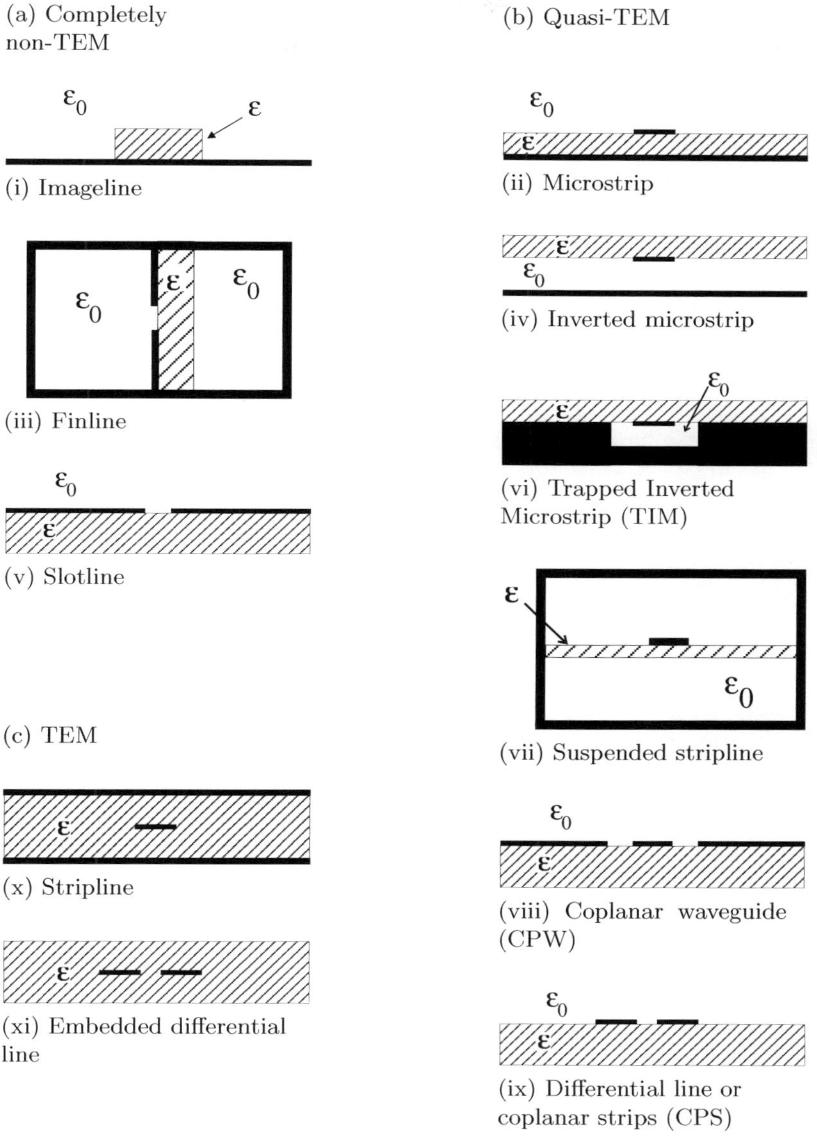

Figure 3.1 Common transmission line structures suited to planar fabrication.

is afforded by dielectric-filled waveguides of approximately semicircular cross-section [38]. Although measurements have been made available at somewhat lower frequencies (28–40 GHz), unloaded Q-factors of about 103 are obtained. The following advantages are also evident:

(a) No radiation losses, especially at bends (several decibels are typical for 90 degree bends in imagelines).
(b) Line-packing density is potentially high. Again, this is due to the inherent shielding.
(c) Tolerance requirements in the manufacturing process will be somewhat less restrictive than for imagelines, since the flowing process involved in the manufacture of these waveguides equalizes irregularities in the longitudinal direction.

Many features, especially parameters determining predictable manufacture, need to be examined before this structure becomes useful in engineering terms.

3.3.2 MICROSTRIP (FIGURE 3.1(ii))

Although microstrip has a very simple geometric structure the electromagnetic fields involved are actually complex. Accurate and thorough analysis requires quite elaborate mathematical treatment, and this is briefly described in Chapter 4. However, simple approaches to the quasi-TEM mode calculations combined with frequency-dependent expressions yield quite acceptable design accuracy for many applications. Accuracies within 1% are generally achievable.

Micromachining enables microstrip lines to be fabricated on a membrane with effectively no dielectric [39–51]. This reduces loss considerable especially at millimetre-wave frequencies. Micromachining also enables the microstrip line to be partially or totally shielded and so eliminating surface waves which can be a significant cause of loss, again at high frequencies. The line is then called a microshield line.

MICs using microstrip can be designed for frequencies ranging from a few gigahertz, or even lower, up to at least many tens of gigahertz.

At higher frequencies, particularly into the millimetre wavelength ranges, losses (including radiation) increase greatly, higher-order modes become a considerable problem, and fabrication tolerances become exceedingly difficult to meet using hybrid MICs. It is probable that the frequency limit for the extensive use of microstrip is in the region of 80 GHz. With ICs fabrication tolerances are much finer than with hybrid MICs and the options available for both microstrip and other transmission structures are extended considerably.

3.3.3 FINLINE ('E-PLANE' CIRCUITS) (FIGURE 3.1(iii))

With finline, a totally shielding rectangular conducting box (like rectangular waveguide but avoiding waveguide modes) has a dielectric substrate fixed, usually centrally, across two of its faces [35,36,38,52–56]. A metal circuit is deposited on one side of the substrate and a slot pattern in this metal forms the finline circuit.

It is essential to maintain an excellent continuous microwave short-circuit to the upper and lower waveguide walls, but this is fairly straightforward to achieve in practice. Since the structure is connected in the E-plane[3], circuits made using it are also called 'E-plane' integrated circuits. As a result, the transmission mode is also far removed from TEM. Some considerable advantages over other structures are obtained, mainly:

(a) low losses (typically a factor about three better than microstrip)
(b) simpler fabrication.

The second advantage, simpler fabrication, is very significant because many applications have been studied for finline at frequencies up to 100 GHz, and thus finline has to compete with structures such as inverted and trapped-inverted microstrip.

Despite some biasing problems, compatibility with solid-state devices is fairly good, especially in the case of beam-lead devices. Satisfactory bandpass filters, with approximately 10% bandwidth, have been built; quadrature hybrids, transitions to waveguide, and balanced mixer circuits have also been designed and manufactured. Discontinuities in finline have been analysed on a modal basis, and the results have been presented in a form appropriate to the design of impedance transforming and filter networks.

3.3.4 INVERTED MICROSTRIP (FIGURE 3.1(iv))

With this arrangement the strip conductor pattern is manufactured in the same manner as for the microstrip. The main difference is the absence of any conductor on the opposite side of the substrate — a ground plane is air-space separated instead. Thus the substrate mainly supports the microstrip lines and most of the field lies within the air between the strip and the ground plane. This ensures that the wavelength is long relative to that in conventional microstrip operated at similar frequencies. It also infers that inverted microstrip can be operated satisfactorily at substantially higher frequencies and this has been achieved [58]. Compared with conventional microstrip, a wider line is obtained for a prescribed characteristic impedance, and this both reduces conductor dissipation and relaxes fabrication tolerances. The inverted microstrip interconnect leads to higher Q components than achievable using microstrip. Thus, for example, lower loss filters can be fabricated using inverted microstrip.

Substrate/ground-plane spacers have to be well separated from the circuit to avoid interference. Some design information is available [59].

3.3.5 SLOTLINE (FIGURE 3.1(v))

This transmission line structure consists of a dielectric substrate metallized on one side only. The metallization has a completely separating narrow slot etched into it to form the slotline. The structure was first proposed by Cohn in 1968 and an extensive account has been given by Gupta *et al.* (See Chapter 6 in Gupta's book reference [57]). Gupta's material includes a wide range of basic design details as well as information on discontinuities, transitions and applications.

---

[3] E-plane refers to the plane of the electric field.

Some clever circuit applications have been devised, but the following disadvantages should be noted for slotline:

(a) Characteristic impedances below about 60 Ω are difficult to realize.
(b) Circuit structures often involve difficult registration problems (especially with metallization on the opposite side to the slot).
(c) The Q-factor is significantly lower than the other structures considered here.

There is another important derivative of slotline. When the conductors are made finite, i.e. restricted such that we have two adjacent strips differentially driven and on one side of the substrate only, this structure is termed coplanar stripline (CPS) — described in Section 3.3.8 on page 57.

3.3.6 TRAPPED INVERTED MICROSTRIP (TIM) (FIGURE 3.1(vi))

This is similar to the inverted microstrip, except that now the ground plane is made into a slot- or 'channel'-shaped section effectively grounding and shielding three planes, away from the strip conductor. This provides for inherent suppression of some higher-order modes which could otherwise propagate.

Both inverted microstrip and TIM structures are potential candidates for applications at frequencies extending up to and through 95 GHz, i.e. for millimetric wave applications. Manufacturing difficulties are, however, considerable — due to the tight fabrication tolerances that must be imposed on conductor/ground-plane spacing and channel dimensions. This is particularly significant with hybrid MICs, but may be less so when fabrication is on-chip (MMICs). Advances in micromachining of interconnects is allowing this structure to be revisited.

A limited amount of design information is available [36] for TIM and it is recommended that considerable further work should be carried out on this structure.

3.3.7 COPLANAR WAVEGUIDE ((CPW, FIGURE 3.1(viii))

Although CPW supports a quasi-TEM mode of propagation the actual shape of the structure is distinctly different from any of those considered so far because the 'active' metallization and the ground planes are formed on one side of the substrate alone. Each 'side-plane' conductor is grounded and the centre strip carries the signal; thus much less field enters the substrate when compared with microstrip. In conventional CPW the ground planes extend indefinitely but in Finite Ground CPW (FGCPW) the extent of the grounds is limited and this results in reduced coupling of adjacent and crossing CPW lines. Considerable design information is available and the structure has been used in many applications ranging upwards in complexity to complete amplifiers and subsystems. Semiconductor diodes or complete MMIC chips may be strapped across or ball-grid bonded on to the intermetallic gaps, meaning that there is often no need to drill the substrate or to etch vias into semiconductor substrates. An important modification is afforded by the conductor-backed CPW in which the opposite side of the substrate (the 'backside') is also metallized and often segmented to decrease losses.

Chapter 6 deals with CPW in considerable detail, and also provides extensive references.

3.3.8 COPLANAR STRIP (CPS) AND DIFFERENTIAL LINE (FIGURE 3.1(ix) AND 3.1(xi)

This simple transmission structure is formed by having just two conductors deposited on the same substrate with no other conducting layer. This is shown in two forms: as a surface line in Figure 3.1(ix) and as an embedded line in Figure 3.1(xi). In both structures the possible existence of ground planes (as in microstrip or stripline) is incidental and ideally these should not influence the field pattern. Essentially, the substrate merely acts as a mechanical supporting element and a quasi-TEM mode forms the main propagating field distribution. In one realization, one of the conductors is grounded, and this form is called a CoPlanar Stripline or CoPlanar Strips (CPS) [57,59,61–69]. In this configuration CPS is used as an area efficient variation of CPW. When neither of the conductors is grounded and the line is driven differentially, the interconnect is called differential line. Differential line is used extensively with RFICs and in critical nets in high-speed digital ICs. The two forms essentially have identical electrical characteristics with differences resulting from interaction with other metallic structures such as ground planes.

For a hybrid MIC this simple approach does not work well because of radiation losses and over-moding. However, the expansion of monolithic technology has rekindled interest in the differential line and work is developing intensively on the design and realization of this technology. This line is also popular for use in long bus lines and clock distribution nets on chip as the signal return path is well defined, the cross-sectional geometry is uniform along the interconnect's length (and hence the interconnect has a controlled impedance), and is on the same metallization layer as the signal path. This interpretation of the current paths — a signal path and a current return path — is not quite correct as neither metallization can be identified as the 'ground' reference for the other. Differential line and CPS are discussed in detail, in Section 6.11 on page 218.

3.3.9 STRIPLINE (FIGURE 3.1(x))

This basic structure predates microstrip and any of the other planar transmission options described in this book.

It comprises a 'completely filled microstrip', i.e. a symmetrical structure somewhat like a coaxial line completely flattened out and rectangularized so that the centre conductor is a rectangular metal strip and the outer grounded metal is simply a rectangular box. This is similar to microstrip in a box, and the entire structure is 100% filled with dielectric, and therefore the transmission is entirely TEM and dependent upon the relative permittivity ε_r explicitly. Therefore the wavelength is simply the free-space value divided by the root of ε_r. Further design criteria associated with stripline are developed in appropriate sections throughout this book.

If the ground 'plane' is made to totally enclose the structure, with air gaps at top and bottom, we have suspended stripline (Figure 3.1(vii)). Although the losses are all relatively low, resulting in a fairly high Q-factor, waveguide modes can easily be excited at higher frequencies. This could possibly be prevented by bringing the top and bottom walls very close together and/or by introducing various structures to act as mode suppressors. In the first place the separations become so small that

Table 3.1 Characteristic impedance (Z_0) ranges for the various structures with a substrate permittivity of 10.

Structure	Z_0 (Ω)
Microstrip	11–110
Inverted microstrip	11–130
Trapped inverted microstrip (TIM)	14–140
Suspended stripline	40–150
Coplanar waveguide (CPW)	40–110
Differential line, Coplanar strips (CPS)	40–110
Slotline	35–250
Finline	10–400
Imageline	≈ 26

there is insufficient space for the incorporation of semiconductor devices, etc., and the introduction of mode suppressors represents a new design problem. In any case, the basic structure is not compatible with shunt-mounted semiconductor devices.

Some work has also been performed on 'modified' suspended striplines [60].

3.3.10 SUMMARY OF INTERCONNECT PROPERTIES

For circuit applications the characteristic impedances and Q-factors achievable with these various structures need to be known. Table 3.1 lists the ranges of characteristic impedance that may be expected. These values are the practically realizable values determined by what can be fabricated given manufacturing tolerances and avoiding multimoding. Z_0 varies as the inverse of $\sqrt{\varepsilon_r}$ although this is modified slightly by the cross-sectional geometry of the line for non-homogeneous media such as open structures.

Each of the quasi-TEM structures is capable of being designed at lower characteristic impedances than those indicated in Table 3.1. The general technique is simply to increase the width of the strip conductor, but this results in a risk of transverse resonance modes (dealt with in detail for microstrip in Chapter 5). At low-to-moderate frequencies microstrip impedances down to just below 10 Ω are thus achievable.

In most cases, the upper limit to characteristic impedance is set by manufacturing tolerances and the unreliability associated with a very narrow strip. For the quasi-TEM lines, especially microstrip, this is an embarrassment, since some circuit applications (e.g. RFICs) can demand $Z_0 > 150\ \Omega$. Unless slotline or finline can then be used, alternative clever circuit/system design is necessary.

The characteristic impedance range of finline is remarkably large — 10–400 Ω — which is more than adequate for nearly all conceivable filter or matching network applications. On the other hand, imageline is restricted to around one value ($\approx 26\ \Omega$) which is a further drawback with this structure. However, several useful circuits have been developed employing various directional couplers, ring resonators, etc., which do not depend upon changes in characteristic impedance for their design and operation [38]. Unloaded Q-factors are of great importance in resonant circuit

Table 3.2 Comparison of typical unloaded Q-factors (Q_u) at a frequency of 30 GHz for the structures shown in Figure 3.1. The substrate permittivity is 10 for all structures except finline, where it is 3.3.

Structure	Unloaded Q, Q_u
Microstrip	250
Inverted microstrip	400
Trapped inverted microstrip (TIM)	450
Suspended stripline	600
Coplanar waveguide (CPW)	200
Differential line, Coplanar strips (CPS)	200
Slotline	200
Finline	550
Imageline	2500

applications (e.g matching networks and filters), and these are indicated in Table 3.2.

Most of the results are typical practical values, but imageline suffers dramatic reductions in unloaded Q-factors when any bending or other shape discontinuity is introduced. Many practical circuits or subsystems are likely to demand a range of characteristic impedances, different operating frequencies and different bandwidths, etc. The integration of multifunction circuits, especially for millimetric wave applications, is likely to require the use of more than one type of transmission structure to achieve the desired level of performance. Transitions would also be required to efficiently couple energy from one structure into another, e.g. waveguide to imageline, imageline to finline, TIM to finline, etc. Extensive advancements will almost certainly be seen in these areas.

3.4 SUBSTRATES FOR HYBRID MICROCIRCUITS

Many facets, mechanical and thermal as well as electronic and economic, influence the decision process leading to the correct choice of a particular substrate for a specific type of hybrid circuit and application. The kinds of questions that should be asked must include:

(a) Can the cost of the substrate under consideration be justified in the light of the application and attendant circuit/system costs?
(b) Is the technology to be thin- or thick-film?
(c) What frequency range or ranges are involved? (This will influence thickness and permittivity.)
(d) Will the surface finish be sufficiently good to keep conductor losses tolerably low — yet maintain metal-film adhesion?
(e) Are the mechanical strength and thermal conductivity sufficient for the application in mind?
(f) Are the substrates readily available with sufficient surface area, considering the circuit complexity and the operating frequencies?

This is not an exhaustive list and other considerations may well arise in certain situations. Before comparing materials it is useful to list a variety of possible substrate materials; these are shown in Table 3.3. We now consider the three major hybrid microcircuit substrates and also a fundamental benchmark substrate — namely sapphire. We will return and consider two of these in much more detail later (printed circuit boards and ceramic substrates in Sections 3.8 and 3.9. on pages 3.8 and 3.9, on pages 71 and 74 respectively.

3.4.1 FR4 ('PRINTED CIRCUIT BOARD')

This substrate material is basically designed for conventional low-frequency printed circuit (or printed wiring) boards or cards. It can be used at much higher frequencies, into the cellular bands, but losses increase rapidly and design is imponderable because of the great variability in basic substrate properties. For RF transmission, or very high speed pulse transmission, it is essential that the permittivity and thickness of any substrate are both known accurately and can be held repeatedly.

FR4 is therefore unsuitable for RF and microwave or very high speed (gigahertz rates) digital applications.

3.4.2 CERAMIC SUBSTRATES

Alumina — a typical ceramic used — is well suited for production circuits functioning at frequencies up to about 40 GHz. The grade to be used depends on a number of factors including the choice of basic fabrication technique: thick-film or thin-film. Comparatively high conductor and dielectric losses and poor reproducibility usually exclude the choice of 85% purity (or worse) material. Since alumina substrates have become very popular for manufacturing circuits to operate at any frequency up to about 40 GHz (although higher is possible), a further table is now presented (Table 3.4) which summarizes some important features regarding alumina substrates.

A wide variety of commercially available substrates exists, one example being MRC Superstrate®. Extensive microstrip measurements indicate that the maximum permittivity of such high-purity substrates must lie somewhere between the limits 10.1 and 10.3.

Ceramic substrates such as High Temperature Co-fired Ceramics (HTCC) are also used as the basic medium for MultiChip Modules (MCMs).

Low Temperature Co-fired Ceramics (LTCC) have become very popular for all ceramic substrate applications including MCMs and single chip packages. They have tightly controlled tolerances, as shown in Table 3.5, and can be stacked in 80 layers or more. As with all fired ceramics, metal patterns are printed on soft un-fired material using a thick film process. (The thick film process is discussed extensively in Section 3.6 on page 66.) Holes for vias are puched in the un-fired layers and are filled in the metallization process. The layers are then stacked, compressed, and then fired when they soften, flow and shrink. Key to the process is the tight control of shrinkage.

Table 3.3 Properties of some typical substrate materials.

Material	Surface roughness (μm)	$10^4 \tan\delta$ (at 10 GHz)	ε_r	Thermal conductivity K (W/cm^2/°C)	Dielectric strength (kV/cm)
Air (dry)	N/A	≈ 0	1	0.00024	30
Alumina:					
99.5%	0.05–0.25	1-2	10.1	0.37	4×10^3
96%	5–20	6	9.6	0.28	4×10^3
85%	30-50	15	8.5	0.2	4×10^3
Sapphire[1]	0.005–0.025	0.4–0.7	9.4, 11.6	0.4	4×10^3
Glass, typical	0.025	20	5	0.01	–
Polyimide	–	50	3.2	0.002	4.3
Irradiated polyolefin	1		2.3	0.001	≈ 300
Quartz (fused) i.e. SiO_2	0.006–0.025	1	3.8	0.01	10×10^3
Beryllia (BeO)[2]	0.05–1.25	1	6.6	2.5	–
Rutile	0.25–2.5	4	100		–
Ferrite/garnet	0.25	2	13–16	0.03	4×10^3
FR4 circuit board	≈ 6	100	4.3–4.5	0.005	–
RT-duroid® 5880	0.75–1[3] 4.25–8.75[4]	5–15	2.16–2.24	0.0026	–
RT-duroid® 6010	0.75–1[3] 4.25–8.75[4]	10–60	10.2–10.7	0.0041	–
AT-1000®	–	20	10.0–13.0	0.0037	–
Cu-flon	–	4.5	2.1	–	–
Si (high resistivity)	0.025	10–100	11.9	0.9	300
GaAs	0.025	6	12.85	0.3	350
InP	0.025	10	12.4	0.4	350
SiO_2 (on-chip)	–	–	4.0–4.2	–	–
polyarylether (SiLK™)	CMP[5]	–	2.65	0.19 @ 25°C 0.23 @ 125°C	– –
silicon oxycarbide (SiCOH)	CMP[5]	–	2.7	–	–
LTCC (typical, green tape™951)	0.22	15	7.8	3	400

[1]Single-crystal. The properties and effects of sapphire are discussed in more detail in Chapter 4.
[2]Beryllia is, however, a dangerous ceramic to handle. Its dust is toxic and must not be machined in any way. It is therefore best avoided.
[3]Average peak-to-valley height differences for rolled copper construction.
[4]Average peak-to-valley height differences for electro-deposited copper construction.
[5]Chemical Mechanical Polishing used for planarization.
[6]From DuPont, see also Table 3.5.

Table 3.4 Some specific observations regarding alumina substrates.

Permittivity: Average values	Within the approx. range $8.2 \leq \varepsilon_r \leq 10.2$. Depends on purity and other factors.
Permittivity: Batch-to-batch variation	Considerable. Therefore has to be taken into account in design. (Worsens with deteriorating purity).
Permittivity: Material dispersion	Significant variation of ε_r with frequency, through microwave frequencies (e.g. 8 to 12 GHz for some aluminas). Some dipolar behaviour, amount depends on impurities.
Dielectric loss ($\tan\delta$)	Generally $< 1 \times 10^{-3}$ (1×10^{-4} for $> 99.5\%$ purity material).
Anisotropy	Present to varying extents. Depends upon method of manufacture (caused by preferential orientation of crystallites).
Surface roughness	See Table 3.3.
Surface flatness	Bowing can often occur to a significant extent with cheaper substrates (not necessarily just low-purity substrates).

Table 3.5 Specific properties of Green Tape™ 951 — an LTCC material from E. I. du Pont de Nemours and Company. See http://www.dupont.com/mcm/greentape for more information.

Parameter	Value	Tolerance
Permittivity (ε_r)	7.8	± 0.1
Conductor thickness (μm)	9	± 2
x-y shrinkage, %	12.7	± 0.3
z shrinkage, %	15	± 0.5
Unfired tape thickness tolerance, %	–	± 7
Fired thickness, AT tape (μm)	97	± 8
Fired thickness, A2 tape (μm)	140	± 10
Fired thickness, AX tape (μm)	216	± 18
Line width tolerance (μm)	–	± 13

3.4.3 SOFTBOARD

This is suitable for experimental circuits at frequencies below a few gigahertz as well as array antennas for frequencies up to and beyond 20 GHz. At the higher frequencies irradiated polyolefin and Cu-flon as well as other materials are used.

3.4.4 OVERALL APPRAISAL — ALTERNATIVE SUBSTRATES AND STRUCTURES

Production circuits for millimetric wave applications from some tens of gigahertz up to perhaps 300 GHz: quartz, or an alternative structure such as finline or imageline. The lower permittivity of quartz allows larger distributed circuit elements to be incorporated.

3.4.5 SAPPHIRE — THE 'BENCHMARK' SUBSTRATE MATERIAL

The weighted average permittivity of sapphire is within the range 10.1 to 10.3. Using the information given in Table 3.3, this average can be calculated quite accurately:

$$\varepsilon_{r(\text{sapphire}\rightarrow\text{average})} \approx \frac{1}{3}(2\varepsilon_{\perp} + \varepsilon_{\parallel}). \tag{3.1}$$

Substituting the given values (Table 3.3) of $\varepsilon_{\perp} = 9.4$ and $\varepsilon_{\parallel} = 11.6$, we obtain:

$$\varepsilon_{r(\text{sapphire}\rightarrow\text{average})} \approx \frac{1}{3}(18.8 + 11.6) = 10.13. \tag{3.2}$$

The very highest grade of polycrystalline alumina has this bulk permittivity. However, surface finish is usually relatively poor — leading to inadequate adhesion of the metal layers and also a deterioration in conductor losses.

MRC's Superstrate® alumina has a permittivity of about 9.8 at microwave frequencies. The surface finish is generally good, and the material finds much application in MICs because film adhesion is also more than satisfactory. Over the last decade a large number of high performance substrates have been developed by several manufacturers.

Although sapphire is usually the most expensive substrate likely to be entertained for hybrid MICs, it does offer some unique advantages:

(a) It is optically transparent. Hence chip devices can be accurately registered for attachment since the underside may be viewed.
(b) It has a fairly high permittivity (around 10). This is almost compatible with high-resistivity silicon. It also means that microstrip components are fairly small due to the reduced wavelength.
(c) Sapphire substrates are highly reproducible; all pieces are essentially identical in dielectric properties and they are not subject to bowing.
(d) Surface finish is optical grade which results in low power loss from surface roughness.
(e) Thermal conductivity is quite high, being about 30% higher than the best alumina.

Against these advantages, and apart from the relatively high cost, we must bear in mind that sapphire substrate areas are limited (usually little more than 25 mm square) and the dielectric anisotropy poses some additional circuit design problems.

3.5 THIN-FILM MODULES

Microwave-grade hybrid circuits place stringent demands on the accuracy and repeatability of the circuit technology. These demands can be met by carefully controlled thin-film manufacturing processes, which are therefore used for the great majority of MICs. 'Clean-room' areas are used for the purpose.

Various metals form the bulk material with gold, silver, copper and aluminium being in common use. Owing to its superior conductivity and also its lower price, copper is preferred — but a final gold film must be evaporated over the copper to provide an environmental protective layer. The necessary thickness of the metal film may be determined by considering the skin depth in the metal at the lowest frequency of interest. Taking this lowest frequency to be about 4 GHz, the skin depth in copper is then approximately 1 μm. At least four skin depths are recommended, and therefore the required thickness should be 4-5 μm. The processes actually used for making the circuit may be visualized, rather crudely, as precision (small-scale) printed circuit processes. Practical techniques involve three main steps:

(a) Evaporation or sputtering a thin 'seed' layer of a suitable metal on to the surface of the substrate.
(b) Evaporating or sputtering a thin layer of conductor metal on to this 'seed' layer.
(c) Electroplating the bulk conductor metal on to layer (b).

Steps (a) and (b) provide, respectively, mechanical and electrical foundation layers on which to electroplate good quality bulk conductor metal. The techniques differ chiefly in the choice of evaporation or sputtering, and in the manner in which (using photolithographic methods) the circuits are defined. In one well-proven technique [70], step (a) comprises approximately 200 Å of chromium, step (b) comprises a layer of copper of similar thickness, and approximately 5 μm of copper is plated in step (c). The thin layers are produced by magnetron sputtering [71] where the combined electric and magnetic fields produce efficient ionization of the metal (copper). The ions are attracted towards the substrate target which behaves as an anode.

Circuit definition may be accomplished either by a plate-through technique or by an etch-back technique and each of these techniques is now described.

3.5.1 PLATE-THROUGH TECHNIQUE

Substrates which are cleaned to a high specification have their surfaces sputter coated as explained above. Next, a photoresist layer, of similar thickness to the final metal film required, is 'spun' on to the surfaces. Exposure with ultraviolet light through a precision contact photomask (with the required circuit pattern) is followed by subsequent processing to yield a pattern of slots in the photoresist layer. The copper is plated through these slots to form the circuit pattern and then the photoresist layer is washed away and the excess sputter coating is etched off. There are further details in the cited reference [70].

3.5.2 ETCH-BACK TECHNIQUE

Again, the process starts by coating the substrates with thin films of metal by sputtering, exactly as explained before, and a photoresist layer is also spun on to the surfaces. In this case, however, the photomask exposure step is arranged so that those larger areas of the surface where metal is not required are left unprotected by the photoresist layer. Thus, at this stage, the circuit pattern is effectively defined by a fine pattern of the photoresist layer across the very thin sputtered layers. 'Etching-back' then removes the unwanted areas of metal which were originally sputtered on. Electroplating yields the desired final thickness. This technique has the following advantages:

(a) A thinner photoresist layer may be used — typically 0.5 μm.
(b) Plating is more uniform.
(c) Microstrip edges are relatively square.

However, for the thinner photoresist layer to work satisfactorily, the substrate surface must be extremely smooth — and free from dust or grit (otherwise the layer becomes perforated). Sapphire substrates, with their optical surface finish, are very good in this respect.

3.5.3 EQUIPMENT REQUIRED

For precise and repeatable circuit production the following items of equipment are recommended:

(a) Class 10000 clean room, achieving class 500 in the main working area.
(b) Coordinatograph for preparing 25 x size of artwork (in the 'cut'n strip' method).
(c) Precision 25 $\times$ reduction camera.
(d) Magnetron sputtering operating in high vacuum, typically of the order of 10^{-6} Torr.
(e) Vacuum spinner, having adjustable speed in the range 1000–7000 revs/min, and timer.
(f) High-intensity collimated ultraviolet light source, for exposing photoresist layers.
(g) Ultrapure water (UPW) system, delivering water of about 18 MΩ resistivity.

Many other details should be considered by anyone contemplating the establishment of a thin-film circuit manufacturing facility, and the paper by Ladbrooke *et al.* [70] treats the topic very well, as do Jensen [72] and Sergent [73].

3.5.4 THIN RESISTIVE FILMS

Resistive films are required for many types of MICs, including circuit elements such as terminations, bias networks and attenuators. These resistive films must have a low temperature coefficient of resistivity and be of good stability. Commonly used resistive materials include nichrome and tantalum. Resistance values within the range 50–5000 Ω typify MIC requirements and such values can readily be realized by the deposition of films using these materials. With abrasive trimming or, alternatively, laser trimming it is possible to achieve resistance tolerances to an accuracy exceeding

1%. The creation of these resistive films demands additional processes of deposition and etching beyond those of the thin-film metallization described above, and this complexity may be obviated by bonding directly chip resistors onto the conducting pattern (i.e. using surface mount or flip-chip components).

3.6 THICK-FILM MODULES

Thick-film technology has become well established in the low-cost manufacture of hybrid microcircuits at lower-than-microwave frequencies and for densely packaged digital subsystems [74]. Large advantages in cost reductions are obtainable compared with thin-film technology, and even more so with monolithic integrated circuit technology, especially where small 'batch production' runs are involved.

Thick-film materials and techniques have been developed for MICs operating at frequencies up to at least 40 GHz [71]. The losses associated with the thick-film conductors have been shown to be (typically) about 20% worse than those associated with comparable thin-film conductors and this limitation, combined with tolerancing difficulties, tends to restrict production circuits to a maximum frequency of around 10 GHz. Circuits such as parallel-coupled and open-circuit stub filters, compensated couplers, attenuators and power levellers have all been satisfactorily realized.

Two methods are commonly encountered for the manufacture of 'thick-film' cheap MICs:

(a) Thick-film patterns are printed and fired on to the ceramic substrate — usually alumina or an LTCC substrate, but occasionally quartz [71, 74–76].
(b) A printed circuit technique is used to etch the desired pattern in the copper cladding of what is usually a plastic substrate (often polyolefin for microwave applications; see Table 3.3).

Either method is substantially simpler and less demanding on both equipment and environment than thin-film technology. Method (b) is very well known and need not be described here. The first method, (a), is used for LTCC, MIC, and single chip package manufacture to an increasing extent and therefore it will be briefly described.

3.6.1 PASTES, PRINTING AND PROCESSING FOR THICK-FILM MODULES

The metal that will ultimately form the microcircuit conductors, often gold or silver but increasingly a copper-based material, is initially available to the circuit manufacturer as a paste or 'ink' contained in a jar. A few millilitres of this paste are placed on a fine-mesh 'screen,' with areas open for the circuit pattern, and some of the paste is squeezed through these open areas and on to the surface of a substrate held rigidly just beneath the screen. Settling, drying, and firing sequences complete the 'thick' deposit, a very high proportion of which is pure metal. Although the process has earned the generic name 'thick film' the actual film thicknesses are only usually about twice that of thin films, i.e. 10 μm or so. The best way to explain the process is to briefly describe each step, and so we start with the pastes.

(a) The conductor pastes have to be special 'fritless' materials, and manufacturers such as Du Pont or Electro-Science Labs provide such pastes. For substantial shelf life pastes should be kept in a refrigerator.

(b) The artwork is prepared using similar procedures to those described for thin-film technology and photographic processing is then used to obtain a positive transparency.
(c) Suitable screens are available, often with dimensions about 15 × 20 cm and made of at least 305 mesh stainless steel or polyester. These are stretched to 'drum tightness' over a rigid frame which will fit into the screen printer. They are also coated with a suitable photoresist layer.
(d) The positive transparency, obtained in step (b), is held in intimate contact (image side) with the coated surface of a screen, and then standard ultraviolet exposure and wash-and-bake processes leave a screen with apertures for the required circuit. All other areas of the screen are then opaque with a durable photoresist layer.
(e) This screen is placed in a printing jig (preferably a machine with well-defined adjustments and controls). The substrate which is to receive the circuit is placed beneath the aperture region of the screen — with accurate registration. This substrate is held firm by vacuum suction and is usually about one-substrate-thickness clear of the base of the screen.
(f) A few millilitres of conductor paste are placed on the screen, between the squeegee and the aperture region.
(g) When the parameters have all been optimized, the aperture region of the screen is wet and the action smooth, a satisfactory wet deposit of paste will be transferred on to the substrate.
(h) This 'wet-circuit' deposit must next be left to settle by placing it horizontally in a clean area for about 15 minutes. Following this it is dried at approximately 1000 °C for about 15 to 20 minutes under an infrared drying unit.
(i) The firing process returns the material of the deposit to a predominantly metallic substance of high electrical conductivity. For the gold-suspension pastes this generally entails about 10 minutes at a temperature set somewhere in the range 900–1000 °C, depending upon the specification of the paste.
(j) Immediately after firing the circuit is, in principle at least, ready for use.

In many instances, however, the precise definition required cannot be achieved with the as-fired circuit, and either laser trimming or an etch-back process is necessary. This is especially a problem in proximity-coupled microstrips that are often required in MICs [55, 71].

As noted above, there exists a requirement for resistive films and these can also be provided in thick-film form. The deposition and processing mainly follows the steps given above for the conductive films except that the firing profile is more critical since, in conjunction with the resistive paste formulation, it controls the final resistivity of the film.

3.7 MONOLITHIC TECHNOLOGY

3.7.1 INTRODUCTION

Conceptually, electronic Integrated Circuits (ICs) comprise the simultaneous, or near simultaneous, processing of all electronic functions within a semiconductor chip. At lower frequencies, and in by far the greater number of electronic applications, the basic semiconductor material used is silicon. At microwave frequencies, up to 38 GHz at least, silicon can still be used and indeed this is sometimes the case for production Monolithic Microwave Integrated Circuits (MMICs). Silicon, or silicon-germanium, Microwave Millimetre-Wave Integrated Circuits (sometimes termed SIMMWICs) are also feasible [77].

Semiconductor substrates of suitable form for MMICs must have high resistivities, in the order of 2000 Ω.cm or more in order to maintain the transmission losses at sufficiently low levels, and also to provide enough isolation between components on the chip. High-resistivity silicon, semi-insulating gallium arsenide (GaAs) and other compound semiconductors meet this requirement. In fact, semi-insulating GaAs provides significantly higher resistivity than the high resistivity silicon (by a factor of well over 100). Indium phosphide (InP) represents another useful semiconductor, particularly for millimetre-wave applications and mixed microwave/optoelectronic.

Other suitable semiconductors include silicon-germanium (SiGe), which is being promoted by several companies, and silicon carbide (SiC) for very high stress and high temperature applications.

Various lithographic processes are used to delineate device geometries and circuit patterns in ICs. For example, with photolithography, diffraction effects limit the resolution of microstrip and other lines to dimensions in the order of a wavelength of light. For substantially higher resolution electron beam lithography is used, particularly for the realization of submicron gates for devices such as GaAs MESFETs or PHEMTs, to operate at the higher microwave and into millimetre-wave frequencies. Combined processing techniques, taking the best advantages of photolithography and electron beam lithography are often used in the fabrication of commercial MMICs. Since the passive circuit patterns use precisely the same lithographic techniques that are employed to define the submicron transistor gates it follows that dimensional tolerances in the circuit patterns using this technology are controlled far more closely than they are in hybrid microelectronic technology.

In this text we are principally concerned with the design of both hybrid and monolithic circuits, but with a specific emphasis on the circuit design itself, i.e. not the active devices. Therefore in our brief discussion of monolithic technology here, we shall exclude any details regarding the processes required for the realization of active devices in MMICs (i.e. including the various transistor types like GaAsFETs, HEMTs and HBTs). Some remarks are, however, in order regarding the substrate requirements.

Selected substrates should be capable of retaining their intrinsic properties during the required annealing cycle in the production process. Chrome-compensated semi-insulating GaAs offers a stable resistivity over the temperature range of 850–900 °C necessary in the fabrication of the microwave devices within an RFIC or MMIC. In these terms the implementation of MMICs in silicon represents a drawback, due to the fact that silicon's intrinsic insulating qualities become degraded at temperatures exceeding 800°C. It is, however, possible to process silicon wafers, using a combination of ion implantation and selective laser annealing and this avoids increasing the temperature of the entire substrate which would otherwise have resulted in the degradation. In spite of the relatively high capital costs associated with wafer fabrication facilities, MMICs provide several distinct technical and economic advantages for an increasing variety of microwave and millimetre-wave applications. To introduce yet another abbreviation, Millimetre-wave MMICs are sometimes termed

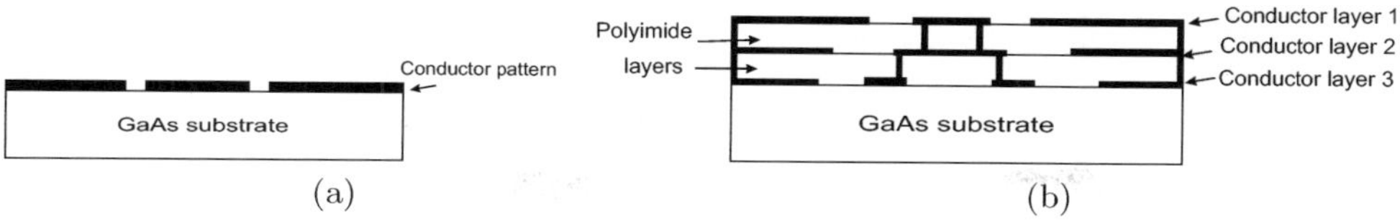

Figure 3.2 Conducting vias connecting between conductor layers in an MMIC.

MMMICs. The main application outlets for MMICs or MMMICs, as distinct from hybrid MICs, are those with relatively high volume requirements. Some microstrip and CPW orientated MMIC design approaches are described in Chapter 11.

3.7.2 MULTILAYER INTERCONNECT

The transmission medium options available to the RFIC or MMIC designer include a selection of those described earlier in this Chapter: microstrip, slotline, differential line, coplanar waveguide, coplanar stripline and (possibly) inverted microstrip.

As described above, all forms of integrated circuits (whether microwave or otherwise) require technologies for interconnecting between several layers. Transistor drains and gates, also power requirements and connections between other devices such as lumped passive elements (C, R, L) or Schottky diodes, all demand conductive interconnections across layers.

Whilst the layers themselves must be formed, for microwave chips, using one or more of these transmission medium options — vias (conducting 'holes' between the layers) are necessary [78]. The general arrangement is shown in Figure 3.2. In this structure (Figure 3.2) several vias are shown, connecting between CPW and microstrip lines located in various regions of the chip. The structure can be continued longitudinally to reduce the resistance of the interconnect and skin effect.

These entire structures, the active and passive devices plus the transmission lines and the vias, must all be modelled using CAE software that is proven fundamentally capable and reliable over the frequency range or time intervals demanded by the circuit function. The demands on such software intensify as frequency increases (or time intervals decrease) towards the high tens of GHz ranges (tens of picoseconds). In particular designers must be aware of the danger of extrapolating models. Simple extrapolation most probably will not correctly model the behaviour of the circuit beyond the strict capability of the basic software.

In general: never believe just what the software models are predicting alone — always seek additional support from measurements, if possible. At the very least, ensure that predicted (computed) data makes sense in terms of engineering fundamentals. When two sets of predictions, derived from two completely independent CAE packages, agree reasonably closely then more confidence may be placed upon either package regarding predicted results.

3.7.3 METALLIZATION

A variety of metals have been used to form the metal layers in MMICs, notably gold, copper and aluminium. Gold represents a relatively expensive solution, but its conductivity is not particularly high, and in any event the process is less environmentally acceptable than other approaches. The major advantage with gold is the self-passivation effect — the metal resists corrosion and will not form alloys with other materials. In monolithic integration its use is restricted to MMICs

Aluminium has traditionally been used for Si-based ICs. It is inert with Si and SiO_2 but its conductivity is markedly worse than copper. Copper is an excellent electrical conductor but has some disadvantages in that it readily forms intermetallic compounds with several semiconductors including silicon, and it has electromigration problems.

Aluminium-copper mixtures are also being utilized because this approach provides the best compromise to exploit the advantages of each metal whilst avoiding the main problems associated with either metal on its own. Copper with buffer layers of other metals is also being used.

For ICs having feature sizes below 0.18 μm the use of copper, copper alloys, or copper with buffer layers, is essential. As the feature size is reduced the resistance of the conductors, now with smaller cross-sections, increases and, as has been noted, the delay of a resistive line depends strongly on the resistance — the higher the resistance the longer the delay.

3.7.4 LOW-K DIELECTRICS

Conventionally silicon-based integrated circuits have used aluminium conductors and silicon dioxide (SiO_2) insulators between the conductors. For higher speeds it is necessary to use insulator materials with permittivities below that of SiO_2 (4.0–4.2)[4] to reduce the self-capacitance of lines and also to reduce the mutual capacitance of neighbouring interconnects.

Zarkesh-Ha *et al.* developed a simple formula for the speed improvement in a digital IC that can be achieved by replacing the conventional Al metallization and SiO_2 dielectric [79]. For long interconnects, so that the interconnect delay dominates timing, the fractional reduction of the critical path delay with the modified material system relative to the Al/SiO_2 system is [79]

$$\frac{\Delta T_{\text{CRITICAL}}}{T_{\text{CRITICAL}}} = \frac{\Delta\rho}{\rho_{\text{Al}}} + \frac{\Delta\varepsilon}{\varepsilon_{\text{SiO}_2}} - \left(\frac{\Delta\rho}{\rho_{\text{Al}}}\frac{\Delta\varepsilon}{\varepsilon_{\text{SiO}_2}}\right) \tag{3.3}$$

where ρ is the resistivity Now the ratio of the copper resistivity to that of aluminium, $\Delta\rho/\rho_{\text{Al}}$, is 39% and, using a relative permittivity of 2.5 for the low-k dielectric, the ratio of the low-k permittivity to that of SiO_2, $\Delta\rho/\rho_{\text{Al}}$, is 36%. Then the ultimate speed improvement is 39% if Cu alone is used, 36% if the low-k material alone is used, but when combined the improvement is 61% less than would be expected by considering the technologies individually. The actual improvement obtained would be somewhat less than this as critical timing is also determined by gate delay.

At the time of writing of this book there are no silicon ICs in production using low-k (or low permittivity) materials — defined as materials with permittivities below 4.0. There are several candidate materials, one or more of which will be adopted during the

[4] The permittivity of SiO_2 depends on the growth conditions and can extend down to 3.8 but this is less common.

lifetime of this book. The candidate materials include polyarylether (manufactured by Dow Chemical, Schumacher, and Honeywell Inc.) which is deposited using spin-on technology; and silicon oxycarbide which forms when trimethylsilane (3MS) (manufactured by Dow Corning Corp.) is used in a chemical vapour deposition (CVD) system. The low-k dielectric materials that are spun-on result in lower manufacturing cost than do the materials requiring CVD processing but there are problems with their hardness.

SiLK^{TM}, a polyarylether from Dow Chemical[5], has a dielectric constant of 2.65 but is relatively soft confining its use as an insulator for the first few metal layers. SiLK^{TM} can be either hot plate or furnace cured. When cured it produces a thin-film of 0.2–1.5 μm thickness with low shrinkage and good planarization properties. The different thicknesses are obtained using formulations of different viscosity. Capping layers using SiO_2, for the top one or two metal layers, are required to obtain satisfactory mechanical strength. Special attention must also given to the metallization patterns to avoid the material being damaged during chemical mechanical polishing (CMP). The capping layer increases the effective permittivity for the upper metal layers to 3.0 from SiLK's 2.65. However lower layers are unaffected as the electric fields here are confined to the immediate region. IBM has announced that they will be using SiLK^{TM} in volume production of 0.13 μm technology silicon ICs beginning in 2001.

Silicon oxycarbide (SiCOH), as formed in a process developed by Novellus[6], has a dielectric constant of 2.75 and is reportedly hard enough so that it does not require capping layers (or a hard mask).

Considerable efforts are being directed at developing ultra-low dielectric constant materials for IC interconnects. The aim is to develop materials with adjustable relative permittivity to 1.5. Materials being investigated include inorganic materials derived from organosilicate polymers and organic thermosetting polymers. The technical challenges include developing a material which does not produce pores, has sufficient mechanical strength that capping layers are not needed, and can be deposited using low cost manufacturing techniques.

3.7.5 MIC AND MMIC APPROACHES COMPARED

An outline comparison of hybrid MIC and MMIC technologies, focusing on their respective features and attributes, is provided in Table 3.6. An excellent detailed coverage of economics, costs and other comparative aspects of MIC versus MMIC technology is provided by Sweet [80].

3.8 PRINTED CIRCUIT BOARDS

PCBs are of two main types:

(a) Organic PCBs: these use organic resins generally reinforced with fibres but also

---

[5] See also http://www.dow.com/silk/products/silk.htm

[6] Novellus is a CVD equipment manufacturer that has developed a process for laying down silicon oxycarbide using trimethylsilane. Similar processes have been developed by Applied Materials and Trikon, also CVD equipment manufacturers.

Table 3.6 MIC/MMIC technology: manufacturing costs and operational capability comparisons.

Technology	Approximate capital cost for fab facility[1] (constant US$)	MM-wave capabilities?	Volume applications?
Thin-film hybrid MICs	10^4–10^5	Yes. To well over 100 GHz	Not usually[2]
Thick-film hybrid MICs	10^3–10^4	No. Maximum perhaps to around 20 GHz	Not usually[2].
Monolithic Microwave ICs (MMICs)	10^6–10^8	Yes. To well over 100 GHz	Yes — well suited to high volume[3].

[1]Order-of-magnitude indications of the start-up capital requirements for an MIC or MMIC fabrication ('fab') facility.
[2]Hybrid MICs are best restricted to low-to-moderate volume applications, on economic grounds.
[3]There are many considerations apart from volume that may influence the decision to realize a circuit in MIC or MMIC form. Monolithic realization has certain specific technical advantages over hybrid approaches.

Table 3.7 A selection of semiconductor substrates (mainly from Ponchak *et al.* [81]).

Substrate	Relative permittivity, ε_r	Loss tangent, $\tan\delta$	Typical thickness, h (mm)
Si[1]	11.9	4×10^{-3} at 30 GHz	0.36
Si[2]	11.9	1×10^{-4} at 30 GHz	0.36
GaAs	12.85	5×10^{-4}	0.5
InP	12.4	5×10^{-5}	0.6

[1]Standard silicon with conductivity approximately 1000 Ω.cm
[2]High resistivity silicon with conductivity approximately 4000 Ω.cm

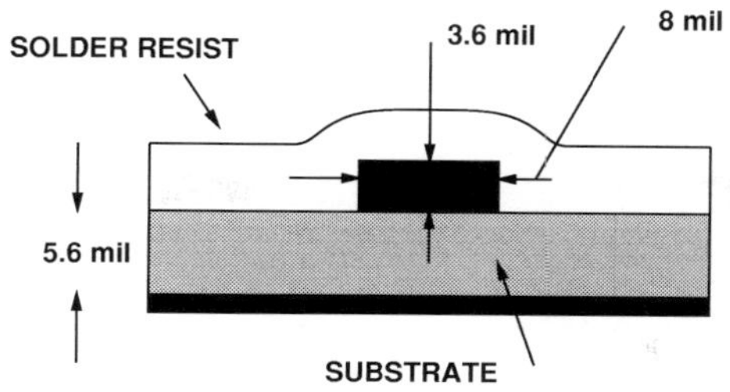

Figure 3.3 Cross-section view of an interconnect on a printed circuit board. (3.6 mil = 88 μm, 5.6 mil = 137 μm, 8 mil = 196 μm)

ceramic particles to form a rigid structure.

(b) Ceramic PCBs: these use ceramics as the dielectric layers. The conductor patterns are formed using a variety of processes. Ceramic PCBs include thin-film and thick-film modules, discussed previously.

In the next two subsections we review the essential properties of PCBs. The reader is referred to Brzozowski [82] for an extensive treatment of PCBs.

3.8.1 ORGANIC PCBs

Organic PCBs are comprised of dielectric layers which are are conductor clad, generally with copper. The dielectric layers are formed from organic resins that are reinforced by fibres to form a rigid structure. The reinforcing material is referred to as a fabric and is generally woven. Typically the copper is patterned, that is selectively coated by photoresist applied using silk-screening or photo imaging. Then the copper is etched. Vias are formed by drilling or punching through one dielectric layer at a time. The final stage of the fabrication process involves compression bonding the individual metallized dielectric layers. The vias are plated to form through holes either as a finished assembly or as individual layers.

The cross-section of an interconnect fabricated this way is shown in Figure 3.3, where the solid black areas are conductors. This is a simple microstrip realization showing dielectric layers (shaded) with a surface conductor. This is coated by a solder resist layer.

The permittivities of a sample of PCB boards are given in Table 3.8. Organic PCBs therefore comprise two materials of different permittivities. E-glass, the most common fabric, has a permittivity of 6.3, and the most commonly used epoxy resin (#5010) has a permittivity of 3.8. The permittivity of the composite material is somewhere between these values depending on the volume ratio of the component materials. The most common PCB material, FR4, uses the E-glass/epoxy combination. More specifics on PCB materials and properties are provided by Brzozowski [82]. Since the fabric has a decidedly planar orientation the permittivity in the board plane differs from that in the direction perpendicular to the plane of the board. Thus there is appreciable anisotropy of ε_r due to prefered orientation of the glass fabric. This problem is addressed using non-woven glass fabrics. In non-woven glass fabrics random microfibres of glass support the epoxy which is usually PTFE. RT/Duroid®

Table 3.8 Permittivity ε_r and dielectric loss tangent $\tan\delta$ of the major organic PCB materials.

Fabric/Resin	ε_r	$\tan\delta$
E-glass/epoxy (FR4)	4.3–4.5	0.013–0.020
E-glass/polyimide	4.1–4.8	0.015–0.025
E-glass/cyanate ester	3.8	0.01
non-woven E-glass/PTFE	2.17–2.33	0.0009–0.0013
woven E-glass/PTFE	2.17–3.20	0.0013–0.009

5870 (with $\varepsilon_r = 2.33$) and RT/Duroid® 5880 (with $\varepsilon_r = 2.2$) are the oldest of these products developed as an alternative to woven glass products. These products are best suited to microwave and broadband applications where it is critical to achieve low loss and low dispersion. The main disadvantage is that the PTFE/microfibre glass boards are soft leading to less precise dimensional control than is achievable using epoxy based boards.

The permittivity of PCBs can be reduced using an alternative glass such as S-glass which has a permittivity of 5.3, and an alternative filler material such as PTFE which has a permittivity of 2.2.

3.8.2 CERAMIC PCBs

The use of a ceramic as the dielectric layer ensures that the PCB is rigid with precisely controlled tolerances including surface flatness. This is particular attractive for microwave circuits where the dimensions define the properties of transmission lines and when lengths of lines are used as circuit elements. In a typical ceramic PCB, PTFE and ceramic are mixed. Depending on the ratio of PTFE and ceramic, and the type of ceramic, these boards can have permittivities ranging from about 2.94 to 10.2, since the most common ceramics are alumina, which has a permittivity of 10.2, and LTCC substrates with varous permittivities but values around 8 are usually chosen. The ceramic mix ensures that there is little or no directional dependence, and so these boards have negligible dielectric anisotropy.

3.9 MULTICHIP MODULES

Multichip Modules (MCMs) are used to provide small, high precision interconnects that can be cost-effectively produced with larger areas than for an IC, have lateral dimensions that are slightly larger than on-chip but smaller than on a PCB. They have higher performance than interconnects on PCBs. In the interconnect performance hierarchy they are between on-chip interconnects and those on PCBs. MCMs include metallization structures that are fabricated using IC fabrication equipment, at the high end, and PCB-like structures at the lower performance end. One of the features that distinguish MCMs from PCBs is that bare (unpackaged) die (i.e. ICs) are generally used and there are usually no active devices on the MCM substrate, although there are instances of this. The intention is to connect two or more chips together using low loss interconnects that can support higher speed digital or analog signals

Table 3.9 Comparison of typical parameters of interconnects on the various MCM substrates.

Parameter	MCM-L μm	MCM-C μm	MCM-D μm
Minimum metal spacing	75	20	15
Minimum trace width	75	15	10
Metal thickness	17–68	10	5
Minimum via diameter	150	20	15

than is possible with PCBs. MCMs also have higher interconnect density than PCBs and have more regular metallization geometries than do ICs. MCMs are nearly always designed to provide well defined signal return paths — interconnects have controlled impedance — and can be properly modelled as transmission line structures. They have multi-level metallization generally with solid or mesh power and ground planes. Low cost versions could have just two metallization layers in addition to a ground plane. There are four dominant types of MCM technology:

(a) MCM-L having a *laminated* PCB-like structure.
(b) MCM-C based on co-fired *ceramic* structures similar to thick-film modules.
(c) MCM-D using *deposited* metals and dielectrics in a process very similar to that used in semiconductor processing.
(d) MCM-C/D having *deposited* layers on an MCM-C base.

In the following sections we will consider the first three MCM types.

3.9.1 MCM-L SUBSTRATES

MCM-L, for MultiChip Module — Laminate, substrates and processing are very similar to that for PCBs. The essential difference is that MCMs, as compared to PCBs, are the choice when performance is a premium and so it is cost-effective to use more exotic epoxies such as PTFE (teflon) and polyimide. These have relative permittivities of around 2.2 when laminated with a glass fabric compared to the permittivity of the most common type of PCB (FR4) of 4.2. More importantly, the mechanical aspects such as surface roughness and layer thickness — critical for defining the electrical characteristics of interconnects — are much better controlled than with PCBs such as FR4 having woven fabrics. Many other materials can be used as the fabric and as the epoxy with varying characteristics. Mechanical characteristics of these materials are very important and the choice of MCM-L substrate involves many tradeoffs. Brzozowski [82] provides a comprehensive treatment of the issues involved.

3.9.2 MCM-C SUBSTRATES

MCM-C, for MultiChip Module — Ceramic, are rigid interconnect substrates which are very similar to the thick-film modules discussed in Section 3.6 on page 66, the functional difference being that MCM-Cs are multi-layer. The base substrate is generally alumina, and layers are alumina, other ceramic, or ceramic/glass blend. The layers are fabricated individually. The base substrate and the ceramic layers are formed from a mix of ceramic powder and organic binders. The material at this stage is soft and flexible and is known as *green tape*. Conductor patterns are formed on the individual layers by silk-screening pastes as discussed in Section 3.6.1 on page 66. Vias, formed by punching holes in the green tape layers, are filled in the silk-screening process. The silk-screen mesh limits the minimum dimensions of lines with the finest meshes yielding a minimum line width of 10 μm and a minimum via diameter of 20 μm. These minimum dimensions limit the use of MCM-C structures in digital systems. The MCM-D structure described below is more suitable for these applications because of the higher wiring densities that can be obtained using smaller dimensions.

As with the thick-film structures considered previously, the green taped layers are stacked to form a multi-layer structure and fired to drive off the organic compounds and form a rigid ceramic structure. The conductors also merge to form good metallic contact between layers at vias. If the firing is done at high temperatures the structure is known as High Temperature Co-fired Ceramic (HTCC) and considerable shrinkage of the structure in the range of 20–40% occurs. More recently, formulations have been developed that allow low firing temperatures and these structures are known as Low Temperature Co-fired Ceramic (LTCC). LTCCs have little shrinkage and thus require less precise process control and are increasingly popular — especially for RF and microwave applications. When used for RF and microwave circuits, multi-layer inductors can be efficiently formed with better characteristics than can be obtained when they are fabricated on-chip with the associated dielectric losses.

3.9.3 MCM-D SUBSTRATES

MCM-D, for MultiChip Module — Deposited, are most like semiconductor integrated circuits but without active devices in general[7]. The most common base substrate is a silicon wafer on which alternate layers of metal and dielectric are deposited. The dielectric layers can be grown in the case of SiO_2 or spun-on in the case of polyimide. Metal layers are deposited typically by sputtering possibly assisted by electro-plating for more rapid build up of conductor. After each layer is added, photolithography and etching are used to define conductor patterns or to open vias in the dielectric. The ultimate geometric limits are those used in semiconductor processing. However, the low loss connections over the long distances needed to connect multiple chips require relatively substantial interconnect cross-sectional dimensions. As with semiconductor processing, the metallic structures are often composed of two or more element-types. Approximate modelling of these structures can be performed using electromagnetic tools to develop the capacitance, resistance and inductance matrices of coupled line structures or by developing distributed transmission line modules. However, the dielectric density varies around the conductors and the multiple-metal structures cannot be modelled properly.

[7] Often previous generations of semiconductor processing equipment can be used, so it is possible to use some active devices.

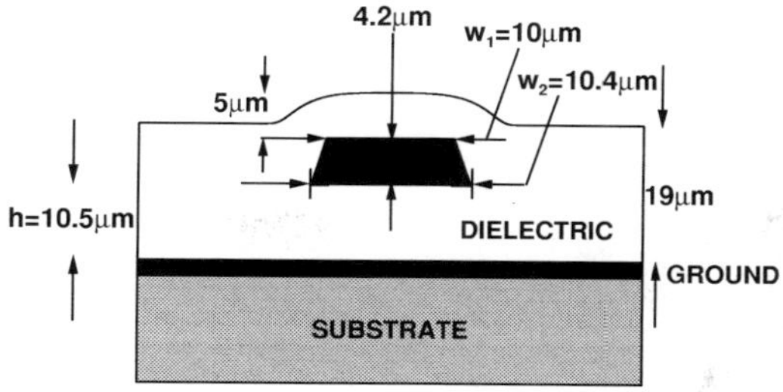

Figure 3.4 Cross-section of the second layer, narrow MCM-D interconnect.

Table 3.10 Parameters of the MCM-D interconnects.

Interconnect	Layer	Height, h μm	Width, w_1 μm	Width, w_2 μm	Capacitance fF/mm
1A	1	5	10	10.4	138
1B	1	5	16	16.4	171
1C	1	5	32	32.4	257
2A	2	10.5	10	10.4	82.4
2B	2	10.5	16	16.4	102
2C	2	10.5	32	32.4	156

3.9.4 CHARACTERIZATION OF INTERCONNECTS ON A MULTICHIP MODULE: A CASE STUDY

Lipa *et al.* [17] undertook an experimental investigation of an MCM-D structure. The MCM had two metallization layers in addition to the ground. The interconnect structure was of microstrip construction and is shown in Figure 3.4. Three different widths on two layers were considered as indicated in Table 3.10.

One of the important things to note is that the cross-section of the interconnect has a low width-to-thickness ratio. This is typical of MCM-D interconnects (and also of on-chip interconnects) where special attention is given to supporting a process that leads to relatively thick interconnect so as to minimize resistance. Thus for the same resistance per unit length, the interconnect widths can be less — resulting in a higher wiring density. This is especially important with the advent of wide buses to increase system performance.

The low aspect ratio of the interconnects means that many fast electromagnetic analyses that assume thin metallization cannot be used. Analytic formulas for interconnect parameters such as characteristic impedance and propagation constant are also derived assuming this thin metallization assumption. The inductance and capacitance matrix extractions considered in the previous chapter for modelling on-chip interconnects are not strictly appropriate as path lengths in MCM-Ds can be substantial compared to the wavelength of the highest significant frequency component of a signal.

The parameters of most interest for MCM interconnects are the loss constant, the delay, and whether or not it is reasonable to use constant per unit length lumped element parameters to describe the properties of MCM interconnects. One of the

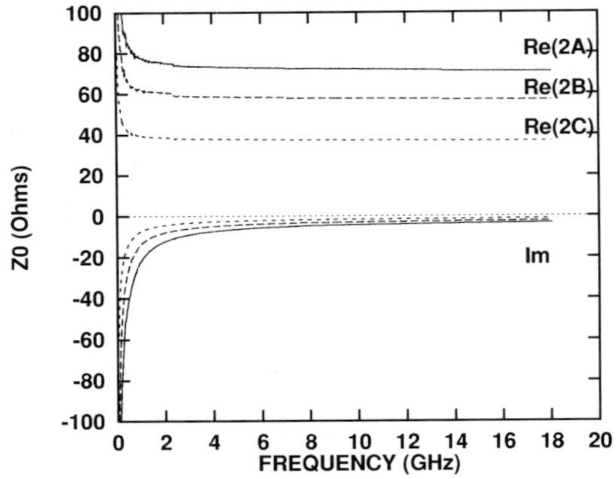

Figure 3.5 Measured characteristic impedance of the microstrip interconnects shown in Figure 3.4 with the line on layer 1 and for three different line widths.

essential transmission line parameters is the characteristic impedance which is shown in Figure 3.5 for the second layer interconnects. As expected, the characteristic impedance of the lines reduces as the width increases. Then for the same power transmission level, currents are lower for narrower lines.

The interconnect loss is captured by the attenuation constant α which is composed of conductive α_C and dielectric α_D attenuation:

$$\alpha = \alpha_C + \alpha_D. \tag{3.4}$$

In the work report by Lipa *et al.*, accurate electromagnetic modelling was used to model the interconnect taking full account of the skin effect and the actual shape of the interconnect [83]. The conductive and dielectric attenuation constants are shown in Figure 3.7; Figure 3.7(a) when the interconnect is on the first layer and Figure 3.7(b) when the interconnect is on the second layer. The layer associated with the measurements is indicated by the primary subscripts 1 and 2. Three different widths were considered, denoted by the secondary subscripts A, B, and C, respectively. As would be expected the conductive attentuation is higher the narrower the line:

$$\alpha_{C,1A} > \alpha_{C,1B} > \alpha_{C,1C}$$
$$\alpha_{C,2A} > \alpha_{C,2B} > \alpha_{C,2C}.$$

Also, the attentuation of the lines on layer 2, with a height above the ground greater than that of the corresponding width lines on layer 1, has lower attentuation:

$$\alpha_{C,1A} > \alpha_{C,2A}$$
$$\alpha_{C,1B} > \alpha_{C,2B}$$
$$\alpha_{C,1C} > \alpha_{C,2C}.$$

This is because the lines closer to the surface have some of the field lines in the air, have a lower permittivity, hence travel faster. Since the signals travel faster, there is less loss as this is largely proportional to the amount of time the signal is exposed to the loss mechanism.

The interesting point to note about the results presented in Figure 3.7 is that for lines on the same layer the dielectric loss, indicated by α_D, depends on the layer number, and not the width of the lines. For the same propagation velocity discussion as before, α_D on layer 1 is greater than that on layer 2.

The dielectric loss makes up about one third of the total loss. This loss is far in excess of the loss due to relaxation loss in the dielectric, and is attributed to the excitation of slab waves and surface waves that travel in the dielectric or at the dielectric/air interface away from the microstrip line. This phenomenon has only recently been identified [84–87].

The attenuation constant can be related to the per unit length lumped element parameters describing the line by considering the propagation constant:

$$\gamma = j\omega\sqrt{LC}\left(1 - j\frac{R}{\omega L}\right)^{\frac{1}{2}}\left(1 - j\frac{G}{\omega C}\right)^{\frac{1}{2}}. \tag{3.5}$$

Approximating the terms in the parentheses by the first two terms in their power series expansion since $R \ll \omega L$ and $G \ll \omega C$ (except at very low frequencies), the conductive attenuation constant is interpreted as

$$\alpha_C = \frac{1}{2}R\sqrt{C/L} \tag{3.6}$$

and the dielectric attenuation constant as

$$\alpha_D = \frac{1}{2}G\sqrt{L/C}. \tag{3.7}$$

The total attenuation coefficient, of course is $\alpha = \alpha_C + \alpha_D$, with the term α_C describing the loss due to the interconnect resistance and α_D representing the loss in the dielectric. Both α_C and α_D also include the effects of radiation loss.

3.9.5 MCM SUMMARY

In summary, the principal advantages of an MCM over a PCB are:

(a) higher interconnect density
(b) finer interconnect geometries so that chips can be placed closer together resulting in shorter interconnect lengths
(c) finer geometries that enable direct chip connect without needing intermediate chip packaging (e.g. using flip-chip or solder bump connections).

In addition to the three principal types of MCMs (MCM-L, MCM-C and MCM-D) MCM-C/D is sometimes used. This substrate begins as an MCM-C with interconnects of relatively large dimensions having low loss and especially suited to distributing power and ground. Additional layers of metallization and dielectric are deposited using the same process as for MCM-Ds. The MCM-D part of the structure provides the higher wiring density required to distribute digital signals where being able to accommodate wide buses boosts performance.

The reader is referred to books by Doane and Franzon [2] and Harper [3] for extensive treatments of MCMs.

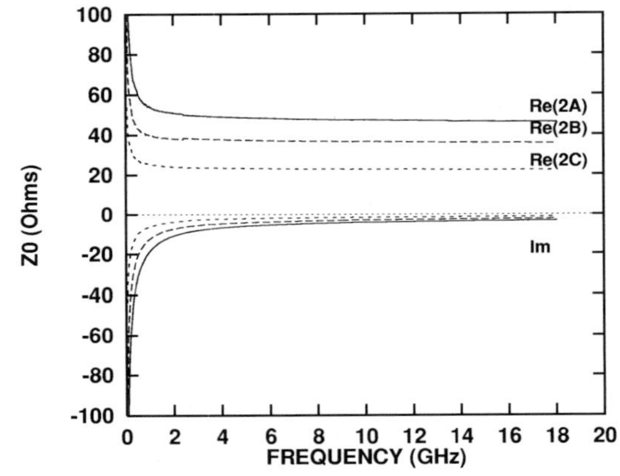

Figure 3.6 Measured characteristic impedance of the microstrip interconnects shown in Figure 3.4 with the line on layer 2 and for three different line widths.

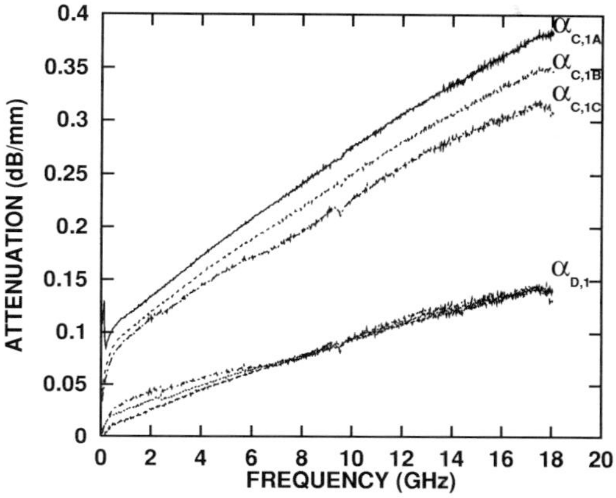

(a)

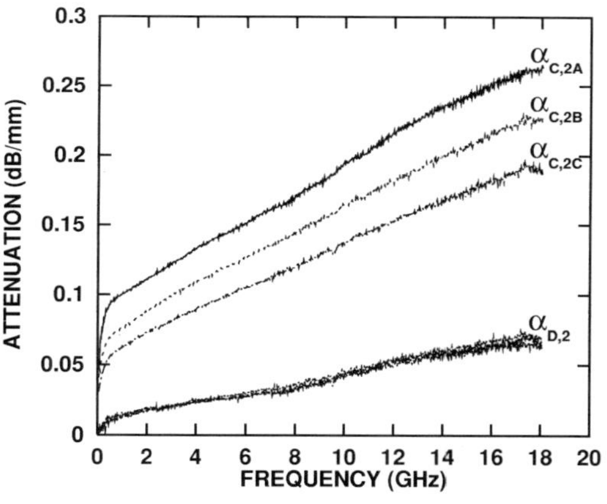

(b)

Figure 3.7 Conductive and dielectric attenuation constants, α_C and α_D respectively, of an interconnect on a two layer MCM-D: (a) first layer metallization; and (b) second layer metallization.

4

Microstrip Design at Low Frequencies

4.1 THE MICROSTRIP DESIGN PROBLEM

As remarked in Chapter 3, microstrip is a particularly useful transmission line medium for implementation in distributed circuit designs at frequencies from below 1 gigahertz through some tens of gigahertz. In this chapter we are concerned with some fundamental aspects of the microstrip structure and with the realization of dimensional design parameters for establishing desired electrical characteristics. This is chiefly for applications through the lower microwave frequencies including clock frequencies up to several gigahertz. When the length of the interconnect is an appreciable fraction of a wavelength (say 1/20 th or more), then it can be regarded as a circuit element. Then the electrical requirement is often to realize a structure that provides maximum signal, or power, transfer.

4.1.1 DIGITAL INTERCONNECT

Digital interconnect comprises two types: low resistance interconnects generally found off-chip and high resistance interconnects generally found on-chip. As discussed in Chapters 1 and 2, high resistance interconnect can often be adequately modelled as RC structures and can form an electromagnetic viewpoint of the field distribution, the integrity of the signal path, and coupling to adjacent structures is of greatest concern. Chip-to-chip interconnection is generally designed with a driver and receiver with specific Thevenin equivalent output and input, respectively, resistances which are generally the same. Thus, a 75 Ω driver can be modelled as a step voltage generator in series with a resistance $R_D = 75\ \Omega$. The receiver resistance would also be around 75 Ω in this case. The characteristic impedance must be 75 Ω for maximum power transfer. (Package inductance and capacitance complete the model.) This provides for the fastest signal transmission. However, as discussed in Chapter 2, when the driver is current limited, frequently the case with drivers connecting gates on-chip, an interconnect with a higher characteristic impedance is desirable. In any case, interconnects must be designed with specific electrical characteristics.

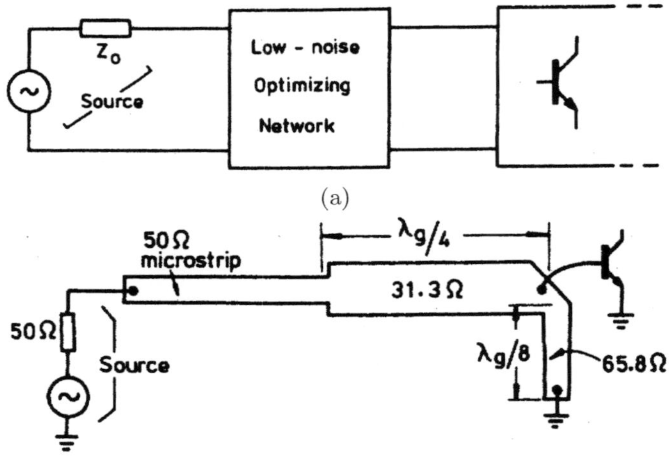

(b)

Figure 4.1 Input network arrangement (a) and microstrip circuit (b) for a microwave bipolar transistor amplifier operating at a frequency of 2 GHz. (Reproduced by permission of Hewlett-Packard Company, Palo Alto, CA.)

4.1.2 A TRANSISTOR AMPLIFIER INPUT NETWORK

One of the many areas where a microstrip design approach is really attractive is that of microwave transistor amplifiers. Further examples are considered in more detail in Chapter 11, but the case of an input network for a bipolar transistor is examined briefly here as a basic example. The design techniques described here are quite accurate for this purpose. The general input arrangement for this type of amplifier is shown in Figure 4.1(a), where the low-noise optimizing network is sometimes referred to as a 'noise-matching' network. This is typical of a microwave receiver 'front-end.' In other cases power matching may be required and a different network would therefore be developed.

A fairly simple microstrip realization of the input network is illustrated schematically in Figure 4.1(b). Detailed design is given in Chapter 11. The signal source is assumed to be connected without any discontinuity through the 50 Ω microstrip to the 'L-shaped' microstrip network. This consists of two lines having characteristic impedances and electrical lengths determined by the network design. The main design problem is then to evaluate the physical widths and lengths of the microstrip lines (this applies to all microstrip designs).

We shall see that the width of a microstrip line is principally a function of its characteristic impedance and the thickness of the substrate. The physical length depends upon the wavelength, which is a function of the width, the substrate permittivity, and of course, the signal frequency. Explicit techniques and formulas will be given here for the calculation of width and length to accuracies within about 1%.

From Figure 4.1(b) it should be clear that various discontinuities are inherent in the structures as follows: where the 50 Ω and 31.3 Ω lines join, at the right-angled

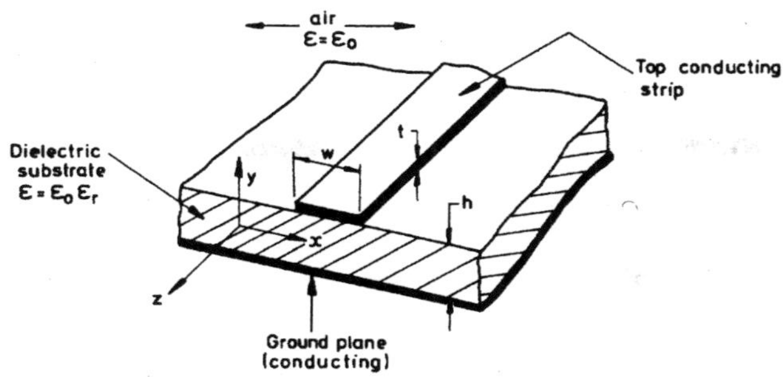

Figure 4.2 The general geometry of a microstrip line, including choice of coordinates.

junction of the 31.3 Ω and 65.8 Ω lines, and (less obviously) at the short-circuit to ground on the end of the 65.8 Ω line. These, and other, discontinuities are described in more detail in Chapter 7, where the details of the mitre to reduce the discontinuity effect of the right-angled bend are also considered.

4.1.3 THE GEOMETRY OF MICROSTRIP

The general geometry of microstrip is shown in Figure 4.2. Materials and techniques for its manufacture were discussed in Chapter 3, where the basic advantages were also described in comparison with other types of MIC and MMIC transmission lines. The most important dimensional parameters are the microstrip width, w, and the height, h, (equal to the thickness of the substrate). Also of great importance is the relative permittivity of the substrate, ε_r. In RF and microwave applications the thickness t of the metallic, top-conducting strip is generally of much lesser importance and may often be neglected. However, microstrip line on-chip and on MCMs are relatively thick as a result of the need to keep resistance down while still achieving high wiring density by keeping the width down. Effects of this thickness are dealt with in Section 4.10 on page 105. We shall consistently refer to the xyz coordinate description, as shown in Figure 4.2.

Some of the particularly useful characteristics of microstrip include the following:

(a) DC as well as AC signals may be transmitted.
(b) Active devices, diodes and transistors may readily be incorporated (shunt connections are also quite easily made).
(c) In-circuit characterization of devices is straightforward to implement.
(d) Line wavelength is reduced considerably (typically one-third) from its free-space value, because of the substrate fields. Hence, distributed component dimensions are relatively small.
(e) The structure is quite rugged and can withstand moderately high voltages and power levels.

In the next six main sections of this chapter we are concerned with basic considerations and microstrip synthesis where the substrate is a homogeneous, isotropic dielectric. Materials such as alumina, which is widely used, approximate to this description. We shall then consider synthesis using a dielectrically anisotropic substrate and a ferrite substrate. The chapter concludes with discussions on the effects of finite thickness and manufacturing tolerances and pulse propagation.

4.2 THE QUASI-TEM MODE OF PROPAGATION

It is clear from Figure 4.2 that microstrip involves an abrupt dielectric interface between the substrate and the air above it. Any transmission line that is filled with a uniform dielectric can support a single, well-defined mode of propagation, at least over a specified range of frequencies (TEM for coaxial lines, TE for waveguides, etc.). Transmission lines that do not have such a uniform dielectric filling cannot support a single mode of propagation; microstrip is within this category. Although this is true, the bulk of the energy is transmitted along microstrip with a field distribution which quite closely resembles TEM; it is usually referred to as 'quasi-TEM.' The detailed field distribution is quite complicated, but the main transverse electric field can be visualized as shown in Figure 4.3.

Gupta *et al.* [88] have used Maxwell's equations to convincingly demonstrate the necessity for longitudinal components of electric and magnetic fields. This is clearly inconsistent with a pure TEM or a TE propagating mode.

Representative views of the magnetic and electric field distributions are given in Figure 4.4. (These are not precisely determined field contours, and they must only be regarded as diagrammatic illustrations of the situation.) The longitudinal components can clearly be seen, and these become increasingly significant as the frequency is raised.

At most transverse cross-sectional planes, taken across the shielded microstrip, the electric (E) and magnetic (H) fields follow the distributions indicated in Figure 4.5. Note the abrupt change in direction of the electric field line as it passes through the air-substrate interface. These fields have been analysed by a number of researchers using various static techniques. We shall not study any of these techniques in detail here, but the results are powerful and significant for the circuit designer wishing to use microstrip. There are two reasons for this:

(a) For the majority of microstrip lines suitable for MICs, the statically derived results are quite accurate where the frequency is below a few gigahertz.
(b) At higher frequencies, up to the limits for the useful operation of microstrip, these 'static' results can still be used in conjunction with frequency-dependent functions in closed formulas. This is developed in detail in Chapter 5.

The earliest work concerning microstrip is generally accepted as that reported by Grieg and Engelmann [89]. Many other workers investigated basic characteristics of microstrip [90–92]. However, the major fundamental work on important and closely related parallel-strip transmission lines was due to Wheeler [93]. This led to analysis and synthesis on the static-TEM basis.

4.3 STATIC-TEM PARAMETERS

The microstrip synthesis problem consists of finding the values of width w and length l corresponding to the characteristic impedance Z_0, and electrical length θ (in degrees or radians) defined at the network design stage.

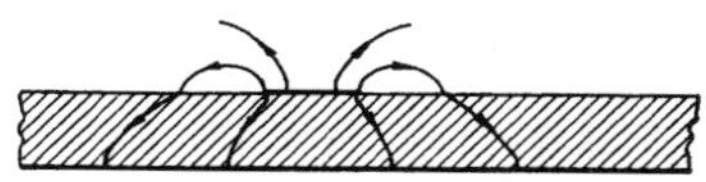

Figure 4.3 Transverse cross-section of microstrip, showing electric field.

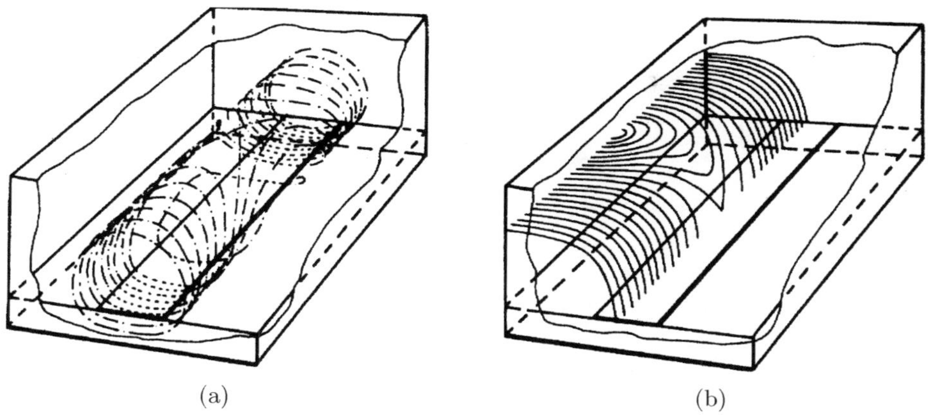

Figure 4.4 Three-dimensional views of approximate magnetic (a) and electric fields (b) surrounding a shielded microstrip line. For simplicity here, only the electric field in the air is shown in (b).

Initially a suitable substrate, of thickness h and relative permittivity ε_r, will have to be chosen. A basis for this choice is given in Section 3.3 on page 52 and the choice also depends upon certain frequency limitations which are discussed in Section 5.9 of Chapter 5. The synthesis actually yields the normalized width-to-height ratio w/h initially, as well as a quantity called the effective microstrip permittivity ε_{eff}. This quantity is unique to mixed-dielectric transmission line systems and it provides a useful link between various wavelengths, impedances and propagating velocities. We shall shortly define the static-TEM effective microstrip permittivity (ε_{eff}) precisely.

4.3.1 THE CHARACTERISTIC IMPEDANCE Z_0

For any TEM-type transmission line the characteristic impedance at high frequencies may be expressed in any one of three alternate forms (see Section 1.7 on page 15):

$$Z_0 = \sqrt{L/C} \tag{4.1}$$

or

$$Z_0 = v_p L \tag{4.2}$$

or

$$Z_0 = 1/(v_p C). \tag{4.3}$$

Both (4.2) and also (4.3) involve the phase velocity v_p of the wave travelling along the line. It is also recalled (Chapter 1) that this phase velocity is given by

$$v_p = 1/\left(\sqrt{LC}\right). \tag{4.4}$$

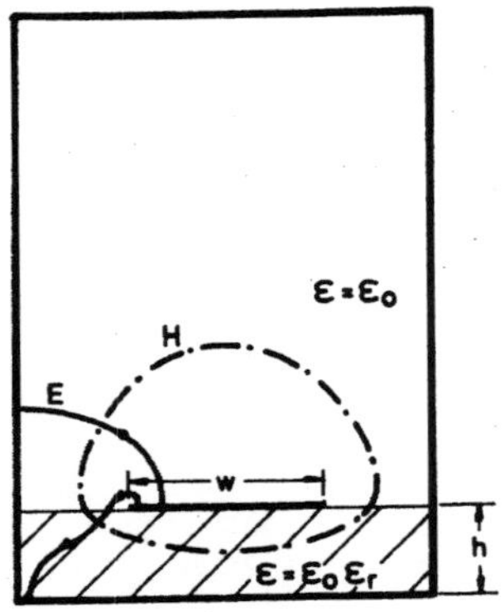

Figure 4.5 A typical transverse cross-section of a shielded microstrip line showing representative magnetic (H) and electric (E) fields.

When the substrate of the microstrip line is (effectively) removed we have an air-filled line along which the wave will travel at c, the velocity of light in free space ($c = 2.99793 \times 10^8$ m/s). The characteristic impedance of this air-filled 'microstrip' Z_{01}, is given by

$$Z_0 = \sqrt{L/C_1} \tag{4.5}$$

where L remains unaltered by the change in dielectric constant and C_1 is the capacitance per unit length for this structure. Alternatively:

$$Z_{01} = cL \tag{4.6}$$

$$Z_{01} = 1/(cC_1). \tag{4.7}$$

Combining Equations (4.1), (4.6) and (4.7) yields the very significant result

$$Z_0 = 1/\left(c\sqrt{CC_1}\right). \tag{4.8}$$

This means that we have the required characteristic impedance if only we can evaluate the capacitances per unit length of the structure, with and without the presence of the dielectric substrate. One way in which this has been achieved will be outlined shortly (in Section 4.6 on page 94).

4.3.2 THE EFFECTIVE MICROSTRIP PERMITTIVITY ε_{eff}

For the air-spaced microstrip line the propagation velocity is given by

$$c = 1/\sqrt{LC_1} \tag{4.9}$$

and, by dividing Equation (4.9) by Equation (4.4) and squaring, we obtain

$$\frac{C}{C_1} = \left(\frac{c}{v_p}\right)^2. \tag{4.10}$$

The capacitance ratio C/C_1 is termed the effective microstrip permittivity ε_{eff}, an important microstrip parameter. From (4.10), ε_{eff} is given by

$$\varepsilon_{\text{eff}} = \left(\frac{c}{v_p}\right)^2. \tag{4.11}$$

Although this result has been derived on a static basis here, it is fundamentally important for microstrip, and will be used again where v_p is taken to be frequency dependent (in Chapter 5).

A useful relationship between Z_0, Z_{01}, and ε_{eff} can be obtained by combining Equations (4.3), (4.7), (4.10) and (4.11). The result is

$$Z_0 = Z_{01}/\sqrt{\varepsilon_{\text{eff}}} \tag{4.12}$$

that is

$$Z_{01} = Z_0\sqrt{\varepsilon_{\text{eff}}}. \tag{4.13}$$

This will be useful in several respects, including the operation of a graphical design technique that is to be described shortly.

Upper and lower bounds can readily be found for ε_{eff}, in the static low-frequency limit, by considering the effects of very wide and very narrow lines as indicated in Figure 4.6. For the very wide lines nearly all of the electric field is confined to the substrate dielectric, the structure resembles a parallel-plate capacitor, and therefore, at this extreme

$$\varepsilon_{\text{eff}} \rightarrow \varepsilon_r. \tag{4.14}$$

In the case of very narrow lines, the field is almost equally shared by the air ($\varepsilon_r = 1$) and the substrate so that, at this extreme

$$\varepsilon_{\text{eff}} \approx (\varepsilon_r + 1). \tag{4.15}$$

The range of ε_{eff} is therefore:

$$\frac{1}{2}(\varepsilon_r + 1) \leq \varepsilon_{\text{eff}} \leq \varepsilon_r. \tag{4.16}$$

It can be convenient to express the effective microstrip permittivity as

$$\varepsilon_{\text{eff}} = 1 + q(\varepsilon_r - 1) \tag{4.17}$$

where the new quantity, the filling factor q, has the bounds

$$\frac{1}{2} \leq q \leq 1. \tag{4.18}$$

Wheeler [93, 94] has evaluated this filling factor, and we shall employ it in the approximate graphical design technique.

4.3.3 SYNTHESIS: THE WIDTH-TO-HEIGHT RATIO w/h

The width-to-height ratio (w/h) is a strong function of Z_0, and of the substrate permittivity ε_r. Wheeler's results [94] are particularly useful in this respect, although it has been found that some modifications are necessary for high accuracy (within 1%) to be achieved [95]. Closed formulas for w/h will be given in Section 4.5 on page 92.

Figure 4.6 Microstrip lines: (a) extremely wide ($w >> h$); and (b) extremely narrow ($w << h$) lines.

4.3.4 WAVELENGTH λ, AND PHYSICAL LENGTH l

For any propagating wave the velocity is given by the appropriate frequency-wavelength product. In free space we have $c = f\lambda_0$ and in microstrip the velocity is $v_p = f\lambda_g$. Substituting these products into (4.11) we obtain

$$\varepsilon_{\text{eff}} = (\lambda_0/\lambda_g)^2 \tag{4.19}$$

or

$$\lambda_g = \lambda_0/\sqrt{\varepsilon_{\text{eff}}} \tag{4.20}$$

where λ_0 is the free-space wavelength.

More conveniently, where the frequency is given in gigahertz and denoted by F the wavelength can be expressed directly in millimetres as follows:

$$\lambda_g = 300/(F\sqrt{\varepsilon_{\text{eff}}}) \text{ mm.} \tag{4.21}$$

The physical length l of a microstrip line to yield a specified electrical length θ (in degrees) is easily determined. In Chapter 1 (Section 1.6 on page 14) we derived the phase coefficient β, and βl is equal to this electrical length in radians, i.e.

$$\beta l = \theta \tag{4.22}$$

and hence

$$2\pi l/\lambda_g = \theta. \tag{4.23}$$

With θ in degrees this gives

$$l = \theta\lambda_g/360. \tag{4.24}$$

Thus, with λ_g evaluated using (4.21), we can simply find l.

4.4 APPROXIMATE GRAPHICALLY-BASED SYNTHESIS

Presser [96] devised a graphical technique for the analysis or synthesis of microstrip lines. The method is quick and useful where errors of a few percent can be tolerated and where a calculator is not readily available. This technique is based upon the work of Wheeler [93, 94], and thus it uses the static-TEM approach which is applicable for frequencies up to a few gigahertz on conventional substrates.

The 'design curves' given by Presser are reproduced in Figure 4.7. Notice that the width-to-height or 'shape' ratio w/h and filling factor q are both plotted to a base of the air-spaced characteristic impedance Z_0. This helps to make the graph fairly universal, as the following design synthesis steps will indicate:

(a) Make the initial, very rough, assumption $\varepsilon_{\text{eff}} \approx \varepsilon_r$ to obtain starting values.

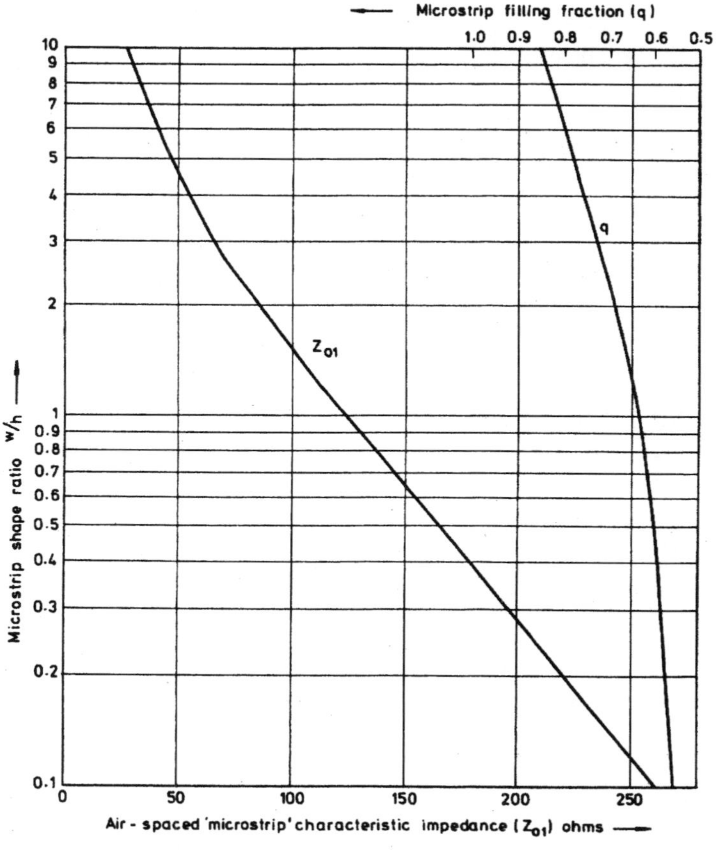

Figure 4.7 General curves for the analysis or synthesis of microstrip. (Reproduced by permission of Microwaves & RF from Presser [96, Figure 2].)

(b) Calculate the air-spaced characteristic impedance approximately, from (4.13), with $\varepsilon_{\text{eff}} \approx \varepsilon_r$ i.e. using

$$Z_{01} \approx \sqrt{\varepsilon_r} Z_0. \tag{4.25}$$

(c) Use the general two-curve graph (Figure 4.7) to find w/h applicable to this Z_{01} and also note the corresponding value of the microstrip filling factor q.
(d) Hence calculate the updated value of ε_{eff} using (4.17):

$$\varepsilon_{\text{eff}} = 1 + q(\varepsilon_r - 1). \tag{4.26}$$

This now ends one iteration. Steps (b) to (d) must be repeated, using the progressively improved values of ε_{eff} (instead of ε_r in steps (a) and (b)). Fast convergence is generally obtained here, and only two or three iterations are usually necessary before ε_{eff} is finally within about 1% of its previous value, i.e. convergence has occurred.

The final value of w/h, and hence the width w, is of course the value appropriate to the final Z_{01}. Lastly, the wavelength must be calculated using (4.21). The static-TEM part of the microstrip design procedure is then complete using this graphical technique.

However, any graphical approach will inevitably involve errors of several per cent and also clearly cannot form part of a Computer-Aided Design (CAD) algorithm. This form of graphical technique is mainly suitable for approximate guidance purposes. Most microstrip MIC design work necessarily proceeds on a CAD basis and demands relatively high accuracy (typically better than 1%). Therefore, closed formulas are very useful for synthesis; in the next section suitable formulas are presented. These closed formulas also provide an insight into trends and sensitivities that cannot be obtained using CAD programs alone.

4.5 FORMULAS FOR ACCURATE STATIC-TEM DESIGN CALCULATIONS

Closed formulas are highly desirable for use in microstrip calculations for the following reasons:

(a) A fairly high accuracy is achievable.
(b) Fast CAD algorithms can be implemented, including the following.
(c) Incorporation of static-TEM formulas in frequency-dependent calculations for higher-frequency design (Chapter 5).

Various researchers have reported formulas for microstrip calculations [93,94,97,98]. Owens [95] carefully investigated the ranges of applicability of many of the expressions given by Wheeler [94], comparing calculated results with numerical computations requiring elaborate algorithms — but known to be very accurate. New changeover values for the shape ratio (w/h) were advocated for greatest accuracy by using alternate expressions dependent upon the w/h range. The formulas are applicable to alumina-type substrates ($8 \ll \varepsilon_r \ll 12$), the most useful expressions being given in the following sections.

4.5.1 SYNTHESIS FORMULAS (Z_0 AND f GIVEN)

For narrow strips (i.e. when $Z_0 > (44 - \varepsilon_r)\ \Omega$):

$$\frac{w}{h} = \left(\frac{\exp H'}{8} - \frac{1}{4\exp H'}\right)^{-1} \tag{4.27}$$

where

$$H' = \frac{Z_0\sqrt{2(\varepsilon_r + 1)}}{119.9} + \frac{1}{2}\left(\frac{\varepsilon_r - 1}{\varepsilon_r + 1}\right)\left(\ln\frac{\pi}{2} + \frac{1}{\varepsilon_r}\ln\frac{4}{\pi}\right). \tag{4.28}$$

We may also use, with slight but significant shift of changeover value to $w/h < 1.3$ (i.e. when $Z_0 > (60 - \varepsilon_r)\ \Omega$):

$$\varepsilon_{\text{eff}} = \frac{\varepsilon_r + 1}{2}\left[1 - \frac{1}{2H'}\left(\frac{\varepsilon_r - 1}{\varepsilon_r + 1}\right)\left(\ln\frac{\pi}{2} + \frac{1}{\varepsilon_r}\ln\frac{4}{\pi}\right)\right]^{-2} \tag{4.29}$$

where H is given by (4.28) (as a function of Z_0) or, alternatively, as a function of w/h, from Equation (4.27):

$$H' = \ln\left[4\frac{h}{w} + \sqrt{16\left(\frac{h}{w}\right)^2 + 2}\right]. \tag{4.30}$$

Another, somewhat simpler, expression for $\varepsilon_{\text{eff}}(Z_0)$ is furnished by Equation (15) of Owens [95], which is

$$\varepsilon_{\text{eff}} = \frac{\varepsilon_r + 1}{2}\left[1 + \frac{29.98}{Z_0}\left[\frac{2}{\varepsilon_r + 1}\right]\left(\frac{\varepsilon_r - 1}{\varepsilon_r + 1}\right)\left(\ln\frac{\pi}{2} + \frac{1}{\varepsilon_r}\ln\frac{4}{\pi}\right)\right]^{-2}. \tag{4.31}$$

We now consider the ranges and formulas for 'wide' strips.

For wide strips (i.e. when $Z_0 < (44 - 2\varepsilon_r)\ \Omega$):

$$\frac{w}{h} = \frac{2}{\pi}\left[(d_\varepsilon - 1) - \ln(2d_{\varepsilon_1} - 1)\right] + \frac{(\varepsilon_r - 1)}{\pi\varepsilon_r}\left[\ln d_{\varepsilon_r} - 1 + 0.293 - \frac{0.517}{\varepsilon_r}\right] \tag{4.32}$$

where

$$d_\varepsilon = \frac{59.95\pi^2}{Z_0\sqrt{\varepsilon_r}}\ . \tag{4.33}$$

With the same slight shift of changeover value as before (i.e. where $w/h > 1.3$ and $Z_0 > (63 - 2\varepsilon_r)\ \Omega$, Owens found that a modified form of an earlier expression due to Schneider [58] gave very accurate results. Owens' formula is

$$\varepsilon_{\text{eff}} = \frac{\varepsilon_r + 1}{2}\frac{\varepsilon_r - 1}{2}\left[1 + 10\frac{h}{w}\right]^{-0.555}. \tag{4.34}$$

Alternatively, where Z_0 is known at first

$$\varepsilon_{\text{eff}} = \frac{\varepsilon_r}{0.96 + \varepsilon_r(0.109 - 0.004\varepsilon_r)[\log(10 + Z_0) - 1]}\ . \tag{4.35}$$

For microstrip lines on alumina ($\varepsilon_r = 10$) this expression appears to be accurate to $\pm$ 0.2% over the impedance range

$$8 \le Z_0 \le 45\ \ \Omega. \tag{4.36}$$

Finally, some accurate expressions for analysis (calculating Z_0) are also given.

4.5.2 ANALYSIS FORMULAS (w/h AND ε_r GIVEN)

For 'narrow' strips ($w/h < 3.3$):

$$Z_0 = \frac{119.9}{\sqrt{2(\varepsilon_r + 1)}} \ln\left[4\frac{h}{w} + \sqrt{16\left(\frac{h}{w}\right)^2 + 2}\right]. \tag{4.37}$$

For 'wide' strips ($w/h > 3.3$):

$$\begin{aligned} Z_0 &= \frac{119.9\pi}{2\sqrt{\varepsilon_r}} \left\{\frac{w}{2h} + \frac{\ln 4}{\pi} + \frac{\ln(e\pi^2/16)}{2\pi}\left(\frac{\varepsilon_r - 1}{\varepsilon_r^2}\right)\right. \\ &\quad \left. + \frac{\varepsilon_r + 1}{2\pi\varepsilon_r}\left[\ln\frac{\pi e}{2} + \ln\left(\frac{w}{2h} + 0.94\right)\right]\right\} \end{aligned} \tag{4.38}$$

where 'e' is the exponential base: $e = 2.7182818\ldots$. These expressions may readily be incorporated in fairly extensive CAD routines that involve frequency-dependent expressions for effective microstrip permittivity $\varepsilon_{\text{eff}}(f)$ (dealt with in Chapter 5).

4.5.3 OVERALL ACCURACIES TO BE EXPECTED FROM THE PREVIOUS EXPRESSIONS

In all cases, the shape ratio w/h will be accurate to $\pm 1\%$. For narrow lines ($w/h < 1.3$), ε_{eff} has the error range $\pm 0.5\%$.

When calculated using (4.34), ε_{eff} is accurate to $\pm 0.25\%$. The expressions for Z_0 result in accuracies to $\pm 1\%$.

4.6 ANALYSIS TECHNIQUES REQUIRING SUBSTANTIAL COMPUTER POWER

Computer-aided design can be used to synthesize microstrip circuits by including an analysis engine in an optimization loop with prescribed goals. From several viewpoints, e.g. initial concept develpment, understanding of sensitivities, and manufacturability, it is preferable to use the types of formulas presented here. However, as pointed out early in Section 4.5, on page 92, the formulas only apply to a restricted range of substrates, and for design work using different substrates (e.g. quartz, plastics, etc.) a fundamental computation is necessary. In a typical approach a family of closed formulas is derived from the results of a design space exploration using computer aided analysis. These computationally derived results are normally verified with experimental characterization of selected structures. It is worth noting that since semiconductor substrates such as High-Resistivity Silicon (HRS) or semi-insulating GaAs have permittivities just 20–30% higher than high-grade alumina, the above formulas should work well provided that the width of a line is much greater than the height.

Since an electrostatic field analysis is performed, it is useful to be able to express the characteristic impedance entirely in capacitive terms [99]. This is already available in the form of (4.8), which is repeated here for convenience:

$$Z_0 = 1/\left(c\sqrt{CC_1}\right). \tag{4.39}$$

Thus, in order to compute Z_0, only the capacitances per unit length, C_1 and C, are required.

Now the capacitance C_x may, in general, be expressed in terms of net charge Q and voltage V_T as

$$C_x = Q/V_T. \tag{4.40}$$

The charge is obtainable by applying Gauss' theorem

$$Q = \oint D \cdot \mathrm{d}s \tag{4.41}$$

in which D is the displacement and s is a surface enclosing the strip conductor. Where v is a point voltage, Laplace's equation may be written as

$$\nabla^2 v = 0 \tag{4.42}$$

and for every voltage v appropriate values of electric field E may be determined. The constitutive relation

$$D = \varepsilon E \tag{4.43}$$

then enables D to be evaluated and substituted in Equation (4.41).

In practice, the computations are performed using some numerical method to solve Laplace's equation. Also, a shielded microstrip is generally considered (the shield can be progressively moved away from the strip to approach the unshielded case, if desired).

Of course, two complete calculations of capacitance, using Equation (4.40), are required: one with the substrate present, yielding C, and a second with the substrate removed, yielding C_1. Then (4.39) gives the characteristic impedance. The effective microstrip permittivity ε_{eff} is also known, via Equations (4.10) and (4.11):

$$\varepsilon_{\text{eff}} = C/C_1. \tag{4.44}$$

Great precision is readily achievable if a sufficiently fine grading of points is set up for the computation.

4.7 A WORKED EXAMPLE OF STATIC-TEM SYNTHESIS

Let us consider the 31.3 Ω series element of the microstrip matching circuit shown for the transistor amplifier in Figure 4.1. We are required to calculate the width and approximate length of this element, and for the moment we shall neglect the effects of the discontinuities present at each end, i.e. the step to 50 Ω microstrip and the bend. These and other discontinuities are dealt with in Chapter 7. The initial specification is, from Figure 4.1:

31.3 Ω $\qquad l = \lambda_g/4 \qquad f$=2 GHz

(We could alternatively have specified $\theta = 90°$, which is identical to $l = \lambda_g/4$.) Also assume that a substrate is chosen having the following specification:

Relative permittivity ε_r= 9.8 $\qquad$ Thickness h = 0.6 mm.

Since the frequency is only 2 GHz the static-TEM methods described here will definitely be of 'sufficient' accuracy (i.e. within $\pm 1\%$ when the accurate expressions are used). At first we carry out a rapid, approximate determination using Presser's graphical technique.

4.7.1 GRAPHICAL DETERMINATION

Referring to step (a) given in Section 4.4 on page 90, we have

$$Z'_{01} = \sqrt{9.8} \times 31.3 \approx 98\ \Omega. \tag{4.45}$$

From the graph of Figure 4.7 we obtain, for this value of Z_{01}

$$q' \approx 0.66. \tag{4.46}$$

From the remaining steps in Section 4.4,

$$\varepsilon'_{\text{eff}} \approx 6.8 \tag{4.47}$$

and hence the improved value of Z'_{01} is

$$Z_{01} \approx \sqrt{6.8} \times 31.3 \approx 81.7\ \Omega. \tag{4.48}$$

Using the graph again, with this new value of Z'_{01} we obtain:

$$\frac{w}{h} = 2.2 \qquad q = 0.7 \tag{4.49}$$

whence

$$\varepsilon_{\text{eff}} = 7.16. \tag{4.50}$$

The reader is advised to check these figures independently.

The frequency of 2 GHz used here is close to that applying to handset connections in PCS mobile communications, although FR4 ($\varepsilon_r = 4.8$) substrates are more usual in this instance. Following the above procedure again leads to $w/h = 4.3$, $q = 0.78$ and $\varepsilon_{\text{eff}} = 3.93$. At the (lower) GSM midband frequency of 960 MHz, almost identical results will apply.

Occasionally very high permittivity substrates are used to reduce the size of resonators at these lower frequencies and in such instances almost all the fields are confined into the substrate. Permittivities typically range from 20 to over 150. At the low end the above procedures must be adopted, but from approximately mid-range upwards the approximation $\varepsilon_r \approx \varepsilon_{\text{eff}}$ may be used.

Attention is now given to the use of accurate formulas.

4.7.2 ACCURATELY CALCULATED RESULTS

An algorithm has been devised that can be executed on a computer to evaluate any desired design quantities using the expressions given in Section 4.5 on page 92. The correct choice of expression is made automatically in the program. For this 31.3 Ω line the results are

$$w/h = 2.225 \quad \text{and} \quad \varepsilon_{\text{eff}} = 7.095.$$

Notice that the approximate graphical technique gives a fairly accurate result for w/h, but involves substantial error in ε_{eff}. This is because q cannot be accurately determined from the graph, e.g. a change of value to $q = 0.69$ yields $\varepsilon_{\text{eff}} = 7.07$, which is actually slightly lower than the computed value.

4.7.3 *FINAL DIMENSIONS OF THE MICROSTRIP ELEMENT*

The accurately calculated results are employed. Since the substrate thickness $h = 0.6$ mm, with $w/h = 2.225$, we find the microstrip width:

$$w = 1.335 \text{ mm}. \tag{4.51}$$

The length is now required. An expression for the microstrip wavelength (4.21) was given earlier:

$$\lambda_g = 300/(f\sqrt{\varepsilon_{\text{eff}}}) \text{ mm} \tag{4.52}$$

where f is in gigahertz.

Since the physical length l is required to be $\lambda_g/4$ we obtain

$$l = \frac{\lambda_g}{4} = \frac{75}{2\sqrt{7.095}} \tag{4.53}$$

so that

$$l = 14.08 \text{ mm}. \tag{4.54}$$

The length calculation was not automatically included in the computer program, because this embodies a frequency-dependent calculation (dealt with in Chapter 5) before microstrip wavelength is determined. Due to the effects of the discontinuities (Chapter 7) the actual length will be slightly less than the calculated value of 14.08 mm.

Physical length determinations at different frequencies and/or on different substrates follow precisely the same procedure as above. For example, at 2 GHz on FR4 ($\varepsilon_r = 4.8$), for the 31.3 Ω line $\varepsilon_{\text{eff}} = 3.93$, and this leads to $l = \lambda_g/4 = 18.9$ mm. Or at the 960 MHz GSM frequency, $l = \lambda_g/4 = 39.4$ mm. Although neglecting end-effect discontinuities, these physical lengths are becoming substantial — often too large for implementation in small-sized mobile handsets. On a ceramic substrate with $\varepsilon_r = 70$, all physical lengths are decreased by the approximate ratio $\sqrt{\varepsilon_{r1}/\varepsilon_{r2}} = \sqrt{70/4.8} = 3.8$. This means that a half-wave linear resonator on the $\varepsilon_r = 70$ material would have a physical length of 20 mm, and this could just about be implemented within a handset.

4.8 MICROSTRIP ON A DIELECTRICALLY ANISOTROPIC SUBSTRATE

Many substrates have been found to possess a degree of dielectric anisotropy. One particularly useful substrate material has turned out to be sapphire, which possesses well-defined dielectric anisotropy and which has been quite intensively investigated [100, 101]. As a result, we only consider sapphire here and before proceeding further it is worth listing the advantages of sapphire, as follows:

(a) The material is extremely pure and the permittivity values are now accurately known (and very stable).
(b) There is negligible batch-to-batch variation. That is to say, one piece of sapphire, taken from anywhere, will be almost exactly like any other piece of sapphire.
(c) An optical surface finish is readily achievable, resulting in very uniform and precisely defined thin metal films for microstrip conductors.
(d) The material is completely optically transparent. This means that the process of mounting chip devices can be accurately registered because the operator can see the underside.
(e) Since the thermal conductivity of sapphire is about 30% higher than that of alumina it may be preferred to the latter for mounting power devices.

Some other advantages were discussed in Chapter 3.

In this section we show that microstrip circuit design on sapphire may proceed in the manner so far described for isotropic substrates, provided a preferred orientation is chosen and a suitable function for an equivalent permittivity is used.

With the nomenclature shown in Figure 4.8, the permittivity tensor $\bar{\varepsilon}$ may be written

$$\bar{\varepsilon} = \begin{bmatrix} \varepsilon_x & 0 & 0 \\ 0 & \varepsilon_y & 0 \\ 0 & 0 & \varepsilon_z \end{bmatrix} \tag{4.55}$$

and this interrelates the electric displacement and electric field vectors thus

$$\bar{D} = \bar{\varepsilon}\varepsilon_0\bar{E}. \tag{4.56}$$

In all practical cases the principal axis of the sapphire is aligned to be very nearly perpendicular to the substrate–air interface, i.e. along the y direction in Figure 4.8. Therefore, Equation (4.55) may be written

$$\bar{\varepsilon} = \begin{bmatrix} \varepsilon_\perp & 0 & 0 \\ 0 & \varepsilon_\parallel & 0 \\ 0 & 0 & \varepsilon_\perp \end{bmatrix}. \tag{4.57}$$

Furthermore, the values are known accurately for sapphire, yielding

$$\bar{\varepsilon}_{\text{sapphire}} = \begin{bmatrix} 9.40 & 0 & 0 \\ 0 & 11.60 & 0 \\ 0 & 0 & 9.40 \end{bmatrix}. \tag{4.58}$$

Thus, with this particular orientation, and the sapphire substrate forming the dielectric between two completely parallel plates with no fringing, only the 11.60 permittivity value affects the TEM propagation which would then apply. Since microstrip is certainly not a symmetrical (infinite) parallel plate and does involve considerable fringing fields, the effects of this anisotropy are significant. The problem was studied fundamentally by Owens *et al.* [100] , who set up special finite difference equations and solved for the potentials around the structure.

The effective microstrip permittivity ε_{eff} is finally found as a ratio of capacitances, as described in Section 4.6 on page 94. However, because of the anisotropy, ε_{eff} depends upon w/h in a different manner to the isotropic dependence. To appreciate this more

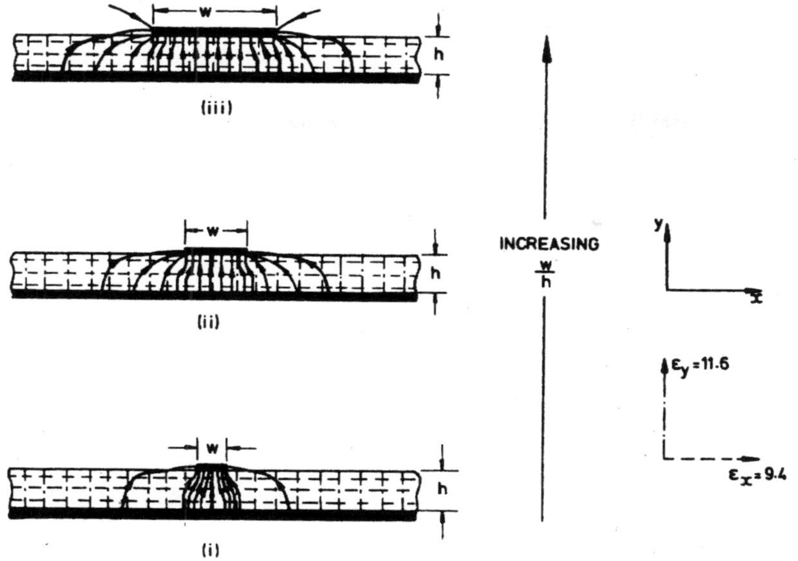

Figure 4.8 Electric field distributions, in the transverse plane, applicable to microstrip on an anisotropic substrate (sapphire) for three shape (w/h) ratios: (i) 'small;' (ii) 'moderate;' and (iii) 'large'.

completely, the electric field distribution is shown in Figure 4.8 for three different w/h values. In case (iii) the fields are almost entirely affected by ε_y (=11.60). In the other extreme shown, case (i) for relatively small w/h, ε_x (=9.40) significantly affects the field because there is a considerable fraction of electric field in the x direction.

To simplify the synthesis of microstrip on a sapphire substrate, a new permittivity concept has been suggested to account for the gross anisotropic effects. This is termed the equivalent isotropic substrate relative permittivity $\varepsilon_{r,\text{eq}}$, and it is, of course, dependent upon the width-to-height ratio w/h. A strict definition of $\varepsilon_{r,\text{eq}}$ is as follows:

> The equivalent isotropic substrate relative permittivity $\varepsilon_{r,\text{eq}}$ is the permittivity of some fictitious uniform and isotropic substrate, supporting a microstrip transmission line, which would yield microstrip static-TEM design parameters having values identical to those for the same line geometry (w/h) on sapphire.

The situation is illustrated in Figure 4.9.

For a known line geometry, w/h, an appropriate value of $\varepsilon_{r,\text{eq}}$ may be calculated and then used as 'ε_r' in any desired microstrip design formulas. Three design curves for microstrip on sapphire, originally published in Owens *et al.* [100], are reproduced here as Figure 4.10.

Quite accurate values of $\varepsilon_{r,\text{eq}}$ may be obtained by a combination of simple considerations and conventional (isotropic substrate) calculation. A cross-sectional model for microstrip on an anisotropic substrate assists this calculation of $\varepsilon_{r,\text{eq}}$, and is shown in Figure 4.11.

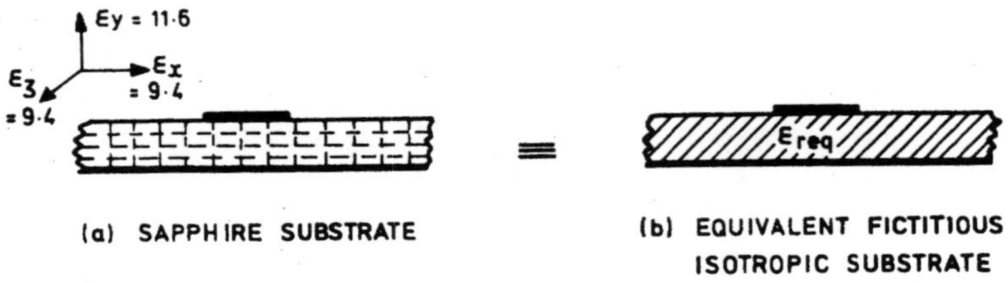

Figure 4.9 Illustrating the concept of equivalent isotropic substrate relative permittivity(ε_r).

The electric field immediately beneath the strip and down to the ground plane is almost entirely affected by ε_y alone. Elsewhere in the substrate, and in the surrounding air, the direction and density of the electric flux is such that ε_x and ε_y have approximately equal influences on the remaining capacitance associated with the microstrip. Thus, in Figure 4.11, we can treat the regions bounded by the magnetic walls separately. The total capacitance C per unit length of the microstrip line can thus be decomposed into separate quantities:

(a) The simple parallel-plate capacitance directly beneath the strip given by

$$C_y = \varepsilon_y \varepsilon_0 \left(w/h \right). \tag{4.59}$$

(b) A fringing capacitance C_f accounting for the remaining flux regions.

As shown in Figure 4.11, there is a hypothetical permittivity quantity ε_i ascribed to the remaining substrate regions. If we calculate the total capacitance C_i per unit length for the microstrip, but on some uniform isotropic substrate of permittivity ε_i, the fringing capacitance is then approximately the difference:

$$C_f \approx C_i - \varepsilon_i \varepsilon_0 \left(w/h \right) \tag{4.60}$$

where C_i may be evaluated using any appropriate conventional technique as described earlier — the method outlined in Section 4.6, for example. At present ε_i is unknown, but it is quite a straightforward task to find a value for this quantity, as indicated shortly.

Finally, the equivalent relative permittivity $\varepsilon_{r,\text{eq}}$ is given by a combination of ε_x and ε_y, weighted by the respective capacitance ratios C_f/C and C_y/C:

$$\varepsilon_{r,\text{eq}} = \varepsilon_i \left(\frac{C_f}{C} \right) + \varepsilon_y \left(\frac{C_y}{C} \right). \tag{4.61}$$

However, the total capacitance $C = C_y + C_f$; therefore,

$$\varepsilon_{r,\text{eq}} = \frac{\varepsilon_i C_f + \varepsilon_y C_y}{C_f + C_y}. \tag{4.62}$$

An optimized value $\varepsilon_i = 10.60$ was obtained by curve fitting to the finite difference theory values making up the discrete points shown alongside curve I of Figure 4.10(a) — where curve II is the actual plot of Equation (4.62) using this value of ε_i.

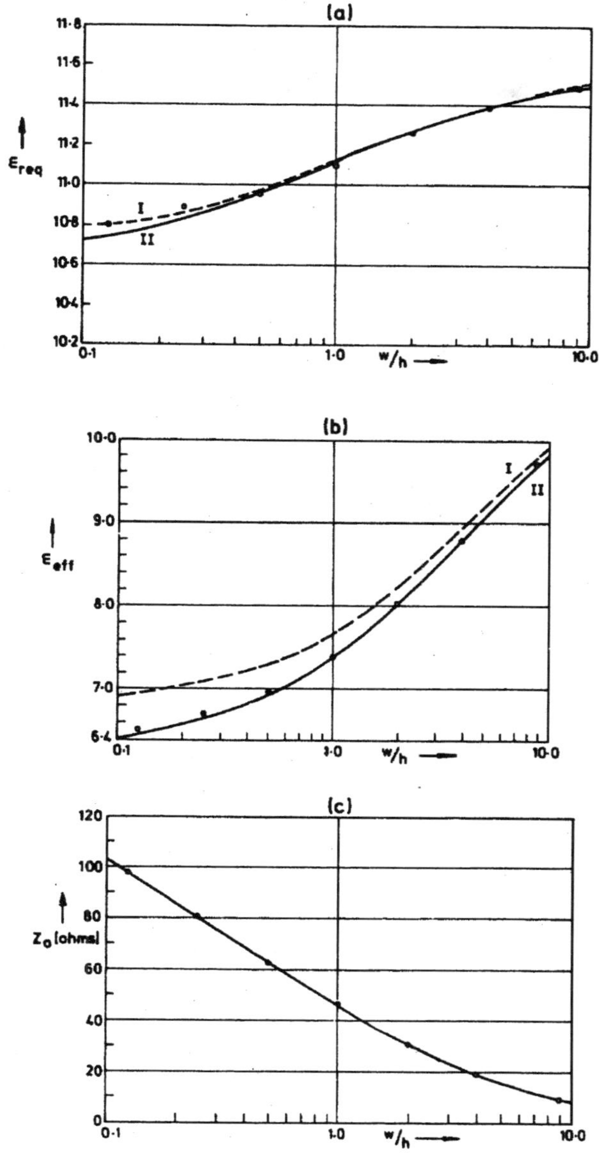

Figure 4.10 Design curves for microstrip lines on an anisotropic substrate of sapphire (C axis perpendicular to the ground plane). (©1976 IEEE. Reprinted, with permission, from Owens *et al.* [100].)

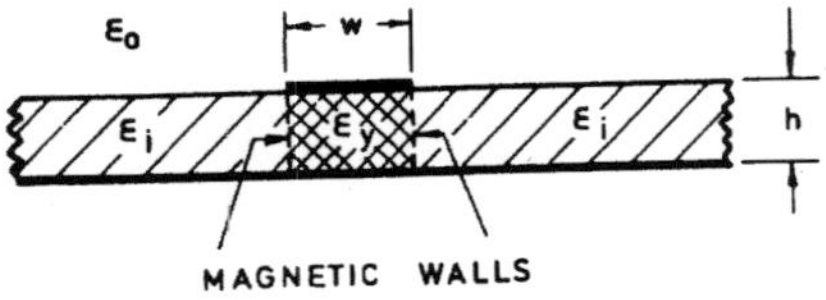

Figure 4.11 A model for microstrip on an anisotropic substrate with C axis perpendicular to the ground plane. (© 1976 IEEE. Reprinted, with permission, from Owens *et al.* [100])

This value can be used as a constant in determining $\varepsilon_{r,\text{eq}}$ for particular microstrip-on-sapphire designs. According to Figure 4.10(a) the worst discrepancy, amounting to 0.65%, occurs at the $w/h = 0.1$ extreme.

As a purely empirical alternative, the following expression gives $\varepsilon_{r,\text{eq}}$ and the only independent variable involved is w/h:

$$\varepsilon_{r,\text{eq}} = 12.0 - \frac{1.21}{1 + 0.39[\log(10w/h)]^2}. \tag{4.63}$$

The accuracy of Equation (4.63) when tested against other theoretical calculations and measurements, with results extrapolated to zero frequency, appears to be $\pm 0.5\%$ throughout the range

$$0.1 \leq w/h \leq 10 \tag{4.64}$$

which, on the sapphire substrate, approximates to the characteristic impedance range

$$10 \leq Z_0 \leq 100\ \Omega \tag{4.65}$$

and this is quite sufficient for a wide variety of circuit design purposes.

When static-TEM analysis is applied to microstrip-on-sapphire designs the effective permittivity evaluated, ε_{eff}, will depend upon the shape ratio w/h. This is clearly shown by the curves of Figure 4.10(b), where curve I has been generated using ε_y as 'ε_r' in the calculations, but curve II is generated using Equation (4.62) for $\varepsilon_{r,\text{eq}}$. The importance of employing the correct permittivity ($\varepsilon_{r,\text{eq}}$) is evident, since only then are the results close to the theoretical predictions (discrete points). For Figure 4.10(c), $\varepsilon_{r,\text{eq}}$ was used in determining the continuous curve. Thus, curve II of Figure 4.10(b) and the curve of Figure 4.10(c) represent design curves for microstrip on sapphire.

There are several other dielectrically anisotropic substrate materials, notably AR-1000®[1], single-crystal quartz and some semiconductor substrates — in particular GaAs. In any design using these materials, the principles outlined above may be developed in the same manner to provide a design basis.

Mariki and Yeh [102] have reported a three-dimensional TLM analysis of microstrip on an anisotropic substrate. Their results are entirely concerned with dispersion for this case, and are considered in rather more detail in Chapter 5.

[1] Formerly known as Epsilam-10®.

4.9 MICROSTRIP ON A FERRITE SUBSTRATE

A number of circuits have been designed and built on various ferrite substrates and it appears that the earliest significant papers appeared as long ago as 1957–58 [103,104]. However, the realization of production circuits using ferrite substrates is expensive, on account of the relatively high cost of the ferrite and its finishing. Further, many circuit functions certainly do not require ferrite for their operation — indeed, greater quality is generally achievable on purely dielectric substrates such as sapphire, alumina, quartz, etc. Where non-reciprocal properties are demanded, ferrite devices may be inserted into the substrate. Therefore only a brief treatment of the design problem is presented here.

For magnetically saturated ferrite, magnetized in the z direction and with an RF field propagating in this direction, the Polder tensor is

$$\bar{\mu} = \mu_0 \begin{bmatrix} \mu & -j\kappa & 0 \\ j\kappa & \mu & 0 \\ 0 & 0 & \mu_z \end{bmatrix} \tag{4.66}$$

where $\mu_z = 1$ and the permeability elements μ and κ are exactly related to precession frequency, microwave frequency, saturation magnetization, DC magnetic field, macroscopic relaxation time, and resonance line width. In the limit of high applied field or high frequency, the ferrite behaves as a dielectric medium with a moderately low loss ($\tan\delta \cong 0.001$ typically) and permittivity range

$$9 \leq \varepsilon_r \leq 16. \tag{4.67}$$

In the partially magnetized state (common in practice) μ is somewhat less than unity, and

$$\kappa = \frac{\gamma(4\pi M)}{\omega} \tag{4.68}$$

where γ and $4\pi M$ are respectively the gyromagnetic ratio and the average magnetizations for the ferrite.

Effective permeability μ_{eff} is an important quantity which is analogous to a bulk substrate relative permeability, and is thus very useful in design. The defining formula is

$$\mu_{\text{eff}} = \left(\mu^2 - \kappa^2\right)/\mu. \tag{4.69}$$

At sufficiently high frequency $\omega >> \gamma(4\pi M)$ and $k \to 0$ so that $\mu_{\text{eff}} \to \mu$. Under these conditions, μ_{eff} is not much less than unity:

$$0.7 \leq \mu_{\text{eff}} \leq 1.0 \qquad (\text{as } \omega \to \infty). \tag{4.70}$$

The quantity κ/μ is usually termed the (ferrite) anisotropy. In the case of microstrip an empirical correction has been given for μ_{eff} [105]. However, the above comments remain approximately true.

It is useful to define an effective microstrip permeability μ_{eff} as being analogous to the effective microstrip permittivity defined earlier in this chapter by (4.17) and repeated here,

$$\varepsilon_{\text{eff}} = 1 + q(\varepsilon_r - 1) \tag{4.71}$$

where the filling factor q, and hence ε_{eff} are functions of w/h and ε_r.

Provided some basic and reasonable assumptions are made regarding a microstrip line (perfect conductors; isotropic, homogeneous, non-gyromagnetic substrate) [105], then its dielectric properties will not affect the magnetic field distribution and its magnetic properties will not affect its electric field distribution. We can then write down a magnetic–electric duality statement:

$$\mu_e \{(w/h), \mu_{\text{eff}}\} = (\varepsilon_{\text{eff}} \{(w/h), \varepsilon_{\text{eff}}\})^{-1} . \tag{4.72}$$

By analogy with (4.71), we can also write

$$\mu_e = \left(1 + q_m \left\{\frac{1}{\mu_{\text{eff}}} - 1\right\}\right) \tag{4.73}$$

where q_m is the magnetic filling factor. This completely defines μ_e. However, although not explicitly indicated here, it must be appreciated that both q_m and μ_{eff} are functions of w/h for the microstrip. For CAD purposes, Pucel and Masse [105] give analytical expressions for μ_e and μ_{eff}. These expressions will not be repeated here, but a graph is given (Figure 4.12) as a design aid.

We also require expressions for the characteristic impedance and the guide wavelength. Using (4.1), $Z_0 = (L/C)^{\frac{1}{2}}$, we obtain

$$Z_0 = \left(\frac{\mu_e L_1}{\varepsilon_{\text{eff}} C_1}\right)^{\frac{1}{2}} \tag{4.74}$$

where L_1 and C_1 are for the air-spaced case. Hence

$$Z_0 = Z_{01}\sqrt{\mu_e/\varepsilon_{\text{eff}}} \tag{4.75}$$

also, using

$$v_p = c/\sqrt{\mu_e \varepsilon_{\text{eff}}} \tag{4.76}$$

we obtain

$$\lambda_g = \frac{\lambda_0}{\sqrt{\mu_e \varepsilon_{\text{eff}}}} \tag{4.77}$$

or

$$\lambda_g = \frac{300}{f\sqrt{\mu_e \varepsilon_{\text{eff}}}} \quad \text{mm} \tag{4.78}$$

(which can be compared to Equation (4.21)).

For synthesis, where w and λ_g are desired, it is recommended that the design is first carried out assuming a non-ferrite substrate (but, of course, with the appropriate permittivity). One or two stages of iteration, using the above techniques and information, should then suffice to obtain quite accurate final values for w and λ_g. This is assisted by the fact that μ_{eff} and μ_e both vary only slowly over the range

$$0.1 \leq w/h \leq 10.0 . \tag{4.79}$$

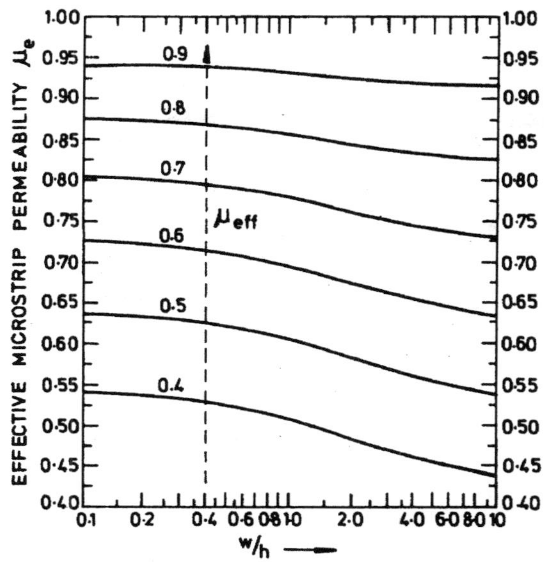

Figure 4.12 Effective microstrip permeabilty μ_e as a function of w/h. The effective relative permeability of the substrate material μ_{eff} is the parameter. (©1972 IEEE. Reprinted, with permission, from Pucel and Masse [105].)

4.10 EFFECTS OF FINITE STRIP THICKNESS, METALLIC ENCLOSURE, AND MANUFACTURING TOLERANCES

4.10.1 EFFECTS OF FINITE STRIP THICKNESS

Referring back to Figure 4.2 we see that microstrip, in general, will have a finite thickness t which must influence the field distribution. For most single microstrip lines the effect of this thickness on the design parameters is very small and may often be neglected. Even in the case of microstrip circuits manufactured using 'thick-film' technology (Section 3.6 on page 66), there is usually no need to allow for thickness when calculating impedance or effective microstrip permittivity, because such films invariably taper towards the strip edges. Notable exceptions are digital interconnects on chips and on MCMs, where the interconnects are kept as narrow as possible to ensure high wiring density.

With some etched microstrip circuits, on plastic substrates, and circuits where the microstrips are designed to carry at least moderate power, the thickness may be significant. An indication of the change in electric field distribution is provided by Figure 4.13, and several researchers have investigated the effects of finite strip thickness. Some straightforward and accurate expressions have been reported [106], which are now stated here, starting with characteristic impedances:

(a) For $w/h \leq 1$:

$$Z_0 = \frac{60}{\sqrt{\varepsilon_{\text{eff}}}} \ln\left(8\frac{h}{w_e} + 0.25\frac{w_e}{h}\right). \tag{4.80}$$

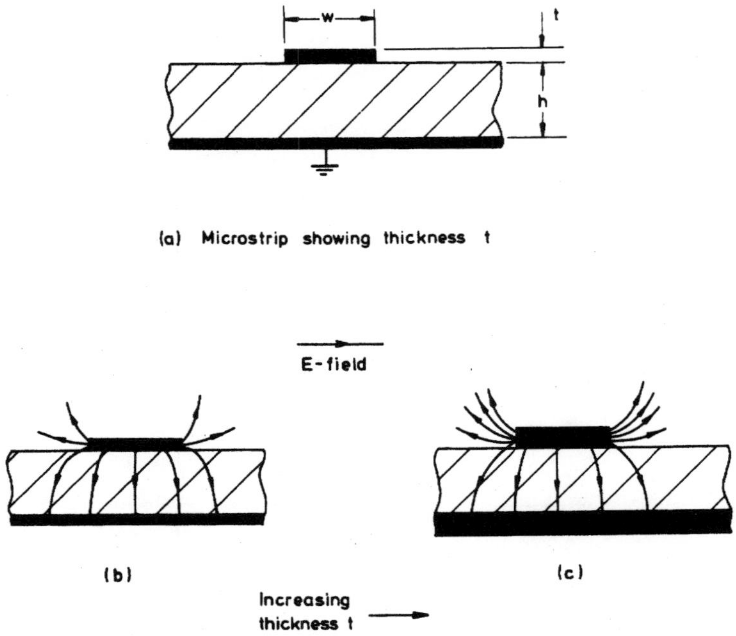

Figure 4.13 Changes in the distribution of electric field (transverse cross-section) as the thickness of microstrip is altered.

(b) For $w/h \geq 1$:

$$Z_0 = \frac{120\pi}{\sqrt{\varepsilon_{\text{eff}}}} \ln\left(\frac{w_e}{h} + 1.393 + 0.667\frac{w_e}{h} + 1.444\right) \tag{4.81}$$

where, in both cases

for $w/h \leq 1/2\pi$

$$\frac{w_e}{h} = \frac{w}{h}\frac{1.25t}{\pi h}\left[1 + \ln\left(\frac{4\pi w}{t}\right)\right] \tag{4.82}$$

and for $w/h \geq 1/2\pi$

$$\frac{w_e}{h} = \frac{w}{h}\frac{1.25t}{\pi h}\left[1 + \ln\left(\frac{2h}{t}\right)\right]. \tag{4.83}$$

Finally, the effective microstrip permittivity ε_{eff} should be evaluated using expressions given previously (Section 4.5 on page 92) and the following term subtracted [106]:

$$\Delta\varepsilon_{\text{eff}}(t) = \frac{(\varepsilon_r - 1)t/h}{4.6\sqrt{w/h}} \tag{4.84}$$

so that the final value is given by

$$\varepsilon_{\text{eff}}(t) = \varepsilon_{\text{eff}} - \Delta\varepsilon_{\text{eff}}(t). \tag{4.85}$$

For microstrip having $t/h \leq 0.005$, $2 \leq \varepsilon_r \leq 10$ and $w/h \geq 0.1$, the effects of this thickness are negligible (within approximately 1% on Z_0 or ε_{eff}). At smaller values of w/h or greater values of t/h, the significance of the thickness increases. These trends are clearly shown in curves published by Schwan [107], although he gave no details of their derivation. It is interesting to compare the trends exhibited in Schwan's results with those predicted by the expressions given here, Equations (4.80) to (4.84). Rather arbitrarily, an alumina substrate is chosen, with relative permittivity 10 and thickness 0.5 mm, and we also assume a strip conductor thickness of 0.032 mm, that is $t/h = 0.064$.

With a shape ratio $w/h = 0.1$, Schwan indicates a 20–30% reduction in Z_0 and approximately a 14% reduction in ε_{eff}, both compared with the $t/h = 0$ values. Careful checks with the formulas given here confirm these trends, but show much smaller changes — approximately 7% for both the permittivity and the impedance. Measurements and other indications tend to support these smaller changes.

The effect of the thickness of the conducting strip is quite significant when considering conductor losses; this is examined in Chapter 5.

4.10.2 EFFECTS OF A METALLIC ENCLOSURE

In many practical applications, MICs entail a metallic enclosure which serves the purposes of hermetic sealing, mechanical strength, electromagnetic shielding, connector mounting, and ease of handling of the module. The presence of conducting top and side walls lowers both the characteristic impedance and the effective microstrip permittivity, which is due to the increased proportion of electric flux in air. The effects of a top cover alone, as well as those due to a complete enclosure, have been reported by several researchers. Closed formulas [108] are given here which predict the effects of a conducting top cover, and show how both the characteristic impedance and effective microstrip permittivity are modified in comparison with the unshielded static-TEM expressions given in Section 4.5. The nomenclature of Figure 4.14 is used throughout.

Characteristic impedance is considered first: when $w/h \leq 1.3$:

$$Z_0(\text{shielded}) = Z_0(\text{unshielded}) - \Delta Z_{0s1} \tag{4.86}$$

and when $w/h \geq 1.3$:

$$Z_0(\text{shielded}) = Z_0(\text{unshielded}) - \Delta Z_{0s2} \tag{4.87}$$

in which

$$Z_{0s1} = 270\left[1 - \tanh\left(0.28 + 1.2\sqrt{h'/h}\right)\right] \tag{4.88}$$

and

$$Z_{0s2} = Z_{0s1}\left(1 - \tanh\left\{1 + \frac{0.48[(w_e/h) - 1]^{0.5}}{[1 + (h'/h)]^2}\right\}\right). \tag{4.89}$$

The effective microstrip permittivity of a shielded (top cover only) microstrip line is given by

$$\varepsilon_{\text{eff}} = \frac{\varepsilon_r + 1}{2}\left(\frac{\varepsilon_r - 1}{2}R\right)\tanh\left[0.18 + 0.237(h'/h) - \frac{0.415}{(h'/h)^2}\right] \tag{4.90}$$

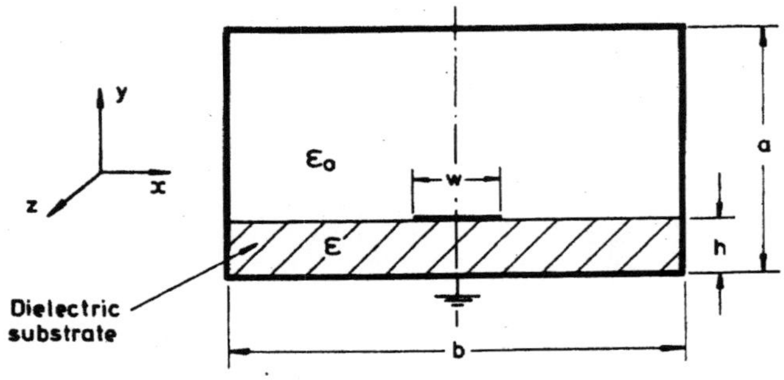

Figure 4.14 Cross-section of shielded microstrip.

in which

$$R = \begin{cases} [1+12(h/w)]^{-0.5} + 0.04[1-(w/h)]^2 & \text{for} \quad w/h \le 1 \\ [1+12(h/w)]^{-0.5} & \text{for} \quad w/h \ge 1 \end{cases} . \tag{4.91}$$

The expressions quoted in Section 4.5 might alternatively be used, but the nature of the shielding-modifying term is then uncertain.

In all these equations, Equations (4.86) to (4.90), w_e has already been defined by Equations (4.82) and (4.83), Z_0(unshielded) can be determined using expressions given previously in Section 4.5, and $b = a - h$, referring to Figure 4.14.

Comparison with the theoretical results due to Kowalski and Pregla [109] shows discrepancies within 1%. However, more extensive experimental investigation is highly desirable and design expressions allowing for the proximity of at least one side wall would be particularly useful.

4.10.3 EFFECTS DUE TO MANUFACTURING TOLERANCES

The effect on the various parameters due to tolerances in the physical dimensions has been quite intensively investigated and reported in detail by Gupta *et al.* [88]. Although often difficult, it is important to control the variations in thickness h and relative permittivity ε_r of non-semiconductor substrates. Some batch-to-batch variations are, however, inevitable, and the effects of these on electrical quantities such as impedance and VSWR must be determined. In addition to the effects of thickness and permittivity variations, the surface finish of the substrate affects the precision with which the strip width can be manufactured.

The maximum excursion in the characteristic impedance is given by

$$\frac{|\Delta Z_0|_{\max}}{Z_0} = \left|\frac{|\Delta w|}{w} S_{Z_0(w)}\right| + \left|\frac{|\Delta h|}{h} S_{Z_0(h)}\right| + \left|\frac{|\Delta \varepsilon_r|}{\varepsilon_r} S_{Z_0(\varepsilon_r)}\right| \tag{4.92}$$

and the maximum excursion in the effective microstrip permittivity is given by

$$\frac{|\Delta \varepsilon_{eff}|_{\max}}{\varepsilon_{eff}} = \left|\frac{\Delta w}{w} S_{\varepsilon_{eff}(w)}\right| + \left|\frac{\Delta h}{h} S_{\varepsilon_{eff}(h)}\right| + \left|\frac{\Delta \varepsilon_r}{\varepsilon_r} S_{\varepsilon_{eff}(\varepsilon_r)}\right| \tag{4.93}$$

where the sensitivity is, in general, defined by

$$S_{Q(p)} = \frac{p}{Q}\frac{\partial Q}{\partial p} \tag{4.94}$$

in which Q is a quantity whose sensitivity is to be determined (Z_0, ε_{eff}) and P is a controlling parameter (w, h, ε_r).

It is particularly important that the demanded accuracy in the manufacture of strip width is evaluated. This is the minimum selected from the two values of Δw obtained by rearranging Equations (4.92) and (4.93) for

$$\frac{\Delta w}{w} S_{Z_0(w)} \tag{4.95}$$

and

$$\frac{\Delta w}{w} S_{\varepsilon_{\text{eff}(w)}} \tag{4.96}$$

respectively. Any of the required sensitivities may be obtained by appropriately differentiating expressions given in Section 4.5.

Gupta *et al.* [88] show that the mismatch (VSWR) due to tolerances in the various quantities (w, h, ε_r) may be directly related to the resultant characteristic impedance change. The following expression is given for VSWR(r):

$$r = \left(1 - \frac{|\Delta Z_0| \max}{Z_0}\right)^{-1} \tag{4.97}$$

where the term in Z_0 is determined from Equation (4.92). Detailed graphs of parameter variations are given for an alumina substrate in the cited reference. In broad terms it is found that the VSWR can be held below 1.05 provided the tolerances are less than approximately 5%.

For similar tolerances the worst VSWR is approximately 1.1 (alumina or plastic substrates). The maximum change in effective microstrip permittivity, from Equation (4.93), is approximately 4.3%, on alumina, when the relative permittivity tolerance is also approximately 4.5%, or the substrate thickness tolerance about 4%, or the microstrip width tolerance is about 8%.

4.11 PULSE PROPAGATION ALONG MICROSTRIP LINES

There are many instances in which microstrip lines are used for carrying signal pulses rather than analog microwave signals in high-speed systems. This area of application tends to be increasing in importance, especially with the pervasiveness of ever higher speed digital circuits in monolithic technology (silicon, SOI, GaAs, InP, SiGe, etc.). Examples of such applications include:

- very high speed computer logic (through GHz clock rates)
- high bit-rate digital communications (e.g. direct pulsing of a laser or LED for optical communications — through tens of Gbit/s)
- high-speed samplers for oscilloscopes or time-domain reflectometers (below 20 ps rise times)
- 'pico-second' resolution counters
- radars.

One of the most significant properties of microstrip in these applications is that of *propagation delay.* The delay time per metre, τ_d, for a microstrip line is given simply

Figure 4.15 A microstrip circuit structure yielding substantial delay.

by

$$\tau_d = \frac{1}{v_p} \quad \text{s/m} \tag{4.98}$$

or, using Equation (4.11),

$$\tau_d = \frac{\sqrt{\varepsilon_{\text{eff}}}}{c} \quad \text{s/m} \tag{4.99}$$

where ε_{eff} and c have the definitions given previously.

The static-TEM effective microstrip permittivity ε_{eff} is determined using any appropriate method as described earlier. As an example, let us consider a 50 Ω microstrip line on high-purity alumina: $\varepsilon_{\text{eff}} \approx 6.7$. We obtain, using Equation (4.99),

$$\begin{aligned} \tau_d &= \sqrt{6.7}/(3 \times 10^8) \quad \text{s/m} \\ &\approx 8.6 \quad \text{ns/m}. \end{aligned}$$

In practice this means that the delay is 8.6 ps/mm and a length of 10 mm produces 86 ps of delay for a pulse. High-speed gates may typically have around 50 ps delay per gate, so that we often require around 5–10 mm of microstrip in systems using such gates. This is usually feasible on non-semiconductor substrates, and even on some semiconductor chips. For instances where such lengths of microstrip are not feasible, one possible way in which to increase the effective propagation delay is to introduce capacitive loading at intervals along the microstrip. The delay then becomes influenced by these capacitances ($v_p = 1/\sqrt{LC}$) and v_p is reduced, i.e. the pulse is slowed, if C is increased. Further, the effective inductance per unit length L may also be increased, and the microstrip takes on the general appearance shown in Figure 4.15.

This structure is, of course, in the frequency domain, a form of lowpass filter, and such circuits are discussed in Chapter 10. The delay associated with such a structure is given by $n\sqrt{LC}$, where n is the (equal) number of sections representing L (or C). Design techniques for the microstrip sections are given in Chapter 10.

Crosstalk between adjacent circuits is a serious problem in pulse systems. Methods for calculating the amount of crosstalk require a study of the coupled microstrip line problem, and are discussed in Section 8.14 on page 307.

4.12 RECOMMENDATIONS RELATING TO THE STATIC-TEM APPROACHES

This section provides a summary of the principal design methods and formulas which have been presented in this chapter. Although the static-TEM approaches, when used directly, exhibit significant errors at frequencies beyond a few gigahertz they form a basis for powerful design expressions which are used at higher frequencies. This is discussed in Chapter 5.

4.12.1 THE PRINCIPAL STATIC-TEM SYNTHESIS FORMULAS

These are given in Section 4.5 on page 92. They are of proven accuracy to within 1% for w/h and 0.5% for ε_{eff}; some are repeated here for convenience.

(a) For narrow strips (i.e. when $Z_0 > (44 - 2\varepsilon_r)\ \Omega$):

$$\frac{w}{h} = \left(\frac{\exp H'}{8} - \frac{1}{4\exp H'}\right)^{-1} \tag{4.100}$$

under the slightly different condition, $w/h < 1.3$:

$$\varepsilon_{\text{eff}} = \frac{\varepsilon_r + 1}{2}\left[1 - \frac{1}{2H'}\left(\frac{\varepsilon_r - 1}{\varepsilon_r + 1}\right)\left(\ln\frac{\pi}{2} + \frac{1}{\varepsilon_r}\ln\frac{4}{\pi}\right)\right]^{-2} \tag{4.101}$$

where, for both expressions,

$$H' = \frac{Z_0\sqrt{2(\varepsilon_r + 1)}}{119.9} + \frac{1}{2}\left(\frac{\varepsilon_r - 1}{\varepsilon_r + 1}\right)\left(\ln\frac{\pi}{2} + \frac{1}{\varepsilon_r}\ln\frac{4}{\pi}\right). \tag{4.102}$$

(b) For wide strips (i.e. when $Z_0 < (44 - 2\varepsilon_r)\ \Omega$):

$$\frac{w}{h} = \frac{2}{\pi}\left\{(d_\varepsilon - 1) - \ln(2d_\varepsilon - 1)\right\} + \frac{\varepsilon_r - 1}{\pi\varepsilon_r}\left\{\ln(d_\varepsilon - 1) + 0.293 - \frac{0.517}{\varepsilon_r}\right\} \tag{4.103}$$

where

$$d_\varepsilon = \frac{59.9\pi^2}{Z_0\sqrt{\varepsilon_r}} \tag{4.104}$$

and, again with the slightly different condition, $w/h > 1.3$:

$$\varepsilon_{\text{eff}} = \frac{\varepsilon_r + 1}{2} + \frac{\varepsilon_r - 1}{2}\left(1 + 10\frac{h}{w}\right)^{-0.555}. \tag{4.105}$$

For the effects of shielding (top cover), finite thickness of strip, and manufacturing tolerances, Section 4.10 on page 105 should be consulted.

The approximate graphical method described in Section 4.4 is useful at least for a 'first run' analysis or synthesis, and is applicable to any substrate. Apart from lack of accuracy, its main drawback is the restricted range encompassed.

4.12.2 MICROSTRIP ON A SAPPHIRE (ANISOTROPIC) SUBSTRATE

This problem was discussed in detail in Section 4.8 on page 97. The main design expression is the empirical formula for equivalent relative permittivity $\varepsilon_{r,\text{eq}}$:

$$\varepsilon_{r,\text{eq}} = 12.0 - \frac{1.21}{1 + 0.39\{\log(10w/h)\}^2}. \tag{4.106}$$

Table 4.1 Adjustments required in w.

$\approx w/h$	$\approx Z_0$ (Ω)	Reduction in w (%)
0.1	110	3
1.0	50	2.5
8.0	12	1

Once more, this expression is only known to be accurate within $\pm 0.5\%$ over the range

$$0.1 \leq w/h \leq 10. \tag{4.107}$$

On the sapphire substrate this approximates the characteristic impedance range

$$10 \leq Z_0 \leq 100\ \Omega. \tag{4.108}$$

The value of $\varepsilon_{r,\text{eq}}$ obtained using the above expression is directly substituted for 'ε_r' in the static-TEM formulas which apply to the more useful isotropic substrate situation. However, w/h is initially not accurately known in design and some iterative CAD routine is therefore required. The designer might start with a rough initial value $\varepsilon_{r,\text{eq}} \approx 11$ for sapphire, compute w/h from a static-TEM expression, evaluate an updated $\varepsilon_{r,\text{eq}}$, etc. Since $\varepsilon_{r,\text{eq}}$ w/h is a fairly slowly varying function, this would be a good, fast-converging approach.

4.12.3 DESIGN CORRECTIONS FOR MICROSTRIP ON NON-SEMICONDUCTOR SUBSTRATES

Once an MIC has been manufactured there is often little that one can do to 'trim' the structure and hence improve upon the performance of the complete circuit. Where the passive (i.e. microstrip) circuitry can be tested in the absence of any active devices, the deviations in actual measured performance may be observed and used to calculate the required artwork/processing adjustments for an improved second attempt. Usually the spread of characteristics and the rejection costs preclude this approach when active devices have been incorporated. Sometimes it may also be possible to etch or laser-trim microstrip elements. Whatever the method, the following guidelines are applicable to microstrip on an alumina-type substrate:

(a) Always start with a slightly lower impedance than actually desired, i.e. larger w/h, if trimming is contemplated. Table 4.1 indicates the approximate magnitude of adjustments for a 1% increase in Z_0.
(b) The physical lengths of sections should also be made, initially, slightly longer than finally required if trimming is contemplated. This is because the operating frequency for these lengths will then be slightly lower than required and the physical length (proportional to wavelength) must be reduced by etching or laser-trimming to increase the operating frequency back to the correct value. The adjustment may be made on a pro rata basis, i.e. with a 1% reduction in length we can expect approximately a 1% increase in frequency.
(c) As a final suggestion the length of a top-cover shield might be adjusted to 'trim' the performance of MICs. See Section 4.10.2 on page 107 for suitable expressions which would yield the values of adjustments required.

5

Microstrip and Stripline at High Frequencies

5.1 THE SCOPE OF THIS CHAPTER

In Chapters 3 and 4 it was explained that microstrip cannot support a pure TEM wave, or any other simple electromagnetic field mode, and the necessary longitudinal field components which exist lead to the propagation of hybrid modes. It was further indicated in Chapter 4 that the 'quasi-TEM' nature of the propagating fields can be approximately calculated by the use of some statically derived expressions. Such expressions have accuracies to within 1% where they are applicable — generally at frequencies below 1 or 2 GHz. They also form an important basis for calculations at higher frequencies.

In this chapter we study the behaviour of microstrip rather more deeply, leading to fairly straightforward design calculations for frequencies above a few gigahertz. The following list gives the principal topics discussed:

(a) dispersion in microstrip
(b) a model leading to design calculations which allow for dispersion
(c) improved accuracy calculations
(d) an example (a specific design calculation)
(e) a frequency-dependent electromagnetic field solution (using a numerical method and requiring a substantial computer program)
(f) characteristic impedance as functions of frequency
(g) operating frequency limitations
(h) power losses and parasitic coupling
(i) stripline design including suspended substrate stripline.

5.2 DISPERSION IN MICROSTRIP

When the frequency of a signal exciting a microstrip line is (say) doubled, the phase constant or wave number $\beta(= 2\pi/\lambda_g)$ is not exactly doubled. Several transmission line structures exhibit this type of behaviour, indicated in Figure 5.1; it is called dispersion. Thus, all microstrip lines are dispersive and it follows that the 'exact' relationship between wavelength and frequency is very complicated. Detailed time and space-variant electromagnetic field analyses have been performed requiring numerical methods and substantial computer programs to solve them. Fortunately, as we shall

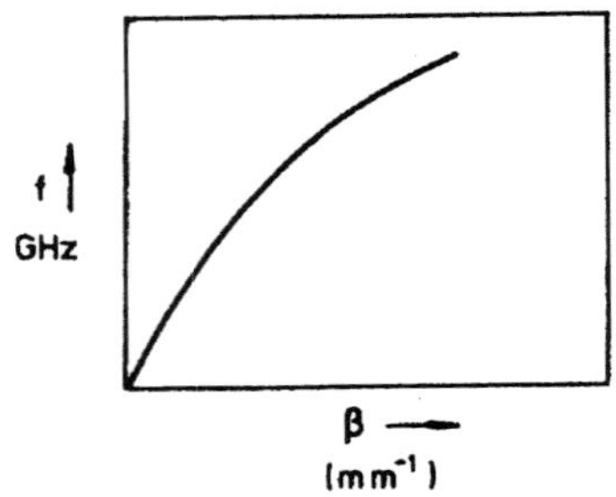

Figure 5.1 The dispersive effect in any general structure or system — non-linearity when frequency f is plotted against wave number or phase coefficient β.

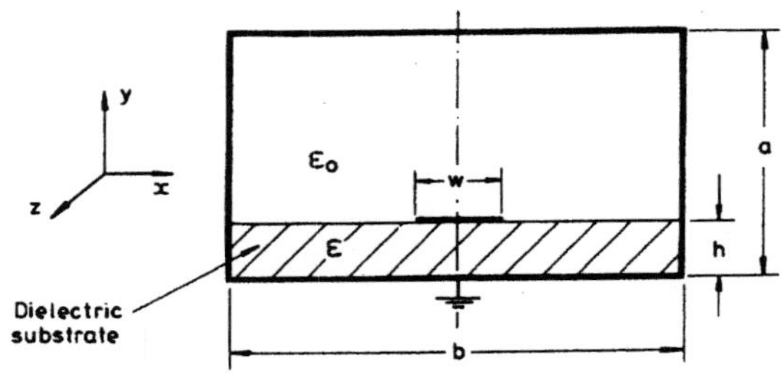

Figure 5.2 Transverse cross-section of shielded microstrip, showing the nomenclature. The strip is the conductor of width w and the ground plane is the bottom, grounded conductor.

see, useful and accurate (typically within 1%) design calculations may be carried out with suitable closed formulas. These formulas also provide intuitive understanding of behaviour as required in the design.

The hybrid mode of propagation along microstrip leads to the dispersive mechanism, and an examination of the types of fields occurring in a shielded microstrip line, itself an important structure, helps to explain this dynamic situation (see Figure 5.2). Another important and revealing aspect concerns the changing current and charge distributions in both strip and ground plane as the frequency increases.

We have performed a series of SONNET® simulations of a 50 Ω alumina-substrate line at three frequencies: 1 GHz, 10 GHz and 30 GHz. The actual structure modelled is shown in Figure 5.2 with a very distant top wall (i.e. $a \gg h$). The results are given for each frequency in Figures 5.3, 5.4 and 5.5 where the longitudinal current and charge distributions on the strip and the ground plane are shown (in the xy plane). The longitudinal current, shown in Figures 5.3(b), 5.4(b) and 5.5(b), is the current that is the largest current component and is the only one considered in a first approximation. The transverse currents are much smaller and their relative levels are indicated in Table 5.1.

The ground plane was modelled realistically using the properties of the ground metallization, but it was necessary to terminate the ground plane in an ideal ground at the top and bottom extremes. Thus the transverse currents can be large at the top and bottom ground plane extremes as these continue into the ideal ground.

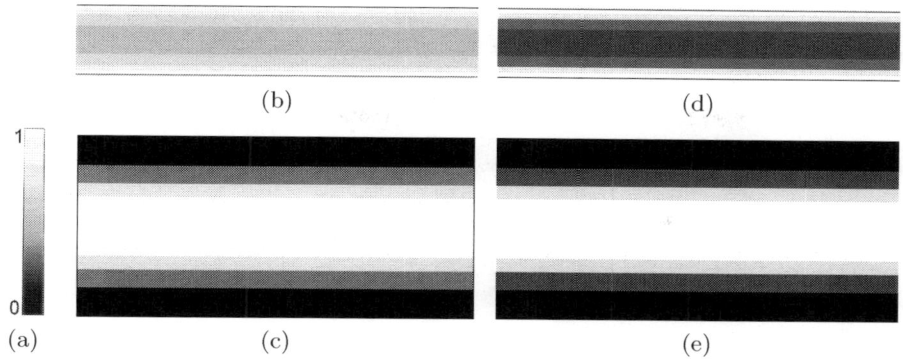

Figure 5.3 Normalized current and charge magnitudes in an alumina microstrip line at 1 GHz: (a): normalized scale; (b) longitudinal current, i_z, on the strip and (c) on the ground plane; and (d) the charge on the strip and (e) on the ground plane.

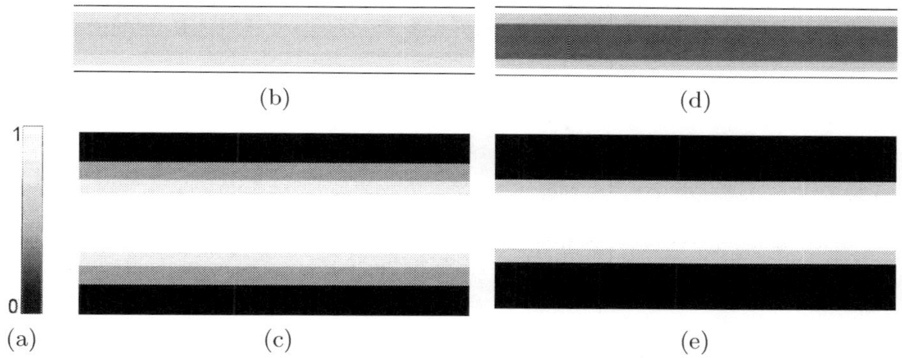

Figure 5.4 Normalized current and charge magnitudes in an alumina microstrip line at 10 GHz: (a): normalized scale; (b) longitudinal current, i_z, on the strip and (c) on the ground plane; and (d) the charge on the strip and (e) on the ground plane.

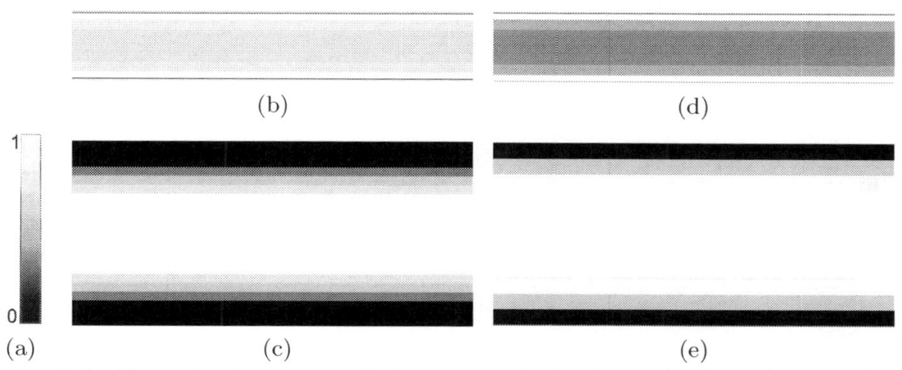

Figure 5.5 Normalized current and charge magnitudes in an alumina microstrip line at 30 GHz: (a) normalized scale; (b) longitudinal current, i_z on the strip and (c) on the ground plane; and (d) the charge on the strip and (e) on the ground plane.

Table 5.1 Peak charge and longitudinal, i_z, and transverse, i_x, current levels for the microstrip line. The currents and charges are normalized to the peak strip levels at 30 GHz.

Freq. (GHz)	Strip Current i_z	Strip Current i_x	Ground Current i_z	Ground Current i_x	Strip Charge	Strip Charge
1	0.84	0.00052	0.10	0.0013	0.52	0.057
10	0.90	0.00081	0.11	0.012	0.61	0.058
30	1.00	0.0052	0.22	0.058	1.00	0.115

In all three instances the charge and current on the strip clearly intensify greatly near the edges, and this is a well known phenomenon.

At 1 GHz, Figure 5.3, both the longitudinal and transverse currents are highly regular and in particular the longitudinal current flows directly down the microstrip. This can be seen by examining the relative levels of the various current components given in Table 5.1. It is seen that at 1 GHz the transverse current is negligible compared to the longitudinal current so that the current on the strip is almost entirely along the strip. The longitudinal current in the ground plane is at its highest level under the microstrip, but is 15% of the peak microstrip current. The transverse current in the microstrip is around 0.06% of the longitudinal current and for the ground current the relative level is around 1%. The transverse current in the microstrip and in the ground is zero along the centre line. This indicates that the current in the strip and the ground has a small circulatory component superimposed on the longitudinal current.

At one decade higher in frequency, Figure 5.4, clear changes in current distributions are evident. In the ground plane the maximum transverse current is now around 11% of the peak longitudinal ground current. However, for the microstrip the relative level is around 1%. When the frequency is increased still further, to 30 GHz, the effects noted here are more dramatic. In Figure 5.5 the current distribution changes dramatically, with a bunching effect now pronounced in both the microstrip and the ground plane. At this frequency the peak transverse current in the ground is in the region of one quarter of the maximum longitudinal current. One other significant change is the ground current is now much mode localized under the strip at the higher frequency. The trend is obvious: as frequency increases the ground current tends to localize under the strip; transverse currents increase in both the strip and the ground; and the strip longitudinal current becomes more concentrated on the edges of the strip. When the ground is lossy, which is the case with the very thin interconnects in digital circuit chips (especially with aluminium metallization), it is the increased localization of the current in the ground plane (or nearby interconnects) that is primarily responsible for resistance increase with high frequency, rather than the skin effect, which is most apparent with the concentration of the current near the edges of the strip [19].

The behaviour of current and charges can be examined further by considering the current and charges on the microstrip line at one instant of time. Figure 5.6 shows the longitudinal and transverse currents for a 30 GHz sinusoidal signal at one instant. The peak levels are the same as for the frequency domain variations shown in Figure 5.5. The strong circulatory motion of the currents in the ground plane is most

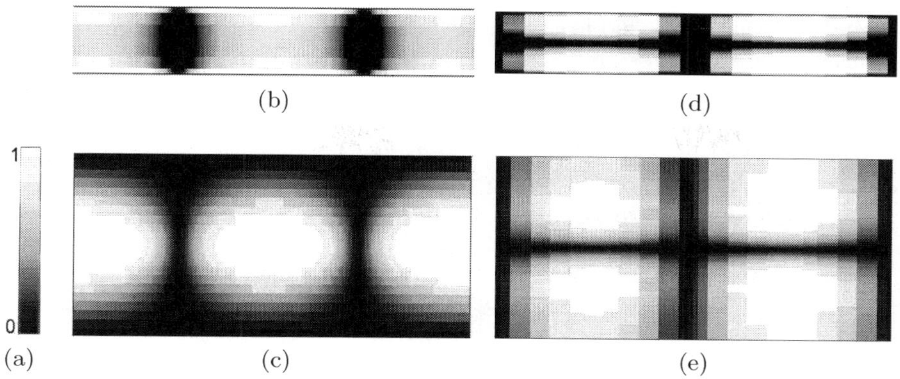

Figure 5.6 Normalized instantaneous current distribution on the microstrip line and in the ground plane: (a): normalized scale; (b) magnitude of the longitudinal current, i_x, on the microstrip and (c) on the ground plane; and (d) magnitude of the transverse current, i_y, on the microstrip and (e) on the ground plane. In all cases an alumina substrate applies and the frequency is 30 GHz.

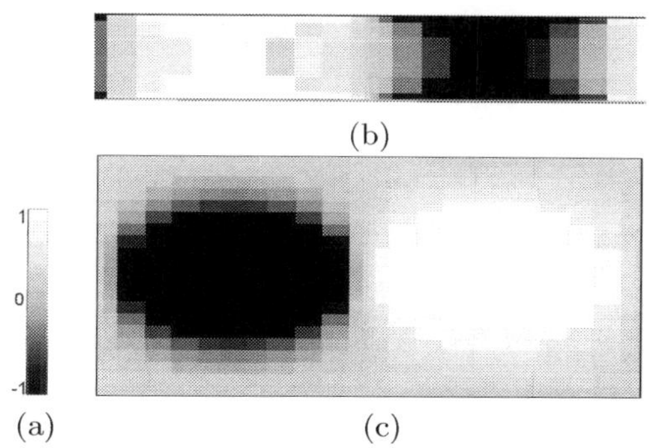

Figure 5.7 Normalized instantaneous charge distribution, q, on the microstrip line and the ground plane at 30 GHz: (a) scale; (b) microstrip; and (c) ground plane.

significant. The current in the ground (as well as on the strip) can be considered as being composed of longitudinal and circulatory components instead of longitudinal and transverse components. Here the circulatory current in the ground is one third of the longitudinal component. The confinement of the current to the region under the strip is more evident than in the frequency domain case. The charge distributions corresponding to the currents of Figure 5.6 are shown in Figure 5.7.

This all amounts to clear evidence of significant dispersion in microstrip.

If the centre strip is removed from the shielding waveguide, Figure 5.2, it reduces to a guide with partial dielectric filling that can support Longitudinal Section Electric (LSE) or Longitudinal Section Magnetic (LSM) modes. Pure TE or TM modes cannot, however, be supported — unlike the conventional unfilled waveguide. When the centre strip is inserted it causes currents to flow in both x (transverse) and z (longitudinal) directions on this strip; these currents serve to couple the LSE and LSM modes so that the final mode configuration is hybrid — what is called the quasi-TEM mode. As the frequency is increased, the mode-coupling efficiency increases and the fields become more concentrated in the region beneath the strip — where the substrate permittivity has already resulted in a relatively large electric field displacement. Since the fields are forced into the dielectric substrate to an increasing extent as the frequency rises we can define a frequency-dependent effective microstrip permittivity $\varepsilon_{\text{eff}}(f)$. This quantity clearly increases with frequency and the wave is progressively slowed down. We may continue to use the more general definition of effective microstrip permittivity given by Equation (4.10), except that ε_{eff} must now be replaced by $\varepsilon_{\text{eff}}(f)$ and v_p by $v_p(f)$ giving:

$$\varepsilon_{\text{eff}}(f) = \{c/[v_p(f)]\}^2 . \tag{5.1}$$

Fundamentally, the dispersion problem then consists of solving the microstrip fields for the velocity $v_p(f)$. The limits of $\varepsilon_{\text{eff}}(f)$ are readily established; at the low-frequency extreme it reduces to the static-TEM value ε_{eff}, while as frequency is increased indefinitely $\varepsilon_{\text{eff}}(f)$ approaches the substrate permittivity itself ε_r. This is summarized as follows:

$$\varepsilon_{\text{eff}}(f) \to \begin{cases} \varepsilon_{\text{eff}} & \text{as } f \to 0 \\ \varepsilon_r & \text{as } f \to \infty \end{cases} . \tag{5.2}$$

Between these limits $\varepsilon_{\text{eff}}(f)$ changes continuously, as shown by the general curve of Figure 5.9.

There have been several approaches to the detailed hybrid-mode analysis of microstrip [110–122], and one example is described later in this chapter. A number of closed formulas based upon various models of microstrip have also been developed, and some of these are briefly discussed and compared in the next section.

5.3 APPROXIMATE CALCULATIONS ACCOUNTING FOR DISPERSION

As explained in Chapter 3 the circuit designer needs to know the wavelength λ_g of the dominant mode in the microstrip whose characteristic impedance Z_0 and aspect ratio w/h are already known (from electrical measurements). This implies that $\varepsilon_{\text{eff}}(f)$is required, since

$$\lambda_g = \frac{c}{f\sqrt{\varepsilon_{\text{eff}}(f)}} . \tag{5.3}$$

It is generally better to find expressions for $\varepsilon_{\text{eff}}(f)$ first, because gross errors in calculating $\varepsilon_{\text{eff}}(f)$ can readily be detected (e.g. it must lie within the limits shown in Figure 5.8).

Fundamentally, this effective microstrip permittivity must be a function of frequency, substrate height, and substrate relative permittivity:

$$\varepsilon_{\text{eff}}(f, h, \varepsilon_r) \tag{5.4}$$

where it should be noted that the substrate height h enters into the expressions because it influences the surface waves, and thus also the complete hybrid field pattern. This can be seen by imagining the effect of increasing the proportion of dielectric substrate in Figure 5.2.

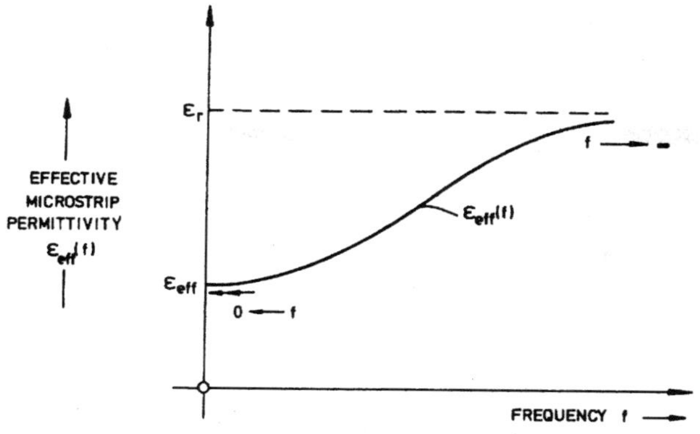

Figure 5.8 Dispersion of microstrip interpreted as effective microstrip permittivity $\varepsilon_{\text{eff}}(f)$ plotted to a base of frequency.

Most of the 'closed-formula' expressions for calculating dispersion also require the static-TEM value of ε_{eff}, and sometimes Z_0 also.[1] These are obtained by the methods given in Chapter 4. Thus, the frequency-dependent effective microstrip permittivity may also be written as a function involving ε_{eff} and Z_0 as follows:

$$\varepsilon_{\text{eff}}(f, h, \varepsilon_r, \varepsilon_{\text{eff}}, Z_0)\,. \tag{5.5}$$

One of the first attempts at developing a closed-formula for dispersion calculations was due to Jain, Makios and Chudobiak [123–125]. Their first formula was derived empirically and showed only a first-order relationship with frequency [123].

This is known to be very inaccurate (clearly it is, at least, incapable of approaching the limit conditions already discussed here and shown in Figure 5.8) and their more comprehensive method must therefore be examined. Their approach consisted basically of analysing the coupling between the TEM mode and a surface-wave mode TM_0.[2] Rather elaborate algebraic expressions are obtained, and either a graphical or a numerical method is required to solve a special effective permittivity which is used. The main expression used takes $\varepsilon_{\text{eff}}(f) > \varepsilon_r$ at high frequencies — which is physically impossible as we have already seen. In fact the full dispersion curves given by Jain, Makios and Chudobiak clearly show $\varepsilon_{\text{eff}}(f)$ progressing beyond the limit of 9.9 set by the substrate permittivity. Otherwise the theory shows quite close agreement with experimental results, except that these were obtained using ring resonators, which give unreliable results and are discussed later. The actual values of ε_{eff} which were used in the work of Jain, Makios and Chudobiak are all too high due to substantial errors in the static-TEM expressions which were used at that time.

A fundamentally different approach was adopted by Schneider [126] who took four basic features regarding microstrip dispersion and identified a rational function which

---

[1] We shall use ε_{eff} to indicate the static-TEM value of the effective permittivity and $\varepsilon_{\text{eff}}(f)$ for the frequency-dependent values.

[2] Note that the hybrid field configuration can be conveniently decomposed into at least two combinations of compound modes, the LSE and LSM mode pair, and the TEM and TM_0 mode pair, for example. At higher frequencies larger combinations of modes in each set must be considered.

satisfied all these features.

The four basic features are as follows:

(a) $\varepsilon_{\text{eff}}(f)$ always increases with frequency.
(b) $\varepsilon_{\text{eff}}(f)$ $(f \to 0) \Rightarrow \varepsilon_{\text{eff}}$ (see Figure 5.8.)
(c) $\varepsilon_{\text{eff}}(f)(f \to \infty) \Rightarrow \varepsilon_r$.

It is also noted that $\varepsilon_{\text{eff}}(f)$ is asymptotic at both limits and the fourth feature concerns phase velocity $v_p(f)$. This suffers inflection at some frequency close to the cut-off frequency f_c of the TE_1 surface wave, Therefore this fourth feature is:

(d) $\partial^2 v_p/\partial f^2 = 0$ when $f = f_c$.

The formulas which Schneider finally obtained were well suited to simple rapid calculations at the time, but Schneider pointed out that this theory can only be expected to yield an engineering approximation and quoted errors of somewhat less than 3% in phase velocity (i.e. about 6% in effective permittivity) calculated by this method compared with other data (for lines having characteristic impedances in the region of 50 Ω). The method relied heavily on satisfying all the discrete conditions, especially (d), and this could be a main reason for the errors that arise. It should be observed that a 3% error in phase velocity, or wavelength, would be unacceptable for the design of many circuits — including filters or matching networks.

Carlin [127] reported a novel and interesting approach that involved setting up and analysing a circuit model of microstrip. His circuit model involved coupled lines where one of the lines propagates only a TEM wave while the other line propagates only a TE wave. The details will not be pursued here, but the final expressions involve frequency f, height h, permittivity ε_r, static-TEM effective permittivity ε_{eff}, and characteristic impedance Z_0 — as well as some constants.

A comparison of calculations based upon Carlin's expressions against several other approaches is given shortly in graphical form (see Figure 5.11). Although more accurate methods are known (which are no more complicated than that due to Carlin) it may well be that fairly minor modifications to the expressions could yield improved accuracy.

In any event, the advances due to researchers including Getsinger, Edwards and Owens, Yamashita *et al.*, Kirschning and Jansen, and also Kobayashi, provide more accurate and wide-ranging sets of expressions for predicting the effects of dispersion. As his approach effectively underlies most of these advances, we start with Getsinger's formulation here.

An apparently more accurate formula, applicable at least to microstrip on alumina-type substrates, has been developed by Getsinger [128, 129]. The full derivation is too lengthy to be given here, but a brief summary of the method is presented. Getsinger's approach begins with proposing an alternative 'model' for microstrip which is arranged for more straightforward analysis than the microstrip itself. Essentially, the model aims to largely overcome the difficulties set up by the 'horizontal" air-dielectric interface in real microstrip and the approach is illustrated in Figure 5.11. It is important to note that the model is not physically realizable. It consists of three side-joined transmission lines, as follows: (i) central parallel-plate line width w and height h filled with the substrate material of permittivity ε_r and two additional parallel-plate lines, (ii) and (iii), attached to the sides of this central line (i). These

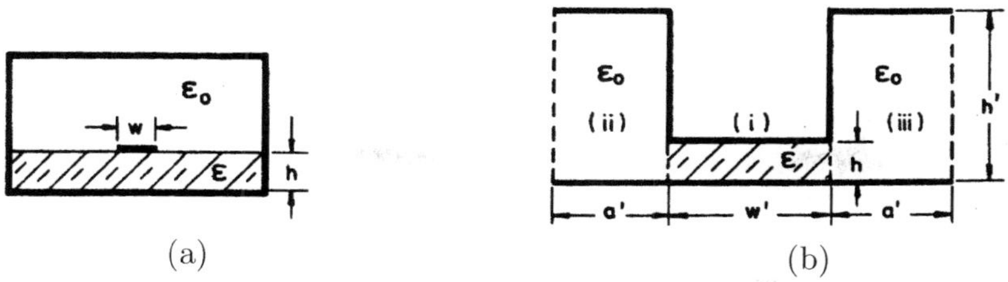

Figure 5.9 Getsinger's model for dispersion analysis: (a) actual microstrip; and (b) microstrip model. (© 1973 IEEE. Reprinted, with permission, from Getsinger [128].)

two additional parallel-plate lines are identical in structure and electrically joined to the central line. Each has width a, height h, and is filled with air having permittivity ε_0.

Line (i) accounts for the substrate fields in the real microstrip. It is much easier to analyze because it is simply a parallel-plate line. Lines (ii) and (iii) allow for the fields which exist in air with the real microstrip. The assumption is made that because the two regions — air filled and dielectric filled — of the model and the microstrip are grossly similar, the two structures will have the same dispersive behaviour for the same mode of propagation. This hypothetical 'microstrip model' involves corners with abrupt dielectric discontinuities, and Getsinger makes it clear quite early in his paper that junction capacitance should strictly be included to allow for the corner errors. However, this would clearly make the analysis overcomplicated, and almost certainly prevent a closed-form expression from being obtained. Therefore, no attempt is made to improve the accuracy by including these junction capacitances.

The first step in the analysis consists of mathematically forcing the model to have the same electrical characteristics at the low-frequency limit, as the microstrip. Thus the static-TEM parameters of each structure are made equal [129]. A transverse resonance analysis of the model (Figure 5.8) is performed with the ultimate aim of obtaining an explicit formula for $\varepsilon_{\text{eff}}(f)$ as a function of known parameters. This transverse resonance technique demands that the sum of admittances looking to the left and right from either air-dielectric interface shown in Figure 5.9 must be zero. Getsinger shows that this sum may be written

$$\frac{\gamma_a}{h'}\tanh(\gamma_a a') + \frac{\gamma_r}{h}\tanh\left(\frac{\gamma_r w'}{2}\right) = 0 \tag{5.6}$$

where the subscripts a and r refer to air and substrate dielectric, respectively.

Also, the propagation coefficients are related by

$$\gamma^2 + \gamma_i^2 = -\omega^2 \mu_0 \varepsilon_0 \varepsilon_r \tag{5.7}$$

where $i = a$ or r as appropriate. These expressions are combined, together with appropriate static-TEM relations, and solved to yield an expression for $\varepsilon_{\text{eff}}(f)$ as a function of ε_{eff}, ε_r, Z_0, f and h'/h. The reader is referred to the details in Getsinger's publications [128, 129]; and only the result is given here:

$$\varepsilon_{\text{eff}}(f) = \varepsilon_r - (\varepsilon_r - \varepsilon_{\text{eff}})\Bigg(1 + f^2\,(2\mu_0 h/Z_0)^2$$

$$\times \left[\frac{\pi^2}{12} \left\{ \frac{(\varepsilon_{\text{eff}} - 1) + (h'/h)^2 (\varepsilon_r - \varepsilon_{\text{eff}})}{\varepsilon_r - 1} \right\} \left\{ \frac{(\varepsilon_{\text{eff}} - 1)(\varepsilon_r - \varepsilon_{\text{eff}})}{\varepsilon_{\text{eff}} (\varepsilon_r - 1)} \right\} \right] \Bigg)^{-1} \quad (5.8)$$

There is one really unknown parameter in this expression and that is h'/h. It turns out that, rather fortuitously, this parameter varies little as a function of Z_0 over quite a wide range. For $Z_0 \geq 35\ \Omega$, $h'/h \approx 3.0$, and as Z_0 falls below 35 Ω the parameter slowly increases — reaching approximately 6.2 at Z_0= 15 Ω. Therefore h'/h can be introduced quite readily into Equation (5.8), although this procedure lacks some precision, and is uncertain for characteristic impedances of less than 15 Ω or substantially greater than 90 Ω. The formula also remains rather complicated.

A somewhat simpler formula, based on the same foundations, may be written in the form [128]:

$$\varepsilon_{\text{eff}}(f) = \varepsilon_r - \frac{\varepsilon_r - \varepsilon_{\text{eff}}}{1 + G(f/f_p)^2} \quad (5.9)$$

where

$$f_p = \frac{Z_0}{2\mu_0 h} \quad (5.10)$$

and μ_0 is the free-space permeability (of universally constant value $= 4\pi \times 10^{-7}$ H/m). The parameter G is purely empirical, thereby giving some flexibility to the formula. G is dependent mainly upon Z_0, but also to a lesser extent upon h, and Getsinger deduced from measurements of microstrip ring resonators on alumina that

$$G = 0.6 + 0.009 Z_0 \quad (5.11)$$

when $h = 0.635$ mm. Fortunately, Equation (5.9) is not particularly sensitive to G, and $G = 1.0$ for the most common characteristic impedance of microstrip lines, that is 50 Ω. For a wide range of lines on alumina-type substrates G has been found to vary over the range

$$0.3 \leq G \leq 1.7\,. \quad (5.12)$$

Substrates where this applies include monocrystalline sapphire, which provides advantages for many MICs, as explained in Chapters 3 and 4.

It has been found that Equations (5.9), (5.10) and (5.11) give rise to discrepancies which increase as Z_0 decreases, and can exceed 3% as Z_0 is reduced below 20 Ω [130]. The work of Getsinger, as outlined above, provides an excellent basis for the development of expressions having an improved accuracy and which may be used in design. These are described in the next section.

5.4 ACCURATE DESIGN FORMULAS

5.4.1 EDWARDS AND OWENS' EXPRESSIONS

When careful measurements are made on microstrip lines over a wide range of frequencies and characteristic impedances, it is found that Getsinger's approach, while more accurate than any of the other methods, requires at least optimization of the G-factor for any given 'family' of microstrip lines on a particular substrate. Edwards and Owens [130], for example, carried out extensive measurements on microstrip lines on monocrystalline sapphire substrates and established (by curve fitting) the following expression for G:

$$G = [(Z_0 - 5)/60]^{1/2} + 0.004Z_0. \tag{5.13}$$

This works well in conjunction with Equations (5.9) and (5.10) for lines having characteristic impedances with the range

$$10 \leq Z_0 \leq 100 \ \ \Omega \tag{5.14}$$

and over the frequency range

$$2 \leq f \leq 18 \text{ GHz}. \tag{5.15}$$

It should be noted that the sapphire substrates were nominally 0.5 mm thick and the equivalent isotropic relative permittivities (see Section 4.12.2 on page 111) were 10.7 to 11.6. This should mean that the formula will also work well for semiconductor substrates such as high-resistivity silicon (HRS), GaAs, InP, etc.

Some careful measurements have also been carried out on microstrips on an alumina substrate [130]. In this case, a suitable expression for the G-factor is

$$G = \left(\frac{Z_0 - 3}{60}\right)^{1/2} + 0.001Z_0 \tag{5.16}$$

which works very well for lines on a 0.65 mm thick alumina substrate (e.g. Alsimag 805, for which $\varepsilon_r = 10.15$) where the characteristic impedance range is

$$30 \leq Z_0 \leq 70 \ \ \Omega \tag{5.17}$$

over the frequency range

$$2 \leq f \leq 18 \ \text{GHz}. \tag{5.18}$$

Equation (5.16) may give slightly high results where the characteristic impedance falls significantly below 30 Ω.

Alternatively, and applicable accurately at least to both alumina and sapphire, Edwards and Owens [130] suggest the following expression (combining their Equations 9 and 11) for dispersion:

$$\varepsilon_{\text{eff}}(f) = \varepsilon_r - \frac{\varepsilon_r - \varepsilon_{\text{eff}}}{1 + (h/Z_0)^{1.33}(0.43f^2 - 0.009f^3)} \tag{5.19}$$

where h is in millimetres and f is in gigahertz.

Based upon extensive measurements carried out with microstrip lines on alumina and sapphire over the impedance and frequency ranges indicated above, and also comparisons with computed results using the analysis due to Itoh and Mittra (to be described shortly), the estimated accuracy of Equation (5.19) is better than ± 0.8%

Even this accuracy, which was the best achieved up to 1976, is still insufficient for many filters and some matching networks. Tolerances and substrate repeatability uncertainties reduce the prediction capability further still, and post-production trimming is then mandatory. The use of monocrystalline substrates with controlled thickness removes some repeatability difficulties and tends to place the onus back on to the accuracy of a suitable dispersion formula again. These comments refer to non-semiconductor substrates.

Since only alumina-type substrates have been tested, Equation (5.19) is only guaranteed to predict with the quoted accuracy, provided the substrate permittivity is within the approximate limits:

$$10 \leq \varepsilon_r \leq 12. \tag{5.20}$$

Measurements reported by York and Compton [134], however, suggest that Equation (5.19) may work well for plastic substrates at frequencies up to 18 GHz (their results indicate discrepancies consistently below 0.9%).

For consistently accurate calculations in cases where the substrates differ markedly from alumina, such as plastics, quartz, many ferrites, etc., it is recommended that new experiments should be carried out. Hence a new curve-fitted formula for G is determined for each case, allowing the use of Equations (5.9) and (5.10) to calculate $\varepsilon_{\text{eff}}(f)$. Figure 5.10 shows a comparison of the five dispersion calculation methods just described. Two characteristic impedances are quite sufficient to indicate the trends: 25 Ω was chosen because, at lower impedances, dispersion is greater; and 50 Ω was chosen because it is a commonly used value. Considerable overall discrepancies are evident although, on these scales, the Getsinger and Carlin curves are fairly close.

A combination of comprehensive measurements, preferably backed by an independent set of theoretical calculations, is needed to decide upon the validity and accuracy of particular formulas. Frequency-dependent analysis, based upon a microstrip field development, is dealt with in outline in Section 5.7 on page 133.

A summary of results obtained by various methods is indicated by the curves of Figure 5.10. The 'polynomial fit' refers to Equation (5.19), Itoh and Mittra's approach will shortly be outlined, and the experimental results were obtained using a technique which is explained in Section 7.7 on page 237. It is clear that consistent results are achieved over this wide frequency range and over the moderately wide range of characteristic impedances (31–71 Ω).

5.4.2 EXPRESSIONS SUITABLE FOR MILLIMETRE-WAVE DESIGN

The design expressions presented so far in this chapter have been well tested and checked against experimental measurements for frequencies up to approximately 18 GHz. This frequency is, however, far below the limits at which microstrip can be used — up to the vicinity of 100 GHz at least — and as outlined in Chapter 3, there are increasing applications available for MICs and MMICs at millimetre-wave frequencies. It has therefore long been recognized that a requirement exists for closed formulas that would enable calculations to be conducted including dispersion in microstrip at these high frequencies.

In 1979 Yamashita *et al.* [131] published an approximate dispersion formula covering the frequency range 1 to 100 GHz. Their expression can be written

$$\varepsilon_{\text{eff}}(f) = \left(\frac{\sqrt{\varepsilon_r} - \sqrt{\varepsilon_{\text{eff}}}}{1 + 4F^{-1.5}} + \sqrt{\varepsilon_{\text{eff}}} \right)^2 \tag{5.21}$$

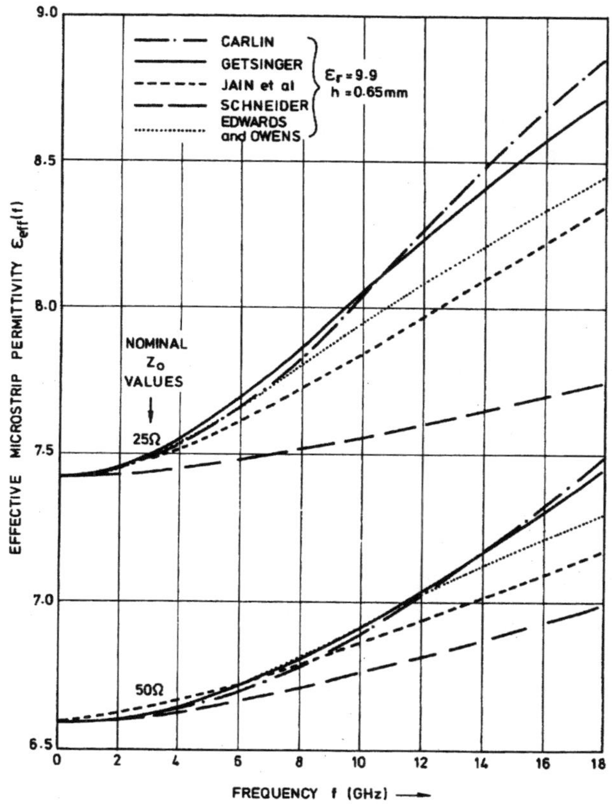

Figure 5.10 A comparison of the results derived from five different dispersion calculation techniques.

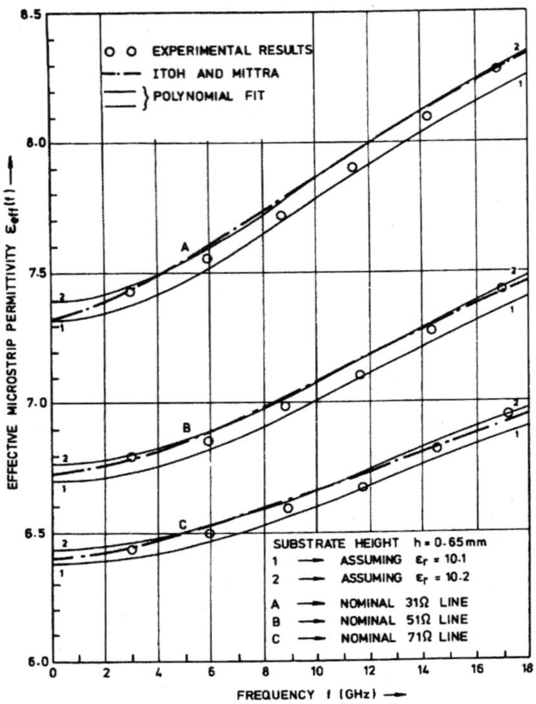

Figure 5.11 Dispersion curves applicable to three microstrip lines on a microwave-grade high-purity alumina substrate.

where

$$F = \frac{4hf\sqrt{\varepsilon_r - 1}}{c}\left[0.5 + \left\{1 + 2\log\left(1 + \frac{w}{h}\right)\right\}^2\right]. \tag{5.22}$$

However, this expression is not particularly accurate — especially in the lower frequency range up to 18 GHz, where the expression due to Edwards and Owens is recommended.

Comparison of Yamashita's design formulas with analytical results determined using electromagnetic solutions shows considerable discrepancies over the full 1 to 100 GHz range.

An improved design formula has been developed by Kirschning and Jansen [132]. Their approach begins with a function that bears a close resemblance to Getsinger's, but they have a new and reportedly more accurate frequency-dependent denominator term. Also, their formula covers a much wider range of permittivities, aspect ratios and frequencies than those considered by Getsinger or Edwards and Owens.

Kirschning and Jansen's basic expression is (with frequency f in GHz and thickness h in cm):

$$\varepsilon_{\text{eff}}(f) = \varepsilon_r - \frac{\varepsilon_r - \varepsilon_{\text{eff}}}{1 + P(f)} \tag{5.23}$$

and the form of the denominator frequency function is

$$P(f) = P_1 P_2 \{(0.1844 + P_3 P_4)\, 10fh\}^{1.5763} \tag{5.24}$$

where

$$\left.\begin{aligned} P_1 &= 0.27488 + \left[0.6315 + 0.525/\left(1+0.157fh\right)^{20}\right](w/h) \\ &\quad -0.065683\exp\left(-8.7513w/h\right) \\ P_2 &= 0.33622\left[1-\exp\left(-0.03442\varepsilon_r\right)\right] \\ P_3 &= 0.0363\exp\left(-4.6w/h\right)\left\{1-\exp\left[-\left(fh/3.87\right)^{4.97}\right]\right\} \\ P_4 &= 1+2.751\left\{1-\exp\left[-\left(\varepsilon_r/15.916\right)^{8}\right]\right\} \end{aligned}\right\}. \tag{5.25}$$

These researchers used computer-based matching in conjunction with the available hybrid mode results (derived from full electromagnetic numerical computations) to force the correct asymptotic behaviour of the function $P(f)$. An accuracy of better than 0.6% is claimed for all frequencies up to 60 GHz (although a full check only appears to have been conducted up to 30 GHz). The validity ranges are very wide, namely:

$$\begin{aligned} 1 \leq &\ \varepsilon_r \leq 20 \\ 0.1 \leq &\ w/h \leq 100 \\ 0 \leq &\ h/\lambda_0 \leq 0.13. \end{aligned}$$

There are further formulations enabling the effects of dispersion in microstrip to be calculated at frequencies through millimetre-wave. Although the last word has almost certainly not yet been stated in the literature on this subject, the position has been reached where reported accuracies are within 1% (of measurements and independent theoretical results over a wide range). Having said this, it must be recognized that care is always required in selecting an appropriate accurate dispersion expression.

The final formulation presented here for dispersion calculation is that due to Kobayashi [133]. This work is based upon the concept of a significant 50% dispersion point at which the effective microstrip permittivity is the arithmetic mean of the substrate relative permittivity and the low-frequency limit value. Kobayashi calculates the frequency for this condition and develops the following dispersion expressions:

$$\varepsilon_{\text{eff}}(f) = \varepsilon_r - \frac{\varepsilon_r - \varepsilon_{\text{eff}}}{1+\left(f/f_{50}\right)^m} \tag{5.26}$$

where

$$\begin{aligned} f_{50} &= \frac{f_{k,TM_0}}{0.75 + \left[0.75 - \left(0.332/\varepsilon_r^{1.73}\right)\right]w/h} \\ f_{k,TM_0} &= \frac{c\tan^{-1}\left(\varepsilon_r\sqrt{\frac{\varepsilon_{\text{eff}}-1}{\varepsilon_r-\varepsilon_{\text{eff}}}}\right)}{2\pi h\sqrt{\varepsilon_r-\varepsilon_{\text{eff}}}} \\ m &= m_0 m_c \\ m_0 &= 1 + \frac{1}{1+\sqrt{w/h}} + 0.32\left(\frac{1}{1+\sqrt{w/h}}\right)^3 \\ m_c &= \begin{cases} 1 + \frac{1.4}{1+w/h}\left[0.15 - 0.235\exp\left(\frac{-0.45f}{f_{50}}\right)\right] & \text{when } w/h \leq 0.7 \\ 1 & \text{when } w/h > 0.7 \end{cases}. \end{aligned}$$

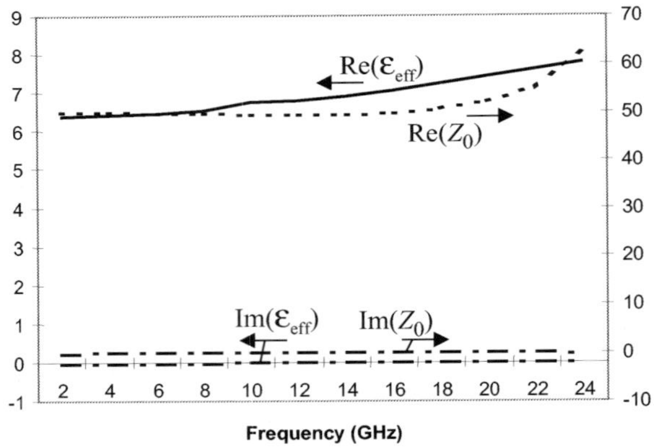

Figure 5.12 Dispersion of lossy gold microstrip on a 635 micron thick alumina substrate ($\varepsilon_r = 9.8$), width = 635 μm, nominal characteristic impedance 50 Ω.

Atwater [135] has reported results of comparative tests on various microstrip dispersion formulas, but it should be noted that Atwater's paper predates the later important Kobayashi work referred to above (Atwater's reference to Kobayashi relates to the 1982 paper). It is, however, clear that Kirschning and Jansen's approach is comparatively accurate within its frequency restrictions.

York and Compton [134] (referred to earlier) report measurements on microstrips ranging in impedance from 35 to 70 Ω, fabricated on both alumina and plastic substrates having thicknesses in the range 0.635–1.605 mm. Although York and Compton's results are limited to frequencies up to 18 GHz it is clear that, out of a total of nine models tested, only those due to Edwards and Owens, Kirschning and Jansen and Kobayashi show deviations well within 1% of measurements.

5.4.3 DISPERSION CURVES DERIVED FROM SIMULATIONS

All the above results apply generally to lossless microstrip lines. It is known however that in practice losses play a significant part in determining the dispersion and a set of SONNET® simulations have been run in order to generate dispersion curves for microstrip on various substrates — taking losses into account. In all cases the 1 to 40 GHz range is covered.

The first curve, applying to a 50 Ω lossy line on a 0.635 mm thick alumina substrate, is shown in Figure 5.12. It is clear that, as expected, the real part of the effective permittivity steadily increases with frequency and this is also true of the characteristic impedance — with a more pronounced increase above 18 GHz. The imaginary parts of each parameter are relatively small. Throughout the 1 to 24 GHz frequency range considered, the effective permittivity is almost identical whether the line has loss or is lossless. However, the presence of losses causes the characteristic impedance to be about 0.13 Ω greater at all frequencies.

In Figure 5.13 the results are shown for a high impedance (nominally 83 Ω) lossy line on high resistivity silicon (HRS), over the wide frequency range 1 to 40 GHz. Copper conductors apply and the silicon is 0.650 mm thick. The conductor width is 70 μm. After an initial shallow dip the real part of the effective permittivity slowly increases

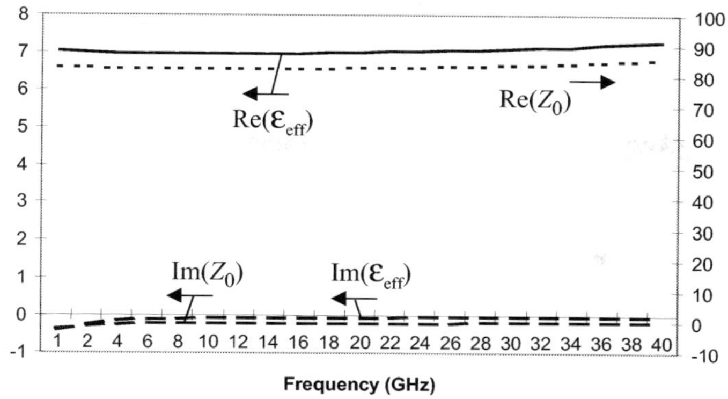

Figure 5.13 Dispersion of lossy copper microstrip on a 650 micron thick high resitivity silicon substrate, width = 70 μm, nominal characteristic impedance 83 Ω.

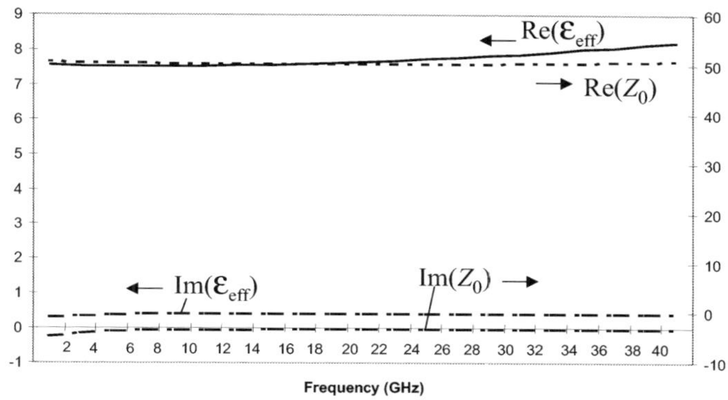

Figure 5.14 Dispersion of lossy copper microstrip on a 500 μm thick GaAs substrate, width = 240 μm, nominal characteristic impedance 50 Ω.

whilst the characteristic impedance continually slowly increases. The imaginary parts of each parameter remain relatively small.

We have also simulated a (nominally 83 Ω) lossy aluminium line on low resistivity silicon. Both the real part of the effective permittivity and also the characteristic impedance are a little lower in value than for HRS (approximately 0.1 to 0.3 Ω for the impedance).

In Figure 5.14 the results are shown for a moderate impedance (50 Ω) lossy line on a GaAs substrate. Copper conductors again apply and the substrate is 0.5 mm thick. The line width is 0.24 mm. The general trends of the data are very similar to that applying to HRS, discussed above.

5.5 EFFECTS DUE TO FERRITE AND TO DIELECTRICALLY ANISOTROPIC SUBSTRATES

5.5.1 *EFFECTS OF FERRITE SUBSTRATES*

The static-TEM problem of microstrip on a ferrite substrate was discussed in Section 4.9 on page 103. The frequency dependence of the effective microstrip permeability is of little practical importance so will not be covered here. However, since ferrite also behaves as a dielectric the frequency-dependent effective microstrip permittivity is important and should be evaluated as recommended in the foregoing discussion of this section.

Kurushin *et al.* [136] have shown that as the frequency increases in the case of microstrip on a ferrite substrate, the conventional dispersion results approach those applicable to the case of a non-ferrite dielectric substrate. Therefore, all the design expressions previously given here continue to apply at high frequencies.

Since ferrite devices are often placed on non-ferrite substrates (alumina, semiconductor), and microstrip is then fabricated on the ferrite device surface itself, it is important to gain some design appreciation of the behaviour of this type of structure. The basic properties of such a structure have been evaluated by Hsia *et al.* [137], who used a spectral domain exponential matrix method. These workers generalized their analysis to accommodate isotropic/anisotropic structures — in terms of both permeability and permittivity. Most of the results presented by Hsia *et al.* refer to a ferrite with permittivity of 12.6 providing the top layer whilst GaAs having a permittivity of 12.9 formed the lower substrate. The magnetic bias field was conventionally disposed normally to the plane of this assembly. Hsia *et al.* give curves showing normalized wave number (corresponding to effective permittivity) and characteristic impedance as functions of frequency over the 5–30 GHz range. Their curves for normalized wave number are reproduced as Figure 5.15, in which $z = b$ refers to the case of microstrip deposited on the lower, non-ferrite, substrate whereas $z = d$ refers to the case where the microstrip is deposited on the upper ferrite superstrate.

5.5.2 *EFFECTS OF A DIELECTRICALLY ANISOTROPIC SUBSTRATE*

Static-TEM calculations associated with microstrip on a dielectrically anisotropic substrate were dealt with in Section 4.12.2 on page 111. In the case of sapphire, probably the most common anisotropic substrate, the measurements of Edwards and Owens [130] seem to indicate that the influence of the sapphire anisotropy on the detail of the dispersion characteristics is negligibly small in practice (up to 18 GHz).

As indicated in his Table II, and also his reported dispersion results, the data derived by Kobayashi [138] confirm excellent agreement with the predictions due to Edwards and Owens, presented in Section 5.6 on page 131. The use of slightly modified versions of Yamashita's dispersion expression also provides very good agreement and it would be useful to study the effects of suitably adjusting Kirschning and Jansen's formulas in an effort to predict dispersion at high frequencies on dielectrically anisotropic substrates.

A more extensive study of the behaviour of microstrip on a dielectrically anisotropic substrate has been performed by Mariki and Yeh [139], as mentioned in Chapter 4.

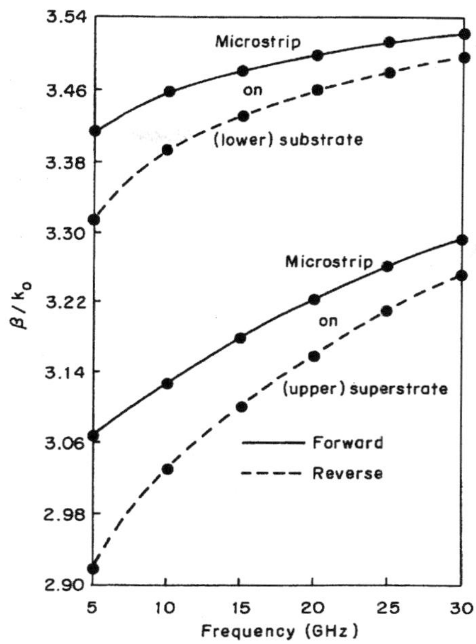

Figure 5.15 Microstrip propagation coefficient as a function of frequency (© 1990 IEEE. Reprinted, with permission, from Hsia *et al.* [137].)

These researchers use a three-dimensional Transmission-Line-Matrix (TLM) analysis to determine the dispersive effects with this type of substrate. In particular the analysis derives the frequency-dependent effective microstrip permittivity and characteristic impedance. Considering a sapphire substrate Mariki and Yeh's results for microstrip permittivity at frequencies up to 21 GHz are very close to those predicted by Getsinger's original model. Characteristic impedance is also shown to increase with frequency — for example, from 37.5 to 54 Ω when the aspect ratio (w/h) is 1.5.

5.6 DESIGNS REQUIRING DISPERSION CALCULATIONS — WORKED EXAMPLES

For the first example we will recall the problem of calculating one element of a transistor matching network, which was calculated for 2 GHz in Section 4.7 on page 95. However, we now design the same element for a frequency of 12 GHz, where dispersion will certainly be significant.

The initial data, as previously, are

$$Z_0 = 31.3\ \Omega, \quad \varepsilon_r = 9.8, \quad h = 0.6\ \text{mm} \tag{5.27}$$

but in this case the frequency f is 12 GHz. We shall also require the effective microstrip permittivity from the static-TEM design completed in Section 4.7 on page 95. This is

$$\varepsilon_{\text{eff}} = 7.09. \tag{5.28}$$

For a frequency of 12 GHz, Equation (5.19) is used to calculate $\varepsilon_{\text{eff}}(f)$:

$$\varepsilon_{\text{eff}}(f) = \varepsilon_r - \frac{\varepsilon_r - \varepsilon_{\text{eff}}}{1 + (h/Z_0)^{1.33}(0.43f^2 - 0.009f^3)}. \tag{5.29}$$

The thickness h is directly substituted in millimetres and the frequency f in gigahertz, giving

$$\varepsilon_{\text{eff}}(f) = 7.621 \tag{5.30}$$

and the wavelength

$$\lambda_g = \frac{300}{f\sqrt{\varepsilon_{\text{eff}}(f)}} = 9.056 \text{ mm}. \tag{5.31}$$

Both values are easily obtained by running a program implementing the appropriate expressions, including Equation (5.19).

From the wavelength we can readily derive the length of the microstrip element, required to be $\lambda_g/4$, i.e.

$$l \approx 9.056/4 \approx 2.264 \text{ mm}. \tag{5.32}$$

Again, this value neglects the effect of discontinuities that will slightly reduce the final length. These effects are dealt with in Chapter 7.

Notice that our new length result, for 12 GHz, is significantly different from the value that would have been obtained had the previous (2 GHz) result been simply linearly scaled for frequency. This would have yielded the erroneous result of 2.346 mm — nearly 4% in error. This error also increases as the frequency is increased. Provided we assume here that we will neglect any change in Z_0 with frequency (dealt with in Section 5.8 on page 137), the microstrip width will remain at the value previously determined: $w = 1.34$ mm.

At higher frequencies the results calculated including dispersion differ more markedly from those calculated using the static-TEM formulas.

For example, let us consider a microstrip line to be designed for use in an MMIC, in this case fabricated on a semi-insulating GaAs substrate. The thickness is 0.127 mm and the relative permittivity is 13.0.

Let us assume that the operating frequency is 40 GHz, which is typical of some MVDS (Microwave Video Distribution Systems), and that the line to be designed has an aspect ratio (w/h) of 2.0.

Calculations using the static-TEM expressions yield the low-frequency limit permittivity as 9.136, whereas calculations using the formulas due to Kobayashi yield a value of 9.74 at 40 GHz. The discrepancy is around 4% — significant in many designs such as filters and multiplexers.

Similar approximate magnitudes of discrepancy occur where plastic (e.g. RT-Duroid®), quartz or many other substrates are considered.

A comparison of calculated results for this 0.127 mm thick GaAs case, using the methods of Yamashita, Kirschning and Jansen and Kobayashi, is provided by the curves shown in Figure 5.16. Discrepancies between the first two methods continually increase, reaching 1.07% at 40 GHz, whereas the differences between Kirschning and Jansen and Kobayashi remain less than 0.25% over the entire frequency range.

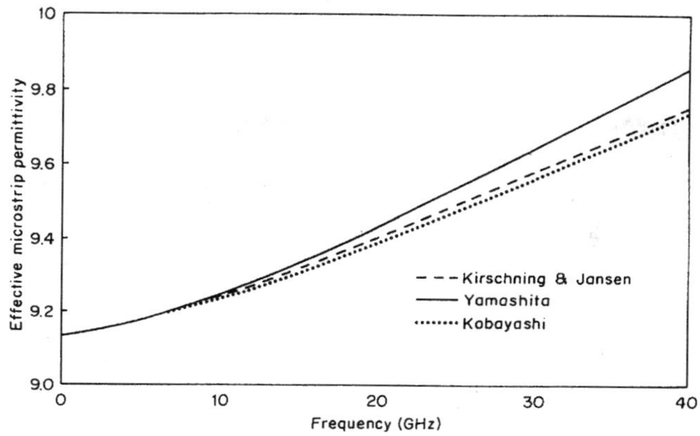

Figure 5.16 Comparative dispersion curves for microstrip on a GaAs substrate ($\varepsilon_r = 13$, $w = 0.254$ mm, $h = 0.127$ mm). (From calculations performed independently by Dr Roger P. Owens, to whom the authors are indebted.)

5.7 FIELD SOLUTIONS

5.7.1 ONE EXAMPLE OF A 'CLASSIC' FREQUENCY-DEPENDENT COMPUTER-BASED FIELD SOLUTION

In the methods for calculating effective microstrip permittivity outlined so far the emphasis has been on approximate approaches using closed formulas. For any of the available approximate formulas it may be said that:

(a) The methods are amenable to calculator-based design.
(b) If used for CAD, the methods are relatively inexpensive because they are very fast.

On the other hand, these approximate approaches have a number of drawbacks. For example:

(i) They often suit only a specified microstrip 'family', e.g. those based on alumina substrates.
(ii) They require accurate static-TEM data to start with, that is ε_{eff} and Z_0.
(iii) Such methods are not fundamentally elegant (some empirical element usually has to be introduced).

Frequency-dependent field solutions requiring 'number crunching' using relatively powerful computer programs have been published by a number of researchers [140–144]. While these are comparatively expensive to run they provide the positive benefits of answering drawbacks (i), (ii) and (iii) above. They generally take up too much computer time and storage to be considered as candidates for inclusion in a larger microwave integrated circuit computer-aided optimization routine. However, such solutions provide an excellent theoretical comparison with measured data and can therefore enable the accuracy of the approximate design formulas to be checked. They can also be readily extended to handle more complex problems such as coupled lines, even involving multiple conductors, etc.

When satisfactory agreement is reached regarding measurements and fundamental theoretical results then 'curve-fitted' closed formulas can be deduced for $\varepsilon_{\text{eff}}(f)$. It should be appreciated that these frequency-dependent field solutions require only the physical input data:

(a) microstrip geometry (w, h, and box dimensions, if applicable), and
(b) the permittivity of the substrate material (ε_r).

Itoh and Mittra [142, 143] in particular, have developed a 'spectral-domain' approach. This approach involves Fourier transforms to convert the field formulation from the actual space domain to a domain having new (transformed) variables.

These researchers set up a superposition of TE and TM fields in terms of scalar potential functions:

$$E_{zi} = j\frac{k_i^2 - \beta^2}{\beta}\psi_i^{(e)}(x, y)\exp(-j\beta z) \tag{5.33}$$

and

$$H_{zi} = j\frac{k_i^2 - \beta^2}{\beta}\psi_i^{(h)}(x, y)\exp(-j\beta z) \tag{5.34}$$

where

$$k_1 = \omega\sqrt{(\varepsilon_r\mu_r\varepsilon_0\mu_0)}, \quad k_2 = \omega\sqrt{(\varepsilon_0\mu_0)} = k_0 \tag{5.35}$$

and β is the unknown propagation constant to be determined. Superscript (e) means 'TE' field and superscript (h) means 'TM' field. The Fourier transform of the scalar potentials is defined by

$$\psi_i^{(P)}(n, y) = \int_{-a}^{a}\psi_i^{(P)}(x, y)\exp\left(j\hat{k}_n x\right)dx \tag{5.36}$$

where $i = 1$ or 2, $p = e$ *or* h, and

$$\hat{k}_n = \begin{cases} (\pi/a)\left(n - \frac{1}{2}\right) & \text{for } E_z \text{ even and } H_z \text{ odd modes} \\ n\pi/a & \text{for } E_z \text{ odd and } H_z \text{ even modes} \end{cases}. \tag{5.37}$$

This procedure leads to a more economical computer program routine. The current distributions on the strip are the crucial features of this particular technique. The final solution is selected only when successively improved current distribution functions, inserted in the program, are observed to give converging sets of output results. Jansen [144] provides an improved and faster solution of this general complete electromagnetic analytical approach.

5.7.2 ANALYSIS OF ARBITRARY PLANAR CONFIGURATIONS

Spectral domain approaches of similar forms to that of Itoh and Mittra have been used elsewhere to analyse the characteristics of MICs and MMICs on a three-dimensional basis [145]. In their work Wu *et al.* [145] develop a space-spectral domain method for analysing arbitrary planar configurations with or without open boundaries.

They express the transverse fields in each layer of a multi-layer structure in a form suitable for application of a Fourier transform and reduce the results of

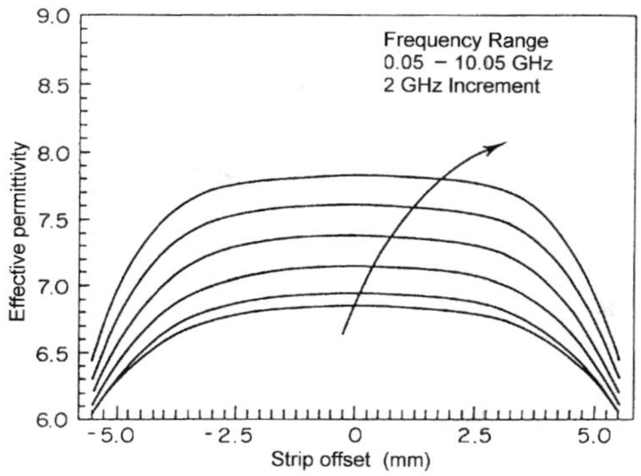

Figure 5.17 Effective permittivity for shielded microstrip. (Reprinted, with permission, from Dumbell *et al.* [146].)

this transformation to a differential solution. Basis functions are introduced for rectangular resonator shapes and, following the application of Galerkin's technique, a characteristic matrix equation system is obtained which has to be solved for the zeros of a determinant. Some interesting microstrip structure shapes are considered and results compare well with measurements. Calculations are reported for a range of rectangular strip-patch microstrip resonators (suspended, grounded and sandwiched), giving resonant frequency as a function of resonator length. Although it is not clear from the paper, it seems that these calculations refer to microstrip on a substrate having a permittivity of 3.82 (fused quartz) and resonant frequencies in the 12–37 GHz range are obtained.

5.7.3 ASYMMETRY EFFECTS

The effects of asymmetrically-disposed shielded microstrip are important in practice and Dumbell *et al.* [146] have reported an analytical technique for this situation. They apply a variational method to analyse asymmetric shielded microstrip both in single-layer and overlaid dielectric forms. In fact, they show that the variation in propagation coefficient as a function of degree of asymmetry can be controlled by the introduction of a dielectric overlay to the extent that microstrip tracks may be located close to conducting box walls and such variations then practically neglected.

Some results obtained by Dumbell *et al.* [146] are shown in Figure 5.17 for the case of a substrate having a relative permittivity of 10.5. This family of curves refers to microstrip with an aspect ratio of 0.866 enclosed in a conducting box 11.43 mm high by 12.7 mm wide. 'Strip offset' is the amount by which the strip is offset from the centre line of the system.

A comparative family of curves, applicable to the same geometry but with a dielectric overlay having permittivity 7.0 and 1.27 mm thickness, is shown in Figure 5.18. Clearly, at the lower frequencies particularly, the variations with strip offset are substantially reduced. Families of curves for the characteristic impedances are also determined by Dumbell *et al.*, and these indicate that variations caused by offset are less effectively compensated by the introduction of an overlay.

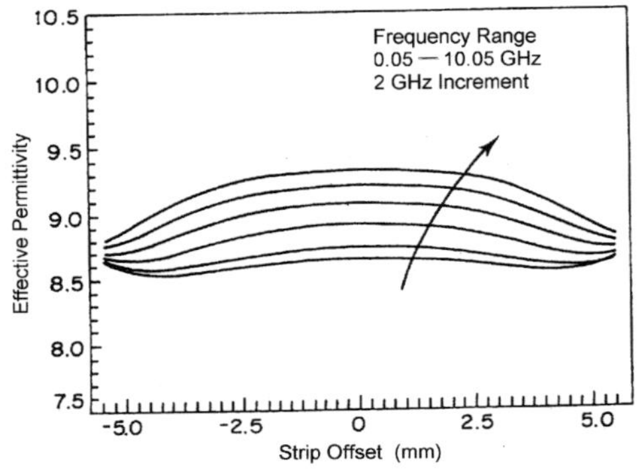

Figure 5.18 Effective permittivity for shielded microstrip with dielectric overlay. (Reprinted, with permission, from Dumbell *et al.* [146].)

5.7.4 TIME-DOMAIN APPROACHES

A length of dispersive transmission line can be used as a 'phase equalizer' to compensate for the phase distortion otherwise resulting in the transmission of asymmetric pulses. This equalization results in the effective reshaping and compression of ultrashort pulses which originate, typically, from high-speed photoconductive switches.

Qian and Yamashita [147] have shown that an optimum length of microstrip or coplanar waveguide can be used to correct the phase distortion. They begin with a Fourier spectrum of a single-sided exponential pulse that suitably quantifies this distortion and they adapt a microstrip dispersion expression (see Section 5.4.2 on page 124). The phase function associated with the single-sided exponential pulse is expanded as a polynomial in which the frequency component at which the amplitude of the spectrum is 10% of its peak value is defined.

The microstrip dispersion expression is expanded using Taylor's series and used to formulate an expresssion for the phase delay. A comparison is finally made between the polynomial derived from the exponential pulse and this dispersion expression series expansion. Qian and Yamashita show that, if the second and third order terms in each series are made equal, then the total phase of the signal after a specific distance has been traversed is close to a linear function of frequency. They report a straightforward formula for the length of line required.

As an example, Qian and Yamashita provide details of a simulation for a 35 ps single-sided exponential pulse propagating along a 50 Ω microstrip line on a GaAs substrate and the resulting waveforms are repeated here in Figure 5.19.

A finite-difference time-domain analysis can be useful for the determination of the broadband frequency response of complex MIC or MMIC components, as reported by Rittweger and Wolff [148]. These workers report results for spiral inductors and microstrip radial stubs obtained using this technique. This numerical technique is three-dimensional and essentially trades a relatively modest requirement for analytical operations against a large numerical demand.

The microstrip radial stub is important in some filter and matching circuit designs and Rittweger and Wolff's results for a specific case are shown in Figure 5.20. It can be seen that, at frequencies below 19 GHz, the measured and calculated results

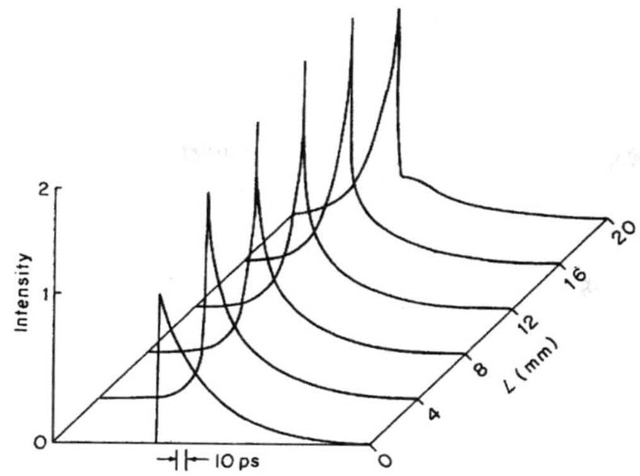

Figure 5.19 Simulation results for the propagation of a 35 ps single-sided exponential pulse along 50 Ω microstrip lines on a GaAs substrate. (©1990 IEEE. Reprinted, with permission, from Qian and Yamashita [147].)

compare favourably. In the 19 to 24 GHz range there is a greater discrepancy in the data although the pattern of variation still compares quite well.

5.8 FREQUENCY DEPENDENCE OF THE MICROSTRIP CHARACTERISTIC IMPEDANCE

There have been many different approaches to the dynamic problem of microstrip characteristic impedance and quite different functions of frequency have been predicted — even extending to opposing trends. With no definitive function available MIC and MMIC designers have continued to use the static-TEM calculation for Z_0.

The problem of characteristic impedance as a function of frequency is not easy to solve satisfactorily, but some background theory and tentative recommendations are presented here in an effort to provide some insight and assistance for the design engineer.

5.8.1 DIFFERENT DEFINITIONS AND TRENDS WITH INCREASING FREQUENCY

Fundamentally, characteristic impedance may be defined in the following different ways:

$$Z_{0,a}(f) = \frac{\mathbf{V}}{\mathbf{I}} \tag{5.38}$$

$$Z_{0,b}(f) = \frac{P}{\mathbf{II}^*} \tag{5.39}$$

$$Z_{0,c}(f) = \frac{\mathbf{VV}^*}{P}. \tag{5.40}$$

Different functions of frequency are obtained, depending on which of these definitions is selected. This is due mainly to different assumptions which have to be made regarding mode coupling as the frequency is increased. This mode coupling, it is recalled, is the fundamental dispersive mechanism for microstrip.

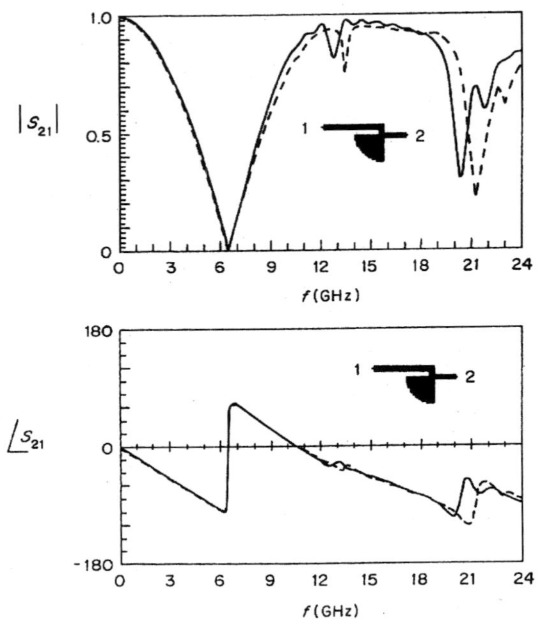

Figure 5.20 Calculated (—) and measured (- - -) S-parameters of the radial stub. Alumina substrate, $\varepsilon_r = 9.768$ (measured), height $h = 0.635$ mm, width $w = 0.61$ mm, gap $= 0.61$ mm, radius $= 3.253$ mm. (© 1990 IEEE. Reprinted, with permission, from Rittweger and Wolff [148].)

Bianco *et al.* [149] further define characteristic impedances in terms of mean voltage $\mathbf{V}$ and centre voltage $\mathbf{V}_c$ to yield a total of five definitions as follows:

$$Z_{0,1}(f) = \frac{\mathbf{V}}{\mathbf{I}} \tag{5.41}$$

$$Z_{0,2}(f) = \frac{\mathbf{V}_c}{\mathbf{I}} \tag{5.42}$$

$$Z_{0,3}(f) = \frac{P}{\mathbf{I}\mathbf{I}^*} \tag{5.43}$$

$$Z_{0,4}(f) = \frac{\mathbf{V}_c\mathbf{V}_c^*}{P} \tag{5.44}$$

$$Z_{0,5}(f) = \frac{\mathbf{V}\mathbf{V}^*}{P}. \tag{5.45}$$

The general theoretical conclusion is that both $Z_{0,2}(f)$ and $Z_{0,4}(f)$ always rise with increasing frequency whereas the remaining impedances all fall as the frequency rises. ($Z_{0,3}(f)$ always exhibits the smallest variation.)

In all calculations based upon Equations (5.38) to (5.45) the power P flowing in the microstrip is evaluated using Poynting's theorem.

Getsinger [150], using the model previously described in this chapter, developed an expression which (like that of Denlinger [122]) has microstrip characteristic impedance varying inversely with the square root of $\varepsilon_{\text{eff}}(f)$. The development is based upon Schelkunoff's definition of wave impedance and Getsinger is very careful about the interpretation of the model for the analysis pursued. This derivation seems very plausible, but it must be considered in the light of alternative definitions and trends that will now be examined.

5.8.2 *USE OF THE PLANAR WAVEGUIDE MODEL (FIGURE 5.21)*

Owens [151,152] developed a radically different approach to the analysis of microstrip with the effective microstrip width concept [153] as a principal feature. Owens obtained the following expression for characteristic impedance:

$$Z_0(f) = \frac{h\eta}{w_{\text{eff}}(f)\sqrt{\varepsilon_{\text{eff}}(f)}} \tag{5.46}$$

where, following the dispersion expressions given earlier (such as Equations (5.9) and (5.10)), the effective microstrip width $w_{\text{eff}}(f)$ may be written as

$$w_{\text{eff}}(f) = w + \frac{w_{\text{eff}} - w}{1 + (f/f_p)^2}. \tag{5.47}$$

(Note that the overall form of Equation (5.47) is very similar to the earlier equations provided by Getsinger.)

In Equation 5.47

$$f_P = \frac{c}{2w_{\text{eff}}\sqrt{\varepsilon_{\text{eff}}}} \tag{5.48}$$

and

$$w_{\text{eff}} = \frac{h\eta}{Z_0\sqrt{\varepsilon_{\text{eff}}}}. \tag{5.49}$$

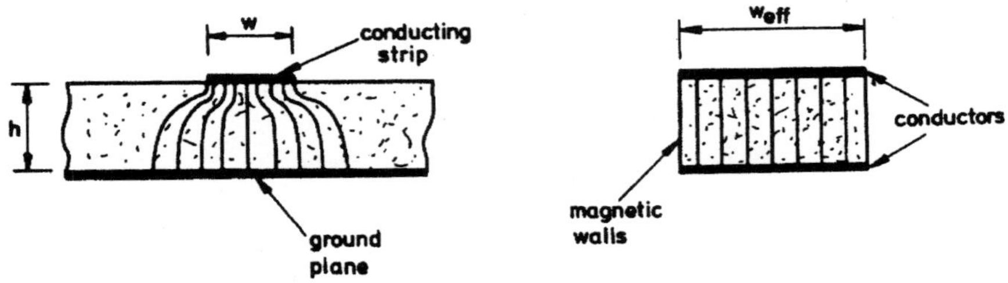

Figure 5.21 Planar waveguide model for microstrip.

In all these expressions η is the characteristic impedance of free space:

$$\eta = \sqrt{\mu_0/\varepsilon_0} = 376.7\,\Omega. \tag{5.50}$$

From Equation (5.47) it is observed that $w_{\text{eff}}(f)$ which appears in the denominator of Equation (5.46) varies as a direct function of $\left[1 + (f/f_p)^2\right]^{-1}$, whereas $\sqrt{\varepsilon_{\text{eff}}(f)}$ is also involved in Equation (5.46). Hence the frequency-dependent effect of $w_{\text{eff}}(f)$ dominates, and this quantity decreases as the frequency increases.

Thus $Z_0(f)$ increases with increasing frequency. For a 50 Ω line the increase is about 10% over the 0–16 GHz range.

5.8.3 A FURTHER ALTERNATIVE EXPRESSION

Another expression has been given by Bianco *et al.* [154], who also utilize the form of the main Getsinger dispersion formula, Equation (5.9), to develop the following result:

$$Z_0(f) = Z_{0T} - \frac{Z_{0T} - Z_0}{1 + G(f/f_P)^2}. \tag{5.51}$$

In this formula Z_{0T} is twice the characteristic impedance calculated for a triplate (stripline) transmission line having the same width w as the microstrip line, but twice the substrate height (i.e. total triplate dielectric thickness is $2h$). Z_{0T} is readily determined from standard design curves [155] or from some good approximate expressions (see Section 5.12.1 on page 156 for a formula for stripline characteristic impedance).

Again we consider a 50 Ω line on an alumina substrate and find that this expression, Equation (5.51), yields an increase of only about 5% over the 0–16 GHz frequency range. Now this order of magnitude (5%) is in fair agreement with a more fundamental calculation by Bianco *et al.* [149] based upon Equation (5.26), the theoretical results due to Krage and Haddad [141] (also Kowalski and Pregla [140]) and the theoretical calculations by Knorr and Tufekcioglu [157]. It may be significant that the latter researchers employed the spectral-domain techniques which have been demonstrably successful in the accurate prediction of $\varepsilon_{\text{eff}}(f)$ (see Section 5.7 on page 133). Mirshekar-Syahkal and Davies [158] also obtain similar results.

For these reasons the use of Equation (5.51) is, rather tentatively, recommended here. Further work is needed to test against the theoretical predictions, and to develop validating measurements. Both recommendations are needed to cover a wide range of nominal characteristic impedance values and, preferably, a range of useful substrates. This testing and validating should be performed in the light of any recent theoretical work.

5.8.4 A DESIGN ALGORITHM FOR MICROSTRIP WIDTH

It is very useful to have the facility for evaluating microstrip width with the aid of static-TEM expressions, such as those given in Chapter 4. The concept, and a routine for achieving this when the frequency-dependent characteristic impedance is known, was first suggested by Getsinger [129]. A modified form of this design algorithm is presented here. The design specification includes ε_r, h, f and $Z_0(f)$, and we require the microstrip width w.

Proceed as follows:

(a) As an initial approximation let $Z_0(f) \approx Z_0$.
(b) Find the approximate microstrip width w using this value of $Z_0(f)$, with ε_r and h.
(c) Determine the frequency-dependent denominator $\left[1 + G(f/f_p)^2\right]$, as described in Section 5.3 on page 118, using this approximate value of $Z_0(f)$. (It may eventually be found that the denominator function in Equation (5.19) gives improved results, but this requires further investigation.)
(d) Evaluate Z_{0T} for triplate having the same strip width w as the microstrip, but twice the dielectric thickness ($2h$). Design curves, or good approximations, may be used.
(e) Rearrange Equation (5.51) for Z_0, letting

$$A = \left\{1 + G(f/f_p)^2\right\}. \tag{5.52}$$

The result of this is

$$Z_0 = A\{Z_0(f) - Z_{0T}\} + Z_{0T}. \tag{5.53}$$

(f) With this approximate value of Z_0, calculate w/h using static-TEM expressions as given in Chapter 4, and hence calculate w (knowing h).
(g) Repeat steps (b) to (f) until negligible difference is observed in the values of w obtained (e.g. it may be desired that the result should be accurate to within 1%).

Further information is available regarding the characteristic impedance as a function of frequency. Brews [159] has provided a theoretical study of microstrip characteristic impedance. His paper is entirely theoretical but he points out that the 'very natural' requirement for complex power P is given by

$$P = \frac{1}{2}\mathbf{I}^*\mathbf{V}. \tag{5.54}$$

This leads to a preferred choice of impedance-defining expressions following Equations (5.43) to (5.45) given above, and it is based upon these definitions that calculations and measurements of $Z_0(f)$ should be made and used.

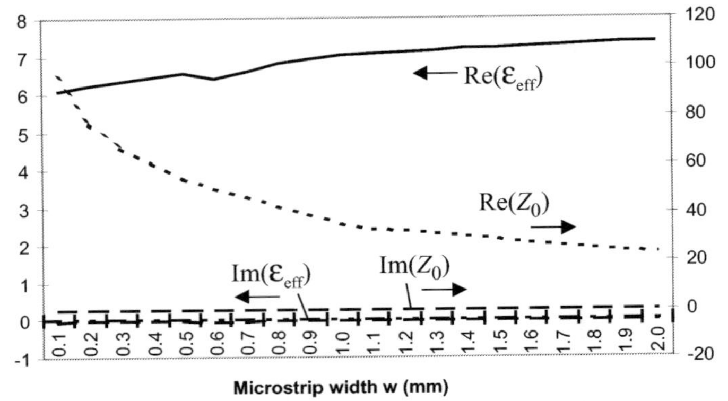

Figure 5.22 Variation of effective permittivity and characteristic impedance with width for a lossy gold microstrip on a 635 μm thick alumina substrate ($\varepsilon_r = 9.8$).

5.8.5 AN EXAMPLE DERIVED FROM A SIMULATION

As an example of some high-frequency results we present here the characteristic impedance and effective permittivity as functions of microstrip width. The data shown in Figure 5.22 were derived from a SONNET® simulation of a lossy microstrip line on a 0.635-mm thick alumina — simulated at 10 GHz.

Apart from the minor decrease, over the 500–600 micron range, the real part of the effective permittivity otherwise exhibits the expected steady increase as width increases (increasing proportion of fields in the substrate). Also as anticipated, the characteristic impedance strongly decreases with increasing width. The imaginary parts of each parameter remain relatively small for all values of width.

These curves can be used as design guidance for microstrip lines on alumina substrates. However, for frequencies other than 10 GHz careful corrections are necessary as described in previous sections here, and in all instances the accurate design formulas should be used.

5.9 OPERATING FREQUENCY LIMITATIONS

As the signal frequency applied to an MIC (or MMIC) is steadily increased some characteristic frequency may be reached at which undesirable effects occur. Two possible spurious effects restrict the desirable operating frequencies. These are

(a) the lowest-order TM mode
(b) the lowest-order transverse microstrip resonance.

In practice, one of these modes will be experienced at some frequency lower than the other, and will thus set the frequency limitation.

5.9.1 THE TM MODE LIMITATION

Vendelin [160] has indicated that the most significant modal limitations in microstrip are associated with strong coupling between the quasi-TEM mode and the lowest-order TM mode. In the cited paper Vendelin gives the main relationships quoted here

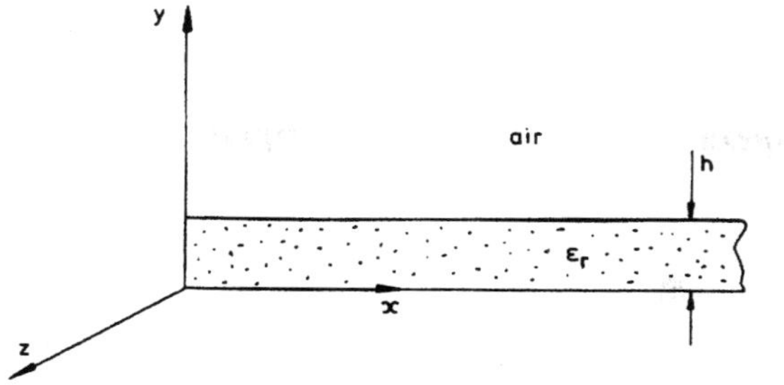

Figure 5.23 An (isotropic) substrate viewed as a dielectric slab, showing the nomenclature.

(although the original analysis is due to Collin — Reference 6 of Vendelin's paper).

We start with the substrate viewed as a dielectric slab, having the coordinate notation shown in Figure 5.23. This slab is a fair approximation for the situation with narrow microstrip lines.

Under these conditions, the eigenvalue expressions are

$$0.5ph = 0.5qh \tan(0.5qh) \tag{5.55}$$

and

$$(0.5ph)^2 + (0.5qh)^2 = (\varepsilon_r - 1)(0.5\beta_0 h)^2 \tag{5.56}$$

where β_0 is the free-space phase coefficient given in terms of the free-space wavelength λ_0 by

$$\beta_0 = 2\pi/\lambda_0. \tag{5.57}$$

Here p is the eigenvalue for the field in the air, and q is the eigenvalue for the field in the substrate (ε_r). It may be convenient to solve Equations (5.55) and (5.56) graphically for p and q, as shown in Figure 5.24.

The following wave equation is then used to find the net phase coefficient for the TM wave:

$$\beta^2 = \beta_0^2 + p^2. \tag{5.58}$$

A frequency for strong coupling between the quasi-TEM mode and the TM mode is identified when the associated two phase velocities are close. The result is given by

$$f_{\mathrm{TEM},1} = \frac{c \tan^{-1}(\varepsilon_r)}{\sqrt{2}\pi h\sqrt{\varepsilon_r - 1}}. \tag{5.59}$$

This is an important relationship. Vendelin points out that the lowest-order TE wave has a cut-off frequency given by the better known expression

$$f_e = c/4h\sqrt{\varepsilon_r - 1}. \tag{5.60}$$

This frequency merely happens to coincide with the $f_{\mathrm{TEM},1}$ result for wide microstrip lines, although the phenomena involved are different. In fact the frequency for the lowest-order TM wave to couple strongly to wide microstrip line is given by

$$f'_{\mathrm{TEM},1} = 75/h\sqrt{\varepsilon_r - 1} \tag{5.61}$$

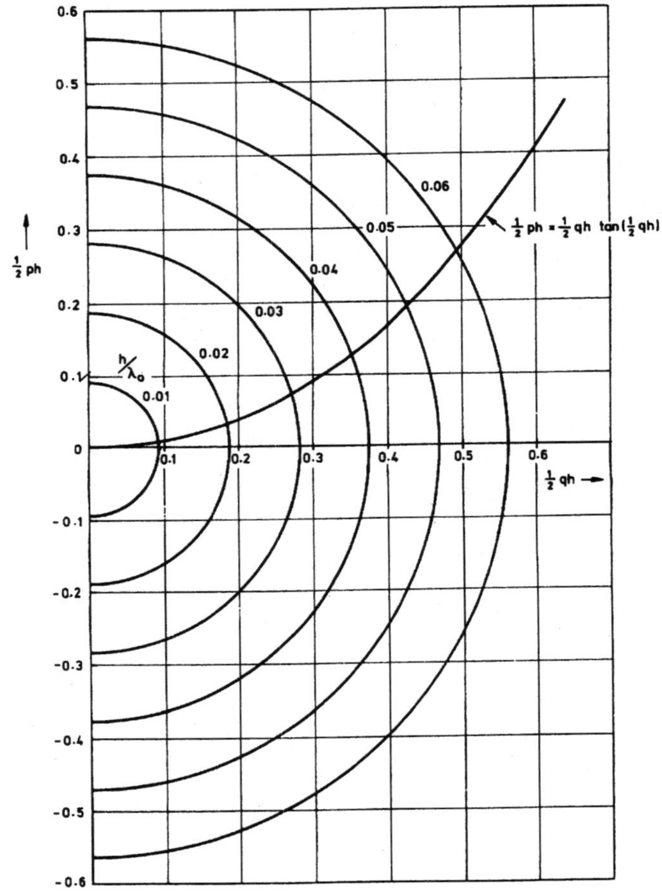

Figure 5.24 Eigenvalue solutions for TM waves.

where $f_{\mathrm{TEM},1}$ is in gigahertz and h is in millimetres.

Returning to the more important case of narrow microstrip lines, where the permittivity ε_r is high (say $\varepsilon_r > 10$), Equation (5.59) tends towards

$$f_{\mathrm{TEM},1} = \frac{c}{2\sqrt{2}h\sqrt{\varepsilon_r - 1}} \tag{5.62}$$

or

$$f_{\mathrm{TEM},1} = \frac{106}{h\sqrt{\varepsilon_r - 1}} \tag{5.63}$$

where $f_{\mathrm{TEM},1}$ is in gigahertz and h is in millimetres. The maximum restriction on usable substrate thicknesses is then easily obtained as

$$h = 0.354\lambda_0/\sqrt{\varepsilon_r - 1}. \tag{5.64}$$

Care should be exercised in the use of these expressions, Equation (5.58) being the more general and applicable to lower-permitivity substrates. It is recommended that substrates should be as thin as possible, Equation (5.64), and that operating frequencies be kept below $f_{\mathrm{TEM},1}$, Equation (5.59).

5.9.2 THE LOWEST-ORDER TRANSVERSE MICROSTRIP RESONANCE

For a sufficiently wide microstrip line a transverse-resonant mode can exist which can also couple strongly to the quasi-TEM microstrip mode. At the cut-off frequency for this transverse-resonant mode the equivalent circuit is a resonant transmission line of 'length' $(w + 2d)$, where d accounts for the microstrip side-fringing capacitance: $d = 0.2h$. The situation is illustrated in Figure 5.25. A half-wavelength must be supported by the 'length' $(w + 2d)$. Therefore the cut-off half-wavelength is

$$\frac{\lambda_{\mathrm{CT}}}{2} = w + 2d = w + 0.4h \tag{5.65}$$

or

$$\frac{c}{2f_{\mathrm{CT}}\sqrt{\varepsilon_r}} = w + 0.4h. \tag{5.66}$$

Hence

$$f_{\mathrm{CT}} = \frac{c}{\sqrt{\varepsilon_r}\,(2w + 0.8h)}. \tag{5.67}$$

Vendelin [160] has indicated that slots, introduced into the metal strip, can suppress the transverse-resonant mode. However, this may not always be practicable (e.g. when short stubs are involved) and it should be checked, by calculating from Equation (5.67), whether the transverse resonance might be excited. Sometimes a change in the circuit configuration will enable the offending wide lines to be avoided altogether.

Under the assumption that transverse resonances are suppressed or avoided in some way, Figure 5.26 shows the cut-off frequency limitation imposed by Equation (5.59) as a function of substrate thickness h for a variety of substrate materials.

Microstrip, compared with media such as waveguide or coaxial lines, offers a rather low unloaded Q-factor. The details of power losses, leading to Q, are discussed shortly. Arbitrarily assuming that a Q-factor of at least a hundred is often desirable, the

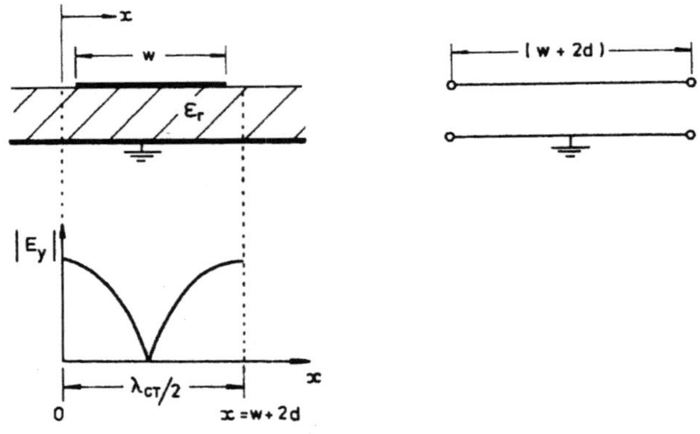

Figure 5.25 Transverse resonance: standing wave ($|E_y|$) and equivalent transmission line of 'length' $(w + 2d)$.

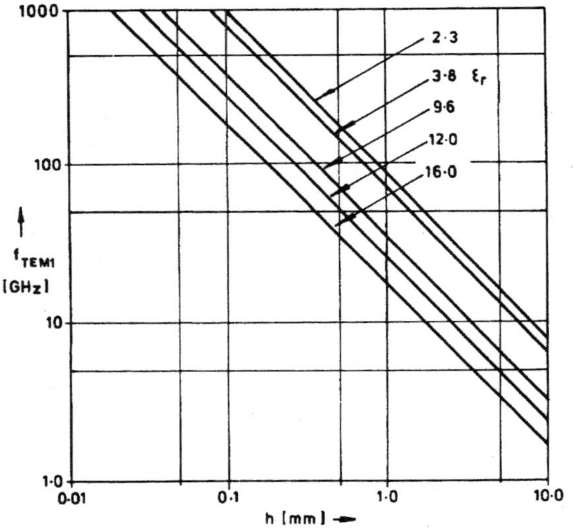

Figure 5.26 TM threshold excitation frequency $f_{\mathrm{TEM},1}$ as a function of substrate thickness h and relative permittivity ε_r. (Reproduced by permission of the *Microwave Journal* from Vendelin [160].)

hatched region of Figure 5.27 then provides a 'useful operating region' knowing the substrate material and the intended operating frequencies.[3]

As an example, consider a microstrip line on an alumina substrate of permittivity 9.6 and thickness 0.65 mm. A vertical ordinate drawn at the $h = 0.65$ mm position shows that we can safely operate at frequencies considerably above 10 GHz and with a 'good' Q-factor of several hundred.

5.10 POWER LOSSES AND PARASITIC COUPLING

Four separate mechanisms can be identified for power losses and parasitic effects associated with microstrip lines:

(a) conductor losses
(b) dissipation in the dielectric of the substrate
(c) radiation losses
(d) surface-wave propagation.

The first two items are dissipative effects whereas radiation loss and surface-wave propagation are essentially parasitic phenomena. Thus, both conductor and dielectric losses can be lumped together for the purposes of calculation and embodied as the attenuation coefficient (α) for a microstrip line.

Considering a fictitious microstrip resonator, which does not radiate or propagate surface waves, the dissipative losses may also be interpreted in terms of a Q-factor as defined by the following expression:

$$Q = \omega_o U/W \tag{5.68}$$

where U is the stored energy and W is the average power lost per cycle.

We now provide the relationship between ε_r and Q.

5.10.1 Q-FACTOR AND ATTENUATION COEFFICIENT

With V_p and I_p, the peak voltage and current at some plane along the line, we may use Equation (5.68) to write, for unit length,

$$Q = \omega_o \frac{\left(LI_P^2/2 + CV_P^2/2\right)/2}{\left(RI_P^2 + GV_P^2\right)/2} \tag{5.69}$$

upon substituting average energy stored.

If $I_P = V_P/Z_0$ is now introduced, we get

$$Q = \omega_o \frac{CV_P^2}{RV_P^2/Z_0^2 + GV_P^2}. \tag{5.70}$$

But $Z_0 = \sqrt{L/C}$, yielding

$$Q = \frac{\omega_o \sqrt{LC}}{R/Z_0 + GZ_0}. \tag{5.71}$$

---

[3] Notice that thinner substrates, while allowing higher frequencies to be used before the onset of restrictions due to TM modes, cause serious degradation of the Q-factor, especially at the lower operating frequencies.

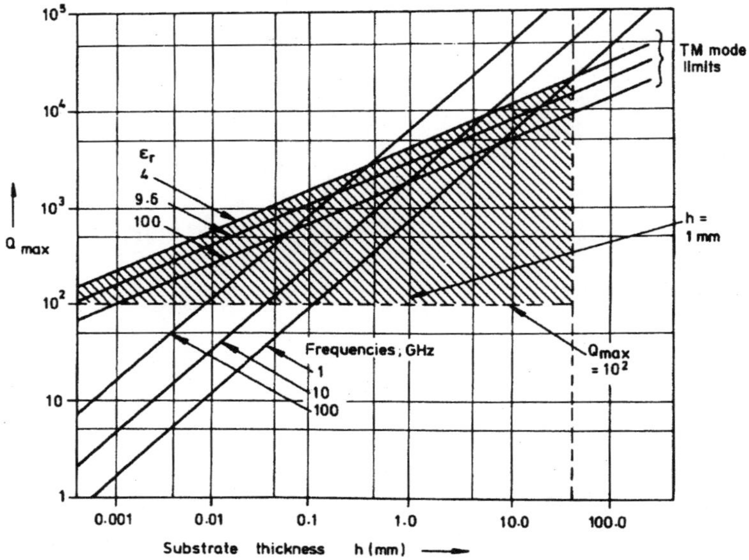

Figure 5.27 Parameters governing the choice of substrate for any microstrip application: acceptable Q-factor, operating frequency, substrate thickness and relative permittivity ε_r. (Reproduced by permission of *Microwave Journal* from Vendelin [160].)

Now, $\beta = \omega\sqrt{LC}$ and $\alpha = R\sqrt{C/L} = R/Z_0 = GZ_0$ so that

$$Q = \frac{\beta}{\alpha\lambda_g}. \tag{5.72}$$

The phase coefficient β is identical to $2\pi/\lambda_g$ and hence Equation (5.72) finally becomes

$$Q = \frac{\pi}{\alpha\lambda_g}. \tag{5.73}$$

In this important result it must be noted that α is in units of Nepers per metre (1 Np = 8.686 dB). Therefore, since α is usually given in decibels per metre, the expression must be divided by 8.686.

To clarify the meaning and use of this expression consider the following example. A Q-factor of 200 is obtained for a microstrip resonator which is $2\lambda_g$ in length and which involves negligible parasitic effects. We require the total loss in decibels for this resonator. From Equation (5.73):

$$\alpha = \frac{\pi}{Q\lambda_g} \quad \text{Np/m} \tag{5.74}$$

and hence, for length l,

$$\alpha(l) = \frac{8.686\pi l}{Q\lambda_g} \quad \text{dB}. \tag{5.75}$$

Substituting $Q = 200$ and $l = 2\lambda_g$, $\alpha(l) = 0.27$ dB for the $2\lambda_g$ length of line. If $\lambda_g = 10$ mm, the attenuation coefficient is then $\alpha = 0.27/20 \approx 0.013$ dB/mm.

Measurements of this Q-factor are best made using ring resonators, where the parasitic effects can be kept very small. (See Chapter 9 for a discussion on measurements.)

5.10.2 CONDUCTOR LOSSES

This topic has been treated in detail elsewhere [58, 161–165] and perhaps the most useful expression has been given by Hammerstad and Bekkadal [97]. This is

$$\alpha_C = 0.072\frac{\sqrt{f}}{wZ_0}\lambda_g \quad \text{dB/microstrip wavelength} \tag{5.76}$$

where the frequency f is in gigahertz and Z_0 is in ohms.

In practice Equation (5.76) yields somewhat low results and account must be taken of surface roughness. Hammerstad and Bekkadal [97] have given the following curve-fitted formula to accommodate this:

$$\alpha'_C = \alpha_C\left\{1 + \frac{2}{\pi}\tan^{-1}\left[1.4\left(\frac{\Delta}{\delta_s}\right)^2\right]\right\} \tag{5.77}$$

where

Δ = r.m.s. surface roughness
δ_s = $1/(R_s\sigma)$ is the skin depth at the appropriate operating frequency
R_s = surface resistance
σ = metal film conductivity.

In order to observe the typical magnitudes involved consider a copper microstrip, where the skin depth at a few gigahertz is 1 μm. Also assume the r.m.s. surface roughness to be of similar magnitude, i.e. 1 μm (typical of rutile, see Table 3.3). Equation (5.77) then shows that the attenuation coefficient becomes

$$\alpha'_C \approx 1.6\alpha_C. \tag{5.78}$$

That is to say, the loss is approximately 60% increased when surface roughness is taken into account. The increase will be still greater for alumina and other substrates where the roughness exceeds 10 μm.

5.10.3 DIELECTRIC LOSS

This is analysed by several researchers, including Gupta *et al.* [166] and Hammerstad and Bekkadal [97]. The first group [166] derive the following expression for dielectric attenuation coefficient, α_D, per unit length:

$$\alpha_D = 27.3\frac{\varepsilon_r(\varepsilon_{\text{eff}} - 1)\tan\delta}{\sqrt{\varepsilon_{\text{eff}}}(\varepsilon_r - 1)\lambda_0} \quad \text{dB/unit length} \tag{5.79}$$

where $\tan\delta$ is the loss tangent for the substrate material and the microstrip wavelength λ_g given by

$$\lambda_g = \frac{\lambda_0}{\sqrt{\varepsilon_{\text{eff}}}}. \tag{5.80}$$

Equation (5.79) can also be written

$$\alpha_D = 27.3\frac{\varepsilon_r\left(\varepsilon_{\text{eff}}-1\right)\tan\delta}{\varepsilon_{\text{eff}}\left(\varepsilon_r-1\right)} \quad \text{dB/microstrip wavelength} \tag{5.81}$$

which is the expression given by Hammerstad and Bakkadal [97].

For microstrip lines on alumina, $\varepsilon_{\text{eff}} \rightarrow \varepsilon_r$ at least very approximately, and the attenuation coefficient is, roughly, $\alpha_D = 27\tan\delta$. In most cases, $\tan\delta \approx 10^{-3}$ (or less) and therefore $\alpha_D \approx 0.027$ dB/microstrip wavelength. This is at least a factor of five smaller than the conductor loss which was calculated earlier.

Conductor losses greatly exceed dielectric losses for most microstrip lines on alumina or sapphire substrates. However, where plastic substrates are used this will by no means always be the case, and (especially) silicon or gallium arsenide substrates result in much larger dielectric losses (about 0.04 dB/mm, that is 0.4 dB/microstrip wavelength at 10 GHz for silicon).

5.10.4 RADIATION

Microstrip is an asymmetric transmission line structure and it is often used in unshielded or poorly shielded circuits where any radiation is either free to propagate away or to induce currents in the shielding. Further power loss is the net result.

In particular, discontinuities such as abruptly open-circuit microstrip (i.e. 'open' ends), steps, and bends will all radiate to a certain extent. Such discontinuities form essential features of a microwave integrated circuit and radiation cannot therefore be avoided altogether. Efforts must be made to reduce such radiation and its undesirable effects. In circuits such as filters, amplifiers, etc., this radiation is an acknowledged nuisance.

Both radiation and surface-wave propagation may be represented as a shunt admittance at the end of an open-circuit microstrip stub or at the plane location associated with some other abrupt discontinuity. This equivalent admittance is

$$Y = G_r + G_s + jB. \tag{5.82}$$

James and Henderson [167] show that, at frequencies where the surface wave is highly trapped in the substrate and with h/λ_0, $w_{\text{eff}}/\lambda_0 \ll 1$ the radiation conductance G_r is approximated by

$$G_r Z_0 \approx \frac{4\pi h w_{\text{eff}}}{3\lambda_0^2\sqrt{\varepsilon_{\text{eff}}}} \tag{5.83}$$

in which w_{eff} is the effective microstrip width defined earlier in Equation (5.46).

The remaining terms in Equation (5.82), G_s and B, represent conductance due to surface-wave propagation and susceptance arising from the many field influences around the discontinuity.

For a 50 Ω microstrip line on 0.65 mm thick alumina the equivalent radiation conductance due to an open-circuit, at 10 GHz, is approximately 30 μS. Under similar conditions a 10 Ω line yields a conductance of about 120 μS.

5.10.5 SURFACE-WAVE PROPAGATION

Surface waves, trapped just beneath the surface of the substrate dielectric, will be propagated away from microstrip discontinuities in the form of a range of TE and radial TM modes. James and Henderson [167] have analysed the problem thoroughly and have given a complicated expression for G_s. The important conclusion is that, for microstrip lines on alumina substrates up to frequencies around 10 GHz, $G_s \ll G_r$ and G_r dominates. However, at higher frequencies G_s also becomes a very significant component. The curves in James and Henderson's paper (their Figures 4 and 6, particularly) are very illuminating in this respect and they should be studied for further information.

When we are concerned with a rectangular microstrip 'patch' resonator, loosely coupled to some source of microwave excitation, there will be two equivalent end admittances, each given by Equation (5.82). In Appendix B it is shown that the 'external' Q-factor due to such end loads may be written as

$$Q_e = \frac{l\pi\rho}{1-\rho^2} \tag{5.84}$$

where l is an integer and ρ is given by

$$\rho = \left|\frac{YZ_0 - 1}{YZ_0 + 1}\right|. \tag{5.85}$$

With the aid of the expressions given in all the preceding sections James and Henderson calculate a total Q-factor of 300 falling to 180, and 60 falling to 20 and lower, for rectangular microstrip patch resonators on alumina and plastic (polyolefin) substrates, respectively. The frequency ranges were 6 to 10 GHz for plastic. Agreement with Q-factor measurements was fairly good, although only 130 was measured for the resonator on alumina at 18 GHz. (Details of the resonators are given in the cited paper.)

Various techniques may be adopted to reduce radiation:

(a) Metallic shielding or 'screening.'
(b) The introduction of a small specimen of lossy (i.e. absorbent) material near any radiative discontinuity [168].
(c) The utilization of compact, planar inherently enclosed circuits (e.g. spurline filters — see Chapter 10, hairpin resonators).
(d) Reduce the current densities flowing in the outer edges of any metal sections and concentrate currents towards the centre and in the middle of the microstrip.
(e) Possibly shape the discontinuity in some way to reduce the radiative efficiency.

Surface-wave propagation may be reduced by (b) above, or by cutting slots into the substrate surface just in front of an open-circuit [169].

5.10.6 PARASITIC COUPLING

It has been pointed out [167] that while screening the circuit removes the radiation it reinforces the surface-wave fields. Thus, we are often left with the option — (b) in the foregoing discussions — of introducing a lossy element to absorb the waves. This may not always be practicable due to space limitations and in any case it further reduces the Q-factor.

In the absence of any such material or other technique the total power in the parasitic waves, $(P_{\text{rad}} + P_{\text{sur}})$, is related to the incident power in the microstrip P by the ratio

$$\xi = \frac{P}{P_{rad} + P_{sur}} = \frac{1}{1 - \rho^2}. \tag{5.86}$$

For a high degree of isolation the circuit must be operated at a relatively low frequency so that $P_{\text{sur}} \ll P_{\text{rad}}$ and then, substituting Equations (5.83) and (5.85) into Equation (5.86):

$$\xi = \frac{1}{4G_r Z_0} = \frac{Z_0 \varepsilon_{\text{eff}}}{640} \left(\frac{\lambda_0}{\pi h} \right)^2. \tag{5.87}$$

Where possible, the following observations then apply for high isolation:

(a) Use relatively high permittivity substrates. Then ε_{eff} is relatively large (e.g. alumina is better than plastic — which should be anticipated on physical grounds).
(b) Use fairly thin substrates (λ_0/h as large as possible).
(c) Employ high impedance (Z_0) stubs, etc., wherever this is feasible.

5.10.7 RADIATION AND SURFACE-WAVE LOSSES FROM VARIOUS DISCONTINUITIES

Although microstrip discontinuities are the principal subject of Chapter 7, the losses originating from them should logically receive some consideration here. Horng *et al.* [170] have reported a generalized method for calculating and clearly distinguishing between radiation and surface-wave losses. They base their method on a rigorous Poynting vector analysis in which the current distribution around the entire discontinuity is a result of a complete electromagnetic solution by moments. These researchers provide self-consistency checks of their results, based upon power conservation calculations.

The losses calculated by Horng *et al.*, due to radiation and surface-wave propagation from the important open-end discontinuity, are shown as a function of frequency in Figure 5.28. It must be appreciated that such losses are often highly significant from the circuit design standpoint and usually they have to be minimized.

5.10.8 LOSSES IN MICROSTRIP ON SEMI-INSULATING GaAs

This specific and important subject has been addressed by several workers, for example Goldfarb and Platzker [171] who compare data measured over the frequency range 0–40 GHz with predictions made using three proprietary CAE packages. For the measurements, microstrip resonators were formed on 0.1 mm thick GaAs substrates.

Goldfarb and Platzker's results for a 20 Ω line having such a structure are shown in Figure 5.29. Even with this relatively wide line operated at 27 GHz the radiation loss is negligible as a consequence of the high permittivity (12.9) and the small thickness

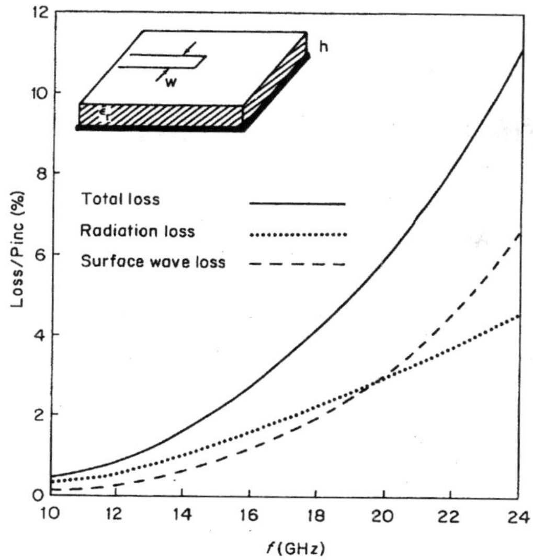

Figure 5.28 Power losses versus frequency for open-end discontinuity ($\varepsilon_r = 10.2$, $w = 24$ mil, $h = 25$ mil). (©1990 IEEE. Reprinted, with permission, from Horng *et al.* [170].)

of the substrate. So these results are dominated by conductor and dielectric losses. It is clear that all three CAE packages yield predictions of loss well within the range of the measurements, except for the losses as frequency approaches zero.

5.11 SUPERCONDUCTING MICROSTRIPS

As pointed out in Chapter 3, a major disadvantage of most planar transmission lines is the dissipative loss (when compared with more traditional media such as all-metallic waveguide or coaxial cable). This forms an advantage for antenna applications, but represents a distinct difficulty in the realization of several types of circuits — notably filters.

Since the principal mechanism responsible for the power dissipation in microstrips is conductor loss it is clear that if this could be reduced then overall loss would also be significantly decreased. One direct method of achieving this is to 'supercool' the structure, typically down to a specific temperature within the range 20–77 K.

By no means incidentally, the action of supercooling in this manner can also benefit the function of other circuit components such as switches by accelerating their performance and thereby rendering them capable of higher-frequency operation. By the end of the 1980s, this area of microwave technology had expanded to the extent that commercial organizations were already in evidence, for example Superconducting Technologies Inc., USA.

Lee *et al.* [172] have reported conductor loss calculations for superconducting microstrips. They describe the complex conductivity of the superconductor approximately, using the 'two-fluid' model. The resulting modelled conductivity is used to determine the conductor loss by means of the Phenomenological loss

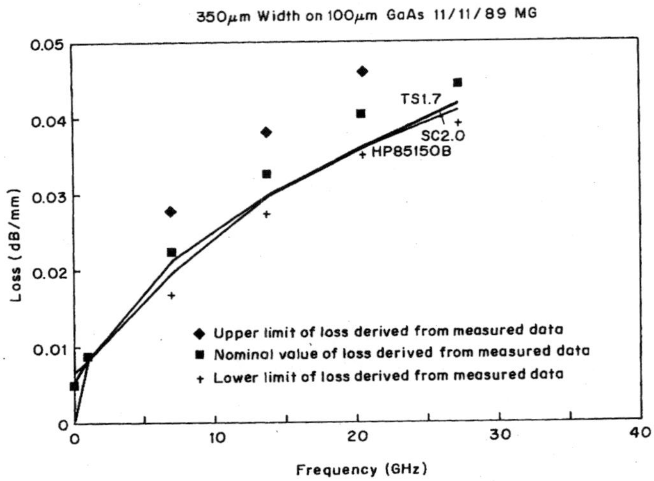

Figure 5.29 Losses in microstrip on a GaAs substrate. (350 μm width on 100 μm GaAs.) (© 1990 IEEE. Reprinted, with permission, from Goldfarb and Platzker [171].)

Equivalent Method (PEM). A problem with this type of analysis is that for complete penetration of the current into the conductor, the current distribution is uniform inside the conductor.

Contrastingly, where the current penetration is relatively moderate then the current distribution becomes distinctly non-uniform inside the conductor. In order to obtain an equivalent strip for calculations Lee *et al.* consider two different superconducting materials to which are separately ascribed the shallow and deep penetrations.

The curves of Figure 5.30 are repeated from the data presented by Lee *et al.* and they show results for both conventional (room temperature) microstrip having aluminium metallization and, alternatively, the superconducting line held at a temperature of 77 K (critical temperature 92.5 K).

The conventional microstrip line is fabricated on a quartz substrate with a permittivity (Lee *et al.* data) of 3.9. In both cases, results using the PEM are compared with results using an independent Monte Carlo method. Clearly, at all frequencies considered here the superconducting line offers a much lower loss regime than conventional (room temperature) metallic microstrip. Even at 100 GHz the superconducting line loses only 0.1 dB/cm in contrast to the conventional line for which a loss of 10 dB/cm is applicable at this frequency.

Phase velocity is also shown as a function of frequency in Figure 5.31. The behaviour associated with the aluminium line exhibits the usual variation of phase velocity, including the marked dispersion which, for this example, is pronounced above about 1 GHz. In the case of the superconducting line, however, the phase velocity is clearly approximately constant with frequency over the entire band considered (DC to 1000 GHz). Any dispersive effects at the higher frequencies are masked by the scale of this graph.

Bhasin *et al.* [173] also reported on the performance and modelling of superconducting ring resonators operating at 35 GHz (Chapter 7 provides information on ring resonators). They showed that, at temperatures below 60 K, the

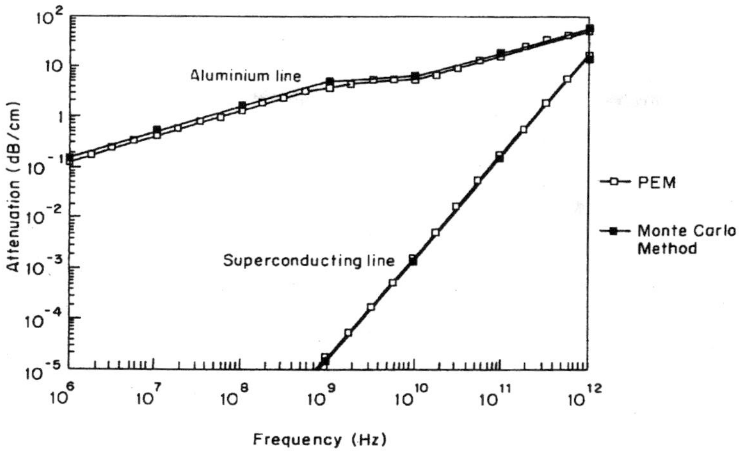

Figure 5.30 Attenuation constants at 77 K for superconducting and aluminium microstrip lines. (Reprinted, with permission, from Lee, Kong and Itoh [172].)

superconducting strip exhibited a performance superior to that of a conventional gold-metallized microstrip. In the superconducting case the unloaded-Q was higher by a factor of approximately 1.5. These researchers also show results for resonant frequency as a function of temperature.

Clearly there is a substantial amount of work required to refine both the models and/or the quality of the metallizations used for the microstrips. The difference between data obtained by theory and measurement approached a factor of two at the lower temperatures and remained about 25% at the higher temperatures for the superconductive case.

Closely related work has also been performed by van Deventer *et al.* [174]. This group used an integral equation approach to calculate the propagation characteristics of shielded superconducting microstrip lines. Phase and attenuation coefficients, as well as characteristic impedance, are presented as functions of frequency (DC to 10 GHz). It is shown that the attenuation coefficient remains below 0.5 dB/m in all cases, up to 10 GHz, except for the relatively extreme case where the substrate height is reduced to 25 μm, when the attenuation increases to 0.83 dB/m at 10 GHz. However, unlike all other situations reported, in this case the phase coefficient stays constant over the entire band, i.e. at these lower microwave frequencies design is relatively straightforward and there is negligible dispersion. Characteristic impedance is also invariant with frequency over this range. Hammond *et al.* [175] have reported the performance of superconducting microstrip resonators. Their measurements indicate that the loaded Q-factor of superconducting, open-ended S-shaped microstrip resonators is always much greater than that of conventional copper or even silver microstrip resonators. The Q-factors are, however, dependent upon the level of input power and significant reductions occur as this input power approaches 1 W. As the power level is increased from -65 to around 0 dBm, the Q-factor falls slowly from around 5500 at the lower power, to approximately 4000 at 0 dBm. Even so, across the entire range of power levels, the Q-factors are about one order of magnitude higher

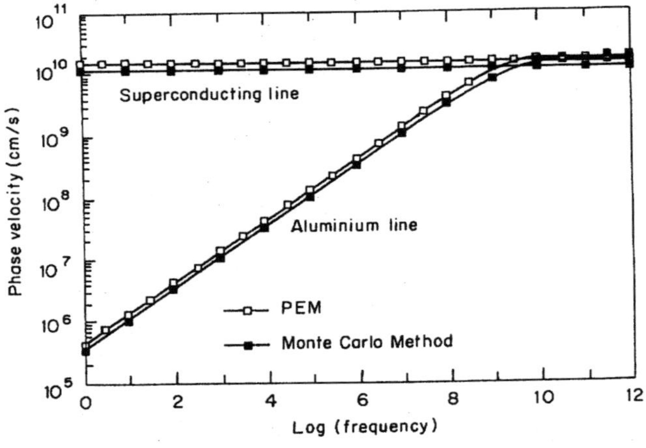

Figure 5.31 Phase velocities at 77 K for superconducting and aluminium microstrip lines. (Reprinted, with permission, from Lee, Kong and Itoh [172].)

than is achievable with conventional resonators held at room temperature.

In summary, the performance achievable with superconducting microstrip renders the approach attractive, at least in situations where the other circuit components also benefit.

5.12 STRIPLINE DESIGN

The main thrust of this text concerns planar transmission structures and interconnects. The problem of stripline design, both filled and also suspended substrate is, however, also significant and this is briefly considered here. In practice, stripline design is generally more straightforward than either microstrip or co-planar waveguide design.

5.12.1 SYMMETRICAL STRIPLINE FORMULAS

The basic structure of stripline simply comprises a centre conductor pattern symmetrically embedded completely with a dielectric, the top and bottom layers of which are coated with the connected conducting ground plane. Unlike microstrip, the entire separation between ground planes is generally denoted by b or $2h$.

As with any transmission structure the designer starts with knowing the desired characteristic impedance and requires the width w of the centre conductor. This can be obtained by iterating on the following formulas which give the characteristic impedance of stripline [156, 262]:

$$\frac{W_{\text{eff}}}{b} = \frac{W}{b} - \left[\frac{(0.35 - W/b)^2}{(1 + 12t/b)}\right] \tag{5.88}$$

$$Z_0 = \left[\frac{30\pi}{\varepsilon_r^{1/2}}\right] \frac{(1 - t/b)}{(W_{\text{eff}}/b) + (C_f/\pi)} \tag{5.89}$$

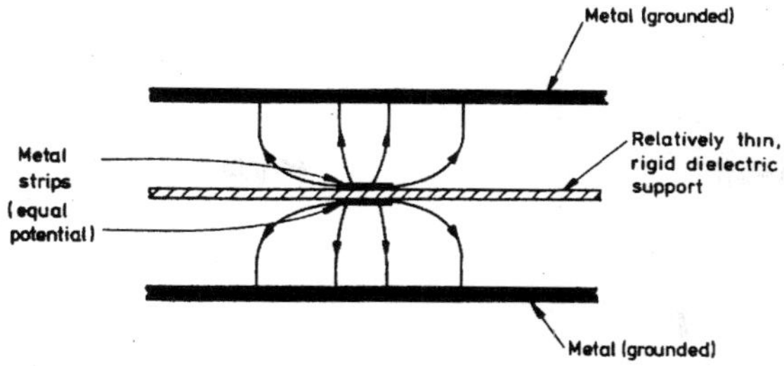

Figure 5.32 High-Q stripline, showing the electric field.

and

$$C_f = 2\ln\left[\frac{1}{1-(t/b)}+1\right] - \frac{t}{b}\ln\left\{\frac{1}{[1-(t/b)]^2}-1\right\} \tag{5.90}$$

The accuracy of these formulas is claimed to be better than 1% for the ranges:

$$0.05 < W/(b-t) < 0.35 \quad \text{and} \quad t/b < 0.025$$

where t is the strip thickness.

Finding the physical length required to realize an electrical length θ (degrees) is also relatively straightforward. We first evaluate the wavelength using

$$\lambda_g = \frac{300}{F\sqrt{\varepsilon_r}}. \tag{5.91}$$

When F is in GHz this automatically provides the wavelength in mm. In Equation (5.91), ε_r is the permittivity of the filling dielectric material and, of course, $\varepsilon_{\text{eff}} = \varepsilon_r$.

Then the physical length is calculated using Equation (4.24), i.e.

$$\lambda_g = \frac{300}{F\sqrt{\varepsilon_r}} l = \frac{\theta\lambda_g}{360}. \tag{5.92}$$

As there is no air-dielectric interface, no progressive mode-coupling occurs and the medium is essentially non-dispersive. Therefore, the above expressions apply to the highest frequencies at which a stripline can be used, and this tends to be limited by the onset of higher-order modes such as the parallel-plate waveguide mode.

Suspended-substrate stripline represents an important variant of high-Q stripline, shown in Figure 5.32.

When one of the metal strips is removed we have suspended-substrate stripline. In practice this means that the structure is not perfectly symmetrical, and a design adjustment is required to account for this fact. Only a very small proportion of the separating space is now dielectric material, and this largely accounts for the high-Q. This feature also means that all the design equations typically substitute 1.03 instead of ε_r because of the mainly air spacing.

Suspended-substrate stripline is frequently used in filter and multiplexer designs, taking advantage of the high-Q.

5.13 DESIGN RECOMMENDATIONS

In the light of the techniques presented in this chapter and also Chapter 4, we can now summarize the main microstrip design recommendations.

Initially, system and network design procedures yield a specific value of characteristic impedance Z_0 and of electrical length θ for each microstrip line. We must determine the correct physical dimensions for the microstrip, that is w and l. It is recommended that the following sequence be used.

5.13.1 RECOMMENDATION 1

Select the substrate such that the TM mode effect is avoided. That is, the maximum intended operating frequency should be less than

$$f_{\mathrm{TEM},1} = c\tan^{-1}(\varepsilon_r) / \sqrt{2}\pi h\sqrt{\varepsilon_r - 1} \tag{5.93}$$

as given by Equation (5.59).

Also ensure that parasitic coupling is sufficiently low, Equation (5.64).

5.13.2 RECOMMENDATION 2

Carry out a 'first-cut' design on the static-TEM basis explained in Chapter 4. This yields w and $\varepsilon_{\mathrm{eff}}$.

5.13.3 RECOMMENDATION 3

Check that the first-order transverse resonance cannot be excited at the highest frequencies using Equation (5.67):

$$f_{CT} = \frac{c}{/\sqrt{\varepsilon_r}\,(2w + 0.8h)}. \tag{5.94}$$

In the event that such a resonance might occur, either redesign (using narrower microstrip lines) or, if possible, incorporate slots in the metal strips.

5.13.4 RECOMMENDATION 4

Calculate the total losses and the total Q-factor, and check that this meets the circuit requirements (Section 5.10 on page 147 gives the methods for this). If the Q-factor is too low a major reappraisal of the design philosophy may be necessary, or superconductivity might be used.

However, as a general indication, many satisfactory microstrip circuits may be designed with Q-factors no greater than about 100.

5.13.5 RECOMMENDATION 5

Evaluate the frequency-dependent effective microstrip permittivity $\varepsilon_{\mathrm{eff}}(f)$, and hence the wavelength $\lambda_g(= c/f\sqrt{\varepsilon_{\mathrm{eff}}(f)})$.

For alumina-type substrates, where $10 \le Z_0 \le 100\ \Omega$ and $2 \le f \le 18$ GHz the recommended expression is the relatively simple Equation (5.19):

$$\varepsilon_{\mathrm{eff}}(f) = \varepsilon_r - \frac{\varepsilon_r - \varepsilon_{\mathrm{eff}}}{1 + (h/Z_0)^{1.33}\,(0.43f^2 - 0.009f^3)} \tag{5.95}$$

where h is in millimetres and f is in gigahertz. The estimated accuracy is $\pm 0.8\%$. (Note that this includes errors in the static-TEM expressions for ε_{eff} and Z_0 i.e. the $\pm 0.8\%$ is overall.)

Where higher frequencies, and/or substrates having permittivities differing substantially from 10, are involved the more elaborate set of expressions due to Kirschning and Jansen or Kobayashi are recommended. These were provided earlier in this chapter and will not be repeated here (see Section 5.4 on page 122).[4] Since the electrical length θ (degrees) is given in the circuit or antenna specification, the physical length l of the microstrip is

$$l = \theta \lambda_g / 360 \text{ (neglecting discontinuities).} \tag{5.96}$$

5.13.6 CHARACTERISTIC IMPEDANCE AS A FUNCTION OF FREQUENCY

This problem was discussed in Section 5.8 on page 137. The practical effects are fairly small, mainly amounting to a small degradation of VSWR in most cases. If initial calculations indicate that $Z_0(f)$ is likely to be significant in a particular design problem the algorithm given in Section 5.8.4 is recommended.

5.13.7 COMPUTER-AIDED DESIGN

The techniques described in Chapter 4, this chapter and also Chapter 6 (CPW) are very appropriate for CAD implementation. Sometimes MIC designers prefer to set up their own CAD facility. This can incorporate the most accurate expressions available for static-TEM and dispersion computations. It can also embody formulas accounting for a number of microstrip discontinuity elements that are described in Chapter 7.

Where there is discrepancy between predicted and measured results, Goldfarb and Platzker [171] recommend a modification of the conductor resistivity model in the simulator, forcing it to follow the observed approximate square law relationship as a function of frequency.

---

[4] If sufficient time and computer power are available, a technique such as that described in Section 4.6 on page 94 might be used.

6

CPW Design Fundamentals

6.1 INTRODUCTION — PROPERTIES OF COPLANAR WAVEGUIDE

Microstrip remains the most popular medium for circuit design at frequencies ranging from several hundred MHz to some tens of GHz. It is also the medium on which most long-haul, high-performance, on-chip digital interconnect is based. This medium is always inherently well grounded by virtue of the ground plane which also ensures that most of the energy is trapped between the strip and this ground plane, i.e. within the substrate. The top surface of the substrate is available for conductor strip patterns that largely define electromagnetic propagation and circuit function. There are, however, difficulties associated with the use of microstrip, notably:

- The relative inaccessibility of the ground plane.
- The difficulty associated with making shunt connections between strip and ground.
- Limitations upon and sensitivity to substrate thickness.
- Increased radiation when thick substrates are used.

The effects of all these difficulties become more severe as the frequency increases. As a result of these effects, microstrip is certainly not necessarily the best choice of transmission medium for broadband millimetre-wave circuit design or for long on-chip interconnect. In integrated circuit technology, lateral dimensions can be defined photolithographically with great accuracy, but the substrate thickness is often not controlled at all well. It is therefore very attractive to use interconnects that are defined principally by their lateral dimensions. In multi-level connections an uninterupted ground cannot always be assured, so providing a signal return path on the same metallization layer assures near-optimum performance but at the price of loosing the number of interconnects that can be provided, since the signal return lines consume routing area.

Coplanar waveguide (CPW) is a term used to describe what really amounts to a class of planar microwave transmission structures that comprises various arrays of conductors configured in the same geometric plane, i.e. coplanar. Quite unlike microstrip, where a distinct and singular ground plane necessarily occupies the entire 'underside' of the structure, CPW has its principal grounding in the form of wide strips adjacent to the active conductor. Thus, the signal characteristics are almost entirely governed by laterally defined dimensions. Before entering into further details regarding CPW geometry options, we start with observing the related, but simpler,

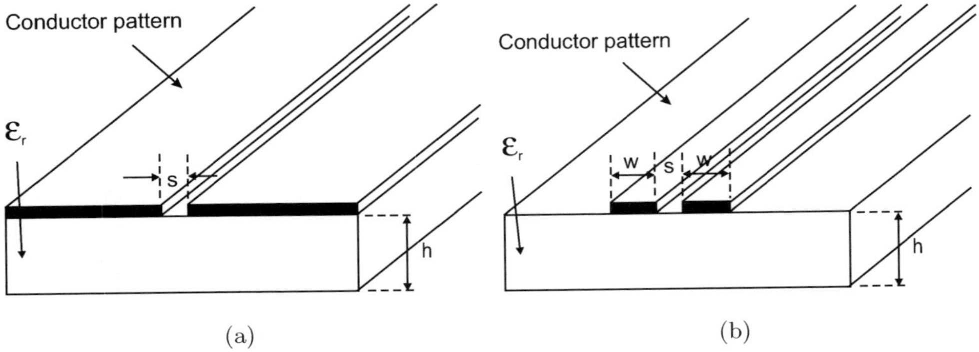

Figure 6.1 Structure of (a) slotline; and (b) differential line or coplanar strips (CPS).

slotline and coplanar strip structures — 3D views of these structures are shown in Figure 6.1.

The slotline structure, Figure 6.1(a), was first proposed by Seymour Cohn in 1969 [176]. Cohn pointed out that the guide mode in this case has a region of elliptical polarization that makes it particularly suitable for applications where the substrate is ferrite. Circulators and isolators — important microwave components — have therefore been realized using slotline. Higher-order moding also occurs with slotline and the behaviour of the first higher-order mode has been investigated [177]. Due to this higher-order moding, care is taken to operate in the dominant mode by ensuring that the height h is much larger than s (so that field lines through to the back of the substrate are negligible) and that s is much less than a wavelength.

Over the intervening years other structures such as finline have been used in what could otherwise have been possible slotline applications. Finline was introduced in Chapter 3. It is like slotline but enclosed in rectangular waveguide, and today is largely confined to applications above 100 GHz. In finline waveguide is used to confine the fields, but at lower frequencies coplanar structures on their own are sufficiently effective for confining the fields locally. This latter situation (confinement of the fields) results in two main attributes: the variation of interconnect characteristics with frequency is small; and unwanted coupling of the signal on one coplanar structure with another is minimal.

Slotline is a special case of the coplanar strip (CPS) structure that has only two conductors separated by a narrow longitudinal slot and grounded at their extremes (away from the slot and theoretically extended to infinity). Both slotline and CPS suffer from relatively high radiation losses when realized on low permittivity substrates which leads to low-Q circuits. The radiation also results in substantial interaction and crosstalk between adjacent circuits, although this is lower than with microstrip. Both slotline and CPS are balanced line structures and are occasionally implemented in applications including radiating elements and transitions. More recently CPS has become important for on-chip implementations on semiconductor substrates — especially for high-speed digital signal transmission, where it is usually referred to as differential interconnect or differential line.

Coplanar waveguide was invented and reported upon by C.P. Wen [178] over three decades ago. The coincidence of the inventor's initials with the acronym for coplanar waveguide (both CPW) may possibly be just that — a coincidence — but it is at least an interesting one. The basic structure of CPW is illustrated in Figure 6.2(a) where the arrangement is assumed to be symmetrical with strip width w and equal longitudinal gaps s. The side conductors are ultimately grounded, theoretically at infinity. A variant of the CPW structure is the Finite Ground CPS line shown in Figure 6.2(b) which results in lower coupling of adjacent lines as the ground (signal return current) is not shared by two or more transmission lines. As originally conceived Wen's design formulations for CPW demanded an infinitely thick substrate, and this practical difficulty doubtlessly impeded its implementation in any general sense. This was in spite of the fact that CPW has the following advantages over microstrip:

- Easier grounding of surface-mounted (or BGA mounted) components.
- Lower fabrication costs.
- Reduced dispersion (for small geometries).
- Decreased radiation losses.
- Availability of closed-form expressions for the characteristic impedance.
- Couplers having higher directivities.
- Photolithographically defined structures with relatively low dependence on substrate thickness.

The current and charge distribution on a CPW line at one time instant is shown in Figure 6.3 for a sine wave travelling from left to right. (The granularity is the result of the necessary discretization of the currents by the electromagnetic simulator.) The charge and the magnitudes of the currents are shown. Just under a full cycle is shown with the longitudinal current on the inner conductor initially (on the left) directed towards the right and the latter in the middle of the figure directed to the left. The adjacent longitudinal currents on the inner and outer conductors are in opposite directions. The small transverse currents are necessary for the charges to rearrange at 100 GHz and the charges establish the electric field between the centre conductor and each outer conductor as well as supporting the longitudinal currents. An expanded view of the currents and charge distributions is shown in Figure 6.4. Note in particular that there is appreciable current on the outer edges of the outer strips.

Throughout the period from 1970 to the present time microstrip design and implementation has been in the ascendancy, and a vast library of algorithms has been assembled for the computer assisted design of circuits using this medium. In contrast, far less has been achieved for CPW in relating the desired electrical parameters to the physical parameters. Gupta *et al.* [180] have reported much design detail for this medium.

The major problem centres upon the characteristic impedance determining parameters associated with CPW *vis-à-vis* microstrip. As explained in the earlier chapters of this book, the characteristic impedance Z_0 of microstrip is very largely determined by the ratio of strip width (usually denoted w) to height h and the substrate permittivity ε_r. This results in a unique value of w for a required Z_0, as h is nearly always fixed by an earlier design decision. However, a similar situation does not, unfortunately, exist with CPW.

Instead, with CPW Z_0 is determined by the ratio of the centre strip width w to the gap width s. This makes the design of a CPW line with a particular Z_0 non-unique because an infinite range of w and s values will result in a specific Z_0 requirement. As

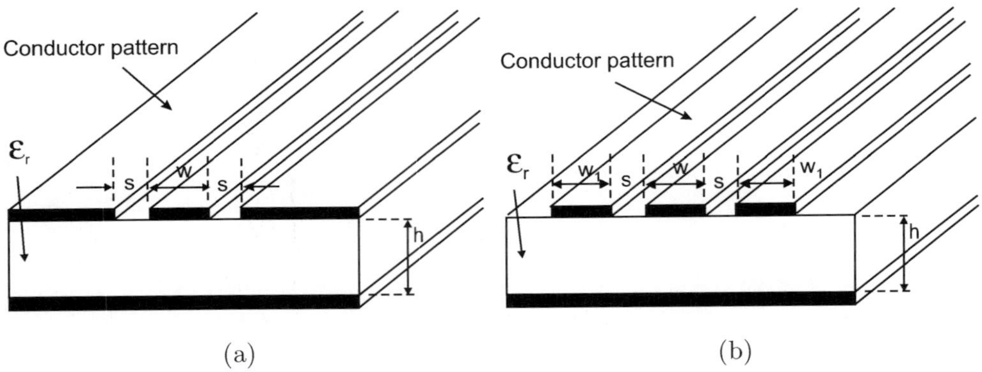

Figure 6.2 Structure of a coplanar waveguide (CPW): (a) conventional; and (b) finite ground CPW (FGCPW).

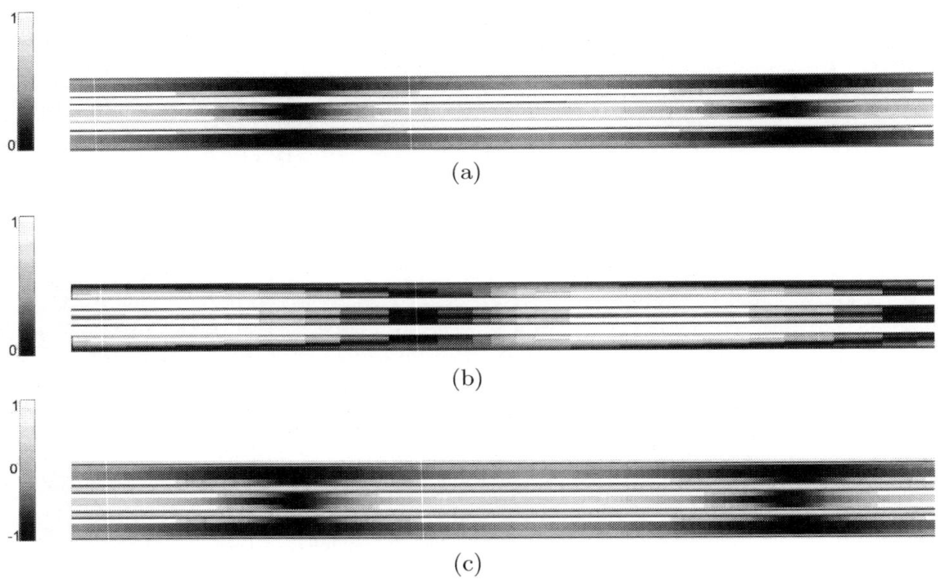

Figure 6.3 Instantaneous normalized current magnitude and charge distribution along a CPW line along with respective normalized scales: (a) longitudinal current; (b) transverse current; and (c) charge. The current and charge are normalized and use the scale shown on the left hand side. The maximum of the transverse current is 500 times smaller than that of the longitudinal current. (The line is 1 mm long, the frequency is 100 GHz, $s = 10\ \mu$m, and $w = 20\ \mu$m.)

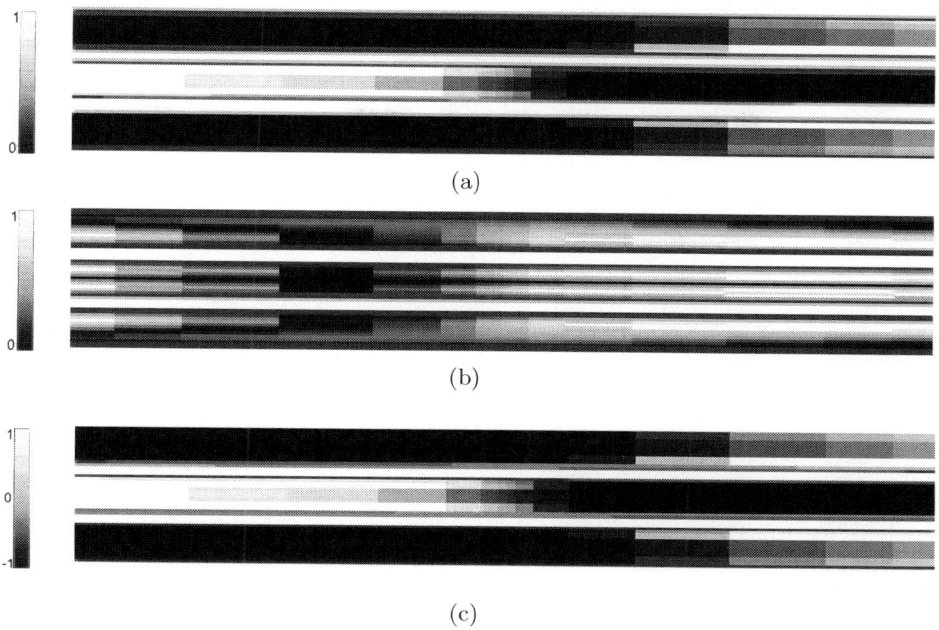

Figure 6.4 Expanded view, current magnitude and charge distribution along a CPW line: (a) longitudinal current; (b) transverse current; and (c) charge.

a result the development of effective simple models for CPW design, such as planar waveguide models, is extremely difficult and much more extensive design algorithms (than for microstrip) are required.

Many researchers have used the notation 's' for the centre strip width and 'w' for the separating gap. We shall use the more common, but opposite and more logical, notation here.

Appreciating that in practice CPW almost always has a ground plane conductor backing, in 1980 Veyres and Fouad-Hanna [181] published analytic equations characterizing the key parameters of this type of CPW. This conductor backing is sometimes termed 'backside metallization' and this must generally apply because MICs or MMICs are usually mounted in a conducting box of some form. There were some errors in Veyres and Fouad-Hanna's analysis, but corrections to these errors were subsequently published by Ghione and Naldi [182,183]. For this conductor-backed structure analytic closed-form expressions were obtained for Z_0 and the phase velocity in CPW and this is actually an improvement over the situation with microstrip.

Design advantages rarely come without attendant drawbacks and this conductor-backed structure is no exception. The following disadvantages apply:

- Spurious modes (notably the microstrip mode) can easily be generated if the separation between the CPW structure and the backing metallization is too close (resulting in field lines between the CPW and the backing metallization).
- For a given Z_0 and substrate thickness the strip width w will always be significantly less than for the corresponding microstrip, in order to maintain the

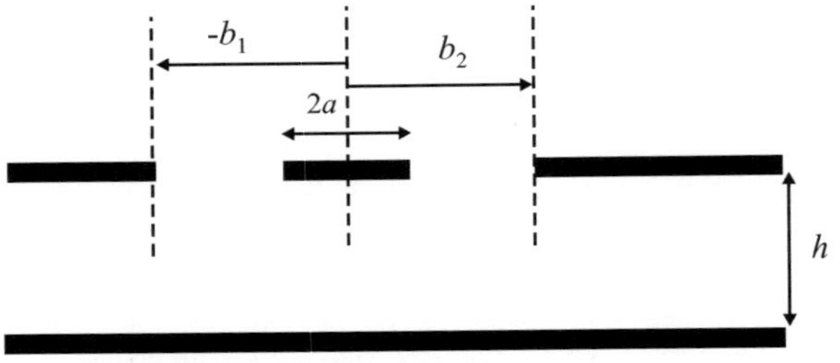

Figure 6.5 Asymmetric CPW with conductor backing.

same capacitance to ground. Therefore, the resistive loss for the CPW line can exceed that of the corresponding microstrip.

Note that with this structure the slotline mode, another spurious mode, can exist, just as without a conductor backing.

In practice, spurious modes can be avoided by careful design and fabrication, whilst the additional loss can be minimized by carefully selecting the substrate thickness.

It can be the case that coplanar is not always precisely symmetrical — the separations between centre strip and the ground planes can differ providing, at almost no cost, an extra dimension of the design space. Amongst the first to investigate the properties of this structure, and its departures from the basic CPW behaviour, were Fan and Wang [184]. Fan and Wang's work was entirely theoretical, and at that time no measured results were reported. The cross-section of the configuration is shown in Figure 6.5. These researchers showed that, for suitable combinations of b_1 and b_2, a virtually frequency independent effective permittivity of 5.6 applied (with a substrate permittivity of 10). The electric field patterns were also reported for this structure.

The key attribute of symmetrical CPW structures (with $b_1 = b_2$) is the relative frequency independence of the effective permittivity, for small geometries, as this leads to almost dispersion free transmission of signals. This attribute is accompanied by a simplified way of connecting devices along the signal path without the use of vias — active devices can be placed under the CPW structure. This is important in many MMIC chip layouts.

6.2 MODELLING CPWs

For an interconnect to be of use in practical design, it must be analyzed and also be capable of being synthesized, i.e. the physical parameters enabling manufacture must be determined, starting with a knowledge of the desired characteristic impedances and electrical lengths. Firstly, expressions for characteristic impedance must be known. It must also be possible to find the wavelength in the medium and this is normally preceded by knowing either the effective permittivity (previously defined for microstrip), the propagation coefficient, or the phase velocity. All of these latter quantities can be derived from knowing just one.

We shall now define these quantities for basic CPW (Figure 6.2(a)), starting with the effective permittivity. The results given were first analysed by Wen [178] and used the Schwartz-Christofell transformation. Rayit [185] provides a good detailed description of this approach and this is followed here. The appropriate form of the Schwartz-Christofell transformation is:

$$\frac{dW}{dZ} = A \prod_{n=0}^{n=\infty} (Z - X_n)(\alpha_n - \pi)/\pi. \tag{6.1}$$

The use of this equation enables the problem to be represented in terms of different planes (or variables) from the physical planes (specifically what are known as the W- and Z-planes). In Equation (6.1) A is a constant that must be evaluated, $\alpha_1, \alpha_2 \ldots\ldots \alpha_n$ are interior angles in a polygon described in the W-plane, and $x_1, x_2, \ldots x_n$ are corresponding points along the real axis in the Z-plane.

Wen, and also Rayit, clearly show that ratios and standard values can then be implemented to solve for the effective permittivities and characteristic impedances. In particular, a ratio a/b is also expressed as the ratio of two complete integrals termed $K(k)$ and $K'(k)$:

$$\frac{a}{b} = \frac{K(k)}{K'(k)} \tag{6.2}$$

in which $K(k)$ is the complete integral of the first kind with modulus k, such that $K(k)$ is given by

$$K(k) = \int_0^{\pi/2} \frac{\mathrm{d}\Phi}{\sqrt{1 - k^2 \sin^2 \Phi}} \,. \tag{6.3}$$

The interrelationships between the quantities are

$$k = \frac{a_1}{b_1} = \frac{w}{w + 2s} \tag{6.4}$$

where w is the width of the centre strip and s is the separation between this strip and each of the coplanar ground planes,

$$k' = \sqrt{(1 - k^2)} \tag{6.5}$$

and

$$K'(k) = K(k'). \tag{6.6}$$

In the initial expressions we are referring to basic CPW, i.e. not conductor-backed, the substrate thickness is considered large enough to be theoretically infinite and the conductor thicknesses are regarded as negligible. The effect of the substrate thickness is considered later.

6.2.1 EFFECTIVE PERMITTIVITY

The effective permittivity of CPW is defined in exactly the same fundamental way as that of microstrip, and for that matter for any form of interconnect structure:

$$\varepsilon_{\text{eff}} = (c/v_p)^2. \tag{6.7}$$

Here c is the free-space velocity and v_p is the phase velocity of the dominant mode in the CPW.

Hilberg [186] gives what are termed exact relations for both the effective permittivity and also the characteristic impedance of CPW. His (Hilberg's) equation for the effective permittivity is:

$$\begin{aligned}\varepsilon_{\text{eff}} &= 0.5(\varepsilon_r+1)\left\{\tanh[1.785\log(h/s)+1.75]\right.\\ &\quad \left.+(ks/h)[0.04-0.7k+0.01(1-0.1\varepsilon_r)(0.25+k)]\right\}.\end{aligned} \tag{6.8}$$

Hilberg quotes this expression as being accurate within approximately 1.5% provided that

$$h/s \geq 1. \tag{6.9}$$

In practice this means that either relatively thin substrates ($h \approx s$, $h \approx w$) or wide gaps ($s \gg w$) are prohibited, but this should be anticipated in any event in almost all practical implementations. In practice, when h/s is near to or less than 1 the microstrip mode can readily be generated with the resultant unpredictable coupling of energy between this mode and the desired CPW mode. In this case, the influence on the effective permittivity is not by any means the major problem.

To quite a close approximation, the distributions in the air and the substrate of the electric fields associated with CPW are reasonably independent of geometry and frequency. Therefore, unlike microstrip and many other interconnect structures, the effective permittivity is not particularly sensitive to the geometry of the structure. Equation (6.8) can therefore be used in a wide variety of CPW 'open' structures.

Conductor-backed shielded or asymmetric CPW require further attention and these will be considered to some extent later.

6.2.2 CHARACTERISTIC IMPEDANCE

The characteristic impedance of CPW is, like the effective permittivity, defined in precisely the same way as for any other transmission medium. A CPW line terminated in a purely resistive load, having the same value as the characteristic impedance of the line, will be perfectly matched.

The basic expression, originally due to Wen [178], is

$$Z_0 = \frac{30\pi}{\sqrt{\varepsilon_{\text{eff}}}}\frac{K'(k)}{K(k)}. \tag{6.10}$$

Many researchers and engineers consider that $K'(k)$ and $K(k)$ have to be determined using tables of elliptic integrals. However, in 1969 Hilberg [179] gave two closed expressions for the characteristic impedance, and these are dependent upon the range of k involved. The first expression covers the lower range of k, $0 \leq k \leq 0.707$, and it is:

$$\frac{K(k)}{K'(k)} = \frac{1}{\pi}\ln\left(2\frac{1+\sqrt{k}}{1-\sqrt{k}}\right). \tag{6.11}$$

The second expression covers the higher range of k, $0.707 \geq k \geq 1$, and this expression is:

$$\frac{K'(k)}{K(k)} = \frac{1}{\pi}\ln\left(2\frac{1+\sqrt{k}}{1-\sqrt{k}}\right). \tag{6.12}$$

It is very important to notice the different k' arguments involved in Equations (6.11) and (6.12), respectively. Again, conductor-backed, shielded or asymmetric CPW require further attention, and these will be considered to some extent later.

6.3 FORMULAS FOR ACCURATE CALCULATIONS

The above expressions provide for reasonable accuracy but are mainly aimed at analysis of CPW; i.e. knowing the structure, the characteristic impedance can be found. However, in practice engineers generally know the characteristic impedance required in a design, and must determine the width w and the spacing s of the CPW. This is termed synthesis.

6.3.1 ANALYSIS AND SYNTHESIS APPROACHES

Equation (6.4), expressing $k = w/(w + 2s)$, shows that we can obtain k by manipulating the analysis equations above. For example, by manipulating Equations (6.10) and (6.11) and simplifying the argument for the logarithm as just x, we have:

$$x = \exp\left(\frac{Z_0\sqrt{\varepsilon_{\text{eff}}}}{30\pi^2}\right). \tag{6.13}$$

A suitable initial starting value for ε_{eff} is simply $(\varepsilon_r + 1)/2$. This expression results from the observation that some of the field is in the substrate ($\varepsilon = \varepsilon_r\varepsilon_0$) whilst some is in the air ($\varepsilon = \varepsilon_0$). As an approximation the effective permittivity must therefore be somewhat greater than just $\varepsilon_r/2$. The parameter k is then easily found from x. It is then necessary to initially select a value for the strip width w at some convenient value and then solve for the required gap s using Equation (6.4). A more complete iteration then returns to Equation (6.8) followed by repeated application of the appropriate impedance expressions.

Specifically, for design on 200 μm thick GaAs substrates, Ghione and Naldi [183] provide a useful chart of Z_0 versus k (their Figure 4), which leads to the required width and separation for the CPW line. Ghione and Naldi use simplifying b $(= s + a)$ and $a = 0.5w$.

By curve-fitting we have developed a simple expression that fits Ghione and Naldi's results to within an accuracy of about 4% for the worst case, and this applies at the low impedance extreme where $b = 0.05$ mm and $a = 12$ mm. In practice this would be unacceptable in any event because the strip width is so large. The expression is:

$$Z_0 = \frac{47.12}{\sqrt{\varepsilon_{\text{eff}}}}\frac{[1 + (b^2/2)]}{\sqrt{0.43a/b}} \ . \tag{6.14}$$

This applies over the ranges

$$0.4 \leq a/b \leq 0.6 \text{ and } 0.05 \leq b \leq 0.35 \tag{6.15}$$

although, as described above, the lower extreme of b should be avoided.

Now, it turns out that ε_{eff} varies only slowly with a/b, but its value as a function of b must be known. Equation (6.8) could be used, although again starting values of the target w and s are required.

For guidance, on GaAs substrates, ε_{eff} takes on the following values for typical values of b:

b (mm)	0.35	0.26	0.05
ε_{eff}	4.2	4.6	5.0

Therefore to a first approximation, values of ε_{eff} may be interpolated between the above data, for intermediate values of b.

For synthesis Equation (6.14) may be used. Explicitly, as it stands we have one equation but two (a and b) unknowns — but it turns out that the term $(1 + b^2/2)$ is always relatively small. For the range of b given above, the maximum value $(1+b^2/2)$ is 1.061. Therefore, a starting value of this term can be, for example, 1.05. Then Equation (6.14) can easily be rearranged for a/b, giving

$$\frac{a}{b} = \frac{1 + b^2 + (b^4/4)}{M} \tag{6.16}$$

where

$$M = \frac{Z_0^2 \varepsilon_{\text{eff}}}{5163.5}. \tag{6.17}$$

Rearranging Equation (6.16) leads to:

$$M\frac{a}{b} = 1 + b^2 + (b^4/4) \tag{6.18}$$

which is a quadratic in b^2. Setting $b^2 = y$ for simplicity, Equation (6.18) becomes

$$y^2 + 4y = 4M\frac{a}{b} - 4 \tag{6.19}$$

which is of course a simple quadratic in y, because the RHS is a known quantity at this stage. Solving for y and then taking the square root of the result yields b.

Clearly there is substantially more work to be completed in order to verify and extend these design equations and procedures. In particular, there is a need for computer simulations to provide far more design data. These equations serve to provide a good initial estimate and, crucially, provide design insight, including trends that are not available from computer simulations.

Once more, the consideration of conductor-backed, shielded or asymmetric CPW requires further attention, and these will be considered later.

6.4 LOSS MECHANISMS

In common with all interconnect structures that include metal parts, CPW exhibits two distinct transmission loss mechanisms: dielectric losses and also ohmic or conductor losses. Radiation losses can also become noticeable with CPW, but this is much lower than with microstrip. It is significant to be able to characterize CPW in terms of these losses because usually their minimization is important.

Both dielectric and conductor losses occur continuously along the length of the transmission structure. We therefore quote these losses in terms of loss or attenuation coefficients α_D and α_C, respectively, and the fundamental units are Nepers per metre (Np/m). Conversion to the more familiar dB/mm or dB per wavelength (dB/λ_g) is straightforward. For example, dB per wavelength is obtained by dropping λ_g in the denominator and multiplying by 27.3.

6.4.1 DIELECTRIC LOSS

The dielectric loss is given as a function of the transmission medium filling factor q, the relative permittivity (ε_r) and loss tangent ($\tan\delta$) of the substrate, the effective permittivity and also the wavelength for transmission along the CPW as

$$\alpha_D = \frac{q\varepsilon_r \tan\delta}{\varepsilon_{\text{eff}}\lambda_g} \quad (\text{Np/m}). \tag{6.20}$$

In practice, the values of the filling factor and the permittivities are minor influences upon this dielectric loss, which always remains dominated by the loss tangent of the substrate material and this tends to be around 0.001 or lower for most substrates, including semiconductors. The effect of the dielectric loss is therefore relatively low in most instances.

Where it is suspected that an unusually dielectrically lossy substrate is being adopted then Equation (6.20) is useful to estimate the extent of these losses. A notable exception is standard silicon substrates for which the resistivity is only about 1000 Ω.cm. A single loss tangent is insufficient to describe the loss of standard silicon because of the appreciable conduction in the substrate. In this case, it is necessary to use an appropriate value of $\tan\delta$ defined at some particular frequency, such as 30 GHz [182]. This problem is avoided by using high resistivity silicon (resistivity of 4000 Ω.cm or more), and in fact this type of substrate was developed for high frequency interconnects.

6.4.2 CONDUCTOR LOSS

The metal conductor resistivity, skin effect and the influence of surface roughness are important parameters influencing loss. Therefore, conductor loss usually dominates the transmission losses in CPW. Rayit [185] and other researchers have reported the unloaded Q-factor Q_u as the means for estimating the effects of the conductor loss:

$$Q_u = \left(\frac{w}{2}\right) Z_0\sqrt{f}. \tag{6.21}$$

The expression appears to work well for CPW up to 2 GHz, and at this frequency the attenuation is approximately identical to that applying to microstrip also having line width s. However, 2 GHz is a relatively low frequency for CPW applications, and it is therefore necessary to provide guidance that allows for design at much higher frequencies.

By solving the Laplace equation for the potential distribution within the CPW cross-section and also by performing measurements, Koshiji *et al.* [187] identified the following features applying to losses:

1. The dielectric attenuation coefficient α_D is principally dependent upon substrate thickness and the geometric ratio w/s has negligible influence.
2. The conductor attenuation coefficient α_C is minimized at a specific (optimum) value of w/s.
3. The conductor attenuation coefficient α_C is dependent upon the relative permittivity ε_r of the substrate, α_C increases with increasing ε_r.

4. As conductor thickness t is varied so also α_C varies, and there is an optimum value of t that minimizes α_C. Generally, however, α_C is reduced by reducing the DC resistance of the interconnect.

Full electromagnetic analyses have been applied to predict the loss associated with CPW and this structure has been compared with microstrip in this regard. The work of Jackson [188], for example, compares conductor loss, dispersion and radiation into parasitic modes for microstrip and coplanar waveguide. In this paper, focused on 0.1 mm thick semiconductor substrates at 60 GHz, it is shown that the dimensions of CPW may be selected so that this structure can provide substantially lower losses and dispersion than microstrip.

Dispersion is principally related to the frequency dependence of the partitioning of the field in the substrate and the covering medium (usually air). For small-geometry coplanar structures the dispersion is smaller relative to the situation with microstrip. Thus, low dispersion relative to microstrip is a fundamental property of small-geometry coplanar structures.

Zhang *et al.* [189] use the Finite-Difference Transmission Line Matrix (FD-TLM) method to compare 50 Ω microstrip and CPW losses. An interesting outcome is that for a 50 Ω coplanar waveguide, differing combinations of s and h yield different losses and these may become greater or less than those applicable to the microstrip line. Frequencies up to 100 GHz were considered.

It is therefore clear that one cannot generalize concerning losses in CPW compared with microstrip — the precise situation depends upon the geometry. For 50 Ω lines it is recommended that the dimensions should be made as large as possible in order to maintain relatively small losses.

Analyses based upon a frequency domain full electromagnetic wave method [190, 191] have shown that on standard 0.1 mm GaAs substrates, CPW yields significantly lower attenuation than microstrip at higher values of characteristic impedance Z_0. On such substrates designers are recommended to choose relatively high Z_0 values, as far as possible. This is fortuitous in many monolithic integrated circuit designs because high Z_0 interconnects have relatively small dimensions, and so are far more economical in their use of substrate area and, for digital circuits, lower drive current requirements. A conceptual understanding of the loss behaviour is that for the same transmitted signal power P, a high impedance (Z_0) interconnect has a lower current $I\left(\propto \sqrt{P/Z_0}\right)$ and hence there are lower I^2R losses.

An extensive study of attenuation in CPW on silicon (Si), gallium arsenide (GaAs) and indium phosphide (InP) substrates has been conducted by Ponchak *et al.* [81]. This study provides detailed design data and expressions applicable to CPW on the substrates identified in Table 3.7, and frequencies in the range from 1 GHz to 40 GHz apply. This table provides a useful reference for the permittivities, loss tangents and some typical thicknesses of these four important semiconductors.

As a result of the much lower losses, the high resistivity silicon is preferred to standard silicon for RF, microwave and millimetre-wave design.

In the cited study [81], the transmission lines were fabricated on double-sided polished semiconductor wafers with top side metallization comprising a 0.02 μm titanium layer followed by gold layers with three specific thicknesses: 0.49, 1.5 and 2.22 μm, respectively. The structures do not have any conductor-backing, i.e. there is no underside grounding. Other features of this study include:

- Aspect ratio $k = w/(w + 2s)$ for the range $0.2 \leq k \leq 0.7$.
- Z_0 range: $35 \leq Z_0 \leq 65\ \Omega$.
- Centre strip width range $10 \leq w \leq 80\ \mu$m.

Side ground plane widths are maintained equal to $4w$ in all cases, ensuring that all results are independent of this width, with no spurious resonances being excited. Also, the substrate thickness h is maintained such that $h \geq 3(w + 2s)$ in order to ensure that the results are independent of this thickness.

Ponchak *et al.* [81] report a substantial array of information on the average RMS error between other reported quantities from five separate sources, their measured data and curve-fitted equations. Their new closed form expressions are:

$$\alpha = af^b \quad \text{(dB/cm)} \tag{6.22}$$

$$a = \sqrt{\frac{\varepsilon_r - 1}{2}} \left[\frac{(45.152)}{(sw)^{0.41} e^{2.127\sqrt{t}}} \right] \tag{6.23}$$

$$b = 0.183(t + 0.464) - 0.095k_t^{2.484}(t - 2.595) \tag{6.24}$$

where w, s and t are in μm, f is in GHz and k_t is given by the following expressions:

$$k_t = \frac{w + \Delta t}{w + 2s - \Delta t} \tag{6.25}$$

$$\Delta t = \frac{1.25t}{\pi} \left[1 + \ln \frac{4\pi w}{t} \right] . \tag{6.26}$$

As the development of the above expressions is based upon measurements, all three loss mechanisms are accounted for: dielectric, conductor and radiation.

The application of Equations (6.22) through (6.24) is quoted by Ponchak *et al.* [81] as being valid for the following parameter ranges:

$$0.2 \leq k \leq 0.7 \quad 10 \leq w \leq 80\ \mu\text{m} \quad 0.5 \leq t \leq 3.0\ \mu\text{m} \quad 1 \leq f \leq 40\ \text{GHz}.$$

The applicability of these expressions to Si substrates requires a significant condition, namely that the Si must be of high resistivity — specifically $\geq 2500\ \Omega$.cm. Otherwise there would be a significant substrate resistivity effect influencing loss. The average error of these expressions is reckoned as 4.7% where the substrate is GaAs with metal thickness 1.58 μm, but may be somewhat higher for Si and InP. The error increases to 6% when all three metal thicknesses are accounted for.

For design, Equations (6.22) through (6.26) may be used judiciously for design guidance. Where a maximum attenuation must be specified, the metal thickness is known and k_t is relatively small to the extent that Equation (6.24) may be reduced to its first term only i.e. $b \approx 0.183(t - 0.464)$. Then Equations (6.23) and (6.24) may be combined in order to determine the $w.s$ product, guaranteeing that this maximum attenuation is not exceeded.

The design approaches given earlier, through synthesis leading to a/b and hence strip width w and separation s, can then be used to finalize the CPW design process.

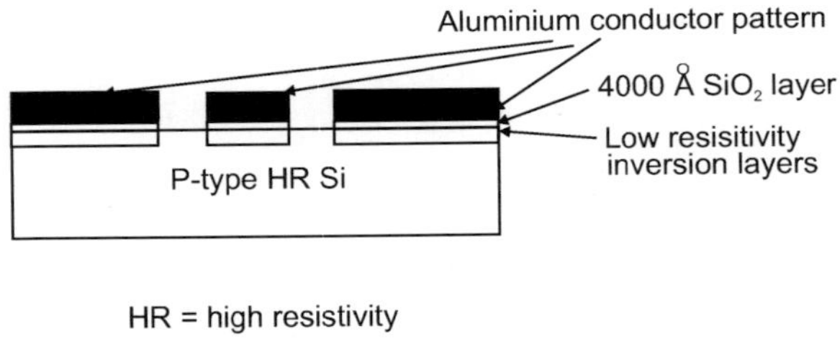

Figure 6.6 CPW with an intervening SiO_2 layer.

6.4.3 RADIATION LOSS

The subject of radiation loss is only covered briefly at this stage. In the latter discussion under 'conductor losses' above it was pointed out that the work of Ponchak *et al.* [81] includes the effects of radiation losses in CPW lines. More details concerning radiation losses are provided within the subsequent sections on discontinuities — notably the pseudo short-circuit.

6.4.4 CPW WITH INTERVENING SiO_2 LAYER

There are disadvantages involved with directly depositing metal layers onto the silicon substrate surface due to surface effects on the silicon and often relatively high losses. This surface effect is attributed to trapped charges in the silicon near its surface. Increasingly, therefore, the metal conductor pattern is deposited on to an intervening layer of silicon dioxide (SiO_2) which has a relative permittivity of around 3.8, similar to that of fused quartz.

Low resistivity inversion layers are frequently required between the SiO_2 layer and the silicon and a cross-section of the entire structure is shown in Figure 6.6. Although aluminium metal is shown, copper is increasingly being used for its lower resistivity which is particularly important for fine interconnects with cross-sectional dimensions of a micron or less. However, the losses inherent in the SiO_2 layer (if isotropic) again increase and become almost unmanageably high — as high as 1.8 dB/mm at high frequencies. By patterning this intervening layer Wu *et al.* [192] showed that these losses can be reduced to nearer 0.3 dB/mm at 30 GHz.

6.5 DISPERSION

6.5.1 FUNDAMENTAL AND THEORETICAL CONSIDERATIONS

In common with all electrical signal transmission structures coplanar waveguide is to some extent dispersive, i.e. the transmission phase coefficient β varies non-linearly with frequency. This effect, however, can be quite small for CPW relative to microstrip.

Using the familiar effective permittivity concept, expressed by Equation (6.8) for low frequencies, we can extend this to account for the frequency-dependent quantity $\varepsilon_{\text{eff}}(f)$ that is now fundamentally defined by

$$\varepsilon_{\text{eff}}(f) = (c/v_p)^2 \tag{6.27}$$

where c is the free-space velocity and v_p is the velocity of propagation of the dominant

CPW mode.

Many closed-form expressions have been developed for microstrip dispersion over the years and Rayit [185] chooses Yamashita's formula, which was first reported in 1979 [131]. This is repeated here in a form suited to the case of CPW dispersion

$$\varepsilon_{\text{eff}}(f) = \sqrt{\varepsilon_r} + \frac{\left(\sqrt{\varepsilon_r} - \sqrt{\varepsilon_{\text{eff}}}\right)}{1 + p\left(f/f_{te}\right)^{-r}}. \tag{6.28}$$

We have found that this expression, used in conjunction with the following analytical formulas, works well for alumina substrates with thicknesses around 0.635 mm. Also, its use is restricted to relatively large geometry CPW — in the region of hundreds of microns in lateral dimension.

In this expression ε_{eff} is the static-TEM effective permittivity, f is the frequency as independent variable as usual, and f_{te} is the cut-off frequency for the lowest order TE mode, given by

$$f_{te} = \frac{c}{4h\sqrt{(\varepsilon_r - 1)}}. \tag{6.29}$$

Parameters p and r depend upon the configuration of the CPW line, and Rayit [185] quantifies these for various aspect ratios. It was found, for design on alumina substrates at least, that $r \approx 1.8$ (compare Yamashita's 1.5), and according to Rayit the quantity p approximately follows the linear equation

$$\log p \approx u \log\left(w/s\right) + v \tag{6.30}$$

in which u and v are functions of the substrate thickness h, as follows

$$u = 0.54 - 0.64q + 0.015q^2 \tag{6.31}$$

$$v = 0.43 - 0.86q + 0.54q^2 \tag{6.32}$$

where

$$q = \log(w/h). \tag{6.33}$$

By curve fitting to Rayit's experimental and theoretical data, it has been found that Equation (6.30) considerably overestimates the dispersion, and in fact the logarithmic relationship does not hold. The relationships always depend upon the geometry, i.e. $w/(w + 2s)$ because the dispersion is inherently geometry dependent, affected by the proportion of the field above and below the metallization layer.

We have found that, for the characteristic impedance range $39 \leq Z_0 \leq 55\ \Omega$ the value for $p = 0.257$ yields an excellent model for dispersion calculations. This value holds up to $Z_0 = 55\ \Omega$, and above this value the following expression is effective:

$$p = 1 - \left(\frac{w}{w + 2s}\right)^{1.5}. \tag{6.34}$$

Examples of calculations using the above expressions are provided in Figures 6.7 through 6.9. The fit is very good in all cases, and may possibly hold above 70 Ω, although this has not been validated.

In general, this approach cannot be expected to work with the smaller geometry CPW applicable to chip-scale semiconductor substrates, because the data fit was

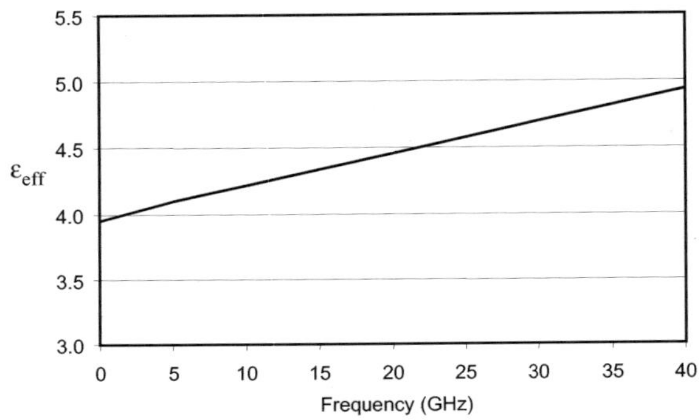

Figure 6.7 Dispersion in a 39 Ω CPW on alumina (ε_r=9.5, h= 0.635 mm).

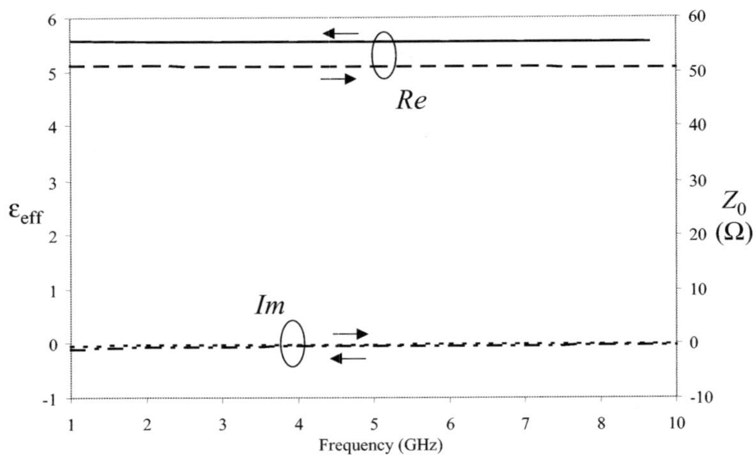

Figure 6.8 Dispersion in a 50 Ω CPW on alumina (ε_r=9.5, h= 0.635 mm.)

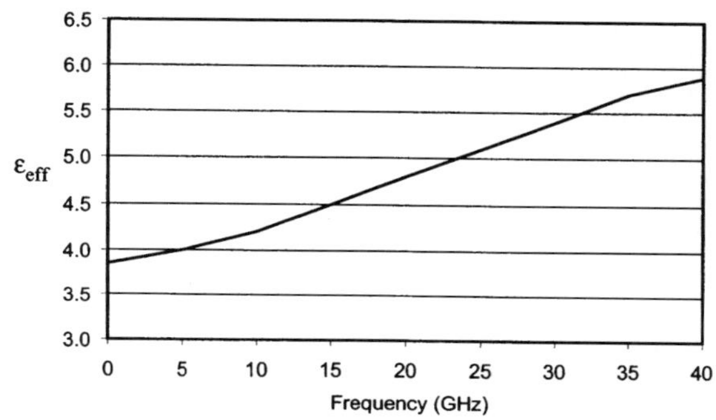

Large Geometry

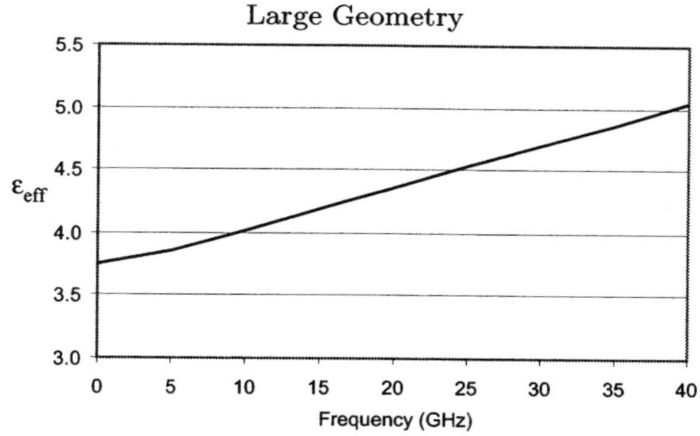

Small Geometry

Figure 6.9 Dispersion in a 70 Ω CPW on alumina (ε_r=9.5, h= 0.635 mm).

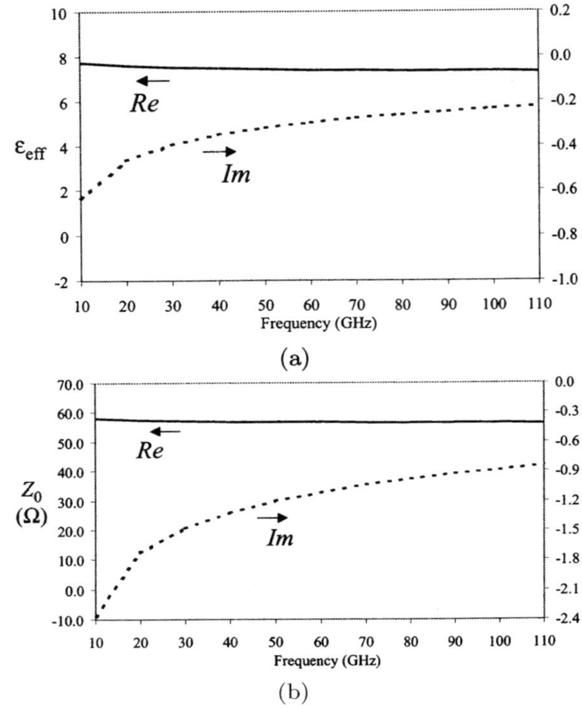

Figure 6.10 Dispersion of CPW on GaAs with no backing: effective permittivity (a) and characteristic impedance (b).

restricted to relatively large alumina substrates. However, for chip-scale substrates it has been shown that dispersion is usually negligible, as indicated by the data generated using commercial electromagnetic simulation software and summarized in the following section.

6.5.2 RESULTS FROM TEST RUNS USING ELECTROMAGNETIC SIMULATION

In this section we present the results from running a variety of representative CPW structures using the SONNET™ simulation program suite. In all instances the structure is metal backed, the thickness of this metal backing is always identical to the top surface metallization and both the substrate and the metal are lossy, unless otherwise stated. Since a major feature concerns the frequency dependence of the effective permittivity $\varepsilon_{\mathrm{eff}}(f)$ and of the characteristic impedance Z_0, most of the following figures show this behaviour for both real and imaginary parts of each parameter, over the 10 to 110 GHz frequency range in most instances.

The first sets of examples are for CPW on GaAs substrates, and in Figure 6.10 the frequency dependent results are shown for such a structure with a gold metal thickness of 2 μm, a CPW width, $w = w_1$ of 20 μm and a separation, s, of 15 μm. The substrate thickness is 500 microns, the dielectric loss tangent is 5×10^{-4} and the

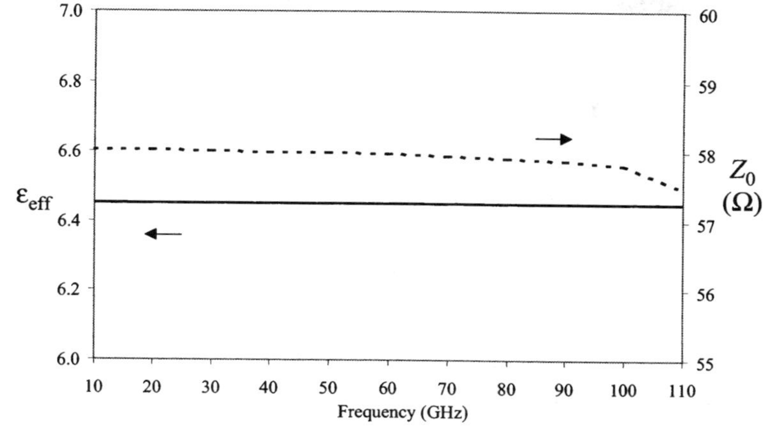

Figure 6.11 Dispersion of CPW on GaAs ignoring dielectric and metallic losses.

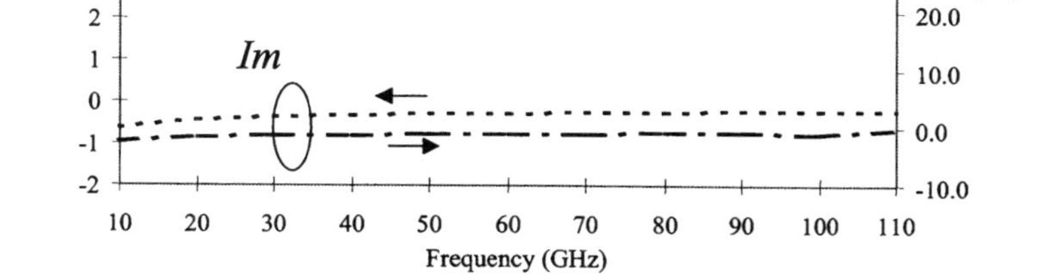

Figure 6.12 CPW on GaAs with lossy gold metallization and ignoring dielectric losses, showing real ε_{eff} and Z_0.

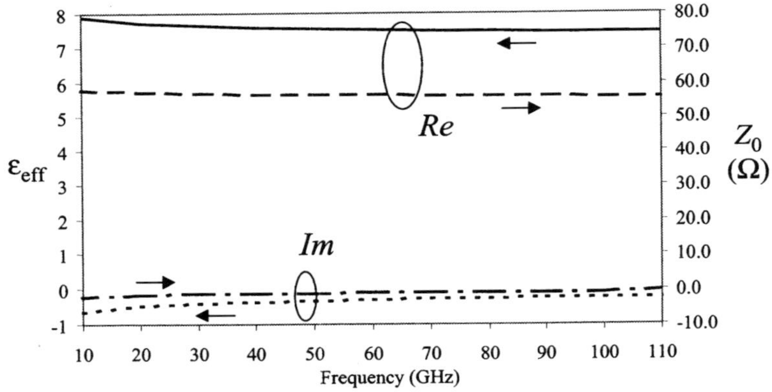

Figure 6.13 CPW on lossy GaAs with 3 μm of covering polyimide.

metal has $R_{dc} = 0.012\ \Omega$/square ($R_{dc} = 1/(\sigma t)$, where σ is the metal conductivity and t is its thickness). The results indicate that both the real and imaginary parts of $\varepsilon_{\text{eff}}(f)$ fall more rapidly with frequency at first and then slowly later. Z_0 also falls slowly with frequency across the 10 to 110 GHz range. These decreases are mainly due to the fact that the losses increase with frequency.

For the same basic structure, but with all losses neglected, Figure 6.12 shows that $\varepsilon_{\text{eff}}(f)$ is practically constant across the entire frequency range. However Z_0, after falling slowly over most of the range, decreases relatively rapidly above 100 GHz. As losses are considered negligible here, the imaginary parts of both Z_0 and $\varepsilon_{\text{eff}}(f)$ are also negligible.

Figure 6.13 has been generated using data for CPW on lossy GaAs, again with the same basic structure as above, but with a 3 μm covering of polyimide commonly used as a passivation layer to protect the semiconductor circuitry. Since this takes up some of the air gap between and just above the lines, as expected, the main effects are to slightly increase $\varepsilon_{\text{eff}}(f)$ and to slightly decrease Z_0 at all frequencies. This decrease in Z_0 can be traced to the increased capacitance per unit length caused by the polyimide layer. The effects of steadily decreasing the substrate thickness h are indicated in Figure 6.14, where are data apply at a constant frequency of 30 GHz. For this structure, with $h = 15\ \mu$m (not shown in Figure 6.9), $Z_0 = 50\ \Omega$ at 10 GHz and its lowest value is 48.5 Ω at 70 GHz.

At relatively small values of h, around and just above the magnitude of the centre conductor width, the real part of $\varepsilon_{\text{eff}}(f)$ decreases steadily whilst Z_0 increases at a similar rate. This is due to the fact that for small values of substrate thickness h, the capacitance per unit length is relatively large (the fields are more concentrated into the substrate). In contrast the imaginary parts of both $\varepsilon_{\text{eff}}(f)$ and Z_0 remain almost constant across the complete range of h. For $h \geq 45\ \mu$m all parameters are practically constant.

Another important parameter, the design sensitivity of which must be studied, is the effect of the widths of the finite ground planes, w_1. This is shown for the basic CPW on GaAs structure in Figure 6.15. Again, all the data apply at a constant frequency of 30 GHz. Although the real part of $\varepsilon_{\text{eff}}(f)$ is almost invariant with w_1 the characteristic impedance varies considerably — mainly decreasing with w_1.

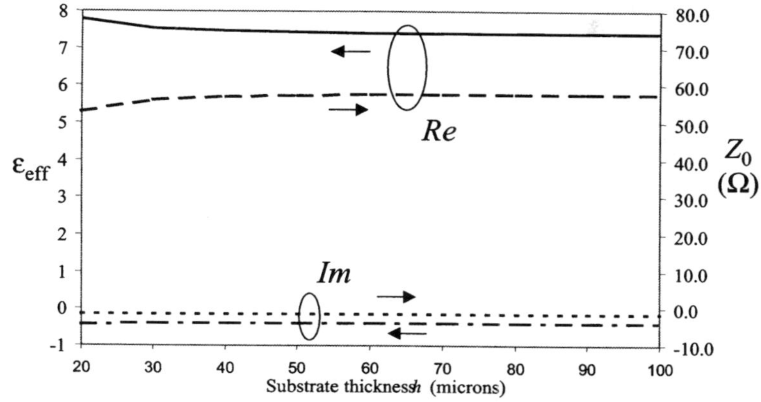

Figure 6.14 CPW on GaAs: variation of parameters with h.

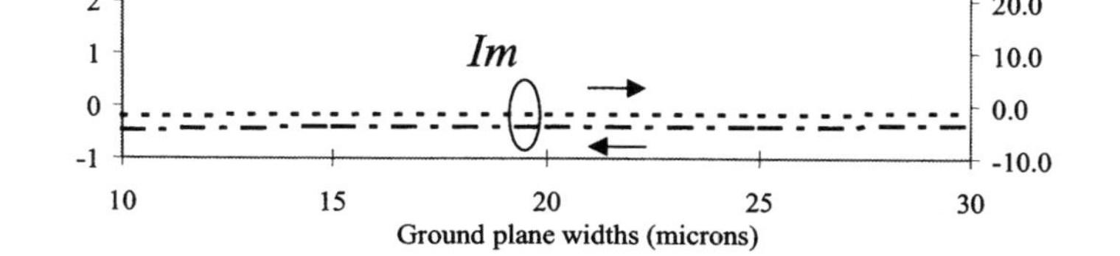

Figure 6.15 CPW on GaAs: variation of parameters with coplanar ground plane width w_1.

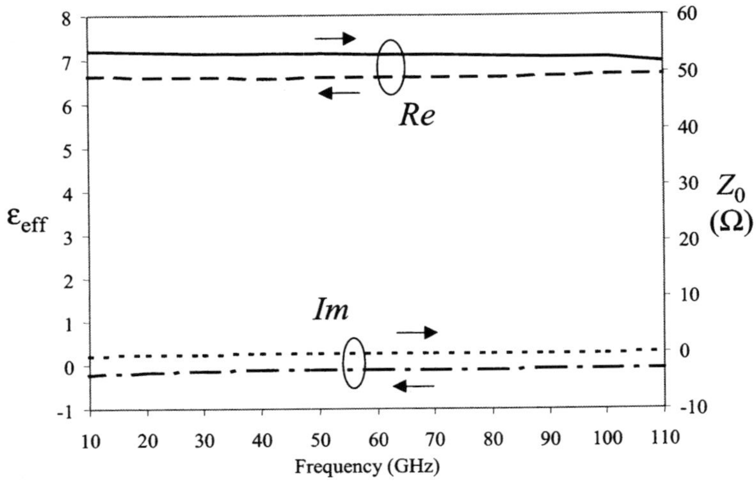

Figure 6.16 CPW on standard silicon with loss included.

Further runs of SONNET™, for values of w_1 above 100 microns, have shown that the parameters settle down to practically constant values (w_1 is then five to ten times the centre conductor width). A design recommendation is then that

w_1 should be at least 5 times w to achieve all the benefits of CPW.

However, compromises are sometimes necessary for on-chip implementations. Referring to the current distribution shown in Figure 6.3, when w_1 is less than that recommended, longitudinal currents will exist on the outside edges of the CPW. The resulting electromagnetic field mean that energy is stored away from the gap, leading to coupling to neighbouring structures if they exist. The effective permittivity ε_{eff} and Z_0 increase as the additional energy is primarily stored in the electric field.

Figure 6.16 shows the results for CPW on a standard silicon substrate with lossy aluminium metallization having $R_{dc} = 0.0185\ \Omega$/square. The metal thickness is 1.46 μm, the CPW centre strip width is 70 μm, and the separation is 40 μm. The thickness of the silicon substrate is 650 μm and the dielectric loss tangent 4×10^{-3}. The results indicate that the real part of $\varepsilon_{\text{eff}}(f)$ is very nearly constant with frequency. However, because of the losses which increase with frequency, the real part of Z_0 falls somewhat at higher frequencies. In both cases, the imaginary parts remain relatively small and change little with frequency.

Apart from the implementation of CPW in MMICs, its use with non-semiconductor substrates is also important and some results have been obtained using SONNET™ for CPW on alumina substrates. The frequency-dependent results for a nominally 50 Ω CPW interconnect are shown in Figure 6.17. The lossy gold metal thickness is 5 microns, centre conductor width and ground plane widths are equal ($w_1 = w$), the separation to the ground planes is 200 μm and the substrate thickness is 635 μm. The relative permittivity of the alumina is 9.8, the dielectric loss tangent is 2×10^{-4} and the metal has $R_{dc} = 0.0048\ \Omega$/square.

Figure 6.17 shows that, for this type of relatively large CPW structure, there is significant dispersion — indicated by the continuous and noticeable increase in the real part of $\varepsilon_{\text{eff}}(f)$ as frequency increases. In contrast, the real part of Z_0 is substantially

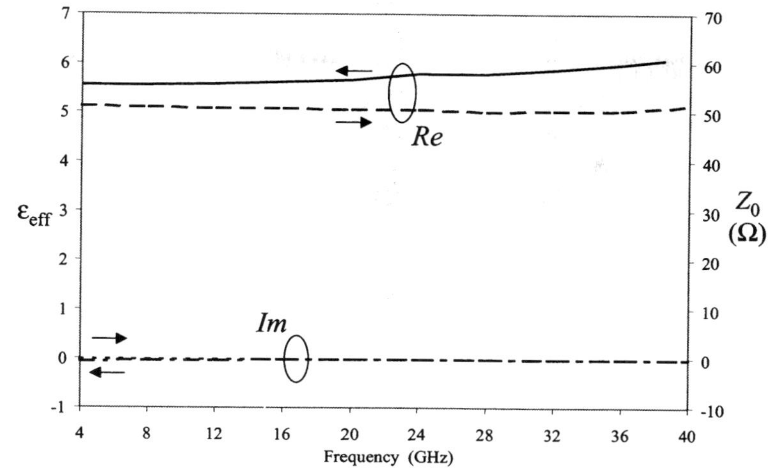

Figure 6.17 50 Ω CPW on alumina: frequency dependent behaviour.

constant with frequency. In both cases, $\varepsilon_{\text{eff}}(f)$ and Z_0, imaginary parts are relatively small. The significant dispersion with this type of structure was also confirmed by Rayit [185], whose results have been referred to earlier here.

From Figure 6.18 it can be seen that the most marked effect is the steady increase in the real part of Z_0. This is expected because the separation is one parameter that influences Z_0 significantly, and provides design freedom over this parameter to some extent. At small separations (s), well below 100 μm, the real part of Z_0 decreases more strongly as the separation decreases. In this instance a CPW structure that is nominally 50 Ω with a separation of 200 μm, becomes only 34.7 Ω when this separation is reduced to 50 μm. Comments on the behaviour of the real part of $\varepsilon_{\text{eff}}(f)$ and also on the imaginary parts of both $\varepsilon_{\text{eff}}(f)$ and Z_0 are provided after the following discussion of the large separation range.

For the same CPW structure on alumina, Figure 6.19 indicates the variations of all four parameters, but now with a much wider range of variation of separation — from 100 to 450 μm. The characteristic impedance Z_0 continues increasing with s, as anticipated and for the same reasons described above. Throughout all values of s, from 50 μm to 450 μm, the real part of $\varepsilon_{\text{eff}}(f)$ slowly increases due to the fields becoming increasingly concentrated into the substrate as the ground planes are steadily taken away from the vicinity of the centre strip. Also, for all values of s, the imaginary parts of both $\varepsilon_{\text{eff}}(f)$ and Z_0 remain relatively small.

For this structure the SONNET^{TM} simulation revealed an interesting feature when the value of s became close to the substrate thickness h. The returned message was '$\varepsilon_{\text{eff}}(f)$ and Z_0 are undefined.' This indicates that multimoding could occur and in this case simulation above $s = 475$ μm was beyond the range of validity. Values from 50 to 475 μm represent the range of validity in this instance.

6.5.3 EXPERIMENTAL RESULTS

For the case of alumina substrates ($\varepsilon_r = 9.5$), commonly used for hybrid MICs, Rayit [185] reports the results of measurements on CPW dispersion. Resonator methods and

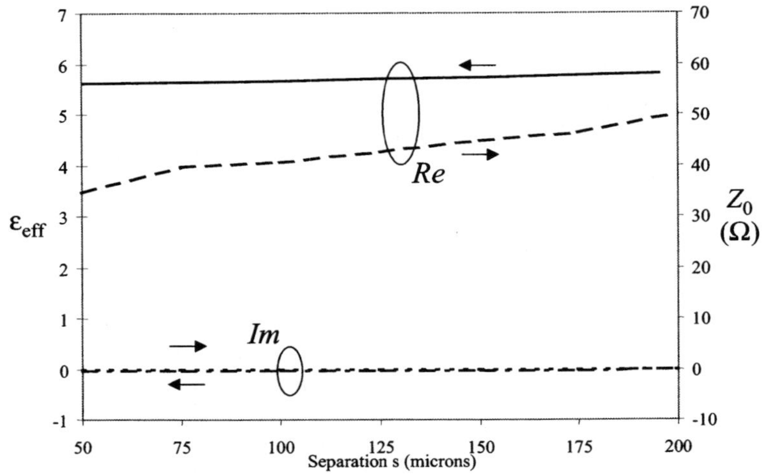

Figure 6.18 50 Ω CPW on alumina: parameter variations with separation (at 30 GHz, small s).

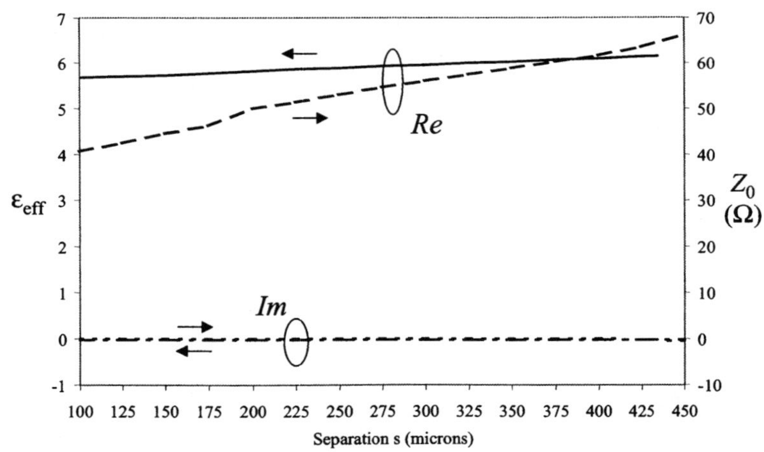

Figure 6.19 50 Ω CPW on alumina: parameter variations with separation (at 30 GHz, large s.)

a scalar network analyser were used to conduct the measurements, employing a very similar methodology to that used for microstrip measurements. The scalar analyser swept the CPW resonators over the band from 2 GHz to 38 or 39 GHz.

Two formats of resonator were employed: the ring resonator and straight or linear resonators. Again, this follows the philosophy used in the case of microstrip, described in Chapter 9. The results show clearly that the measured results track the calculations using Equations (6.28) to (6.33) closely. Characteristic impedances ranged from 39 to 70 Ω. There is a significant difference in the amounts of dispersion resulting from changes in the magnitudes of the relative physical dimensions to realize identical specific characteristic impedances and this leads to an important design axiom:

Smaller physical dimensions always result in lower dispersion.

This comes about because, for a given frequency range, the mode-coupling efficiency is reduced as the physical dimensions of the strip are reduced (cf: the discussion on microstrip dispersion in Section 5.2 on page 113). The principal mode coupling is the coupling of the CPW mode to the microstrip mode as well as to surface waves. The better the fields are localized, by having small w and s dimensions here, the less likely will the modes be excited.

It is unlikely that any analytic expressions could accurately reflect this feature as the coupling of modes can be due to small discontinuities. It is suggested that more research is required to develop, if possible, suitable closed formulas that will predict the dispersion as a function of the physical dimension regime. An electromagnetic simulator approach similar to that used in Section 6.5.2 on page 178 is necessary to predict the CPW behaviour over a wide range of different practical configurations.

Although the above comments were developed for CPW on alumina substrates, the same important generalities apply to any substrate — including all semiconductors. CPW lines on MMICs, then, generally exhibit very low dispersion as a direct consequence of the small dimensions, usually in the tens of microns range. However, more extensive measurements on CPW lines on various important thick substrates, including GaAs, InP and Si, are required over extensive frequency ranges (at least from 1 to 110 GHz) in order to validate the applicability and accuracy of the modelling expressions.

Papapolymerou *et al.* [193] have established that, for Finite-Ground-Plane CPW (FGCPW) on GaAs or quartz, dispersion is essentially negligible over the frequency range from 2 GHz to 120 GHz. Effective permittivity fluctuated between 7.0 and 7.2, being at the higher values at each end of the frequency range. The maximum physical dimensions used were 50 μm by 64.5 μm. Similar behaviour can be anticipated for other semiconductor substrates such as Si or InP — or non-semiconductor substrates such as aluminium nitride (AlN) or SiO_2 provided always that the physical dimensions (geometry) are small — in the tens of microns range.

6.6 DISCONTINUITIES

In common with all types of interconnect structures, even the simplest circuits implementing coplanar waveguide involve discontinuities of various forms. Unlike microstrip, however, static analyses leading to the determination of equivalent circuit elements to account for discontinuity effects apply up to relatively high frequencies.

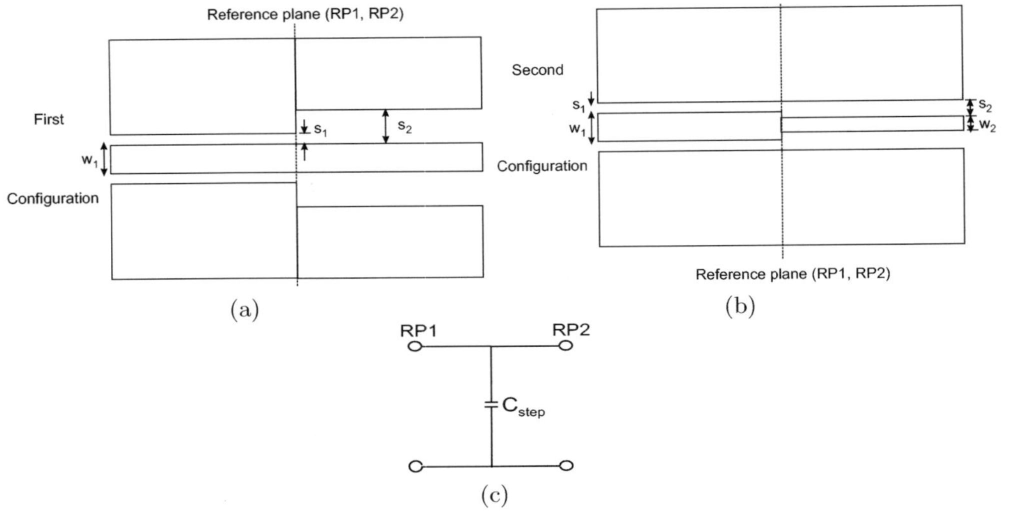

Figure 6.20 Step changes in (a) centre conductor width; (b) separation; and (c) equivalent circuit.

This applies to all CPW structures, but again the extremely small dimensions of MMIC implementations render the discontinuity effects relatively small and even less frequency-dependent.

In this section we deal with several significant discontinuities: step changes in width (i.e. impedance steps), the CPW open-circuit (or pseudo-open-circuit), symmetric series gaps, short-circuit, right angled bend — and T-junctions. Whilst the ultimate aim is to establish suitable equivalent circuits and to evaluate C and L parameter values, an important intermediate aim is also to determine S-parameters, and to link these to circuit parameters (e.g. admittance or impedance parameters) in most instances.

6.6.1 STEP CHANGES IN WIDTH AND SEPARATION

This discontinuity corresponds to a step change in characteristic impedance — but with CPW there exist two distinct structures that apply. These are indicated in Figures 6.20(a) and 6.20(b) where it can be seen that in one instance the step physically occurs in the adjacent ground planes whereas in the other situation it is the centre strip that experiences the physical width change. In either case, regardless of the step realization, a capacitance represents the electrical effect and this applies up to high frequencies. This capacitance is shown as C_{step} in Figure 6.20(c), located at the single reference plane (RP1 = RP2).

The simulated longitudinal and transverse current magnitude distributions of the two types of step discontinuities are shown in Figures 6.21 and 6.22.

Experimental data applying to CPW steps have been reported by several researchers, for example Simons and Ponchak [194] and others have reported theoretical analyses [195–198]. Rayit [185] argues that three main apparent conclusions can be drawn from a study of all these papers:

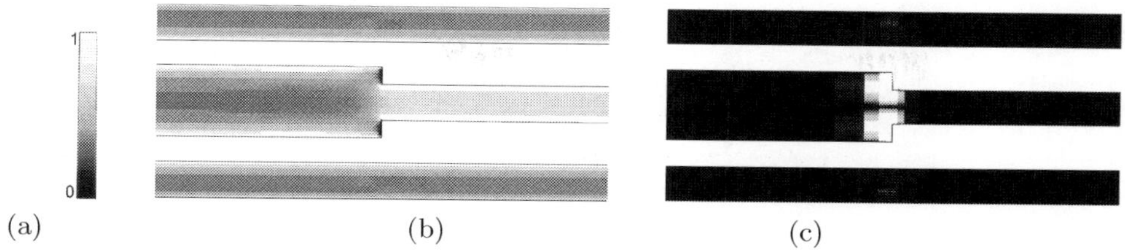

Figure 6.21 Normalized current distribution on the conductors of a CPW impedance step realized using a step of the width of the internal conductor: (a) normalized scale; (b) magnitude of the longitudinal current; and (c) magnitude of the transverse current. (Frequency is 30 GHz).

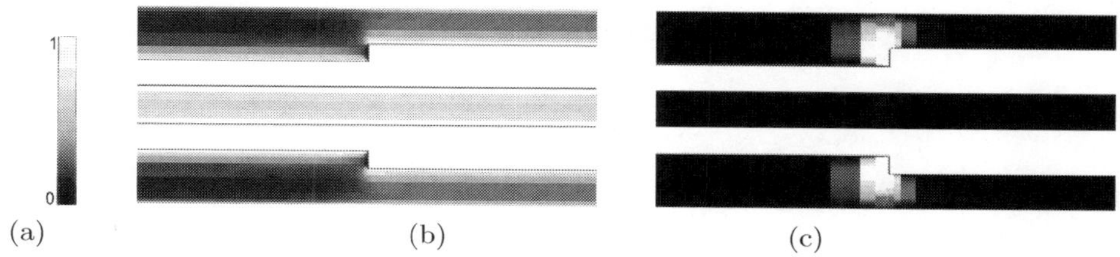

Figure 6.22 Normalized current distribution on the conductors of a CPW impedance step realized using a step of the conductor separation: (a) normalized scale; (b) magnitude of the longitudinal current; and (c) magnitude of the transverse current. (Frequency is 30 GHz.)

- C_{step} CPW is larger than the equivalent C_{step} microstrip.
- C_{step} CPW is of sufficient magnitude for fairly straightforward measurement extraction following S-parameter de-embedding.
- A substantial variation of C_{step} exists, considering the geometries available with typical CPW lines.

Ghioni and Naldi [183] provide the basic conformal mapping technique required to determine the capacitance per unit length of each adjacent coplanar line

$$c(k) = 2\varepsilon_0\varepsilon_{\text{eff}}\left[\frac{K}{K'}(k_3) + \frac{K}{K'}(k)\right] \tag{6.35}$$

where as usual $k = w/(w+2s)$ and

$$k_3 = \tanh\left(\frac{\pi k y}{4h}\right)\Big/\tanh\left(\frac{\pi y}{4h}\right) \tag{6.36}$$

with the effective permittivity being given by

$$\varepsilon_{\text{eff}} = 1 + (\varepsilon_r - 1)\left[\frac{K}{K'}(k_3)\right]\Big/\left[\frac{K}{K'}(k_3) + \frac{K}{K'}(k)\right] \tag{6.37}$$

and $K(k)/K'(k)$ is the elliptic integral ratio. The step width change is therefore equivalent to a step change in the height of a parallel plate waveguide and for this

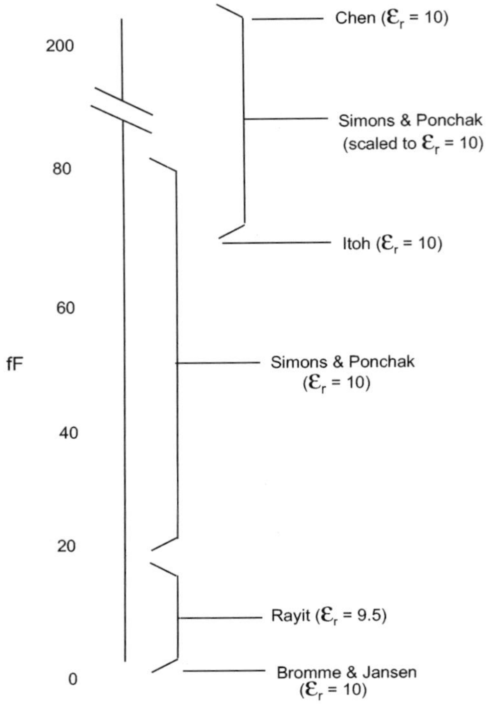

Figure 6.23 Estimates of the capacitance associated with step changes in width.

structure the capacitance per unit width is well known, using this conformal mapping approach

$$C_{\text{step}}(\alpha) = \frac{\varepsilon_0}{\pi}\left[\left(\frac{\alpha^2+1}{\alpha}\right)\ln\left(\frac{1+\alpha}{1-\alpha}\right) - 2\ln\left(\frac{4\alpha}{1-\alpha^2}\right)\right] \tag{6.38}$$

where $\alpha = w_2/w_1$ for $\alpha < 1$. It must, however, be recognized that the width of the equivalent parallel plate waveguide changes on each side of the step, and therefore upper and lower bounds on C_{step} are obtained by alternately setting the width values to w_1 and w_2.

Bracketed ranges of values applying to these upper and lower capacitance bounds are indicated in Figure 6.23, where the results from six sources are provided (all for alumina substrates or similar). A vast range of results are apparent here: from almost zero up to over 200 fF. Whilst, obviously, the value of C_{step} will depend linearly upon the substrate relative to the permittivity — this effect is in any case scaled in Figure 6.23, and yet the range remains extremely large. Rayit's work was the most thorough and up-to-date amongst these data and the range in this instance is from almost zero up to about 10 fF. This is no more than one fifth that of most previously reported data.

Current flow disturbances, critically along the outer edges of the conductor tracks, cause series inductance to be required to fully characterize these step discontinuities. Inductance values can be ascertained analytically, but we concentrate on the more significant capacitance here. Rayit [185] simulated this structure (using the SONNET™ simulator and obtained extensive data that were also supported by

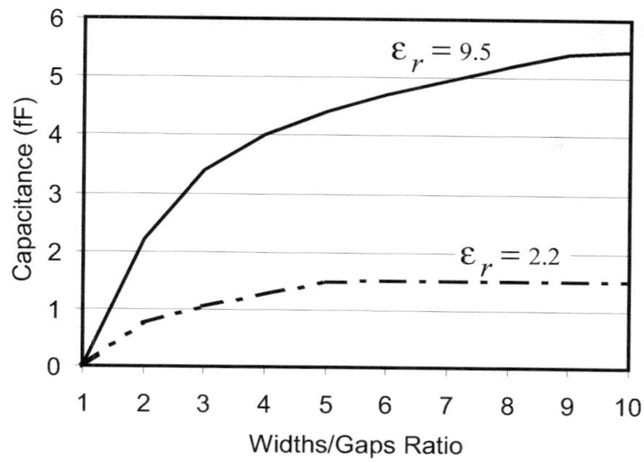

Figure 6.24 Step capacitance versus width/gap ratio for two substrates.

measurements. Here k_1 is the ratio applying to the CPW of width w_1 and k_2 is the ratio applicable to the CPW having width w_2. The overall data are shown in Figure 6.24 from which it should be clear that, as expected, values increase non-linearly with k_2/k_1 but approximately linearly with ε_r (as parameter).

Once more, although these data were obtained for alumina ($\varepsilon_r = 9.5$) the results applying to practically any other substrate can be deduced quite accurately by appropriately scaling according to the relative permittivity of the substrate (i.e. 11.9 for Si, 12.85 for GaAs, etc.). For example, a step ratio of 6 on alumina has a capacitance of 4.7 fF. The capacitance applying to the same line structure (including thickness), but on a GaAs substrate is then $C_{\text{step}} = 4.7(12.85/9.5) = 6.3$ fF. The result has been rounded down to allow for the fact that more of the electric field is concentrated in the substrate as the permittivity rises, and therefore the sensitivity of C_{step} to geometric changes decreases. This approach can be used for most other substrates.

6.6.2 OPEN-CIRCUIT

Quasi-open-circuits occur in many circuit elements and their equivalent circuits must be known. The basic structure is shown in Figure 6.25, where it can be seen that, in contrast to the microstrip open-circuit, the CPW version is complicated by the fact that the adjacent ground plane must be separated from the 'open' end by a distance g.

The problem is twofold: first to determine the equivalent capacitance C_{oc}; and then to represent this by an equivalent extension of the CPW line, with an extra length Δl. The process of determining these parameters starts with the ABCD matrix for this structure

$$\begin{bmatrix} A & B \\ C & D \end{bmatrix} = \begin{bmatrix} 1 & 0 \\ 1/Z_1 & 1 \end{bmatrix} \tag{6.39}$$

from which

$$z_{11} = \frac{A}{C} = Z_1. \tag{6.40}$$

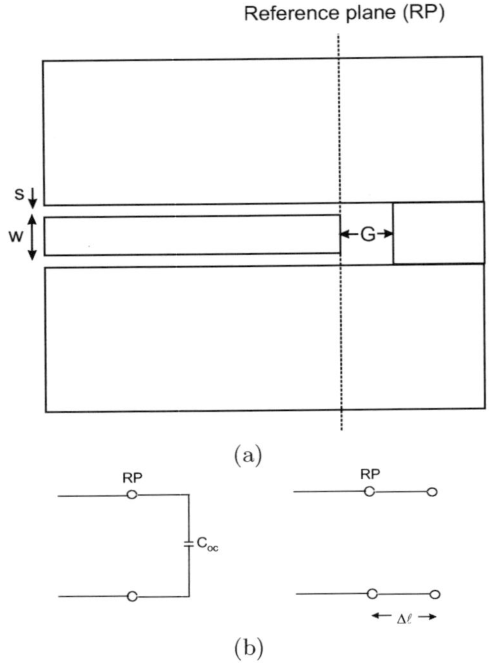

Figure 6.25 Quasi-open-circuit (a) and alternative equivalent circuits (b).

Therefore, with a 50 Ω characteristic impedance

$$Z_1 = \frac{1 + s_{11}}{1 - s_{22}} \cdot 50\Omega. \tag{6.41}$$

Now Z_1 is the capacitive reactance of the open-circuit discontinuity so that

$$Z_1 = \frac{1}{j\omega C_{\text{oc}}} = \frac{-j}{\omega C_{\text{oc}}}. \tag{6.42}$$

By simulation methods Rayit [185] used the above approach for both C_{oc} and Δl. The capacitance can be calculated using

$$C_{\text{oc}} = k \left[36.12 + \frac{1.819}{(g/h)} - \frac{0.02497}{(g/h)^2} \right] \tag{6.43}$$

where

$$k = w/h = w/(w + 2g) \tag{6.44}$$

and the values are found to be almost completely frequency independent at least up to over 20 GHz on alumina substrates. This frequency-independence probably also applies to most other substrates.

6.6.3 SYMMETRIC SERIES GAP

This discontinuity is closely related to the quasi-open end, amounting to a gap in series along the centre conductor. The structure and its 'pi', or π, equivalent circuit

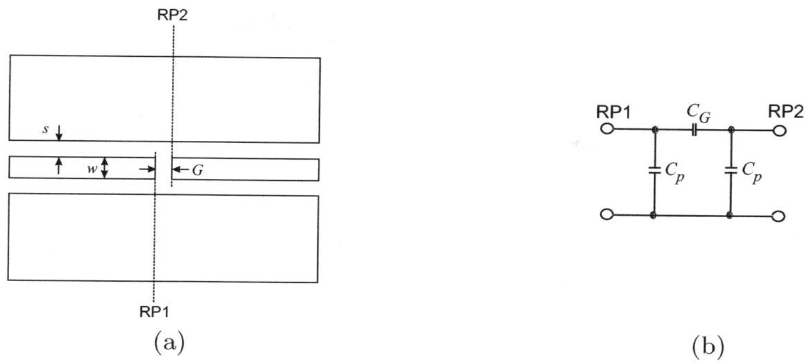

Figure 6.26 Symmetric series gap layout and capacitive π equivalent circuit are reference planes 1 and 2, respectively.

of capacitances is shown in Figure 6.26. The grounded capacitances account mainly for fringing fields from the ends to ground, whilst the series capacitance accounts chiefly for the end-to-end field.

Again, using a similar methodology to that employed with the quasi-open end, we begin with the appropriate ABCD matrix in order to analyse this structure:

$$\begin{bmatrix} A & B \\ C & D \end{bmatrix} = \begin{bmatrix} 1 & 0 \\ 1/Z_1 & 1 \end{bmatrix} \begin{bmatrix} 1 & Z_2 \\ 0 & 1 \end{bmatrix} \begin{bmatrix} 1 & 0 \\ 1/Z_1 & 0 \end{bmatrix} \tag{6.45}$$

which becomes

$$\begin{bmatrix} A & B \\ C & D \end{bmatrix} = \begin{bmatrix} 1 & Z_2 \\ 1/Z_1 & (1 + Z_2/Z_1) \end{bmatrix} \begin{bmatrix} 1 & 0 \\ 1/Z_1 & 1 \end{bmatrix} \tag{6.46}$$

or

$$\begin{bmatrix} A & B \\ C & D \end{bmatrix} = \begin{bmatrix} (1 + Z_2/Z_1) & Z_2 \\ (\frac{2}{Z_1} + \frac{Z_2}{Z_1^2}) & (1 + Z_2/Z_1) \end{bmatrix}. \tag{6.47}$$

The determinant of the ABCD matrix, Δ, is now taken in order to find further parameters explicitly:

$$\Delta = \left(1 + \frac{Z_2}{Z_1}\right) \cdot \left(1 + \frac{Z_2}{Z_1}\right) - Z_2 \left(\frac{2}{Z_1} + \frac{Z_2}{Z_1^2}\right) = 1 \tag{6.48}$$

and

$$z_{12} = z_{21} = \frac{1}{C}. \tag{6.49}$$

Further manipulations enable both z_{11} and z_{12} to be expressed in terms of Z_1 and Z_2. Then finally, equation substitutions and rearrangements lead to both Z_1 and Z_2 being expressed entirely as functions of z_{11} and z_{12}

$$Z_1 = (z_{11} + z_{12}) \tag{6.50}$$

and

$$Z_2 = \frac{(z_{11} + z_{12})^2}{z_{12}} - 2(z_{11} + z_{12}). \tag{6.51}$$

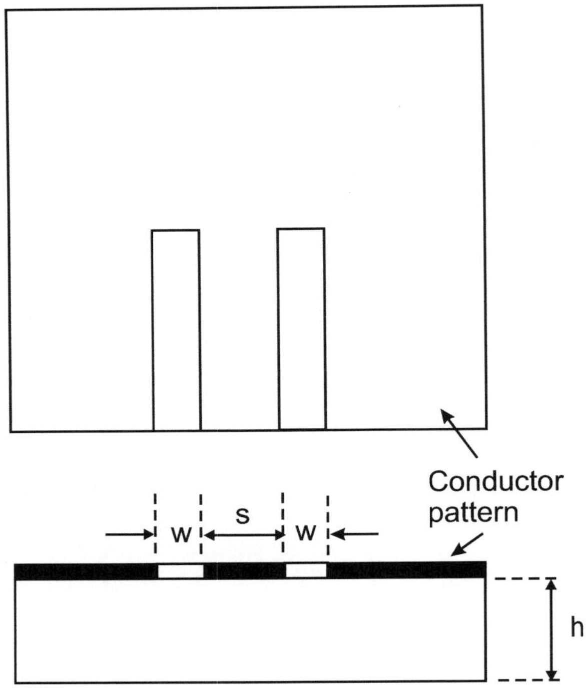

Figure 6.27 Pseudo short-circuit: layout and cross-section.

Simulations enable these expressions to be used to determine the capacitances in the pi-network representing the gap. It is generally found that these capacitances are largely independent of frequency, although C_g, the series capacitance, rises slightly with frequency. Typical values lie in the 20 to 60 fF range.

6.6.4 COPLANAR SHORT-CIRCUIT

This structure is an important and unusual example of a 'termination' in CPW. Unlike other transmission structures, CPW inherently involves the adjacent ground planes and therefore, assuming that the planar structure is preserved, any 'short-circuit' must also inherently be of a pseudo short-circuit nature. This is indicated in Figure 6.27 where the layout is shown at the top of the diagram, and the cross-section is shown beneath. It should be clear that a fairly complex field distribution exists in the vicinity of the pseudo short-circuit.

This structure is also shown in Figure 6.28, where the reference plane, equivalent load (L) and end extension are also indicated. The load is predominantly inductive and may often be represented as an inductor. This is because the abruptly-ended line is mainly characterized by substantial current disturbances.

The current distributions on a finite ground CPW are shown in Figure 6.29, where the substantial current disturbances can be seen.

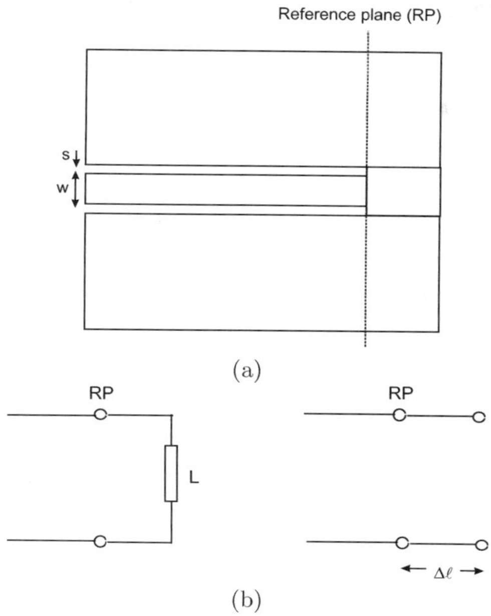

Figure 6.28 Pseudo short-circuit: layout and alternative equivalent circuits.

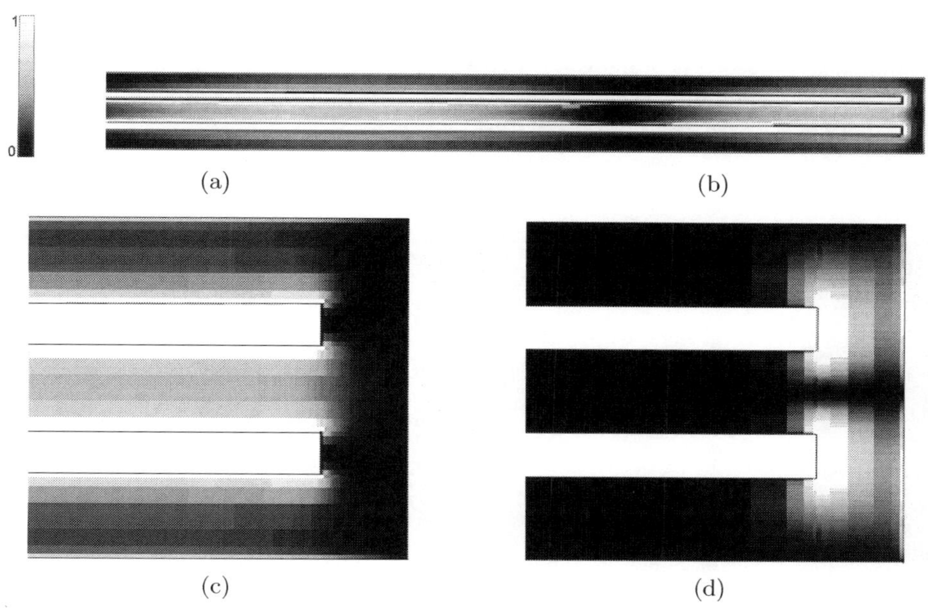

Figure 6.29 Normalized magnitude of the current distribution on the conductors of a CPW pseudo short-circuit: (a) scale; (b) short at 100 GHz showing standing wave pattern of longitudinal current; (c) magnitude of the longitudinal current at 30 GHz in greater detail; and (d) magnitude of the transverse current at 30 GHz.

Figure 6.30 Right-angle bend: layout (a) and equivalent circuit (b).

6.6.5 *RIGHT-ANGLE BENDS*

Although mitring (or chamfering) will always significantly reduce the effects of this discontinuity, we start by considering the direct, unmitred, right-angle bend and once more the appropriate ABCD matrix is the starting point for the analysis. The structure and its equivalent circuit are shown in Figure 6.30.

The ABCD matrix is

$$\begin{bmatrix} A & B \\ C & D \end{bmatrix} = \begin{bmatrix} 1 & Z_1 \\ 0 & 1 \end{bmatrix} \begin{bmatrix} 1 & 0 \\ 1/Z_2 & 1 \end{bmatrix} \begin{bmatrix} 1 & Z_1 \\ 0 & 1 \end{bmatrix} \tag{6.52}$$

so that

$$\begin{bmatrix} A & B \\ C & D \end{bmatrix} = \begin{bmatrix} (1+Z_1/Z_2) & (2Z_1+Z_1^2/Z_2) \\ 1/Z_2 & (1+Z_1/Z_2) \end{bmatrix}. \tag{6.53}$$

In similar fashion to that employed when the series gap was considered, once again we take the determinant of this matrix because this leads, eventually, to explicit expressions for Z_1 and Z_2.

The determinant of Equation (6.53) is

$$\Delta = AB - CD = \left(1+\frac{Z_1}{Z_2}\right)\cdot\left(1+\frac{Z_1}{Z_2}\right) - \frac{1}{Z_2}\left(2Z_1+\frac{Z_1^2}{Z_2}\right) \tag{6.54}$$

which expands simply to $\Delta = 1$ (because all other terms cancel), but

$$z_{11} = Z_1 + Z_2 \tag{6.55}$$

and we can then write

$$Z_2 = z_{12} (= z_{21}) \tag{6.56}$$

also

$$Z_1 = z_{11} - z_{12} \tag{6.57}$$

thus providing the impedances associated with the equivalent circuit elements, in terms of extractable ABCD-related (Z_{mn}) impedance matrix elements.

The bend capacitance and inductance are extracted using this procedure and the results for the inductance are shown, for CPW on 0.635 mm thick alumina, in Figure 6.31. In this graph k is the usual ratio as defined before — see Equation (6.4).

These parameters are almost invariant with frequency from 400 MHz to at least 20 GHz.

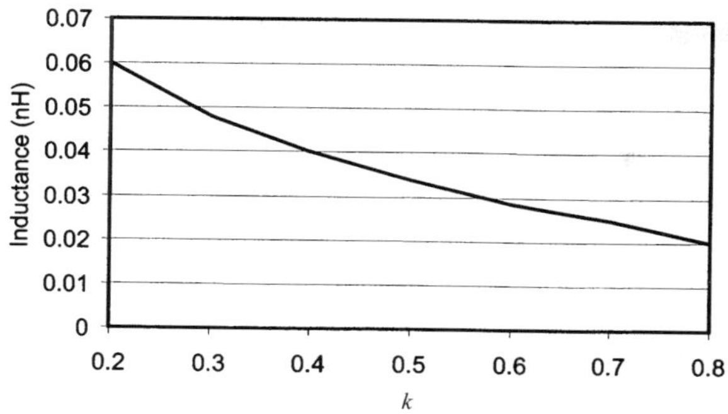

Figure 6.31 Right-angle bend inductance as a function of parameter k.

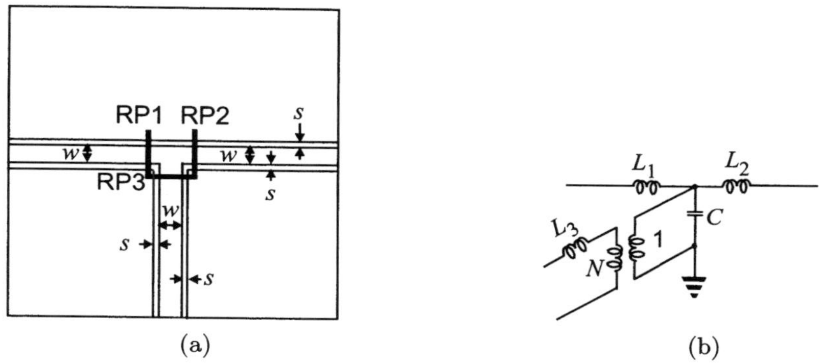

Figure 6.32 T-junction: layout (a) and lumped equivalent circuit (b).

Mitred (chamfered) bends

When optimally mitred the bend capacitance is generally at least halved compared with the unmitred case. Such a bend also exhibits a greatly improved reflection coefficient (S_{11}) with a typical improvement of at least 16 dB compared to an unmitred bend — over the 2 to 20 GHz frequency range (see, for example, the section under Microstrip Discontinuities, Chapter 7 of this book). These comments apply generally to almost every planar transmission structure.

6.6.6 T-JUNCTIONS

This type of discontinuity occurs in a wide variety of circuits. The physical structure and equivalent circuit model are shown in Figure 6.32.

Air bridges (or earthing bonds — see the following section) are absolutely necessary in order to suppress the unwanted odd mode in most CPW circuits and T-junction discontinuities are a major case in point in which these air bridges need to be implemented.

As with the microstrip (and other) T-junctions, the equivalent circuit demands a transformer to identify the effect of the coupled arm on the main transmission route.

Considering a 49 Ω line on an alumina substrate, the characteristic behaviour of the T-junction transformer ratio is shown in Figure 6.33 (data derived by Rayit [185]).

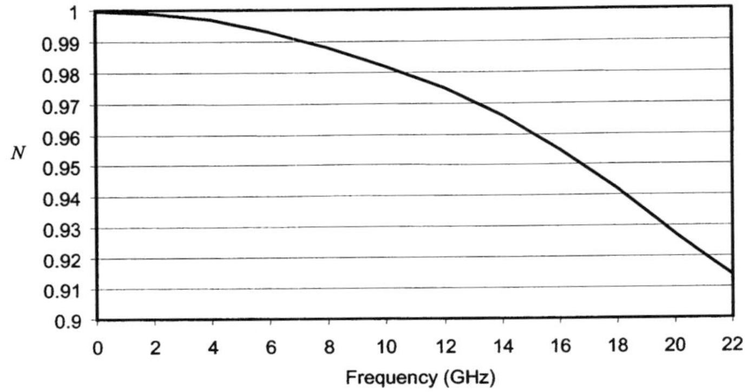

Figure 6.33 T-junction transformer ratio N as a function of frequency.

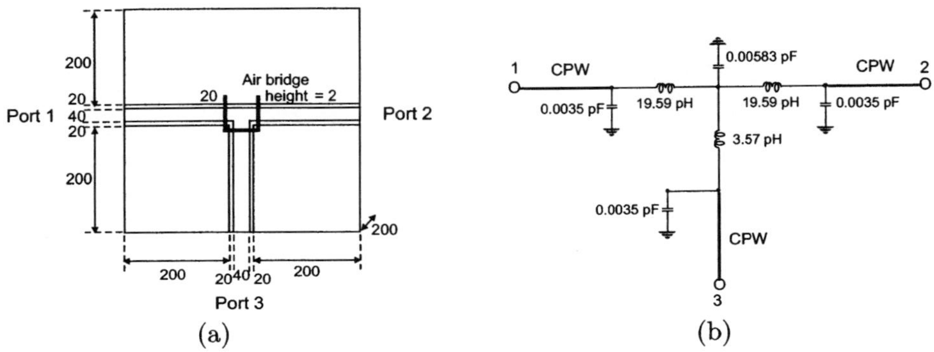

Figure 6.34 The T-junction (a) structure analysed by Mirshekar-Syahkal [199] together with its equivalent circuit (b). (All dimensions are in microns.)

At 20 GHz the value of this ratio has clearly decreased from its unity (DC) value to 0.92. Rayit has analysed and measured a wide range of T-junctions on alumina.

Mirshekar-Syahkal [199] has computed the equivalent circuit element values for a range of CPW discontinuities, including T-junctions, using the quasi-static spectral domain method. A structure analysed by Mirshekar-Syahkal is shown in Figure 6.34, together with a simplified version of the equivalent circuit. This circuit is on an InP MMIC substrate and it is conductor-backed.

6.6.7 AIR BRIDGES.

In order to ensure that only the fundamental CPW mode propagates over the frequency range of interest, it is usually necessary to directly interconnect coupled sections of CPW lines using conductive straps. The electrical lengths of these straps, commonly known as 'air bridges' must be kept much less than one-fifth of a guide wavelength so that they appear as localized short-circuits. These air bridges suppress the unwanted odd CPW mode.

Several researchers have reported on the characterization of these air bridges as well as the effects of the coupled slotline mode and shunt stub realizations. Sewell

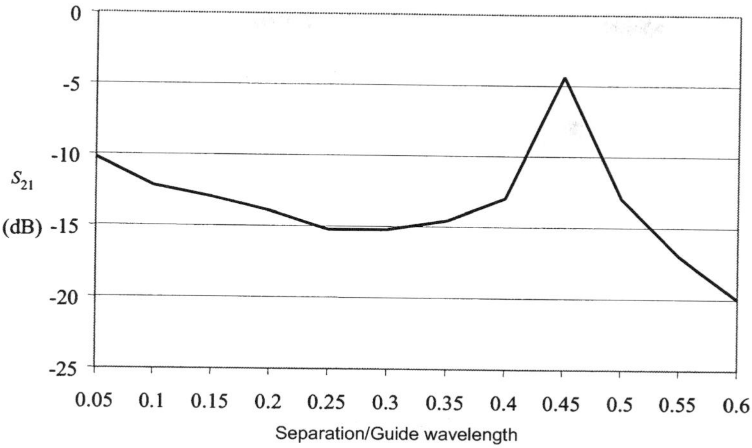

Figure 6.35 Effects of the coupled slotline mode on air bridges (Sewell and Rozzi [200]).

and Rozzi [200] presented calculated characterizations of air bridges under millimetre-wave conditions, considering both single and coupled bridges. Using a mode spectrum technique they showed that, as should be anticipated, the S_{11} and S_{21} scattering parameters are strongly sensitive to the strap separation/wavelength ratio for coupled bridges. Sewell and Rozzi's main results, converted to dB and at 40 GHz and 50 GHz frequencies, are shown in Figure 6.35.

Lee, Liu and Itoh [201] described the effects of the coupled slotline mode and air bridges on CPW and non-leaky coplanar waveguide discontinuities. They focused on a T-junction discontinuity and showed both spectral-domain calculated and also measured results over the 4 to 9 GHz band. For the symmetrical structure considered the thickness $t = 0.635$ mm, centre line width $w = 1$ mm, separation $s = 0.34$ mm. Lee *et al.* also examined an asymmetrical T-junction.

In general, the behaviour of the S-parameters with frequency is highly non-linear. Both the magnitude and phase of S_{11} and S_{21} exhibit sharp dips in response that are almost harmonically related. The magnitude typically dips by up to 9.5 dB, and the phase changes abruptly by more than 180° at 8 GHz for the asymmetrical case.

Hettak, Dib and Omar [202] reported upon a new class of miniaturized and radiation-free CPW shunt stubs that they implemented within the centre conductor. A major aspect of this approach was to reduce the number of air bridges actually required to realize shunt stubs considering that the manufacture of multitudes of these bridges is potentially expensive.

Unlike conventional asymmetric shunt stubs, Hettak, Dib and Omar's design exhibits longitudinal symmetry. Their design has a pseudo short-circuit CPW followed by a re-entrant 'serpentine' shaped nominally quarter-wave stub set longitudinally within the centre conductor. Two air bridges directly connect from the pseudo short-circuit end to the outer fingers of the shunt stub configuration. These researchers also show a two-fold version comprising this entire structure reversed and repeated in cascade.

Experimental results are shown by Hettak *et al.* for 50 μm separations (s), 0.175 μm width (w), and 50 μm fingers with 25 μm gaps manufactured on a substrate with

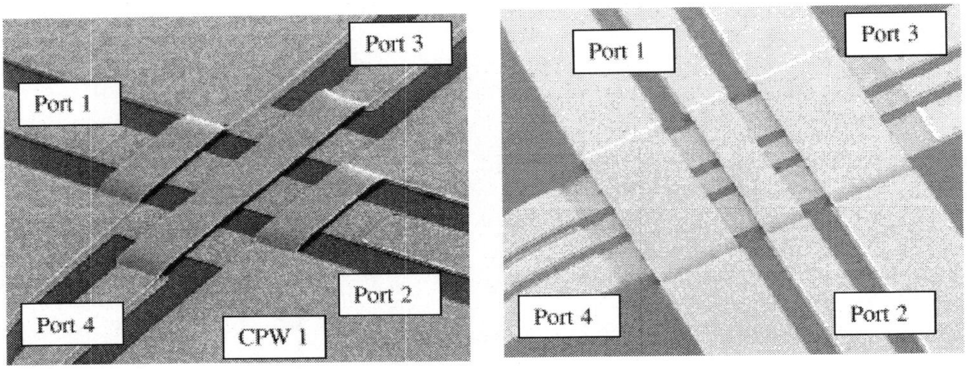

(a) (b)

Figure 6.36 SEM micrographs of CPW cross-over junctions: (a) conventional CPW; and (b) FGCPW. The lines have the same dimensions. (From Ponchak and Tentzeris [203].)

permittivity 9.9 and thickness 0.254 mm. The 10 to 50 GHz band was swept and the required response around 30 GHz was obtained.

6.6.8 CROSS-OVER JUNCTIONS

CPW cross-overs are often essential and then air-bridges must be used to connect the side grounds to ensure the integrity of the CPW mode. It has been shown that FGCPW cross-overs have much lower coupling (typically 15 dB less) than CPW cross-overs [203]. These two types of cross-overs are shown in Figure 6.36. The reason for this difference in coupling can be seen by examining currrent flow in the ground plane. In the conventional CPW cross-over, Figure 6.36(a), ground currents are shared by the two CPW lines, particularly near the intersection of the lines. This results in coupling due to the common impedance effect. However in the FGCPW cross-over, Figure 6.36(b), a common current path does not exist and so there is no common impedance coupling.

6.7 CIRCUIT ELEMENTS

Taken alone, stretches of CPW line, or associated discontinuities, do not of course constitute circuit elements. Rather, they are essential and fundamental aspects that must be identified and quantified before any computer-modelling can be undertaken. In this section we consider some passive circuit elements that can be formed using the CPW medium. The significance of vias in MMICs should first be stressed. This has been reported, for example, by Strohm *et al.* [78].

6.7.1 INTERDIGITAL CAPACITORS AND STUBS

There are many ways in which a lumped (or quasi-lumped) capacitor may be realized in CPW format. The series gap, discussed in Section 6.6.3 on page 190, is arguably the simplest example of a 'fully integrated' series capacitance. However, as in the case

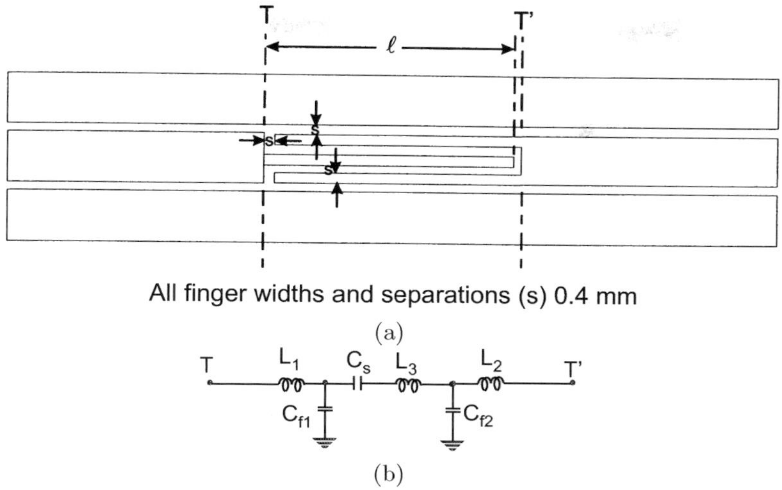

Figure 6.37 Interdigital or stub structure (a) and its equivalent circuit (b). (Here the reference planes are denoted by T and T′, another common convention.)

of microstrip it becomes impracticable to realize even modest values of capacitance this way, because submicron gap widths are required, rendering manufacturing all but impossible.

Another possibility is the introduction of flip-chip capacitors that may be attached across relatively wide series gaps. This approach enables relatively large capacitances to be achieved, but there are at least two fundamental objections:

(a) This can only realistically apply to hybrid MICs, since the real estate available with MMICs makes this an impractical approach with such technology.
(b) The Q-factor of chip capacitors is usually relatively low.

Yet another method for the realization of capacitances in CPW is with interdigital structures. A serpentine shape of gaps across the centre conductor, interspersed with conductor fingers ('digits') results in a dominantly capacitive series element. Capacitances ranging up to several hundred femtofarad and approaching picofarad are achievable this way, and the approach is applicable to MMICs. It is vital to ensure that the maximum length of any finger is kept less than one-fifth of a guide wavelength at the maximum frequency of operation, otherwise such a finger would behave as a distributed element. This is relatively easy to achieve with MMICs, but considerably more difficult with the larger dimensions applicable to hybrid circuits (MICs).

Such a structure is shown in Figure 6.37, where the most basic single-finger example is indicated. In common with most of the discontinuities described, the structure has inductive as well as capacitive effects, and these features are reflected in the equivalent circuit.

Rayit [185] supplies empirical expressions that work well for calculating the capacitances and inductances (l is the finger or stub length in all instances):

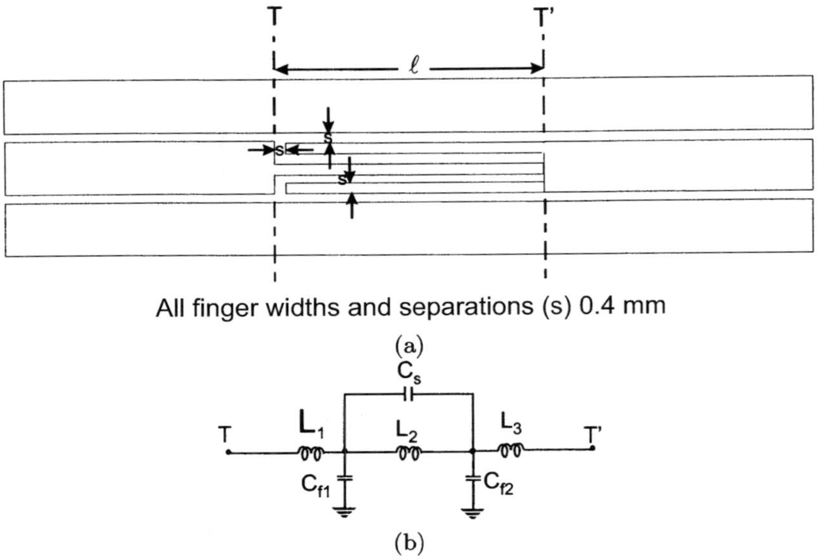

Figure 6.38 Through-connected interdigital stub: structure (a) and its equivalent circuit (b). (Here the reference planes are denoted by T and T'.)

$C_s = 2.03 \times 10^{-4}l + 8.65 \times 10^{-3}$	$L_1 = 1.57 \times 10^{-4}l$
$C_{f1} = 1.58 \times 10^{-4}l + 2.3 \times 10^{-2}$	$L_2 = 1.67 \times 10^{-4}l$
$C_{f2} = 10^{-5}l + 9.7 \times 10^{-5}$	$L_3 = 2.05 \times 10^{-4}l$

In these expressions the length is directly substituted in micrometers (e.g. '50' for 50 μm) and then the capacitances compute directly in pF and the inductances compute directly in nH.

These data apply to a high-permittivity Duroid® substrate ($\varepsilon_r = 10.2$). The calculations are very straightforward, easily incorporated into a computer program, and can be scaled to enable calculations to proceed for substrates where the permittivity is other than $\varepsilon_r = 10.2$.

An interesting and useful alternative structure is realized when the finger is directly connected to the opposing CPW centre conductor, instead of the final series gap, and this arrangement is shown in Figure 6.38. Such a circuit element now behaves in lowpass (or even quasi-resonant) mode because there is a direct inductive path between input and output (T to T$'$).

Rayit's empirical expressions for the equivalent circuit parameters are:

$C_s = 1.51 \times 10^{-4}l + 8.7 \times 10^{-2}$	$L_1 = 3.53 \times 10^{-4}l - 1.2 \times 10^{-4}$
$C_{f1} = 3 \times 10^{-5}$	$L_2 = 1.86 \times 10^{-4}l - 6.12 \times 10^{-4}$
$C_{f2} = 6.9 \times 10^{-5}l + 2.5 \times 10^{-3}$	$L_3 = 3.02 \times 10^{-4}l - 2.09 \times 10^{-4}$

Precisely the same comments made concerning the open interdigital structure apply equally here. To increase the capacitances without entering into highly dimension-critical conductor geometries high-permittivity dielectric materials may be attached across the gaps of these structures. This is achievable with hybrid MICs,— but

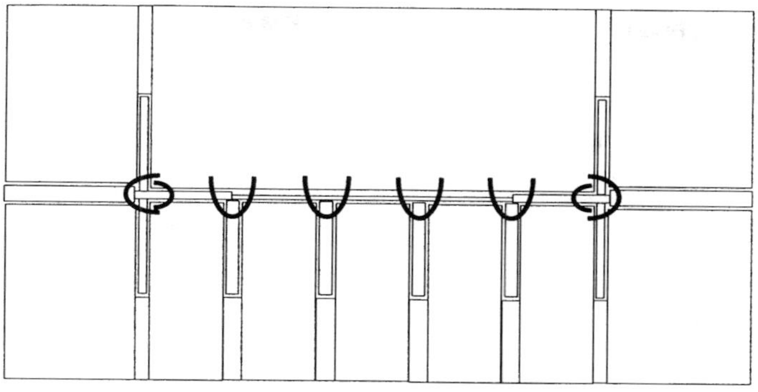

Figure 6.39 Layout of a 30% bandwidth bandpass filter on an alumina substrate.

almost certainly not so with MMICs, due again to the extremely small dimensions. In this case, however, a layer of polyimide may be introduced onto the surface during processing.

6.7.2 FILTERS

The fundamental approaches described in the previous sections of this chapter can now be used in most CPW designs, including filters.

Frequency filters represent good, critical circuit realizations with which to test the adequacy of design models. Such circuit elements are also important in their own right within more complex circuit configurations. Air bridges are also repeatedly introduced in order to suppress the unwanted odd mode (see Section 6.6.7 on page 196).

All four types of basic filter — lowpass, highpass, bandpass and bandstop — are capable of being realized in CPW topology. At this point we concentrate on a bandpass filter, originally investigated by Rayit [185] and manufactured on an alumina substrate.

The basic layout (top view) is shown in Figure 6.39. The aim is to design a 30% bandwidth filter and the shunt shorted stub following the approach due to Malherbe [204]. A Chebyshev response is to be realized. Inherently the filter requires modelling for several discontinuities — including the T-junction. The measured and simulated (incorporating the modelling) responses are shown in Figures 6.40 and 6.41, where it can be seen that several differences occur in practice. Considering the transmission characteristic (S_{21}) first, working upwards in frequency, below 2 GHz the responses are substantially above (lower in dB) the design values. This is of little consequence because the worst case is -53 dB in any case. Both lower and upper skirts (dB per GHz) are actually superior to the design goal.

As practically always, for any type of filter, the in-band loss is inferior to the ideal design goal. In this case, an in-band loss of 1 to 2 dB occurs. Above 8GHz spurious responses become noticeable and the worst result in this case is that at 9.5GHz (-29.5 dB). At this level such a spurious response would not usually be a problem in the remainder of the system.

The return loss or reflection (S_{11}) characteristic (Figure 6.41) demonstrates a generally good response with the exception of the -15 dB level at 5.9 GHz — almost

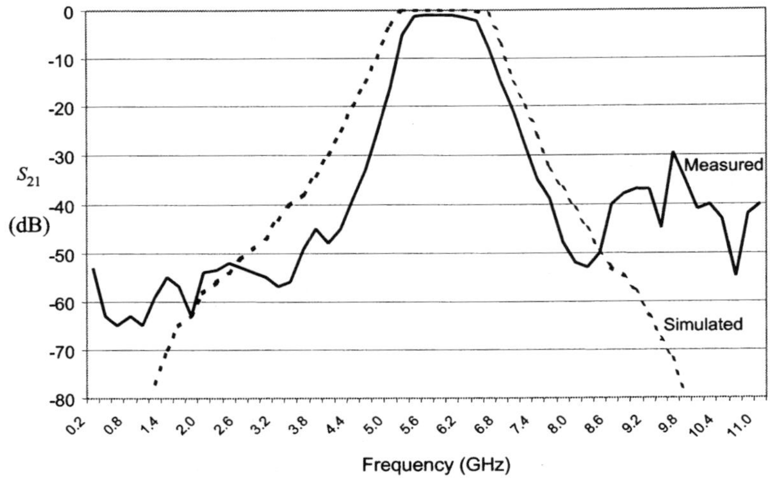

Figure 6.40 Transmission response (S_{21}) of the bandpass filter.

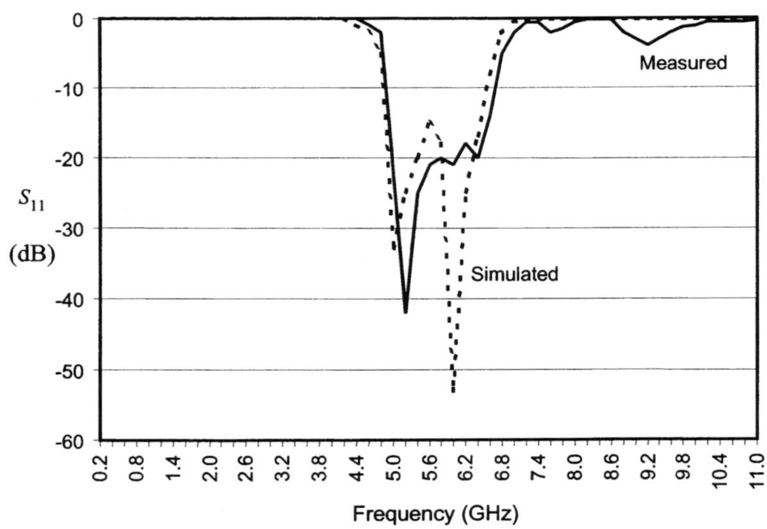

Figure 6.41 Return loss (S_{11}) of the bandpass filter.

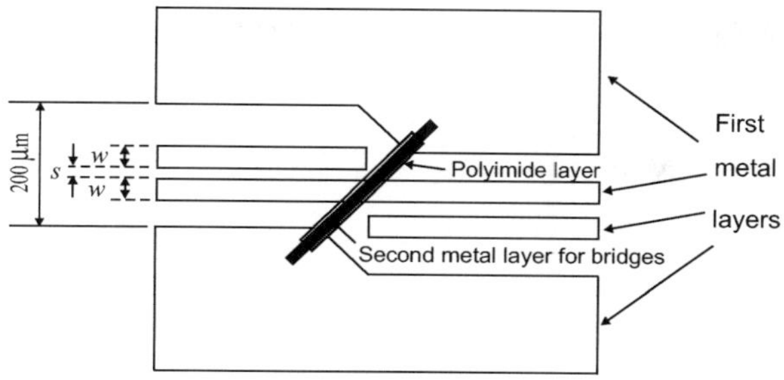

Figure 6.42 Basic structure of an MMIC-implemented bandpass filter (Mernyei *et al.* [206]).

precisely the frequency at which the response should theoretically be excellent, at −42 dB.

It is considered that some of the reasons for these discrepancies and undesirable features in the responses could well lie in a need for still more attention to the air bridges (earthing bonds).

A bandstop filter design, based upon a combination of the interdigitated structures introduced earlier here, has been put forward by Omar and Chow [205]. This design embodies top and bottom conductor shields in place of air bridges, and the characteristics exhibit marked peaking and dipping in the frequency responses. A major feature is that the results with the shields are very nearly as good as those using the air bridges.

Filter designs implemented with smooth substrate surfaces, notably semiconductors, should yield much improved in-band losses.

Mernyei, Aoki and Matsuura [206] have reported MMIC bandpass filters operating in the 17 and 27 GHz bands and having 10 GHz passbands. As shown in Figure 6.42 the basic structure comprises cascaded, coupled, open-ended CPW lines with metal-layer bridges, rather than air bridges deposited on polyimide layers to bridge the line ends. These researchers use the standard filter design procedure to realize the filters on GaAs substrates. In both cases, however, the passband loss is around 5 dB — which is surprisingly high, since just over 2 dB might have been anticipated.

It would also be very interesting to explore the effects of using conductor-backed and/or conductor shielded CPW structures for the realization of filters and other circuit elements.

6.7.3 COUPLERS AND BALUNS

Several forms of couplers have been reported using CPW and its variants. Modelling the attenuation of the basic coupled coplanar waveguide (CCPW) structure is the first prerequisite, and this has been reported by Ghione and Naldi [207]. They show that the net losses are always small, amounting to less than 1 dB/mm for either the odd or the even mode. As the amount of coupling increases so the loss also decreases for the even mode case, down to only a very small fraction of a dB per mm. In contrast, for the odd mode, whilst the loss decreases at first (minimizing around −10 dB of coupling) as the coupling increases further so the loss rises rapidly — almost

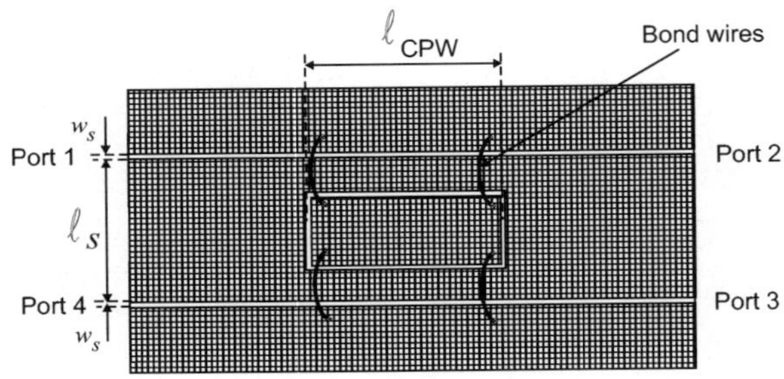

Figure 6.43 Basic structure of a uniplanar, 2-branch directional coupler (Ho, Fan and Chang [208]).

monotonically as the coupling exceeds −5 dB of coupling. This behaviour is similar to that of microstrip.

Uniplanar hybrid-ring and branchline couplers suitable for MMIC implementation have been reported by Ho, Fan and Chang [208]. This team show CPW-slotline transitions using CPW pseudo short-circuits and radial stubs that exhibit well-behaved passband characteristics within the 0.5 to 6 GHz frequency range. The following types of hybrid-ring couplers are considered: rat-race, phase-reverse and cross-over. The rat-race in particular exhibits 3 dB power dividing, and a particularly sharp isolation response at 3 GHz. The cross-over coupler exhibits approximately 4 to 5 dB of power division across the 2 to 4 GHz band.

The circuit configuration of Ho, Fan and Chang's uniplanar two-branch directional coupler is shown in Figure 6.43. In this structure the substrate is RT-Duroid® 6010.8 (ε_r = 10.8 and thickness 1.27 mm), the CPW centre strip width is 0.71 mm and the separations are 0.1 mm. The length of the CPW series arm is 10.39 mm and the length of the slotline shunt branch is 7.77 mm. Whilst the coupling is 3 dB at 2 GHz and 2.3 dB at 4 GHz, this decreases to 0 dB over the band 2.8 to 3.6 GHz. The phase varies from 71° at 2 GHz to 89° at 4 GHz.

Fan, Ho, Kanamaluru and Chang [209] have studied broadband uniplanar magic-T, hybrid-ring and de Ronde's CPW-slot couplers. They report upon the performance of radial-stub compensated hybrid magic-Ts, and show that the phase and amplitude characteristics are well behaved over the 1 to 8 GHz band. A reduced-size reverse-phase hybrid-ring coupler is also reported. In this design the reverse phase is accomplished by implementing two radial stubs in series within one arm. De Ronde's CPW-slot directional coupler comprises a rectilinear four-port circuit with a central quarter-wave slot having radial stubs at each end. With a CPW characteristic impedance of 47.2 Ω, Ho *et al.* provide measured data showing transmission varying from 4 dB at 2.2 GHz through 1.5 dB (at 4 GHz) and coupling of 8 dB at 2 GHz ranging to 8.5 dB at 4 GHz (5.7 dB through the centre frequencies). Isolation and return loss are both generally better than −12 dB over the entire band from 2 to 4 GHz.

In another paper [210] Ho, Fan and Chang report on new uniplanar CPW hybrid-ring couplers and magic-Ts. Extensive use is made of radial stubs and back-to-back baluns.

Uniplanar double-Y baluns are reported by Gu and Wu [211]. These researchers

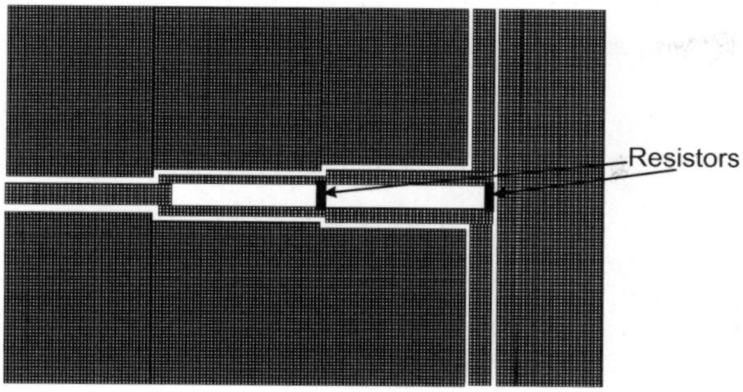

Figure 6.44 2-section in-phase CPW power divider (Fan and Chang [212]).

describe a network technique for the design of broadband circuits of this type using CPW, slotline and coplanar stripline (CPS). Design equations are developed for the resistive and reactive components of the impedances of stubs required in these structures. From these equations curves are plotted showing these resistive and reactive components as functions of the electrical lengths and normalized Z_0 values. From these curves the best choice of stub electrical lengths is determined, consistent with a good impedance match for the baluns.

The back-to-back circuit layout of the double-Y baluns is shown, with re-entrant stubs, and the results of measurements over the frequency range 3 to 30 GHz are given for both the CPW-slotline and FGCPW-CPS baluns. In both cases, the insertion losses are generally better than −4 dB over 3 to 22 GHz and dipping to −5 dB at higher frequencies. VSWR remains below 2 throughout, being below 1.5 over 13 to 30 GHz for the CPW-slotline and below 1.7 over 3 to 28 GHz in the case of the FGCPW-CPS balun.

6.7.4 POWER DIVIDERS

Incident RF power can readily be separated into two or more outgoing transmission lines using virtually any type of structure — planar or otherwise. Therefore, CPW realizations of such power dividers are feasible and various researchers have developed this theme.

Fan and Chang [212] have published the results of their work on uniplanar power dividers using coupled CPW and asymmetrical coplanar slotlines for MICs and MMICs. They report power dividers fed by slotlines with sections of coupled CPW (CCPW) feeding the two CPW output ports. As with most configurations of this general type, balance resistors are employed at the separation T point.

An example of a Fan and Chang two-section, in-phase CCPW power divider is shown in Figure 6.44. The structure is basically similar to that described above in that mm-dimensions apply, an RT/Duroid® substrate is used and the frequency range is 1 to 6 GHz. The input and output CPW lines have widths of 0.62 mm and gaps of 0.33 mm. The intermediate matching sections, having lengths of 10.8 and 10.7 mm, respectively, implement strip widths of 0.17 and 0.52 mm and gaps of 0.12 and 0.18 mm. Central gaps are 1.2 mm and 0.8 mm and the isolation resistances are 651 Ω and 62.5 Ω, respectively.

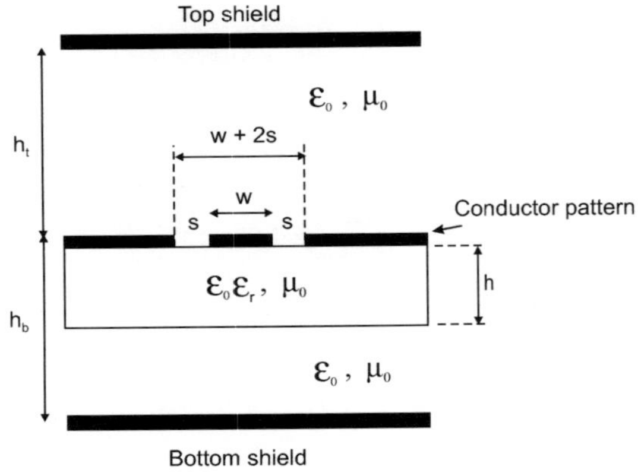

Figure 6.45 CPW with top and bottom metal shields.

For this circuit the insertion loss is less than 3.5 dB over the band and the isolation is greater than 24 dB. Over the 1–5 GHz band 3–4 dB of coupling is achieved from each output port. Return loss is always greater than 10 dB and exceeds 20 dB over the more restricted 1.8–4 GHz range.

6.8 VARIANTS UPON THE BASIC CPW STRUCTURE

6.8.1 CPW WITH TOP AND BOTTOM METAL SHIELDS

As described in the main sections earlier in this chapter, air bridges are traditionally necessary in order to maintain the dominant mode in CPW. Alternatives have, however, been investigated and a significant example comprises the introduction of top and bottom metal shields. In practice, such shields will probably be in place in any event as part of the packaging structure. The cross-section of such an arrangement is shown in Figure 6.45.

Omar and Chow [205] explored this type of structure and used it to implement a band-reject filter. They reported response results for both this structure and also the same pattern of band-reject filter, but implemented using air bridges. A response frequency range of 2 to 40 GHz applied. Using a very thin substrate Omar and Chow showed that, for close proximity shields ($h_{t,b} \approx 100\ \mu$m) the filter responses were very similar regardless of the approach. This demonstrated the utility of applying top and bottom metal shields in place of the more awkward and relatively unsophisticated air bridges.

6.8.2 MULTILAYER CPW

Current crowding at the edges of signal and ground conductors in CPW represents a substantial problem in power amplifiers, for example. Budimir *et al.* [213], reported an interesting solution to this problem. Figure 6.46(a) indicates the traditional 'open' CPW and the lower diagram (b) shows the new multilayer structure with the so-called 'inverted V' cross section.

In this multilayer structure the 'standard' CPW metallization is conductor layer

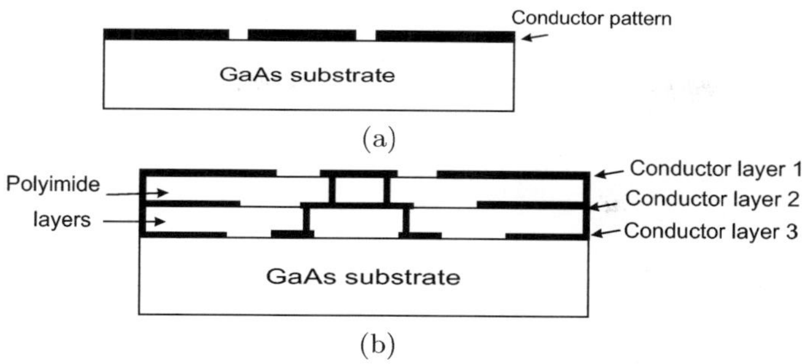

Figure 6.46 Conventional (a) and 'Inverted-V' CPW (b) cross-sectional structures (after Budimir *et al.* [213]).

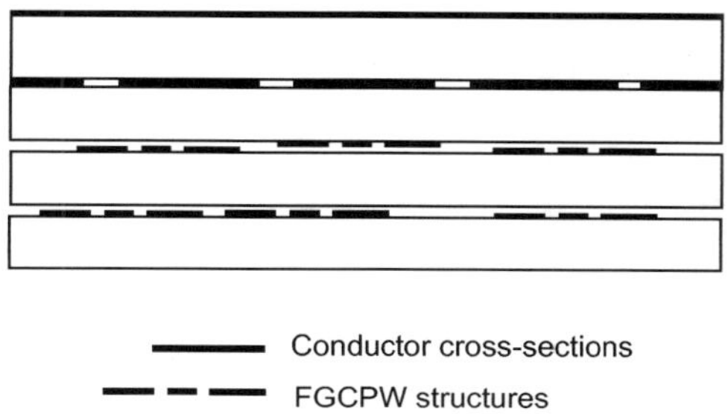

Figure 6.47 A 3-D multilayer stack of FGCPW structures (after Herrick, Henderson and Katehi [214]).

1. However, this layer is deposited upon a polyimide layer with conductive vias connecting to conductor layer 2 — itself deposited upon another polyimide layer. This process is repeated twice, with extending central conductor widths until the final structure, directly on the GaAs substrate, is actually separated into two physical conductor tracks.

The current flow is therefore much more evenly spread through the V-shaped conductor structure. Comparing a conventional CPW the current crowding is reduced by a factor of 4 on the worst affected layer. Budimir *et al.* focused upon investigations in the 1 to 20 GHz band.

Herrick, Schwartz, Henderson and Katehi [48,214], from the University of Michigan research team, have reported wafer effects in Si-micromachined multilayer guiding structures operating at frequencies as high as W-band (75 to 110 GHz). Issues concerning circuit isolation and integrity form the driving force behind this endeavour, especially concerning high-density ICs. Herrick *et al.* examine two CPW architectures embodying Si-micromachined cavities. In all instances, 100 μm thick silicon substrates are used. Iordanescu, Bartolucci, Simion and Dragoman also considered CPW stubs and filters fabricated on thin membranes and contrasted their performance with the components on standard substrates [44].

The main 3-D multilayer structure cross-section is shown in Figure 6.47, in which six CPWs can be seen, in two layers, and also a further layer of active conductors.

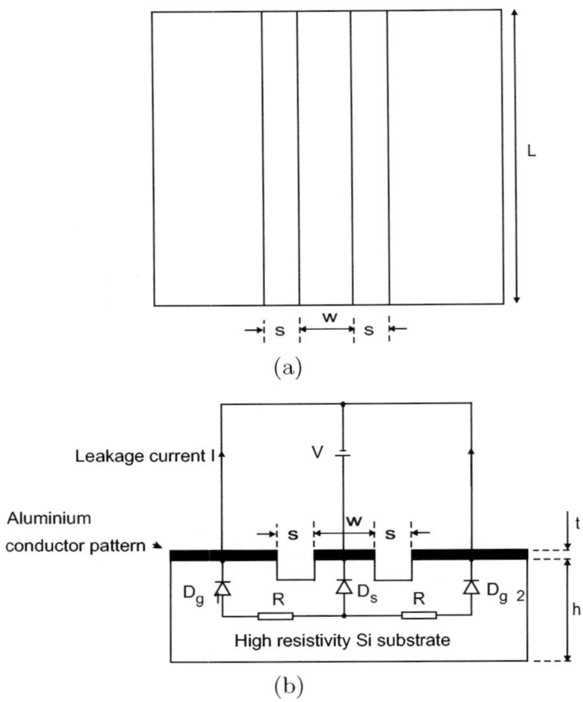

Figure 6.48 Trenched CPW on an HRS substrate (after Yang *et al.* [215]).

Both theoretical and experimental results are indicated and FGCPW is implemented. Very small but significant dispersion (around 1 to 2%) is evident over this wide frequency band, for both the conductor-backed CPW and also the conductor-backed micromachined CPW. Losses remain below 0.35 dB/mm for both cases.

From 75 to 85 GHz the cross-coupling (crosstalk in time-domain terms), at −27 to −45 dB, is broadly similar for both the conductor-backed CPW and also the conductor-backed micromachined CPW. However, at frequencies exceeding 85 GHz the conductor-backed micromachined CPW is considerably superior — with cross-coupling between −45 and −58 dB. This means that the cross-coupling is approaching (and ultimately goes into) the noise floor above 95 GHz. Herrick *et al.* show that the introduction of 40 μm air cavity spacings significantly decrease FGCPW-to-FGCPW coupling, ensuring that this always remains better than −18 dB. In particular, without this air cavity for relatively close spacing (less than 100 μm), the coupling exhibits peaks as low as −12 dB in the 90 to 93 GHz region.

6.8.3 TRENCHED CPW ON A SILICON MMIC

A further novel approach to the problem of RF loss reduction has been offered by Yang *et al.* [215]. This was applied to silicon MMIC structures and it involved the etching of trenches down into the exposed silicon in the CPW gaps. This structure is shown in Figure 6.48 where, in the cross-section (b), the equivalent circuit comprising diodes and resistors is also indicated.

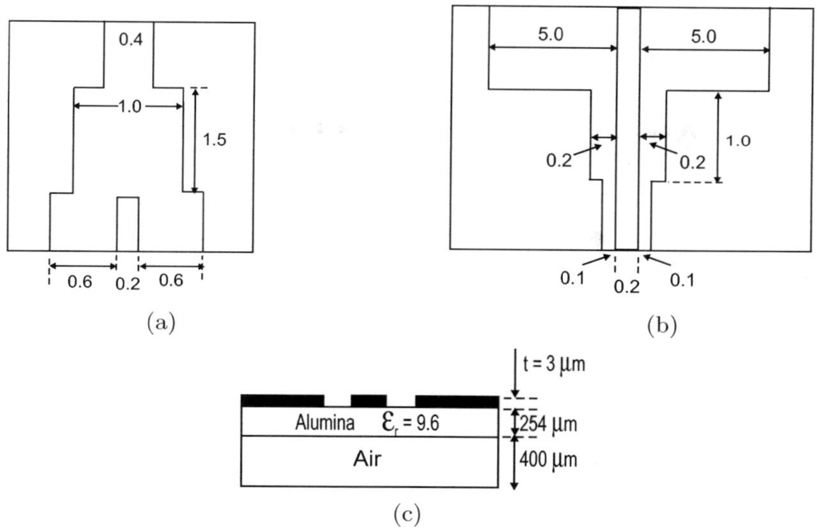

Figure 6.49 CPW transitions: (a) CPW-to-finline; (b) CPW-to-microstrip; and (c) cross-section of the CPW line for both structures (after Jin and Vahldick [216]).

By silicon etching, trench depths with the following values were used (all micrometres): 0, 3, 6, 9 and 12. Electric field and current flow distributions are reported. Yang *et al.* provide C-V curves versus bias voltage, and also characteristic impedance curves versus bias voltage. S-parameters are also provided. Characteristic impedance is also shown as a function of frequency up to a maximum of 40 GHz, and it is clear that this parameter always decreases with frequency. The decrease is greatest for trench depths in the 6 to 9 μm range. The reduction in RF losses varied from 0.5 to 1.8 dB over the 2 to 40 GHz range, for a 12 μm deep trench.

6.8.4 TRANSITIONS BETWEEN CPW AND OTHER MEDIA

It is often vital to be able to make transitions between CPW and other available transmission media. This enables the designer to take advantage of the differing properties of each technology. Therefore, transitions between CPW and microstrip, finline, slotline, etc. are needed and various approaches can be used in each case.

An example of a CPW-finline transition (Figure 6.49(a)) and also of a CPW-microstrip transition (Figure 6.49(b)) are shown. These designs were originally reported by Jin and Vahldick [216] and they operate over the 0 to 40 GHz band. A typical cross-section is also shown, and this may be scaled for any plane along either structure. Top view dimensions are in millimetres and cross-sectional dimensions are in micrometres. In both cases, a nominally quarter-wave central section is required to complete the design. This section is 1.5 mm long for the CPW-finline transition and 1.0 mm in length for the CPW-microstrip transition.

For the CPW-finline transition the transmission factor is 0.9 to 1.0 over the band and the worst-case value for the reflection coefficient is 0.47 at 40 GHz. For the CPW-microstrip transition, the transmission factor is always approximately 0.95 or better

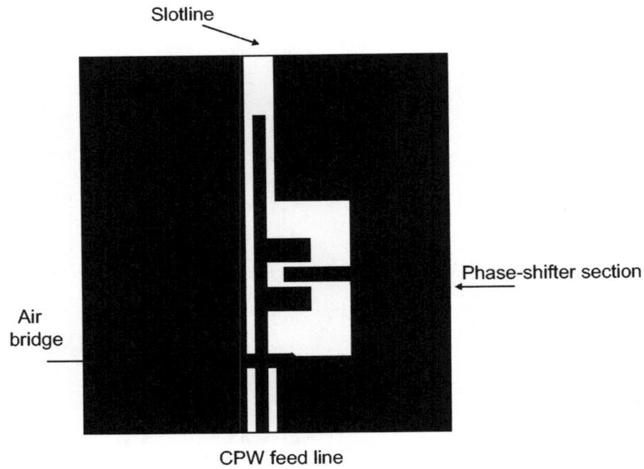

Figure 6.50 CPW-to-slotline transition (after Ma, Qian and Itoh [219]).

over the band, and the reflection coefficient is always less than 0.31.

Ellis *et al.* [217], again from the University of Michigan research team, have reported work on a broadband CPW-to-microstrip transition for millimetre-wave packaging. This development relates to a three layer structure based upon silicon and operating in W-band (75 to 110 GHz). The bottom layer is CPW that feeds the next (microstrip) layer — in turn feeding a microstrip patch antenna top layer. A microstrip to slotline power divider is implemented, and the complete transition has an insertion loss of approximately 0.3 dB and a return loss better than 10 dB over the entire band.

Lin and Chen [218] announced a form of lumped element CPW-slotline transition suitable for the 1 to 6 GHz band. This team used a lumped 'ring-type L' virtual short-circuit as the main element of their design instead of the more conventional quarter-wave transformer. With this design the insertion loss is less than −3 dB and the return loss is better than −10 dB over the 2 to 4 GHz band.

A CPW-slotline transition tailored for feeding a Vivaldi antenna is reported by Ma, Qian and Itoh [219]. The basic layout of this transition is shown in Figure 6.50, in which it can be seen that an air bridge and a phase shifter form major elements. An RT-Duroid® substrate is used ($\varepsilon_r = 10.2$ and thickness 0.635 mm) and the design covers the 1 to 30 GHz frequency range. The CPW and slotline modes are separated by employing a Finite-Difference Time-Domain (FDTD) analysis approach. Over the 11 to 18 GHz band the measured and calculated insertion loss is less than 3 dB. In the same paper [219], Ma *et al.* show this transition used in a CPW-fed Vivaldi antenna, a CPW power divider and a CPW-slotline coupler.

Kaneda, Qian and Itoh [220] report on a broadband microstrip-to-waveguide (rectangular) transition that employs a quasi-Yagi antenna. This uses a microstrip-to-CPS (coplanar stripline) balun, and the antenna is inserted into the E-plane of the waveguide. Over X-band (8.4 to 12 GHz) this transition exhibits near zero insertion loss and better than −8 dB return loss.

Further reported work on transitions increasingly emphasizes applications to antennas. For example Simons, Lee and Perl [221] report the results of work on techniques for exciting Linearly Tapered Slot Antennas (LTSA) using CPW. Their

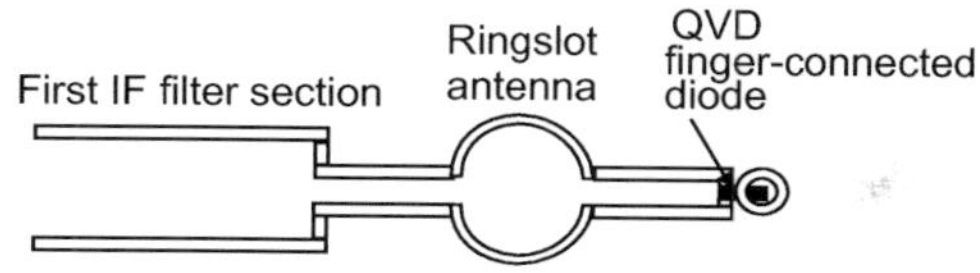

Figure 6.51 A 30–40 GHz bandstop filter within a mixer and a ring-slot antenna (after Vaupel and Hansen [226]).

work covers the band from 16 to 24 GHz, and the return loss achieved is always better than −10 dB. From the same basic team, Simons and Lee [222] report on CPW aperture-coupled patch antennas with FGCPW feeds. It is shown that this type of approach is superior to the use of microstrip for this form of application.

A hybrid FGCPW/CPS design for multilayer circuits and antenna feeds is reported by Zhu and Wu [223, 224]. This covers a 2 to 10 GHz bandwidth and comprises FGCPW-to-CPS transitions. The CPS optimally feeds a dipole antenna, and a measured return loss of −45 dB at 3.35 GHz is obtained.

Nadan *et al.* [225] have performed work on the miniaturization and optimization of a filter/antenna multilayer and multi-function module that embodies a composite ceramic foam substrate with a permittivity of 1.07 (i.e. just 7% greater than air). The characteristic impedance of the CPW is approximately 46.48 Ω, and the characteristic impedance of the microstrip is between 17 and 21 Ω, which results in a fairly wide microstrip line but is fully realizable. This approach is suitable for complex antenna arrays. With this design the insertion loss is better than −12 dB over the 10 to 11 GHz band, with dips as low as −27 dB.

The final example provided here is that reported by Vaupel and Hansen [226]. They report an electrodynamic analysis of coplanar MICs and MMICs that mainly covers the frequency range to 100 GHz. A CPW-slotline analysis is performed for coplanar MMIC applications. A hybrid (E-shaped structure) bandstop filter is shown that implements CPW with air bridges — optimized for the 30 to 40 GHz band. A submillimetre-wave 650 GHz mixer for satellite remote sensing (an ESA project) represents an unusual application that is also reported in the same paper by Vaupel and Hansen. The basic layout of this mixer circuit is shown in Figure 6.51, indicating the first I.F. filter section, the ring-slot antenna and the feed to the mixing diode.

6.9 FLIP-CHIP REALIZATIONS

With the ultimate aim of manufacturing high frequency MultiChip Modules (MCMs) problems surrounding the introduction of flip-chip MMICs are being increasingly addressed by many workers.

Gye-An Lee and Hai-Young Lee [227], for example, have investigated the suppression of energy leakage in GaAs flip-chip to main substrate structures. Two GaAs flip-chip MMICs, side-by-side, are shown above the main substrate in the upper diagram of Figure 6.52, and more detail of one flip-chip *in situ* is indicated in the lower diagram. The main (required) CPW mode is indicated by the curved field line, a TM_0 surface wave mode by the solid line and a TM_0 parallel-plate mode by the dotted line. Lee and Lee compare the leaky and non-leaky scenarios for the surface

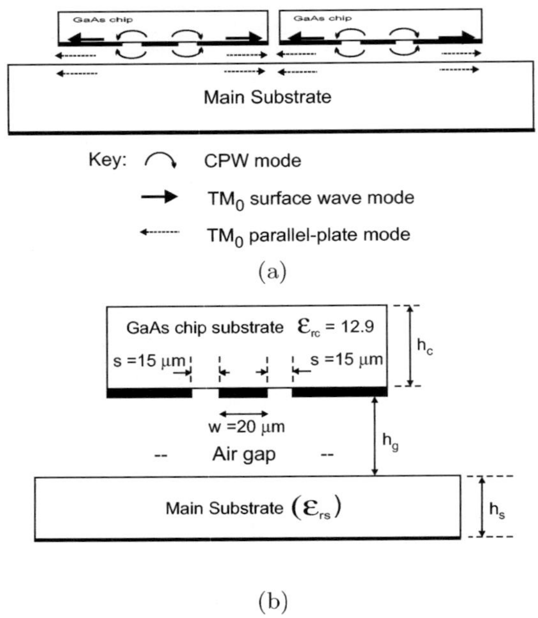

Figure 6.52 Energy leakage due to the various modes occurring in flip-chip structures: (a) modes; and (b) detail (after Lee and Lee [227]).

wave and parallel-plate modes over the frequency range 10 to 100 GHz. An alumina main substrate applies in all cases here, and it is found that the structure becomes leaky at frequencies above 51 GHz. Compared with any structure employing wire bonds or vias, the parasitic inductance of the metal bumps used for interconnection from the flip-chips is relatively small.

Further issues surrounding the millimetre-wave characteristics of flip-chip CPW interconnects for MCMs are addressed by Heinrich, Jentzsch and Baumann [228]. This team study the problem over the 1 to 70 GHz frequency range. Reflection effects are considered in detail, and it is found that the bump-pad length for the flip-chips is a significant parameter. Parasitic resonant and parallel-plate modes are reported upon, although the parallel-plate mode may be suppressed by patterning the conductor backing (see later). Heinrich *et al.* show that the important bump pad structure effect may be represented by a simple, single shunt capacitance model.

Hirose *et al.* [229] describe a flip-chip MMIC design using CPW at W-band. Primarily designed for the 77 GHz automotive radar application, this group report on GaAs amplifiers in two-stage and three-stage formats. For the three-stage amplifier a gain of exceeding 10 dB over 75 to 85 GHz is achieved.

Appreciating that the optimization of flip-chip interconnects is highly significant, especially at millimetre-wave frequencies, Jentzsch and Heinrich have investigated this issue [230]. This is actually a development on the work described in reference [228], and the main approaches considered are staggered bumps and underpasses for compensating the effects of the bumps on the high frequency performance of the flip-chips. Important approaches include: the use of staggered bumps, the addition of close-proximity high-impedance lines and the use of 90-degree underpasses to minimize cross-coupling.

In practice both the RF and also the mechanical characteristics of any circuit structure are equally important. It is of no use having an excellent electrically-performing chip if normal operating fatigue causes failure. With these issues as driving forces Feng *et al.* [231] have reported on the RF and mechanical characterization of GaAs flip-chip interconnects in CPW circuits with epoxy underfill. A simple lumped element equivalent circuit forms an appropriate model for the CPW with flip-chip interconnects. This has the usual C-L-C configuration, with the addition of resistance in series with L for modelling the losses.

Phase-shift measurements with and without the epoxy underfill indicate that this underfill has a significant effect on performance over the frequency range from 1 to 40 GHz. Whilst effects of the the substrate reduce the fatigue life of such flip-chips, the opposite is true for the effects of the underfill encapsulant, where a significant increase in the fatigue life applies.

In a key paper on three-dimensional vertical interconnect technology Goverdhanam, Simons and Katehi [232] report upon the design considerations, fabrication and comparative performance of both CPW and CPS as suitable media for 3-D interconnects. In this case, silicon technology applies and the results are equally applicable to both microwave and also very high speed digital circuits. As these applications are all on-chip, all the dimensions are in the orders of microns or tens of microns. The three main structures described are:

- CPS with cross-over underpass.
- CPS with a vertical interconnect embodying a 180° phase shift.
- CPW with a vertical interconnect embodying a 180° phase shift.

Goverdhanam *et al.* also show measured results for CPS loss in dB/mm versus the strip width. It is clear that this loss increases steeply as the strip width w decreases — most noticeably below about 40 μm. The loss at 30 GHz approaches 1.4 dB/mm when $w = 30$ μm.

In many of these configurations, particularly when using MCM-D technology, parasitic modes arising from CPW discontinuities can become an issue. Soliman *et al.* [233] address this issue and apply their approach to a 3-dB power splitter. A filter structure, due originally to Rittweger (hence named Rittweger's filter) is first studied — implemented in CPW with air bridges and thin-film tunnels to suppress the parasitic modes. The losses and S-parameters over the frequency band 10 to 40 GHz are shown for this configuration. A notable result is that the losses peak rapidly around 25 GHz when tunnels are used, and around 28 GHz when air bridges are implemented.

A 3 dB power splitter in MMIC technology is also shown. With this circuit, using a thin-film bridge, the return loss is better than -12 dB across the 20 to 30 GHz band.

FGCPW is also being applied to the very important and potentially high volume application of 77 GHz automotive radars, as indicated by a benchmark paper by Kerssenbrock and Heide [234]. These researchers point out that cost and mass production issues continue to act as major factors influencing the widespread adoption of miniature radars in this market.

Kerssenbrock and Heide show that flip-chip technology can be effectively applied to the development of a productionized near-distance automotive radar module. The basic transmission technology used is FGCPW, with the line itself deposited on an alumina substrate ($\varepsilon_r = 9.8$, h = 0.635 mm). There is no backside metallization and hence no possibility of parallel-plate moding. A glass layer is interposed between the alumina and the FGCPW structure.

A 77 GHz flip-chip (FC) phase-locked VCO has been built using this technology. This oscillator is used in conjunction with a 12 dB-gain amplifier, an FC power divider, a balanced mixer within a CPW rat-race hybrid and, finally, FC patch transmit and receive antennas. This approach results in a fully uniplanar structure that is potentially economic to manufacture in production volumes. There is no backside metallization, there are no transitions and there are no vias.

This approach may also represents a milestone for further developments in FGCPW applications across a wide spectrum of applications.

6.10 MIXERS, MICROMACHINED STRUCTURES AND OTHER CPW ISSUES

6.10.1 MIXERS AND FREQUENCY DOUBLER

It is well known that mixers form a vital component in any superheterodyne receiver and integrated mixer circuits, including those implemented using CPW, are therefore of considerable importance. A comparatively early example of a mixer incorporating CPW was due to Cahana in 1989 [235]. This was a double-balanced mixer implemented with both CPW and slotline. Another, submillimetre-wave, example was considered towards the end of Section 6.8.4, namely Vaupel and Hansen's 650 GHz mixer [226] for satellite sensing applications.

Most mixers are required to operate at considerably lower frequencies, but the output of CPW-related research results in this area is rather sparse compared with that in other application sectors. One example is from Hsu, Nguyen and Kintis [236] [236] who have reported a CPW mixer design and results. Their mixer operates from 7.1 to 10.5 GHz, and it is implemented on RT/Duroid® 6010. The RF input is fed by a 50 Ω CPW, and this is followed by a CPW-to-slotline transition to a broadening tapered slotline that leads to the hybrid mixing junction. Two Schottky diodes are connected to adjacent ground planes at the end of this tapered slotline to provide the mixing function. The Local Oscillator (LO), at 7 GHz, is fed into the mixing junction via a bandpass filter formed with cascaded CPW resonators (see earlier in this chapter). Finally, the IF output is delivered to the output CPW port via a lowpass filter (LPF) formed with a cascade of nine high-low-high CPW sections. The conversion loss is 6 to 10 dB across the band and the LO-to-IF isolation is better than 20 dB. RF-to-IF isolation is better than 38 dB.

Use can be made of junctions between CPW and other, non-planar, media and an interesting example of this approach is provided by Thiel and Menzel [237]. They report on milimeter-wave diode-based circuits — specifically a frequency doubler operating with an input frequency band of 20 to 25 GHz and providing an output frequency band of 40 to 50 GHz. A 30 GHz balanced mixed is used as the main frequency doubling element and the circuit is fundamentally based upon a CPW-to-finline transition.

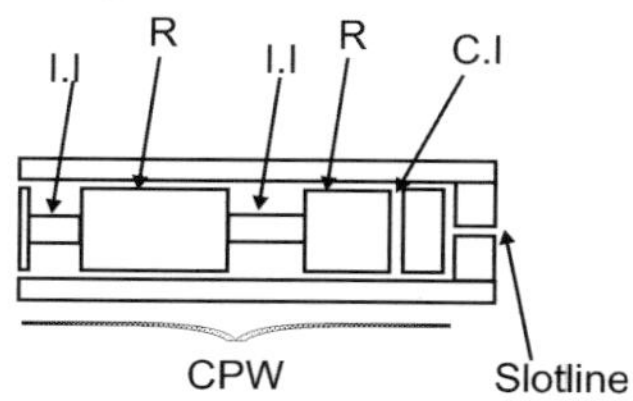

Figure 6.53 CPW-to-finline matching network. I.I = inductive inverter, R = resonator, C.I = capacitive inverter (after Thiel and Menzel [237]).

Since the characteristic impedance of the CPW is 50 Ω but that of the finline is 110 Ω, a matching network is clearly necessary between these elements and the layout of this is shown in Figure 6.53. This comprises the following:

- A cascade of five low-high-low CPW sections (in similar form to the LPF implemented by Hsu *et al.* in the mixer described above), followed by:
- A capacitive coupling gap forming an inverter (C.I.).
- A quasi-lumped short section of CPW.
- Another coupling gap (i.e. another inverter).
- The final output slotline that directly feeds the finline.

The input power is 0 dBm for optimum operation of this circuit and the conversion loss is better than 18 dB over the 40 to 50 GHz output band.

6.10.2 GaAs FET CHARACTERIZATION AND SPECIALIZED RESONATORS

This section is concerned with two relatively unusual aspects of CPW application. The first such application concerns GaAs FET characterization in a quasi-monolithic silicon environment which has been reported by Wasige *et al.* [238]. In this approach discrete GaAs FET chips are embedded in a high-resistivity silicon substrate and air bridges, deposited using thin film technology, are implemented for the interconnects. CPW feeds are used to connect to the transistors that are characterized up to 40 GHz. By using the air bridges instead of the more conventional bond wires parasitic inductances are reduced by more than half the original (wire) values. The air bridges are typically 6 μm in height and 130 μm long.

Another interesting application is the use of Electromagnetic BandGap (EBG) [1] resonators in conjunction with varactor tuning to achieve electronically tunable high-Q results. This has been studied by Yun and Chang [239]. CPW resonators are implemented and the PBGs have reflecting end faces forming Fabry-Perot resonators. Microstrip line and slotline structures of this form are also investigated. Tuning ranges of 7 to 11 GHz and also 6 to 12 GHz are achieved and loaded Q-factors are in the 240 to 300 range.

---

[1] Some would prefer that the term electromagnetic (or photonic) bandgap crystal not be used, but it does usefully summarize the integrated concept of a periodically varing dielectric and the design objective of creating a bandpass or bandstop characteristic reminiscent of a forbidden energy level. The alternative terms Electromagnetic BandGap (EBG), Electromagnetic Crystal, or synthetic periodic substrate are preferred.

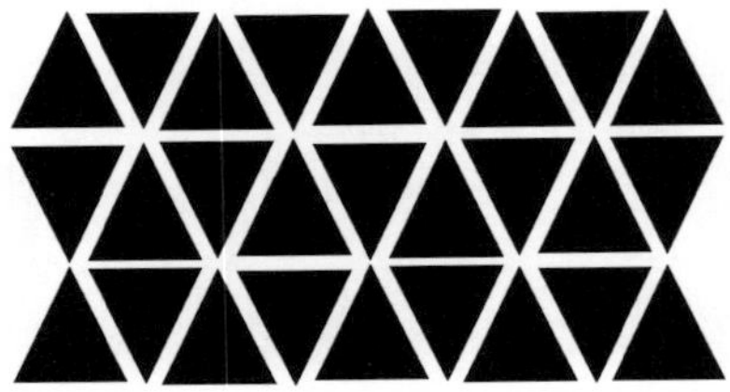

Figure 6.54 Patterned backside metallization suppressing unwanted surface or parallel-plate modes (after Hesselbarth and Vahldieck [243]).

6.10.3 MICROMACHINED STRUCTURES

Increasingly the application of micromachining is making substantial advances possible in many areas of advanced chip technology, including coplanar waveguides and related structures. For example, Milanovic *et al.* [240, 241] have published the results of their work on micromachined CPW in CMOS technology. These workers show that an etched V-shaped pit can readily be formed beneath the CPW structure and excellent control over the very small separating gaps (s) is achieved.

Milanovic *et al.* provide the results of measurements over the 1 to 40 GHz range. Before etching the line loss is as high as 3.8 dB/mm, but following etching this loss is reduced to 0.4 dB/mm — approximately an order of magnitude reduction. However it is pointed out that the very thin metal strip layers available in standard CMOS lead to relatively high series impedance that can only be addressed by widening the strip widths. This in turn means a take-up of valuable real estate on the chip.

Another aspect of micromachining is the prospect of high-Q millimetre-wave structures and this is explored by Guillon *et al.* [242]. In this work a silicon substrate is used and a micromachined silicon Dielectric Resonator (DR) is implemented operating in the whispering gallery modes. Coupling into these modes is achieved by means of CPWs that are formed on a dielectric membrane, itself supported by the silicon substrate.

Loaded Q-factors ranging from 500 to 2400 are measured at 35 GHz and an oscillator is implemented at this frequency. Unloaded Q-factors of 2040 at 77 GHz and 95 GHz have been predicted by finite element 2D simulation.

6.10.4 LEAKAGE SUPPRESSION AND 50 GHz INTERCONNECT

As mentioned earlier in this chapter, the excitation of surface-wave modes or parallel-plate modes can become a problem with many uniplanar structures, especially when these are conductor-backed. Any approach that can minimize or suppress these modes — 'leakage suppression' — therefore represents an important advantage. As discussed above, Soliman *et al.* [233] report extensive results using air bridge technology to suppress parasitic modes.

Another (simpler) approach comprises patterned backside metallization and this has been reported by Hesselbarth and Vahldieck [243]. In this technique the backside metal layer is made discontinuous in that a pattern of interleaved triangles is formed, as shown in Figure 6.54. The planar dimensions must be kept less than $\lambda_g/5$ at the highest frequency of intended operation to ensure that resonance is avoided. Over the 1 to 10 GHz frequency range Hesselbarth and Vahldieck's results indicate that, using a shorted CPW test line, without the patterned metallization the unwanted transmission is around −8 dB whereas by patterning this reduces it down to −60 dB.

Both CPW and CPS are finding increased use for very high speed on-chip (and off-chip) digital technology and also for millimetre-wave applications. As far as possible silicon technology always remains the first choice, and therefore interconnect design at these high frequencies is increasingly important.

Work has been reported on 50 GHz interconnect design in standard silicon technology, notably by Kleveland, Lee and Wong [244]. This group of researchers studied CPWs with characteristic impedances ranging from 40 to 90 Ω, and it was found that coupling through the silicon substrate resulted in substantially increased losses above 30 GHz for the widest lines (40 microns) tested. Kleveland *et al.* point out that the actual results obtained are counter-intuitive because in some frequency ranges the lowest loss was achieved using relatively narrow lines deposited on a low-resistivity (standard) substrate. Full compatibility with standard CMOS was obtained.

There are two recommendations for achieving low-loss CPWs on silicon:

1. Reduce the line widths and spacing in order to avoid substrate coupling, and:
2. Use low-resistivity substrate or metal/poly/diffusion in order to provide a low resistance for the lateral ground connection.

The subject of leakage from planar transmission lines as a multimoding phenomenon has received considerable attention [245–254]. In 1993 it was shown that a leaky dominant-like mode can exist with most printed-circuit transmission lines [246, 247]. Fortunately, the transition from the bound TEM or quasi-TEM mode to a leaky mode can be predicted [245], and occurs when spacings, gaps and widths become an appreciable part of a wavelength. The general guideline is to keep planar transmission line dimensions small.

CPW lines have an additional multimoding phenomenon to contend with. This is the excitation of slotline modes, or the equivalent problem, of having a slotline signal induced by signals on other interconnects that produce a field configuration compatible with the slotline mode. The solution is to enforce ground equalization using bond wires or air bridges to regularly connect the two ground conductors of CPW together, especially at discontinuities. It has been shown that air bridges are far superior to bond wires for implementing this [251, 254].

6.10.5 LIGHT DEPENDENCE OF SILICON FGCPW

An attractive prospect may be afforded based upon the fact that incident light significantly influences the losses of FGCPW fabricated on semiconductor. This illumination dependence has been reported by Spiegel and Madjar [255], whose results show that the conductivity of the silicon is markedly affected by the intensity of the light.

Spiegel and Madjar state that for silicon FGCPW lines losses can be decreased by removing the passivation layer between the centre conductor and the ground plane. As most of the electric field lies within the separating slots (s), the reduction in the surface charges results in a substantial drop in effective substrate conductivity. Under illumination the line attenuation increases by a factor of 6, but the insertion phase

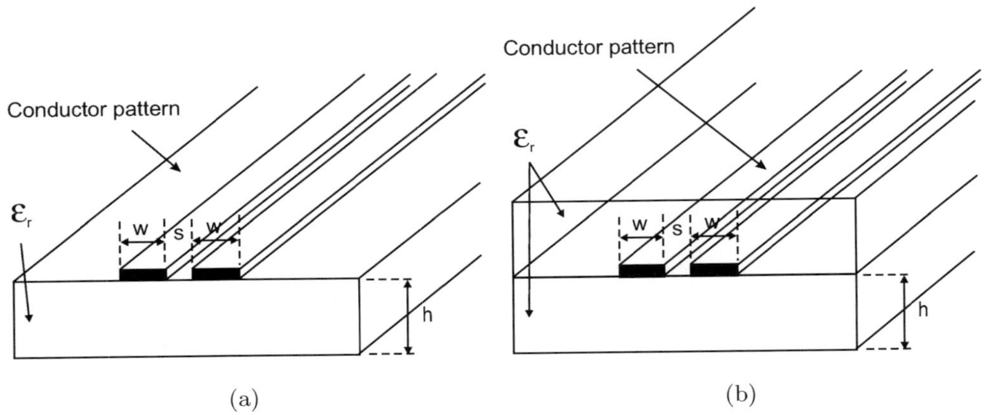

Figure 6.55 Coplanar strip (CPS) or differential line structures on (a) the surface of a dielectric or (b) embedded [243].

shift remains constant at frequencies exceeding 1 GHz. Silicon FGCPWs behave as variable attenuators with constant insertion phase (unlike PIN diodes or PHEMTs), controlled by: light illumination; temperature; or applied voltage. Applications for this effect include: (i) gain control at microwave or millimetre-wave frequencies for phase tracking systems; (ii) Automatic Gain Control (AGC); (iii) power control devices for transmitters; and (iv) temperature control devices.

6.11 DIFFERENTIAL LINE AND COPLANAR STRIP (CPS)

Two coplanar strip (CPS) structures are shown in Figure 6.55. Whether these structures are referred to as CPS or differential line depends on how they are driven. We will talk more about this soon but often CPS is used as the unifying term. The CPS line of Figure 6.55(a) is formed by two conductors on top of a dielectric of sufficient thickness that the metal backing has negligible effect on the field structure. The structure in Figure 6.55(b) is known as embedded CPS and is encountered when multiple dielectric and metal layers are used. From the extensive studies presented in the previous sections we conclude that the dielectric thickness needs to be two to three times greater than the maximum value of w or s.

The two-conductor line of Figure 6.55 is a differential line or balanced line when neither conductor is ground. This is also approximated when the signals are push-pull related and a virtual ground (or AC ground) is formed. Then the signal on one conductor is, at one time, above virtual ground and the signal on the other is then below. Differential line is commonly used with RFICs as the signal path on the IC is almost always differential since differential circuits on-chip remove many of the distortion and noise-coupling concerns. At frequencies in the low gigahertz regime, transmission line structures are rarely used on-chip. However regular differential transmission lines are used off-chip to connect to the RFIC. It s common to mount RFICs on LTCC (Low Temperature Co-fired Ceramic) so that filters, baluns and matching can be realized on the LTCC which has a much lower area cost than the RFIC.

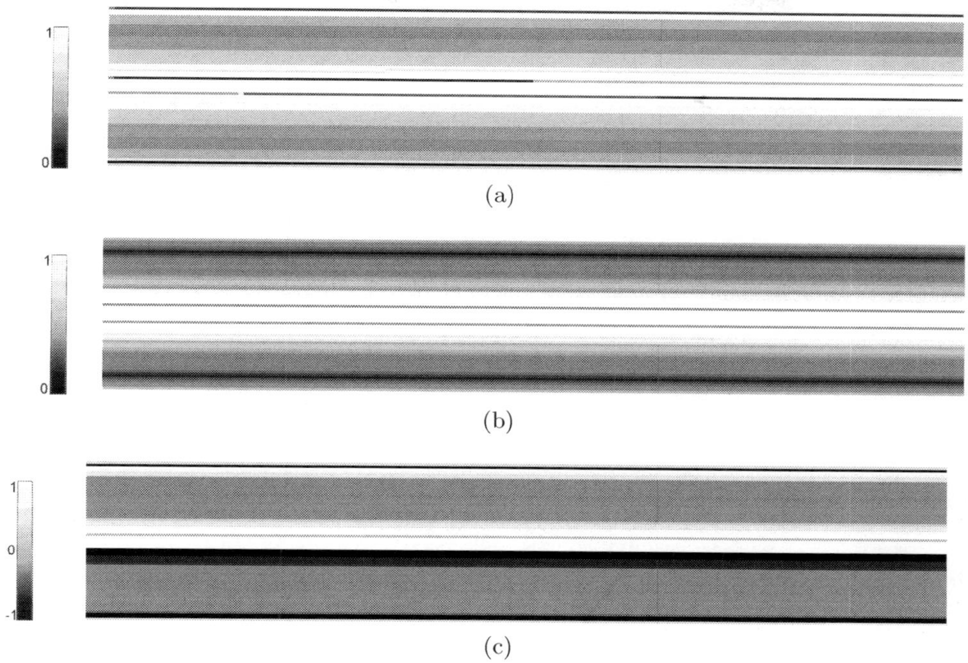

Figure 6.56 Instantaneous normalized current magnitude and charge distribution along a differential line: (a) longitudinal current; (b) transverse current; and (c) charge. The current and charge are normalized and use the scales shown on the left hand side. The peak transverse current is 1500 times smaller than the peak of the longitudinal current. (The the frequency is 10 GHz, $s = 2.5$ μm, and $w = 10$ μm).

When one of the conductors of the transmission line is grounded the line is known as Coplanar Strip Line or CPS. CPS is a good structure to use, instead of CPW, when space is at a premium. CPS and differential line have the same electrical characteristics as long as the metal backing (or ground planes) are sufficiently removed. The attraction of both CPS and differential line is that the signal return path is clearly defined.

The discussion and results presented below relate directly to differential line which is the much more common utilization of this conductor structure. In Figure 6.56 the normalized current magnitude and charge distribution on a differential line is shown. These profiles are calculated at 10 GHz for an electrically short section of line so that there is very little longitudinal variation in the profiles. As with all of the other transmission line structures charges accumulate along the inside edges of the conductors. The concentration however is greater than for any of the other structures that we have considered. Lower level charges also accumulate at outside edges of the conductors. These results indicate that the width, w, and the separation, s, will both be important in determining the characteristics of the differential line.

As mentioned above, differential line is a particularly important transmission line medium where RFICs are used — both for on-chip connections at frequencies 10 GHz and higher and for connections to the differential input and outputs of the RFIC. In

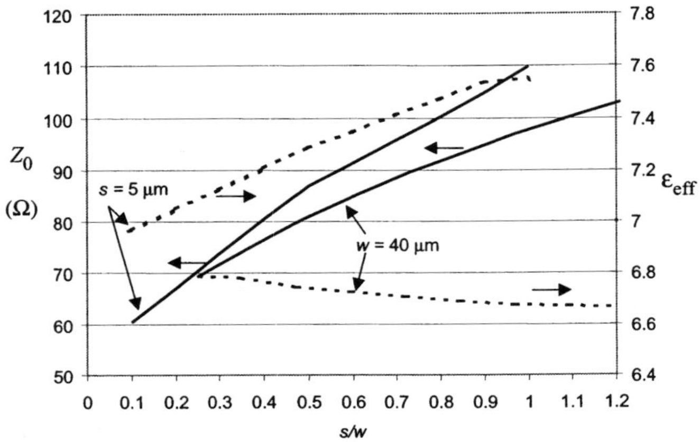

Figure 6.57 Characteristic impedance and effective relative permittivity of open differential line fabricated on high resistivity silicon ($\varepsilon = 11.9$) for various s/w ratios. Frequency = 10 GHz, height above ground plane, $h = 650$ μm.

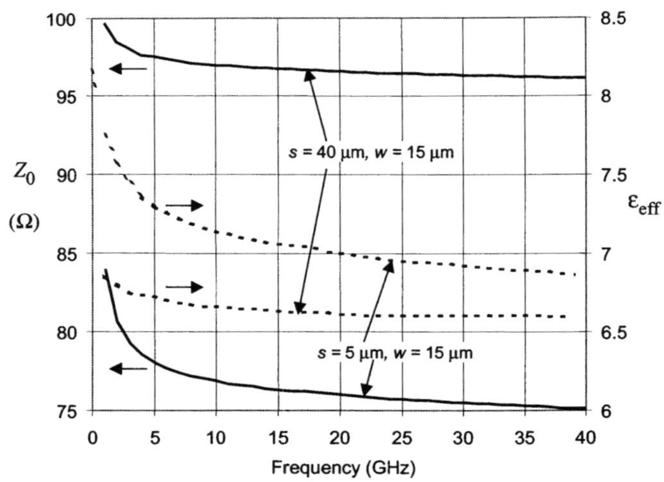

Figure 6.58 Frequency dependence of characteristic impedance and effective relative permittivity of open differential line fabricated on high resistivity silicon ($\varepsilon = 11.9$). Height above ground plane, $h = 650$ μm.

Figures 6.57 and 6.58 curves of characteristic impedance and effective permittivity are presented for a differential line fabricated on the surface of high resistivity silicon. In the first chart, Figure 6.57, the characteristic impedance and the effective permittivity are plotted against the separation-width ratio s/w for two different families of lines. An important point to note here is that the effective permittivity of the lines varies as s/w varies. This results from the field lines fringing more into the air region for larger s values.

Dispersion data are shown in Figure 6.58 for two differential lines fabricated on the surface of high resistivity silicon. The energy in the field concentrates more in the air region at high frequencies and this is the main reason for the impedance variation with frequency.

A more common usage of differential line is as an embedded structure as shown in Figure 6.55(b). Using the common approach based on conformal mapping Crampagne and Khoo [256] derived the characteristic impedance of the line:

$$Z_0 = \frac{240\pi}{\sqrt{2\varepsilon_0(\varepsilon_{r1} + \varepsilon_{r2})}} \frac{K(k)}{K'(k)}. \tag{6.58}$$

This equation allows the dielectrics above and below the strips to be different with the top dielectric having a permittivity of ε_{r1} and the bottom dielectric to have a permittivity of ε_{r2}. In Equation (6.58) $K(k)$ is the complete elliptic integral of the first type; $K'(k)$ is its complementary function, and

$$k = \frac{s}{s + 2w}. \tag{6.59}$$

So k contains the dimensional information, and the important thing to note is that Z_0 depends on the geometrical ratio $k = w/(w + 2s)$, that is, on the ratio s/w as $w/(w + 2s) = 1/(1 + 2s/w)$. Note that this is not a simple dependence between geometry and Z_0, as k is the argument of an elliptical integral function. However it does mean that lines with the same k value, or equivalently the same s/w ratio, will have the same characteristic impedance (within the accuracy limits of the analysis). The characteristic impedances at 2.45 GHz of differential lines embedded in an LTCC are presented in Figure 6.59 for various s/w ratios. Now that the effective permittivity cannot vary (it is equal to the permittivity of the medium) it is seen that Z_0 is essentially determined by the s/w ratio. This is in contrast to the result with the open differential line.

Inevitably there will be a ground plane near the differential line. The analysis and results above assume that the line is sufficiently separated from the ground plane. To investigate the effect of ground plane separation, the characteristic impedance of the LTCC-embedded line was calulated for various separations. The results are plotted in Figure 6.60 as a function of h/s. Ideally, the presence of the ground plane should have no effect on the characteristics of the line and, above $h/s = 3$, the effect is negligible and is only about 1.5% at $h/s = 2$. However values of h/s near or less than 2 should be avoided so that the microstrip (or parallel-plate) mode, with electric field line from the conductors of the differential line to the ground planes, is not excited [257]. Values of h/s greater than 3 are strongly preferred but compromises are necessary given the maximum package thickness and tolerances on the dimensions of fabricated

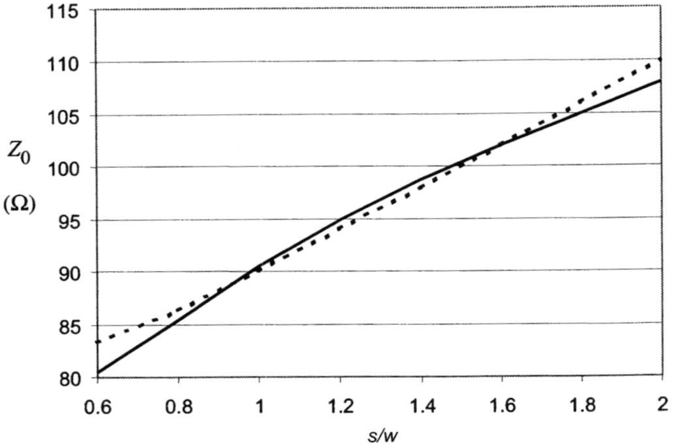

Figure 6.59 Characteristic impedance and effective relative dielectric constant of an embedded differential line fabricated in LTCC with $\varepsilon = 7.3$ for various s/w ratios. Frequency = 10 GHz, height from the line to each ground plane, $h = 340\ \mu$m. Results derived using SONNET^{TM} simulations.

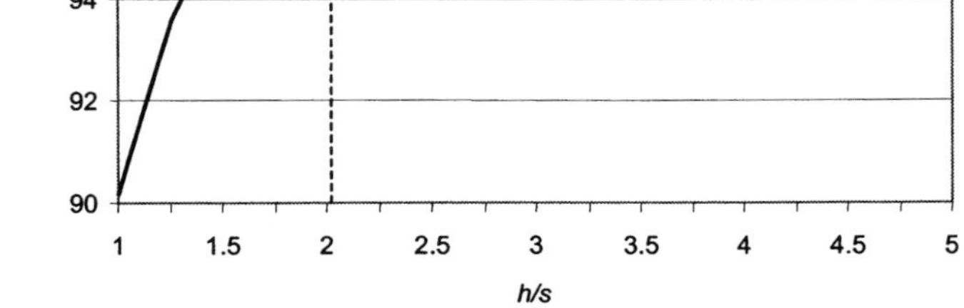

Figure 6.60 Characteristic impedance of an embedded differential line fabricated in LTCC with $\varepsilon = 7.3$ for various h/s ratios. Frequency = 2.45 GHz, Separation $s = 100\ \mu$m, and width $w = 70\ \mu$m. Results derived using SONNET^{TM} simulations.

lines. The ground plane separation should not be used as a parameter to adjust the characteristic impedance of the line.

Synthesis of dfferential line requires that we arrive at the physical dimensions of the line from the electrical requirement, in this case Z_0. The desired synthesis equations are obtained by inverting Equation (6.58):

$$\frac{K(k)}{K'(k)} = \frac{\sqrt{2\varepsilon_0(\varepsilon_{r1}\varepsilon_{r2})}}{240\pi} Z_0. \tag{6.60}$$

The value of k is obtained from $K(k)/K'(k)$ using the following polynomial expression which was fitted by Crampagne and Khoo [256]:

$$k = \frac{w}{w+2s} = \sum_{i=0}^{N} a_i (K/K')^i \tag{6.61}$$

where the polynomial coefficients a_i were fitted over two ranges:

For $0 \leq K/K' < 0.5$:

$$\begin{array}{lll} a_0 = 0.00913 & a_1 = -0.1352 & a_2 = +0.2689 \\ a_3 = -0.1367 & a_4 = +5.523 & a_5 = -5.087 \end{array}.$$

For $0.5 \leq K/K' \leq 1$:

$$\begin{array}{lll} a_0 = -0.0656 & a_1 = -0.3211 & a_2 = 1.676 \\ a_3 = 0.5984 & a_4 = -1.849 & a_5 = 0.6705 \end{array}.$$

One conclusion of the results reported here is that the range of characteristic impedances available is limited. There is a relatively small geometrical dependence (the s/w ratio can be adjusted) but there is a strong dependence on permittivity. The characteristic impedance of embedded differential line (as with all lines) is proportional to the inverse of the square root of the permittivity. So a graphical approach to designing an embedded line with a specific characteristic impedance is as follows. Suppose a line is to be designed on LTCC with a characteristic impedance, Z_0 of 50 Ω. Using Figure 6.59 it is seen that $s/w = 1$ yields a nominal characteristic impedance $Z_0' = 90$ Ω for a nominal relative permittivity $\varepsilon_r' = 7.4$. Then the material required to realize the Z_0 has a permittivity

$$\varepsilon_r = \left(\frac{Z_0'}{Z_0}\right)^2 \quad \text{so that} \quad \varepsilon_r' = \left(\frac{90}{50}\right)^2 7.3 = 23.7 \tag{6.62}$$

LTCC is available in a range of permittivities but many other factors go into selecting the electrical characteristics including the length of quarter-wavelength long lines (required for filters and baluns) and the not insignificant physical tolerances which result in large variations in electrical length if high dielectric constant materials are used. Thus materials with permittivities of 6–10 are often chosen for RFIC applications so that the available range of characteristic impedances for differential lines is 80 to 110 Ω as shown in Figure 6.59.

6.12 SUMMARY

Coplanar waveguide is the transmission medium of choice above 20 GHz or so although this threshold can be a little higher for on-chip lines as they are so small. CPW confines the electromagnetic fields in a more localized manner than does microstrip, thus reducing spurious coupling, radiation and dispersion. The electrical characteristics of CPW are defined by the lateral dimensions which can accurately be defined photolithographcally which is a significant advantage when the thickness of a substrate cannot be accurately controlled as with semiconductors. CPW-like structures are becoming common for long high speed digital interconnect and RFIC connections. Shield lines are placed between neighbouring signal lines thus creating Ground Signal Ground (GSG) configurations. The concept here is to provide a precisely defined signal return path and to reduce capacitive coupling of signal lines. Coplanar strip (CPS) lines are also used as long high-performance digital interconnect media where the desirable attribute is the differential signalling. Differential lines, having the same structure as CPS, are commonly used as off-chip transmission lines connecting RFICs. This is also a desirable attribute in some microwave and millimetre-wave circuits where a balanced line (the microwave term for differential line) is required to drive balanced structures such as antennas.

7

Discontinuities in Microstrip and Stripline

7.1 THE MAIN DISCONTINUITIES

Virtually all practical distributed circuits, whether in waveguide, coaxial lines, or any other propagation structure, must inherently contain discontinuities. A straight, uninterrupted length of any transmission structure (strictly speaking, with a perfectly invariant cross-section) would be genuinely continuous, i.e. it would not contain any discontinuities. In fact such straight lines, taken alone, are of little engineering use, and in any case junctions are essential. In the region of bends, abruptly stopped open-circuits, width changes and transitions discontinuities occur in the transmission line.

Although such discontinuities give rise to only very small capacitances and inductances (often < 0.1 pF and < 0.1 nH) the reactances of these become particularly significant at the high microwave and into millimetre-wave frequencies. The performance of amplifiers, for example, has been shown to be considerably affected by the microstrip discontinuities [258]. As with dispersion, we can very often neglect these discontinuities when the frequencies involved do not exceed a few gigahertz. Above approximately 10 GHz they are definitely very significant.

Several forms of discontinuity emerge from circuit requirements:

(a) foreshortened open-circuits
(b) series coupling gaps
(c) short-circuits through to the ground plane
(d) right-angled corners or 'bends' (unmitred and mitred)
(e) step width changes
(f) transverse slit
(g) T-junction
(h) cross-junctions.

An example of a microwave transistor amplifier layout [259] is shown in Figure 7.1, and at least three of the discontinuities can be readily identified. Some of them are numbered on the diagram, in accordance with the quoted list. Many other circuits, such as filters, mixers and oscillators, involve several discontinuities. All technologies, whether hybrid MIC or MMIC, inherently involve transmission discontinuities.

It is beyond the scope of this treatment, being oriented towards circuit design, to

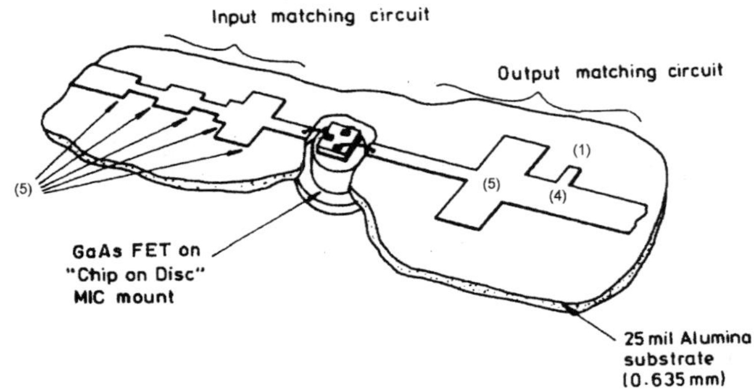

Figure 7.1 Layout of a simple hybrid-MIC microwave amplifier using a GaAs MESFET, showing several discontinuities in the microstrip lines (DC bias filters are not shown).

consider the techniques in any detail which have been used to evaluate discontinuity capacitances and inductances.

In the first twelve sections of this chapter we consider discontinuity modelling based upon equivalent capacitances and inductances. Later, in Section 7.13 on page 250, non-linear frequency-dependent effects are considered.

These techniques have received a good, comparative assessment in detail elsewhere (e.g. Gupta *et al.* [88]), and we shall only note the approaches briefly here, starting with capacitance evaluation techniques as follows:

(a) matrix inversion method
(b) variational method
(c) Galerkin's method in the spectral domain
(d) use of line sources with charge reversal.

Of all the techniques the last (d), adhered to by Silvester and Benedek [260], appears to have been the most successful. This is probably because it avoids the subtraction of two large and similar numbers.

Discontinuity inductance evaluation has been performed by Thomson and Gopinath [261]. Their method starts with magnetic vector potential and uses Maxwell's equations and divergence. Inductance is finally obtained after integrating the current density distribution.

At frequencies much beyond 10 or 12 GHz, the lumped C and L description of most discontinuities becomes less and less meaningful. Discontinuities at these higher frequencies are probably best described by means of scattering matrices, although in many cases considerable success has been enjoyed by using the lumped LC approach at frequencies up to around 18 GHz.

Radiation from various discontinuities gives rise to equivalent resistive elements. These will not be discussed in any detail here since they enter mainly into the province of those interested in antennas rather than circuits, except where the reduction in Q-factor is significant

In some cases techniques have been developed which compensate, at least over a particular range of frequencies, the existing discontinuity effects. Where this is the case less detailed attention is given here to the properties of the uncompensated discontinuity. Although 'quasi-static', the effectiveness of results obtained from

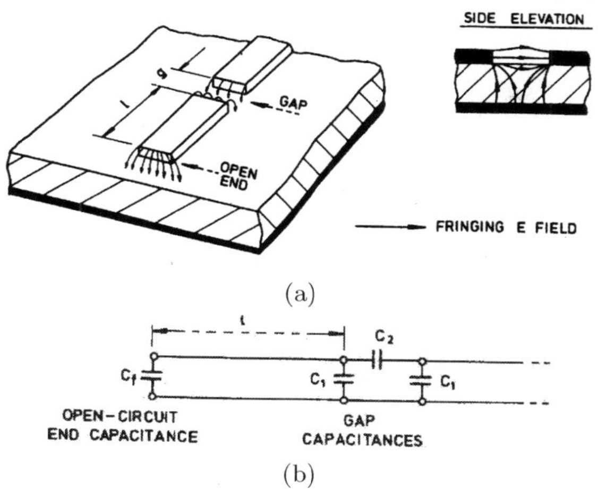

Figure 7.2 Microstrip open-circuits: (a) layout and electric fields; and (b) equivalent lumped capacitive network.

these various previously mentioned approaches has been adequately vindicated by incorporating the methods in several full circuit designs and checking with careful measurements. The calculation of the reactive effects due to most discontinuities will now be shown, as far as is practicable.

7.2 THE FORESHORTENED OPEN CIRCUIT

There are essentially three phenomena associated with the 'open-circuit':

(a) There will be fringing fields extending beyond the abrupt physical end of the metallic strip.
(b) Surface waves will be launched from the end of the strip.
(c) Energy will be radiated from the open end.

Item (a) may be accounted for by assuming some equivalent capacitance to be connected at the open end, and this feature nearly always dominates the 'end-effect'. Items (b) and (c) require equivalent shunt conductance at the open end of the line. In most cases, in this work it will not be necessary to quantify these last two phenomena, but their minimization may be carried out on a practical basis in circuits involving microstrip. This has already been discussed in Chapter 5, and James and Ladbrooke [169] have indicated how substrate slots may be used to minimize surface-wave interactions.

The physical appearance of both a foreshortened open-circuit and a series gap are shown in Figure 7.2, where equivalent lumped capacitor networks are also given.

Several quasi-static field calculations of the capacitance C_f have been documented. The most reliable, judged by comparison with many independent measurements, seems to be the work of Silvester and Benedek [260]. While being thorough, the resulting curves of results need further interpretation and a curve-fitting formula can be deduced for microwave circuit design. In their paper, Silvester and Benedek give

Table 7.1 Coefficients for (7.1). (©1972 IEEE. Reprinted, with permission, from Silvester and Benedek [260].)

ε_r	1.0	2.5	4.2	9.6	16.0	51.0
1	1.110	1.295	1.443	1.738	1.938	2.403
2	−0.2892	−0.2817	−0.2535	−0.2538	−0.2233	−0.2220
3	0.1815	0.1367	0.1062	0.1308	0.1317	0.2170
4	−0.0033	−0.0133	−0.0260	−0.0087	−0.0267	−0.0240
5	−0.0540	−0.0267	−0.0073	−0.0113	−0.0147	−0.0840

a formula which amounts to a series expansion fit to their theoretical data. The expression is

$$\frac{C_f}{w} = \exp\left[2.2036 \sum_{i=1}^{5} K_\varepsilon \left(\log \frac{w}{h}\right)^{i-1}\right] \quad \text{pF/m} \tag{7.1}$$

where the coefficients K_ε are given in Table 7.1.

These results may readily be plotted out as a set of curves giving K_ε as a function of ε_r and with i as parameter. Thus, coefficients K_ε for other materials may be found, within a fairly narrow range of permittivities but including all the important semiconductors with their permittivities lying between 9.6 and 16.0. A set of first-order linear approximation equations can be simply obtained from these curves by curve-fitting so that the formula (7.1) can be used in an iterative CAD routine.

An empirical formula that yields, quite simply, the equivalent end-effect length is given shortly, but first we must properly define this quantity.

7.2.1 EQUIVALENT END-EFFECT LENGTH

In many aspects of circuit design it is very useful to pretend that the microstrip line is longer than it actually is, to account for the end-effect. Then we can deal with fully distributed (microstrip) structures throughout, and do not need to work separately in terms of lumped capacitance. The concept is illustrated in Figure 7.3. We can now define the equivalent end-effect length, l_{eo}:

> l_{eo} is the equivalent extra length of microstrip line, having all the propagation parameters applicable to the main line, continuing from the 'physical' end plane to the final electrically open end plane.

Notice that the 'line' of length l_{eo} must be an exact continuation of the main transmission line — same Z_0 and hence the same w/h and identical ε_{eff}.

It is a simple matter to obtain the end-effect length as a function of C_f, Z_0 and ε_{eff}. The input reactance to the extra length of line is given by a standard open-circuit terminated line result presented in Chapter 1:

$$X_{eo} = -jZ_0 \cot \beta l_{eo}. \tag{7.2}$$

Also, the capacitive reactance due to C_f is

$$X_f = \frac{1}{j\omega C_f}. \tag{7.3}$$

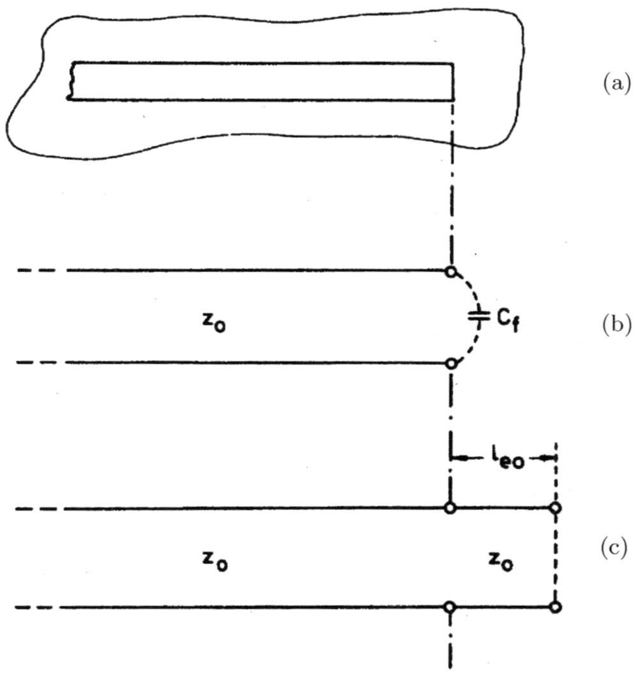

Figure 7.3 Development of the equivalent end-effect length concept: (a) physical transmission line; (b) transmission line with equivalent end fringing capacitance C_f; and (c) transmission line with equivalent extra transmission line of length l_{eo}.

For equivalence we must equate Equations (7.2) and (7.3), giving

$$\frac{1}{\omega C_f} = \frac{Z_0}{\tan \beta l_{eo}} \tag{7.4}$$

Since $l_{eo} \ll \lambda_g$ we may apply the small-angle approximation: $\tan(\beta l_{eo}) \rightarrow \beta l_{eo}$ so that Equation (7.4) then becomes

$$\frac{1}{\omega C_f} \approx \frac{Z_0}{\beta l_{eo}} \approx \frac{Z_0 \lambda_g}{2\pi l_{eo}}. \tag{7.5}$$

Writing c as the free-space velocity, this may be written as:

$$\frac{1}{\omega C_f} = \frac{cZ_0}{\omega\sqrt{\varepsilon_{\text{eff}}} l_{eo}}. \tag{7.6}$$

Then, finally,

$$l_{eo} \approx \frac{cZ_0 C_f}{\sqrt{\varepsilon_{\text{eff}}}}. \tag{7.7}$$

Frequency-dependent effects are discussed in Section 7.13 on page 250. For most microstrip lines, certainly those on alumina substrates less than 1 mm thick, the static-TEM expressions given here appear to work quite accurately over the approximate frequency range:

$$2 \ll f \ll 20 \text{ GHz}. \tag{7.8}$$

An alternative empirical expression, yielding the length extension directly, has been given by Hammerstad and Bekkadal [97]:

$$l_{eo} = 0.412h \left(\frac{\varepsilon_{\text{eff}} + 0.3}{\varepsilon_{\text{eff}} - 0.258} \right) \left(\frac{w/h + 0.262}{w/h + 0.813} \right). \tag{7.9}$$

It appears that, over a wide range of materials and microstrip width-to-height ratios (w/h), this expression can often give errors of 5% or more. Where such errors are acceptable, which is frequently the case for this quantity, Equation (7.9) may be used, because it involves less computing than the more accurate Equation (7.7).

7.2.2 UPPER LIMIT TO END-EFFECT LENGTH (QUASI-STATIC BASIS)

End-effect length is usually normalized to h, to obtain l_{eo}/h, which is the quantity usually employed as a design parameter. The graph of Figure 7.4. shows l_{eo}/h plotted as a function of w/h for the particular case of a substrate having $\varepsilon_r = 11.0$. The maximum value, shown as a broken line, is determined on a static capacitance basis for a line embedded totally in a free-space environment. Cohn [262] has given an exact expression for the fringing capacitance due to a foreshortened semi-infinite line situated at a distance $b/2$ above a ground plane, where the entire structure is completely filled with dielectric of relative permittivity ε_r. Cohn's expression may be reinterpreted for the microstrip case and calculated for l_{eo}/h. This yields the result

$$\left(\frac{l_{eo}}{h} \right)_{\text{max}} = \frac{2}{\pi} \ln(2). \tag{7.10}$$

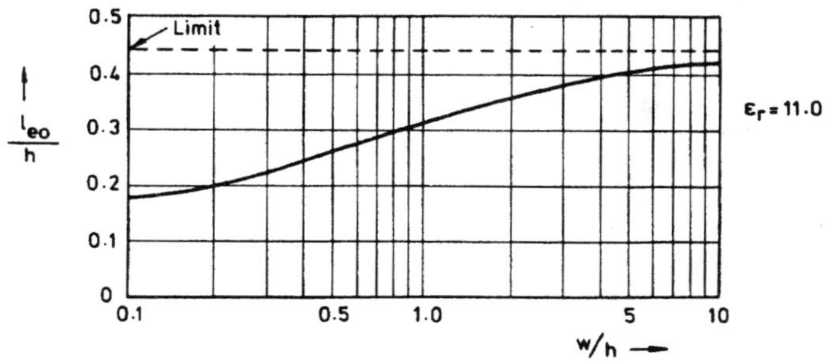

Figure 7.4 Normalized end-effect length (l_{eo}/h) as a function of shape ratio w/h.

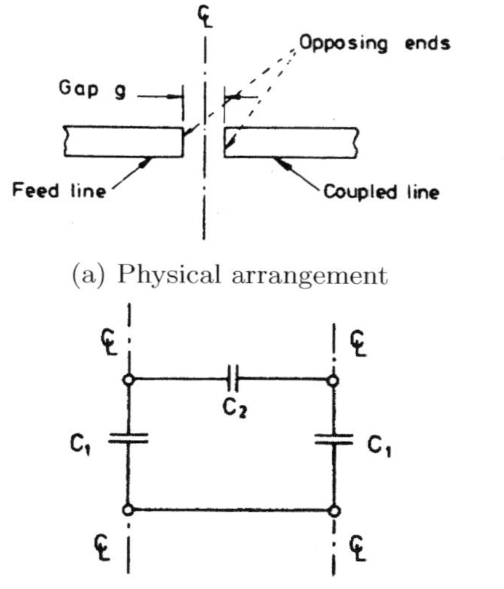

Figure 7.5 The series gap and its equivalent lumped circuit.

Recent work has suggested that this statically derived 'limit' may not be absolute. It does, however, give an indication to the designer of the approximate maximum value likely for open-circuit equivalent length.

7.3 THE SERIES GAP

The physical appearance of the microstrip series gap is also given in Figure 7.2. This discontinuity is important in some forms of end-coupled filters, coupling to resonators, and other circuits. For convenience here the nomenclature is repeated in Figure 7.5.

For energy to be coupled across the gap the 'open-circuit' microstrip ends have to be equally and oppositely charged. Significant capacitance therefore exists across the gap, and is denoted C_2. Grounded capacitors C_1 represent the fields fringing directly down to ground from each end of the line at a symmetrical gap.

An extensive analysis of the gap allowing for the necessary inversion to account for opposite potential signs on each side leads to comprehensive results that work for tightly coupled gaps. Since this is rarely encountered in MIC practice, only the result for loosely coupled gaps is given here. By analogy with the development of Equation (7.7), for the foreshortened open-circuit, the expression for the gap end-effect line extension may be written

$$l_{eg} \approx \frac{cZ_0(C_1 + C_2)}{\sqrt{\varepsilon_{\text{eff}}}}. \tag{7.11}$$

This now means that, where we know C_1 and C_2, we can determine the line extension l_{eg}. This is fairly satisfactory in the case of simple resonators and some other structures where the degree of coupling, leading to the gap separation g, has been rather arbitrarily set.

However, this situation is quite unsuitable for synthesis, e.g. of a filter. In such a case we require ultimately to determine g itself from a starting point of filter characteristics. (For the present purposes, suffice it to say that we can obtain the gap capacitance information from prototype filter characteristics.) Garg and Bahl [263] have given curve-fitted expressions for the gap capacitances as follows:

$$\frac{C_o}{w} = \left(\frac{g}{w}\right)^{m_o} \exp(k_o) \text{ pF/m} \tag{7.12}$$

$$\frac{C_e}{w} = \left(\frac{g}{w}\right)^{m_e} \exp(k_e) \text{ pF/m} \tag{7.13}$$

applicable for $\varepsilon_r = 9.6$ and over the range

$$0.5 \leq \frac{w}{h} \leq 2 \tag{7.14}$$

where

$$C_o = 2C_2 + C_1 \tag{7.15}$$

and

$$C_e = 2C_1. \tag{7.16}$$

The indices and arguments are defined by the following set of equations:

$$\left.\begin{aligned} m_o &= \tfrac{w}{h}\left\{0.619 \log\left(\tfrac{w}{h}\right) - 0.3853\right\} \\ k_o &= 4.26 - 1.453 \log\left(\tfrac{w}{h}\right) \end{aligned}\right\} 0.1 \leq \tfrac{g}{w} \leq 1.0 \tag{7.17}$$

$$\left.\begin{aligned} m_e &= 0.8675 \\ k_e &= 2.043\left(\tfrac{w}{h}\right)^{0.12} \end{aligned}\right\} 0.1 \leq \tfrac{g}{w} \leq 0.3 \tag{7.18}$$

$$\left.\begin{aligned} m_e &= \left[1.565/(w/h)^{0.16}\right] - 1 \\ k_e &= 1.97 - \tfrac{0.03}{w/h} \end{aligned}\right\} 0.3 \leq \tfrac{g}{w} \leq 1.0 \ . \tag{7.19}$$

Garg and Bahl [263] also showed that values can be calculated for C_o and C_e for other values of substrate permittivity ε_r by using the following expressions:

$$C_o(\varepsilon_r) = C_o(9.6)\left(\frac{\varepsilon_r}{9.6}\right)^{0.8} \quad \text{and} \tag{7.20}$$

$$C_e(\varepsilon_r) = C_e(9.6)\left(\frac{\varepsilon_r}{9.6}\right)^{0.9} \ . \tag{7.21}$$

These relations hold over the range

$$2.5 \leq \varepsilon_r \leq 15 \tag{7.22}$$

and an accuracy of 7% is quoted. It is noted that semiconductor substrates are accommodated.

In order to obtain g we select Equation (7.12), because C_o involves both C_1 and C_2 gap capacitances, and rearrange as follows:

$$\ln\left(\frac{g}{w}\right) = \frac{1}{m_o}\left[\ln\left(\frac{C_o}{w}\right) - k_o\right]. \tag{7.23}$$

Having obtained a first value for g, it is desirable that a second value be found using Equation (7.13). Should the two values differ significantly then the final value of g will be that which most closely satisfies the two relations.

7.3.1 ACCURACY OF GAP CAPACITANCE CALCULATIONS (C_1 AND C_2)

Several approaches have been made to the problem of gap capacitance calculation and the results have differed significantly. In a critical paper by Costamagna [264] these various approaches were compared and recommendations were made. Costamagna argued that the results of Benedek and Silvester [265] should be used — but should be reduced by a certain amount in the case of C_2 (Costamagna's C_b, shown in his Figure 2). In critical design situations, great care should therefore be exercised over the determination of C_1 and C_2.

7.4 MICROSTRIP SHORT-CIRCUITS

For hybrid MICs operating at lower microwave frequencies, up to 2 or 3 GHz, a short wire bonded to the microstrip and ground plane and passed through an unmetallized hole in the substrate is known to provide a fairly good short-circuit. At these frequencies discontinuity parasitics (inductance and capacitance) present negligible reactive effects. As the frequency is increased beyond 2 or 3 GHz, the equivalent reactance associated with such a wire becomes increasingly significant, so that the structure departs noticeably from a short-circuit.

This equivalent reactance is also frequency-dependent, and thus different effective lengths of the microstrip line are associated with the same structure at different frequencies. This means that the location, along the line, of the effective short-circuit plane also varies significantly with frequency.

Some interesting work has been performed by Owens [152], primarily in an effort to characterize shunt holes in microstrip lines where active devices such as transferred-electron devices may be connected. One consequence of this work was the discovery that a hole can be produced which has a practically frequency-independent reactance, so that its effective short-circuit plane remains stationary over a wide frequency range.

Extensive microstrip resonator measurements were employed and it was established that the holes may either contain a brass rod or merely be metallized around their 'cylindrical' surfaces to obtain essentially identical results. For fairly low impedance (18 Ω) microstrip lines on alumina substrates (ε_r= 10.1, h= 0.635 mm), it was found

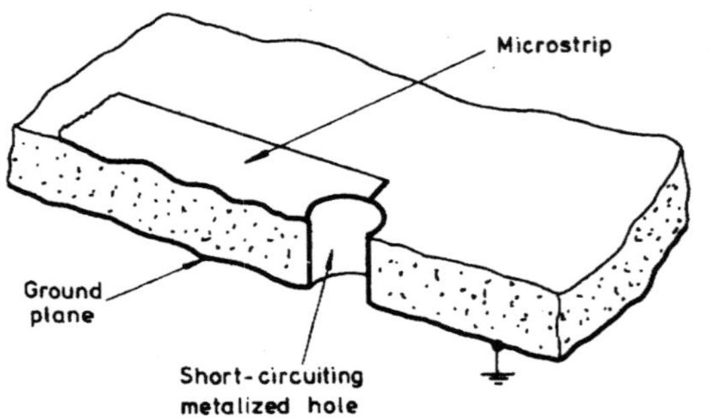

Figure 7.6 A shunt metallized hole which may form a short-circuit in a microstrip line.

that a 1.52 mm diameter hole ultrasonically drilled down through these 3 mm wide lines gave

$$\text{Reactance} = X < \pm j0.5\ \Omega \tag{7.24}$$

over the frequency range:

$$4 \le f \le 18\ \text{GHz}. \tag{7.25}$$

This represents a good *broadband* short-circuit. Holes yielding this performance had microstrip continuing on both sides of them.

Where such a hole appears at the otherwise foreshortened end of the strip, Owens gives the following condition for a broadband short-circuiting hole:

$$\ln\left(\frac{w_{\text{eff}}}{\pi d_e}\right) \approx \left(\frac{\pi d_e}{w_{\text{eff}}}\right)^2 \tag{7.26}$$

where $d_e = 0.03 + 0.44d$, d = actual hole diameter, and w_{eff} = effective microstrip width (see Section 6.8.2)

In the case of the 3 mm wide line this leads to a hole diameter of 1.30 mm, which is substantially smaller than the previous optimum value of 1.52 mm, when the hole was not at the end of a microstrip line. Figure 7.6 illustrates the appearance of the structure.

A 50 Ω line on this substrate material would have a width of approximately 0.6 mm, and many circuits, including filters, require such lines. Assuming, very roughly for present purposes, a pro rata reduction in the hole diameter, we find that 0.26 mm is required for a good broadband short-circuit. This would be extremely difficult, if not impossible, to drill (except, may be, using a laser). It would, however, be feasible using MMIC technology.

It is quite difficult to accurately and repeatably locate these holes, or 'shunt posts,' relative to the microstrip circuit on a hybrid MIC. Microstrip-like structures can be precisely defined and located photolithographically, but this is considerably more difficult with shunt structures such as these short-circuits.

However, with techniques such as computer-controlled laser drilling these short-circuits may become more important, even with hybrid MICs, and some recent work is briefly described in Section 7.13.5 on page 261. The availability of via technology, essentially short-circuits and inherent with MMICs, provides for the precision positioning of such shorts. In particular, Strohm *et al.* [78] have reported accurate and well-defined via hole realization in high-resistivity silicon (HRS) using an advanced silicon etch process. Precise vertical sidewalls are achieved for circular geometries, and an SEM is shown of a 61 μm diameter, 150 μm deep gold-plated via. The inductance is 22 pH and measurements are reported between 50 MHz and 50 GHz.

This via hole technology is not only highly amenable to modelling, but also, if repeatable under production conditions, the technology opens up wider design freedom in MMICs.

7.5 FURTHER DISCONTINUITIES

The remaining sections deal with bends, steps, slits, T-junctions and cross-over junctions. In these cases fringing field effects, leading to equivalent capacitive elements, occur as we have already seen for other discontinuities. However, current flow disturbances are now produced, and these give rise to the concept of equivalent circuit inductances.

For bends and T-junctions, arrangements have been developed which, at least partially, compensate for the discontinuity effects. These are described here because they are significant to the MIC circuit designer, certainly up to around 18 GHz. Sometimes it may not be practical to apply the compensation, and for this reason, for example, the uncompensated bend equivalent circuit parameters are studied.

7.6 THE RIGHT-ANGLED BEND OR 'CORNER'

In many cases where quite complex circuits are required on a single substrate it is necessary to feed, between circuits, using lengths of microstrip that include bends. Another important case comprises hairpin lines such as resonators that embody double bends. These bends usually pass through an angle of 90°. Most often the line does not change width and such a bend, together with its equivalent circuit, is shown in the diagrams of Figure 7.7.

The capacitance arises through additional charge accumulation at the corners particularly around the outer point of the bend where electric fields concentrate. The inductances arise because of current flow interruption. This is considerable, especially bearing in mind that most of the current flows in the outer edges of microstrip. The results of some simulations showing the current densities are provided here.

An alternative equivalent circuit simply treats the inductances as two lengths of equivalent, inductive, microstrip lines. Capacitance data have been determined theoretically by Silvester and Benedek [266] and inductance data has been obtained by Thomson and Gopinath [261]. Fairly extensive measurements have also been conducted.

Gupta *et al.* [88] have given closed formulas for the evaluation of bend capacitance: For $w/h < 1$:

$$\frac{C_{\mathrm{BEND}}}{w} = \frac{(14\varepsilon_r + 12.5)w/h - (1.83\varepsilon_r - 2.25)}{\sqrt{w/h}} \text{ pF/m.} \tag{7.27}$$

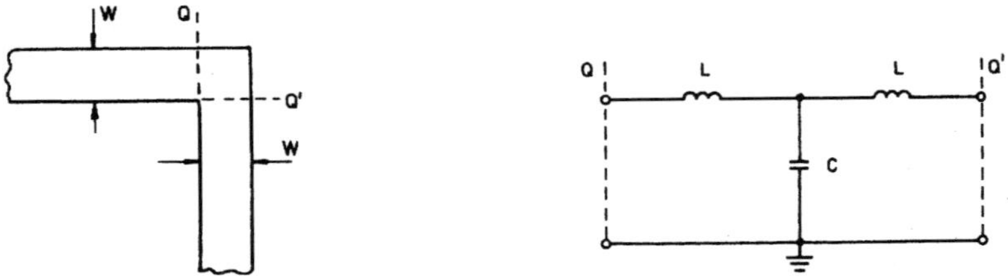

Figure 7.7 Right-angled microstrip bend: (a) structure and nomenclature; and (b) equivalent circuit. (©1973 IEEE. Reprinted, with permission, from Silvester and Benedek [266].)

For $w/h > 1$:

$$\frac{C_{\text{BEND}}}{w} = (9.5\varepsilon_r + 1.25)\frac{w}{h} + 5.2\varepsilon_r + 7.0 \text{ pF/m} \tag{7.28}$$

and inductance:

$$\frac{L}{h} = 100\left[4\sqrt{\frac{w}{h}} - 4.21\right] \text{ nH/m.} \tag{7.29}$$

The accuracies of Equations (7.27) and (7.28) are quoted as within 5% over the ranges: $2.5 \leq \varepsilon_r \leq 15$ and $0.1 \leq w/h \leq 5.0$. The accuracy of (7.29) is quoted as about 3% for the range: $0.5 \leq w/h \leq 2.0$, compared with the results due to Thomson and Gopinath. The permittivity range quoted above is well sufficient to encompass most types of substrates, including semiconductors, although the range applying to the inductance is more restricted than that for the capacitance, because inductance is almost always less significant. This is not a severe limitation and the expression can be used somewhat outside the quoted range without serious overall loss of modelling accuracy.

To illustrate the seriousness of the problem, as an example let us consider a microstrip line of 0.75 mm width and fabricated on a 0.5 mm thick substrate whose permittivity is 9.9. The bend equivalent circuit capacitance and inductance are 0.135 pF and 0.031 nH respectively, shown on the equivalent circuit of Figure 7.8. Although these components may have 'small' values, their reactances, even at 10 GHz, have a significant effect on the electrical properties of the complete network. For the inductance, at 10 GHz:

$$\omega L = 2\pi \times 10^{10} \times 0.031 \times 10^{-9} = 2\ \Omega$$

and for the capacitance:

$$\frac{1}{\omega C} = \frac{1}{2\pi \times 10^{10} \times 0.135 \times 10^{-12}} = 120\ \Omega. \tag{7.30}$$

The 2 Ω reactance in series in a line having a characteristic impedance of 50 Ω will have a significant effect (greater with lower impedance lines), and the 120 Ω shunt capacitive reactance will likewise have a pronounced influence on circuit response (greater for higher impedance lines).

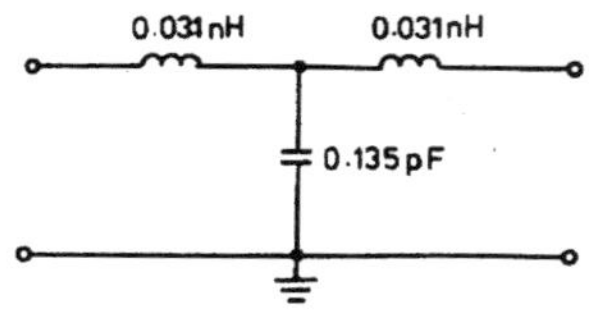

Figure 7.8 Lumped equivalent circuit for the microstrip bend calculated in the text. (© 1973 IEEE. Reprinted, with permission, from Silvester and Benedek [266].)

7.7 MITRED OR 'MATCHED' MICROSTRIP BENDS — COMPENSATION TECHNIQUES

Several techniques have been investigated for the compensation of microstrip bends, greatly reducing the effect of the capacitance and hence improving the VSWR. In particular, Anders and Arndt [267] have reported a moment method to calculate the appropriate capacitances and inductances for both curved and mitred bends. Their results indicate that, at least up to a frequency of about 10 GHz, a mitred bend produces as good as, or better, performance than curved bends. This applies to a wide range of bend angles, from 30° up to 120°. At least 70% mitring is recommended for an acute-angled bend of 120°, i.e. one which acutely bends back upon itself. Guidance for the design of such mitred bends, but for an angle of 90°, is also available for the structure shown in Figure 7.9.

The equivalent circuit shown in Figure 7.9(b) is for the region between planes P and P' and the curves of Figure 7.9(c) relate to measured results [268]. Although these curves apply to the specific instance defined, the considerable reduction of susceptance B (and therefore also capacitance C) would be expected in widely different substrates and structures. As can be seen, the equivalent line-length parameter l_c increases with enhanced mitring. Owing to this, and extreme line narrowing in the centre of the bend, the degree of mitre is generally restricted to around

$$1 - \frac{b}{\sqrt{2}w} \approx 0.6 \tag{7.31}$$

so that $b \approx 0.57w$.

It is tentatively recommended that approximately this amount of mitre be applied in the majority of practical cases, where the substrate permittivity roughly approximates that of alumina. We present SONNET® simulations here that indicate the greatly improved current distributions applying to the mitred bend.

From Figure 7.10(b) the greatly improved current distribution when an optimum mitre is applied is clearly evident, comparing the uncompensated case Figure 7.10(a). The physical origins of what leads to an optimum mitre can be seen by examining Figures 7.11(a) and 7.11(b), where the magnitudes of the horizontal and vertical current components are plotted. It can be seen the the mitred edge of the microstrip bend acts like a reflector. The overhang of the reflector (the fact that the mitred edge extends beyond the microstrip (that is the projection of the mitre is greater than the microstrip width) corresponds very closely to the effective width of the microstrip used in the planar waveguide model for microstrip. The planar waveguide model was developed by Wheeler [270] and is a parallel-plate waveguide with magnetic side walls.

Optimally mitred microstrip bends are re-addressed in Section 7.13.3 on page 257.

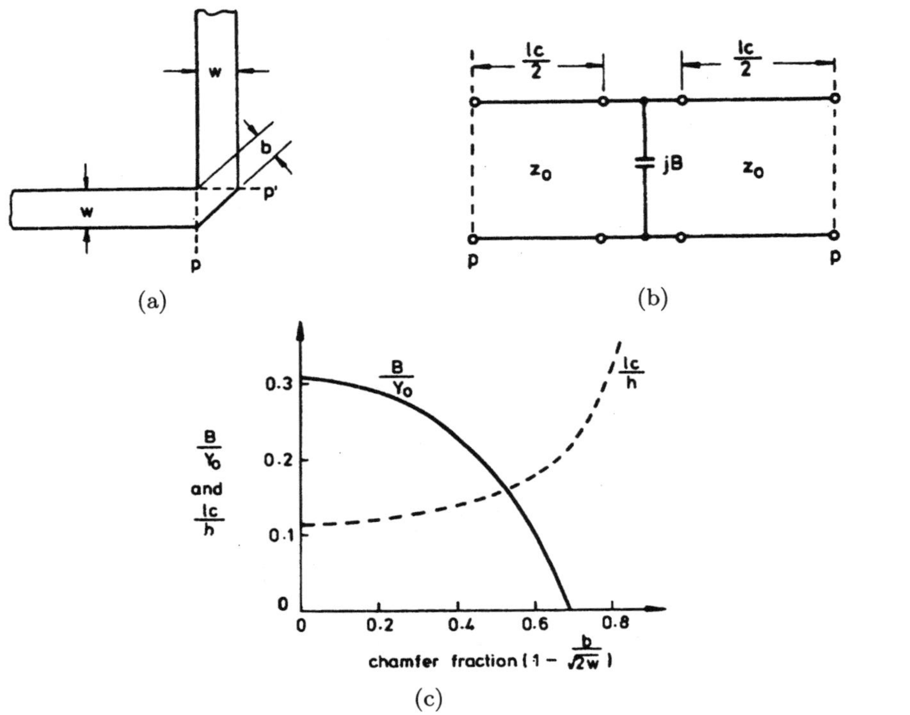

Figure 7.9 Mitred, right-angled bend together with its equivalent circuit and parameter variations (as a function of the amount of mitre): (a) structure and nomenclature; (b) equivalent circuit; and (c) parameter trends. (Reproduced by permission of the Institution of Electronic and Radio Engineers , from Easter *et al.* [268].)

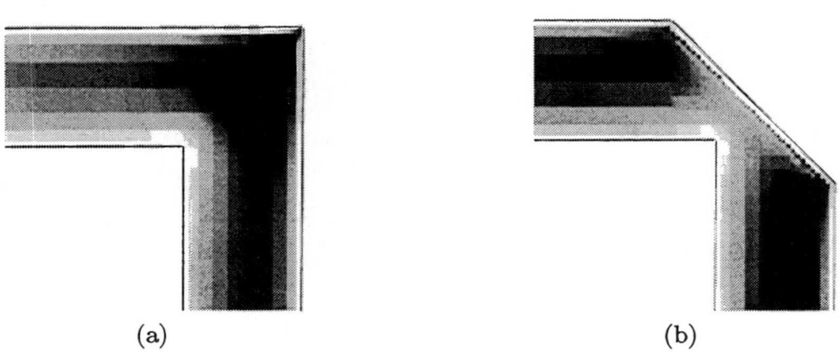

Figure 7.10 Magnitude of the current densities on (a) a right-angled bend, and (b) an optimally mitred bend (in both cases on alumina substrate at 10 GHz).

(a) (b)

Figure 7.11 Magnitude of the (a) horizontal and (b) vertical current densities on a right-angled bend under the same conditions as those shown in Figure 7.10.

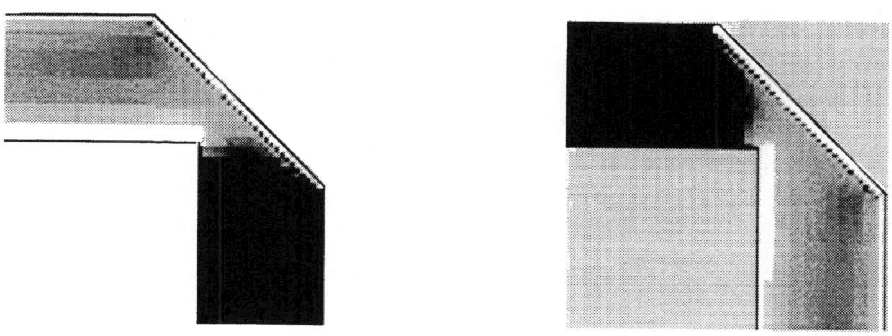

(a) (b)

Figure 7.12 Magnitude of the (a) horizontal and (b) vertical current densities on an optimally mitred right-angled bend under the same conditions as those shown in Figure 7.10.

Figure 7.13 Structure and equivalent circuit of the symmetrical microstrip step (change in width): (a) structure and nomenclature; and (b) equivalent circuit. (©1972 IEEE. Reprinted, with permission, from Benedek and Silvester [265].)

7.8 STEP CHANGES IN WIDTH (IMPEDANCE STEPS)

It is necessary for microstrip lines to experience a width change in many circuits. This is clearly so even for the simple case of a quarter-wave transformer, and several such changes are demanded in transistor amplifier circuits such as that shown in Figure 7.1. Couplers, filters, mixers and oscillators provide further examples.

These forms of discontinuities have been fairly widely investigated and characterized [265]. Like the microstrip bend, the shunt capacitance due to corners is the dominant equivalent circuit parameter and edge-current disturbances occur again which give rise to equivalent inductances.

7.8.1 THE SYMMETRICAL MICROSTRIP STEP

This is illustrated in Figure 7.13(a), with an equivalent circuit in Figure 7.13(b) and parameter variations in Figure 7.14.

We previously interpreted a discontinuity capacitance in terms of an equivalent extra length of the microstrip (for the foreshortened open-circuit, see Section 7.2 on page 227, Figure 7.3). This approach can be used again for the step, where the principal influence of the capacitance is to increase the effective length of the wider line (width w_2). Since the changes and values are all quite small, the fully foreshortened open-circuit result can be used in a modified form as given by the following expression:

$$l_{es} \approx l_{eo}\left(1 - \frac{w_1}{w_2}\right) \tag{7.32}$$

where l_{eo} is obtained from (7.7) or (7.9).

Gupta *et al.* [88] have shown how to account approximately for the inductances. Where L_{m1} and L_{m2} are the inductances per unit length of the appropriate microstrips, having widths w_1 and w_2, the discontinuity inductances are given respectively by the following relations:

$$L_1 = \frac{L_{m1}}{L_{m1} + L_{m2}} L \tag{7.33}$$

and

$$L_2 = \frac{L_{m2}}{L_{m1} + L_{m2}} L. \tag{7.34}$$

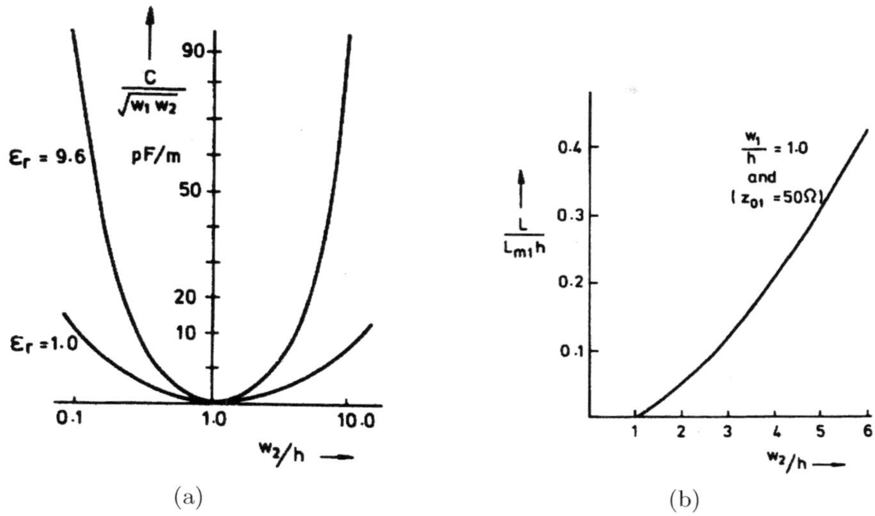

Figure 7.14 The symmetrical step, parameter variations: (a) capacitance; and (b) inductance. (©1972 IEEE. Reprinted, with permission, from Benedek and Silvester [265]; (b) Thomson and Gopinath [261].)

The equivalent extra lengths of microstrip are then given by

$$\frac{l_{es1}}{h} = \frac{l_{es2}}{h} = \frac{L}{(L_{m1} + L_{m2})h}. \tag{7.35}$$

Garg and Bahl [263] obtained design formulas for both C and L. The following expressions for C were derived by curve-fitting earlier theoretical results:

For $\varepsilon_r \leq 10$; $\quad 1.5 \leq w_2/w_1 \leq 3.5$:

$$\frac{C}{\sqrt{w_1 w_2}} = (10.1 \log \varepsilon_r + 2.33)\frac{w_2}{w_1} - 12.6 \log \varepsilon_r - 3.17 \quad \text{pF/m}. \tag{7.36}$$

For $\varepsilon_r = 9.6$; $\quad 3.5 \leq w_2/w_1 \leq 10$:

$$\frac{C}{\sqrt{w_1 w_2}} = 130 \log(w_2/w_1) - 44 \quad \text{pF/m}. \tag{7.37}$$

The error involved in applying (7.36), which applies to relatively slight steps $((w_2/w_1)\,\text{max} = 3.5)$, can be as high as 10%. The error in (7.37) is claimed to be less than 0.5%.

Garg and Bahl's expression for the inductance is

$$\frac{L}{h} = 40.5\left(\frac{w_1}{w_2} - 1.0\right) - 75\frac{w_1}{w_2} + 0.2\left(\frac{w_1}{w_2} - 1.0\right)^2 \text{nH/m}. \tag{7.38}$$

This formula is quoted as having an accuracy of better than 5%, provided that

$$w_1/w_2 \leq 5.0 \text{ and } w_2/h = 1.0 \;. \tag{7.39}$$

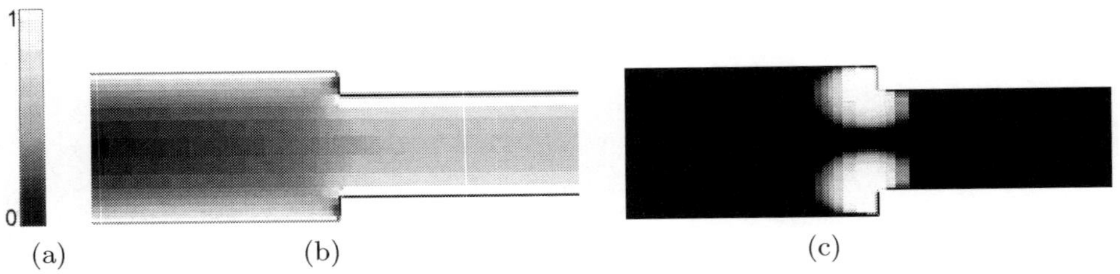

Figure 7.15 Magnitude of the current densities for symmetrical step in width: (a) normalized scale (b) longitudinal current; and (c) transverse current. The signal is 10 GHz and the microstrip line is on alumina.

Figure 7.16 Structure and equivalent circuit of the asymmetrical microstrip step. (© 1972 IEEE. Reprinted, with permission, from Benedek and Silvester [265].)

Once more it should be stressed here that because the discontinuity quantities themselves represent second-order design corrections these error bounds on the values are usually of little if any significance in practice.

We present SONNET® simulations here to show the current distribution around a step.

From Figure 7.15 it is clear that significant current flow disturbances occur in the vicinity of a step in width.

7.8.2 THE ASYMMETRICAL STEP IN MICROSTRIP

The geometry of this is shown in Figure 7.16, together with the appropriate equivalent circuit. As a first approximation the equivalent circuit parameter variations may be taken as the same as those for the symmetrical step. The actual parameter values will be about half of the values obtained for the symmetrical step.

7.9 THE NARROW TRANSVERSE SLIT

A narrow slit, cut transversely into the microstrip width, yields a predominantly series inductance effect. This structure has been investigated by Hoefer [269], and it may be used to compensate for excess capacitance at discontinuities or to fine-tune lengths of microstrip such as stubs. The appearance of the slit and its equivalent circuit are shown in Figure 7.17.

In Hoefer's analysis, dispersion and capacitive effects are neglected and a non-magnetic substrate is taken. A dipole field perturbation method, originally developed by Wheeler [270] is used in the analysis. A parallel-plate waveguide model with

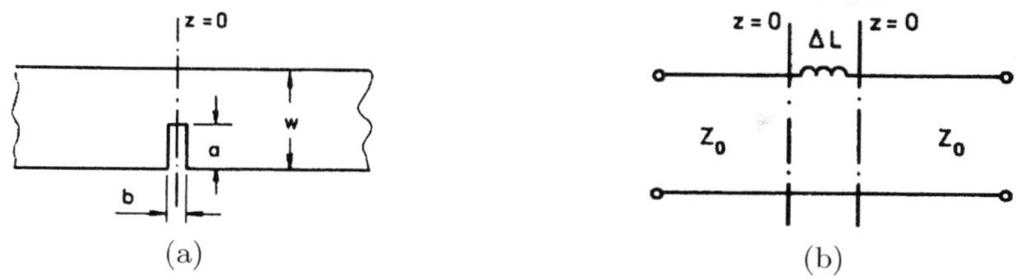

Figure 7.17 The narrow transverse slit: (a) actual slit; and (b) equivalent circuit. (©1977 IEEE. Reprinted, with permission, from Hoefer [269].)

magnetic side walls is also employed. The reader is recommended to see that paper for further details, since only the resulting expressions and their use will be described here. The final formula for the inductance is

$$\frac{\Delta L}{h} = \frac{\mu_0 \pi}{2}\left(\frac{a'}{A}\right)^2 \tag{7.40}$$

where

$$\frac{a'}{A} = 1 - \frac{Z_{0(\mathrm{air})}}{Z'_{0(\mathrm{air})}}. \tag{7.41}$$

For design, the procedure is then as follows:

(a) Calculate the characteristic impedance of the air-filled microstrip line of width w — giving $Z_0(\mathrm{air})$.
(b) Calculate the characteristic impedance of an air-filled microstrip line of width $(w - a)$ — giving $Z'_0(\mathrm{air})$.
(c) Evaluate a'/A with these results, (a) and (b), using Equation (7.41), and hence calculate ΔL using Equation (7.40).

As an example a slit of relative penetration $(a/w) = 0.5$ can be calculated. Considering such a slit in a 50 Ω microstrip line on 0.635 mm thick alumina ($w/h = 1$ when $\varepsilon_r = 9.6$), the inductive reactance of this slit is $j3.7$ Ω at 10 GHz.

There are two important limitations to this type of calculation. In the case of a relatively small slit width b, the capacitance may be significant. Using the calculation techniques for series gaps in microstrip, as described in Section 7.3 on page 231, the capacitance of the slit may be approximately determined as $(a/w)C_2$. For the example calculated above, the capacitive reactance is approximately $-j320\,\Omega$, which would represent a negligible shunt reactance effect in many applications.

Another limitation occurs as the slit is progressively widened. Eventually, the structure looks like a short microstrip line of width $(w - a)$ asymmetrically stepped to the main microstrips of widths w. Hoefer [269] suggests that a lumped inductance, independent of b, is approximately valid while $b < h$.

Experimental results for an RT-duroid® 5880 substrate, with $w/h = 2.0$ and a range of a/w values, show good agreement with the theory.

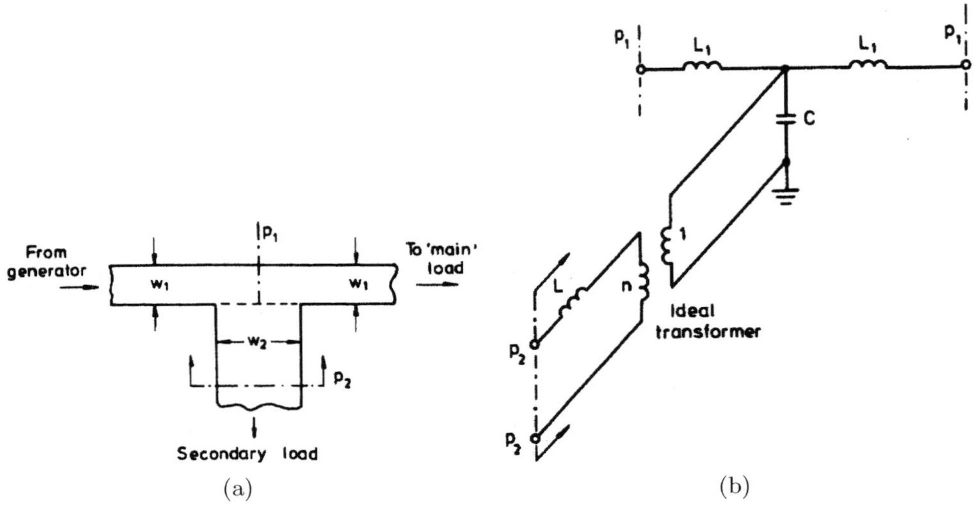

Figure 7.18 The T-junction in microstrip: (a) structure and nomenclature; and (b) equivalent circuit. (©1973 IEEE. Reprinted, with permission, from Silvester and Benedek [266].)

7.10 THE MICROSTRIP T-JUNCTION

This junction necessarily occurs in a wide variety of microstrip circuits. Examples range through stub-matching elements, stub filters, branch-line couplers, and microstrip antenna element feeds, together with several other conceivable types of circuits. The appearance and a fairly elementary equivalent circuit are shown in Figure 7.18. The inductances and capacitance are included for identical reasons to those described previously in respect of other discontinuities. However, one new element has been introduced here: the transformer.

It is necessary to determine the effect, on the main circuit through from generator to main load, of a secondary load connected to the shunt branch arm (width w_2). At very low frequencies, any such load will, of course, appear directly connected across the through circuit, i.e. when

$$f \to 0, \; n \to 1.0 \; . \tag{7.42}$$

At microwave frequencies a load impedance connected at some point along the shunt branch will appear both transformed by the ratio n_2 and located at some specified electrical length from the physical junction. The inductances, the capacitance and the transformation ratio have all been deduced experimentally and, in a number of cases, theoretically [266–268]. It is not intended that all the information be provided here and only the capacitance and transformation ratio results are given (Figure 7.19). Note that, for moderate-to-high impedance branch lines, the junction capacitance becomes reduced below that of either microstrip line capacitance per unit length. Thus, capacitance is 'taken out' of the junction and the equivalent circuit capacitance is negative.

The small difference of 0.1 in the permittivities should not matter much in practice if

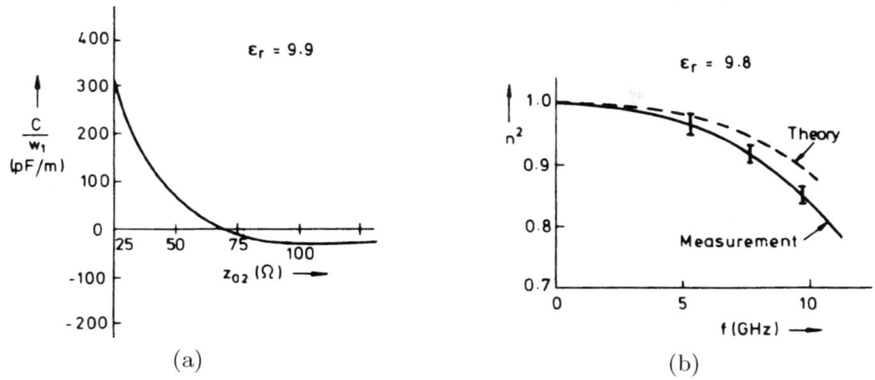

Figure 7.19 Parameter trends for the T-junction ((a)©1973 IEEE. Reprinted, with permission, from Silvester and Benedek [266]; (b) reproduced by permission of the Institution of Electronic and Radio Engineers, from Easter *et al.* [268].)

the two sets of results were both required when carrying out a design on an appropriate alumina substrate. It is observed that a substantial discrepancy exists between the measured and theoretical results for n^2, and it is not clear whether this is due mainly to experimental or to theoretical error. Judging by the given error bounds (only three points), it may be that Vogel's theory [268] is not fully adequate here. The fact that the discrepancy grows with increasing frequency is also unfortunate, since design at frequencies well above 10 GHz is generally increasingly important.

Formulas for the capacitance and for the inductances are given by Garg and Bahl [263] but not for the transformation ratio n. Semi-empirical expressions have also been given by Hammerstad and Bekkadal [97] for shifts in the locations of the electrical reference planes, as well as n, and it is these results that are more useful for circuit design than the equivalent lumped parameters.

The reference planes and hypothetical widths associated with Hammerstad and Bakkadal's method are shown in Figure 7.20, where the effective width (see Equation(5.49), Section 5.8.2 on page 139) of either line is

$$w_{\text{eff}1,2} = \frac{h\eta}{Z_{0(1.2)}\sqrt{\varepsilon_{\text{eff}\,1,2}}}. \tag{7.43}$$

The transformation ratio n is given by

$$n^2 = \left\{ \frac{\sin[\pi(w_{\text{eff}1}/\lambda_{g1})(Z_{0(1)}/Z_{0(2)})]}{\pi(w_{\text{eff}1}/\lambda_{g1})(Z_{0(1)}/Z_{0(2)})} \right\}^2 \left\{1 - [\pi(w_{\text{eff}1}/\lambda_{g1})(d_2/w_{\text{eff}1})]^2\right\}. \tag{7.44}$$

The displacement of the reference plane for the primary line (1) is

$$\frac{d_1}{w_{\text{eff}2}} = 0.05\frac{Z_{0(1)}}{Z_{0(2)}}n^2. \tag{7.45}$$

The displacement of the reference plane for the secondary arm (2) is

$$\frac{d_2}{w_{\text{eff}1}} = 0.5 - \left[0.076 + 0.2\left(\frac{2w_{\text{eff}1}}{\lambda_{g1}}\right)^2\right.$$

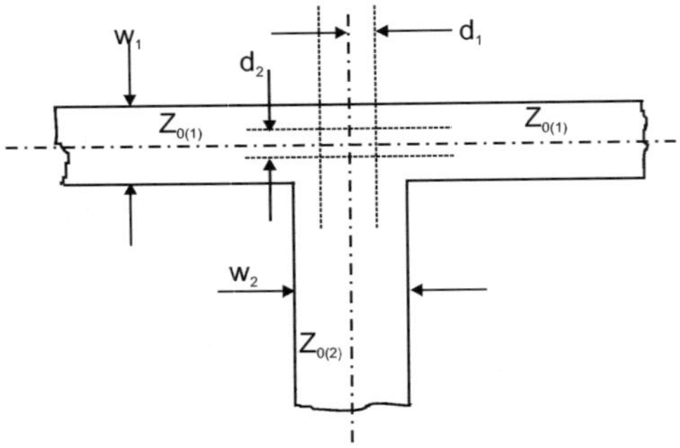

Figure 7.20 T-junction: reference planes and hypothetical widths. (Reproduced by permission of *MicroWaves* from Dydyk [273].)

$$+0.663 \exp\left(-1.71\frac{Z_{0(1)}}{Z_{0(2)}}\right) -0.172 \ln\left(\frac{Z_{0(1)}}{Z_{0(2)}}\right)\Bigg] \frac{Z_{0(1)}}{Z_{0(2)}}. \tag{7.46}$$

Also, the shunt capacitance is determined by the following expressions:

For the condition $Z_{0(1)}/Z_{0(2)} \leq 0.5$:

$$\frac{\omega C \lambda_{g1}}{Y_{0(1)} w_{\text{eff1}}} = \left(\frac{2w_{\text{eff1}}}{\lambda_{g1}} - 1\right) \frac{Z_{0(1)}}{Z_{0(2)}}. \tag{7.47}$$

For the condition $Z_{0(1)}/Z_{0(2)} \geq 0.5$:

$$\frac{\omega C \lambda_{g1}}{Y_{0(1)} w_{\text{eff1}}} = \left(\frac{2w_{\text{eff1}}}{\lambda_{g1}} - 1\right) \left(2 - 3\frac{Z_{0(1)}}{Z_{0(2)}}\right). \tag{7.48}$$

Actual accuracies for these expressions are not quoted, but their discrepancies rise when

$$\frac{2w_{\text{eff1}}}{\lambda_{g1}} > 0.3 \tag{7.49}$$

i.e. when the wavelength becomes small enough (frequency large enough) for this condition to occur. Hall and James [271] have reported measurements at 17 GHz on T-junctions, and they conclude that Hammerstad's numerical results, as he suggests, are too large at such high frequencies. In particular, a discrepancy of a factor of two or more was found for d_2, where the measurements always showed smaller results.

It would be useful to attempt to modify Hammerstad's expressions so that a better fit is obtained at higher frequencies. One might, for example, note that if w_{eff} in Equation (7.46) is replaced by the frequency-dependent value given by Equation (5.47) then the magnitude of the term $(w_{\text{eff1}}/\lambda_{g1})$ rises more strongly with frequency, hence reducing d_2 as desired. Some further investigations are required in this area.

Another limitation to the accuracy of the expressions is imposed by the impedance ratio $Z_{0(1)}/Z_{0(2)}$. When this exceeds about 2.0 the calculated values of d_2 become too large and an improved accuracy is obtained when the ratio $Z_{0(1)}/Z_{0(2)}$ is replaced by its inverse, that is $Z_0(2)/Z_0(1)$. Some useful work has been carried out to compensate T-junctions, and thereby improve the performance of circuits using them. This is discussed in the next section.

7.11 COMPENSATED T-JUNCTIONS

As we have seen in the preceding discussion, direct T-junctions inherently involve substantial discontinuities that significantly influence the behaviour of the structure in any circuit. On many occasions the implementation of the direct T-junction, without modification, is unavoidable. However, in a similar manner to the mitred compensation of the right-angled bend, it is possible to introduce adjustments around a T-junction that will partially compensate for the effects of the direct discontinuities.

One simple way to at least compensate the capacitance of a T-junction is to introduce a slit across the width of the main-through microstrip line, opposite the branch arm. The characteristics of such a slit were discussed in Section 7.9 on page 242. Kompa [272] has shown further that a comparatively wide 'slot' ($b \approx w_2$) in this position, i.e. a rather wider slit, strongly affects the transmission of higher-order modes in the main line and leads to enhanced skirt sensitivity of filters composed of stubs with such modified T-junctions. To the best of our knowledge this kind of approach has not yet been developed further, and more investigation could prove very fruitful.

Dydyk [273] has described a T-junction compensation technique that appears to work well in the case of a branch-line microstrip coupler circuit. Dydyk's aim was to modify the microstrip lines in the vicinity of the junction in order to compensate for reference plane shifts, at least over a specified range of frequencies. The treatment of the junction can exclude any discussion of radiation loss with little error in circuit performance results, at least up to a frequency of 17 GHz. The compensated T-junction takes on the form shown in Figure 7.21. A microstrip branch-line coupler designed using these principles yielded the following performance over the frequency band 16–18 GHz:

Insertion loss	3.5–4.0 dB over entire band
Isolation	$>$ 20 dB over entire band
Return loss	10 dB (at 17.5 GHz), otherwise 10 to -25 dB

The very 'flat', almost frequency-independent, insertion loss and isolation are attributed to the compensated junction design. Dydyk also describes an SPDT switch designed according to these principles.

7.12 CROSS-JUNCTIONS

There are occasions where microstrip lines are required to separate in the shape of a cross, forming a four-port circuit. Examples can be seen in the arrangement indicated in Figure 7.1. In general, a cross-junction may be symmetrical or asymmetrical — where the lines forming the cross do not all have the same widths.

In attempting to design a single stub to be fabricated in microstrip it can be found that the width of the stub line is wide enough to sustain transverse resonance modes at the higher frequencies of operation — see Section 5.9.2, on page 145, and Equation (5.64). One way to overcome this problem is to use two stubs placed on each side of the microstrip instead of the original one. The impedance of each new stub will be twice that of the originally intended single stub, and thus the widths will be greatly reduced. A cross-junction is therefore formed.

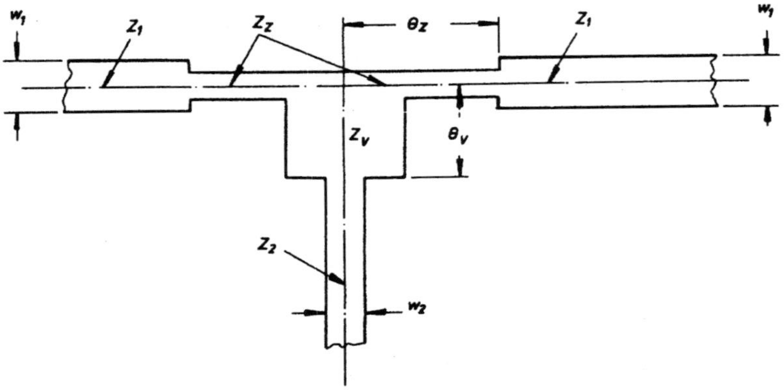

Figure 7.21 A compensated form of T-junction. (Reproduced by permission of *MicroWaves* from Dydyk [273].)

Theoretical and experimental agreement is not good, especially for some inductance parameters associated with the equivalent circuits of cross-junctions. As a result of this it does not appear sensible to give design formulas which merely fit theoretical curves. Instead, some useful engineering design information is given for the asymmetric cross. The asymmetric cross is the most important form for practical microstrip circuits, and is shown in Figure 7.22 together with its equivalent circuit. Microstrip resonator measurement techniques, described in Chapter 9, have proved to be very powerful in the characterization of certain specific cross-junctions. Akello [274] has used such techniques to show that appropriate T-junction inductance parameters can be judiciously applied to the cross. The capacitance C_T presents prediction difficulties (theoretical treatment is available for symmetrical crosses only). Akello indicates that C_T can be approximated to within 8% by using the simple relationship

$$C_T \approx \frac{3}{4} C_m \tag{7.50}$$

where C_m is the capacitance per unit length of the (uniform) microstrip line on port 4 of the cross-junction. This appears to work quite well for the range:

$$1.32 \leq \frac{w_4}{h} \leq 3.0 \ . \tag{7.51}$$

The capacitance C_m may be calculated from expressions given early in Chapter 1. Specifically, $Z_0 = 1/(v_p C_m)$, whence:

$$C_m = \frac{c}{Z_0 \sqrt{\varepsilon_{\text{eff}}}}. \tag{7.52}$$

(Z_0 and ε_{eff} are well known at an early design stage.)

Akello [274] has tabulated values of all the various equivalent circuit elements and has shown how to apply them in practical situations. In particular, computer simulation was used to indicate that discontinuity reactances attributable to steps and cross-overs are significant, and can be incorporated into response prediction

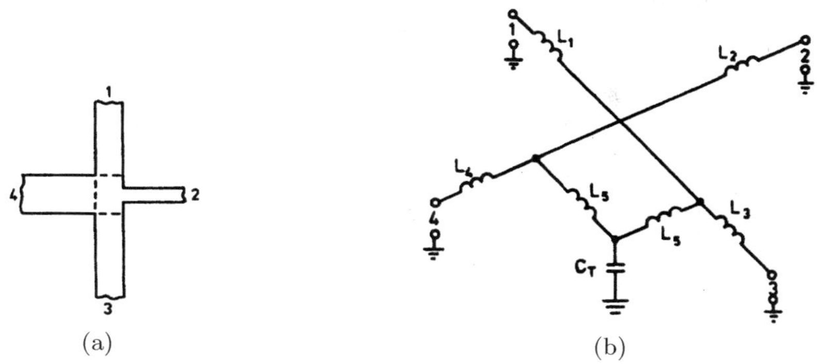

Figure 7.22 The asymmetric cross-over junction: (a) structure and (b) equivalent circuit. (Reproduced by permission of the Institution of Electrical Engineers from Akello *et al.* [274].)

programs. X-band (8–12 GHz), narrowband, and wideband single-stage and double-stage transistor amplifiers are considered. The details and results may be seen by consulting the cited reference.

It would be useful to see whether some compensation technique, perhaps of the type described in Section 7.11 on page 247 (Dydyk), could be developed for cross-junctions. Further information on cross-junctions is given in Section 7.13.3 on page 257.

Examination of the coupling effects that occur with cross-junctions illustrates the origin of cross-talk in complicated interconnection networks. In Figure 7.23 the magnitudes of the currents occur when one a strip on one metal layer, the top metal, crosses a strip on another metal layer which in turn is above a ground plane. The top layer shown in Figure 7.23(b) is driven from the left with a 30 GHz sinusoidal signal. The total curent on the top interconnect is little affected by the cross-junction discontinuity, but it clear that the distribution across the strip is affected and this occurs following the cross-over for a short distance. The most significant effect that occurs is that current is induced on the bottom strip and this has a peak magnitude that is 23% of that of the peak current on the top strip. The origins of this phenomenon can be understood by considering the instantaneous currents and charges on the strip. These are shown in Figure 7.24 at one instant in time, again with the top strip driven on the left by a 30 GHz signal. In the region of the cross-over the signal on the top strip is transitioning from positive to negative as can be seen from the current profile shown in Figure 7.24(c). In the region of the cross-over the charges on the top strip are negative, see Figure 7.24(d). Thus, there is an electric field which in the absence of the bottom strip begins on the ground plane and terminates on the top strip. However, with the top strip and in the region of the cross-over, some of the electric field begins on positive charges on the bottom strip and terminates on the negative charges on the top strip. The positive charges on the bottom strip are drawn in from the regions of the bottom strip above and below the cross-junction, see Figure 7.24(f). This movement of charge constitutes current flow towards the cross-junction, see Figure 7.24(e). Thus the signal on the top strip effectively induces generators on the bottom strip, one for each part of the strip away from the cross-junction.

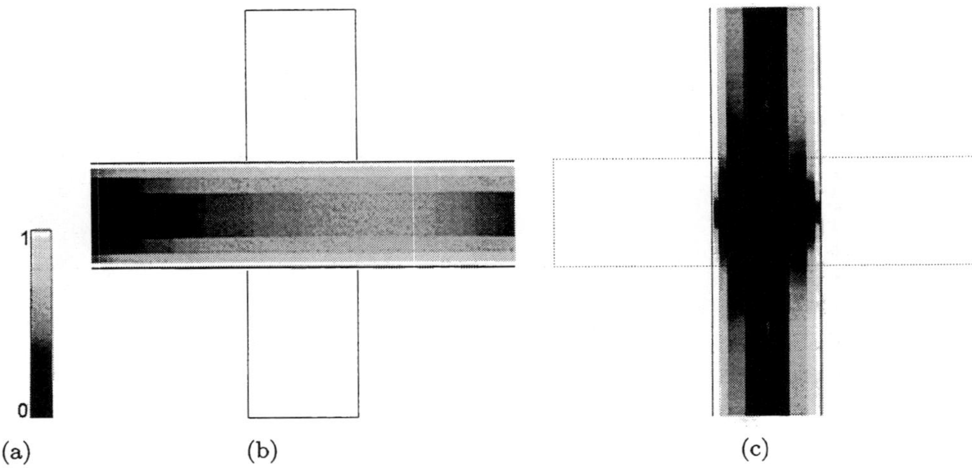

Figure 7.23 Magnitude of the current densities on the strips of a symmetrical cross-over junction: (a) normalized scale (b) current on the top metal; and (c) current on the lower metal layer (between the top metal and the ground plane.

Many coupling effects can be viewed in the same way that the cross-junction discontinuity can be understood by interrupting the electric field and so inducing charges. The inducement of charges constitutes current flow and thus a signal on the victim nets.

7.13 FREQUENCY DEPENDENCE OF DISCONTINUITY EFFECTS

At frequencies much above 18–20 GHz it appears that the essentially quasi-static techniques described in this chapter begin to break down significantly in a number of cases. Eventually, for millimetre and submillimetre design, it is unlikely that these techniques would provide any useful design data, at least not directly.

7.13.1 OPEN-CIRCUITS AND SERIES GAPS

We have already seen here that the quasi-static results provide quite accurate design information, particularly for the capacitive effects, at frequencies approaching 18 GHz. For example, the commonly encountered open-circuit (see Section 7.2 on page 227) has been fairly extensively investigated, both theoretically and experimentally. More effort has been expended in this regard than for any other discontinuity — for the good reason that the open-circuit termination has many applications in both circuits and microstrip antennas. With an emphasis on the latter, James and Henderson [167] published a paper describing detailed research results. The values and trends shown in Figure 10 of their paper are of particular interest. There is a consistent trend for alumina substrates amounting to a peaked behaviour with frequency; two of the curves reported by James and Henderson are reproduced here as Figure 7.25, with the horizontal axis denormalized for the case of a 0.65 mm thick alumina substrate. Some experimental data are also shown which originate from unpublished results obtained

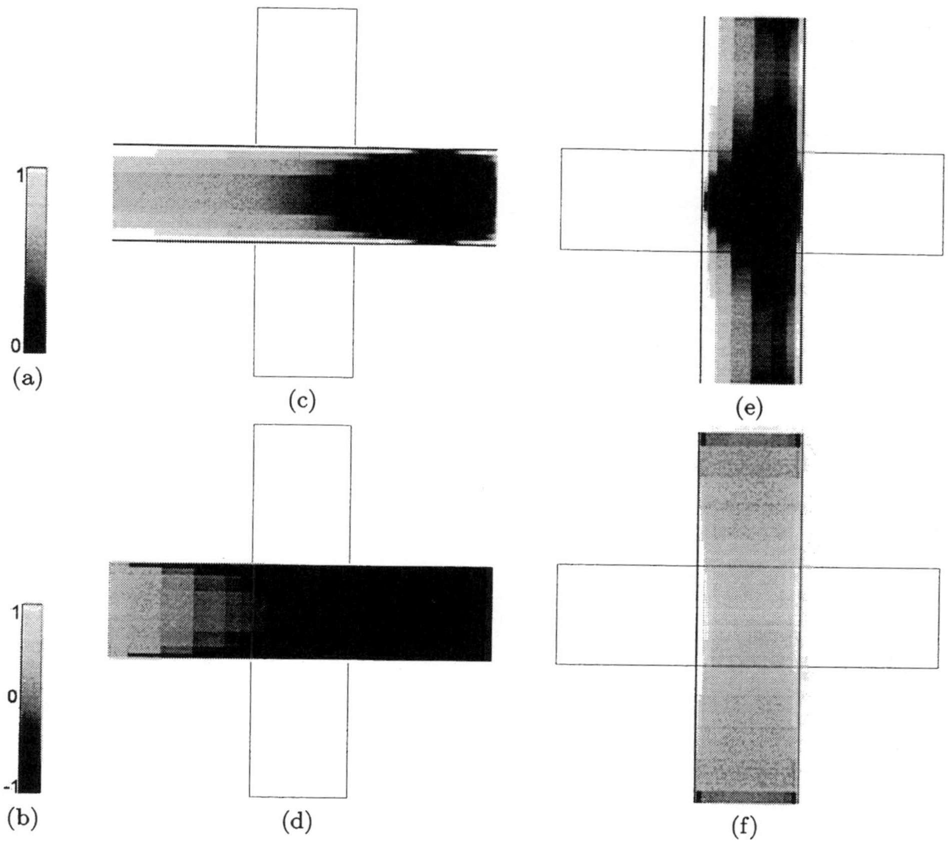

Figure 7.24 Instantaneous current and charges on the strips of a symmetrical cross-over junction: (a) normalized current scale (b) normalized charge scale (c) current on the top metal; (d) charge on the top metal (e) current on the lower metal; and (f) charge on the lower metal.

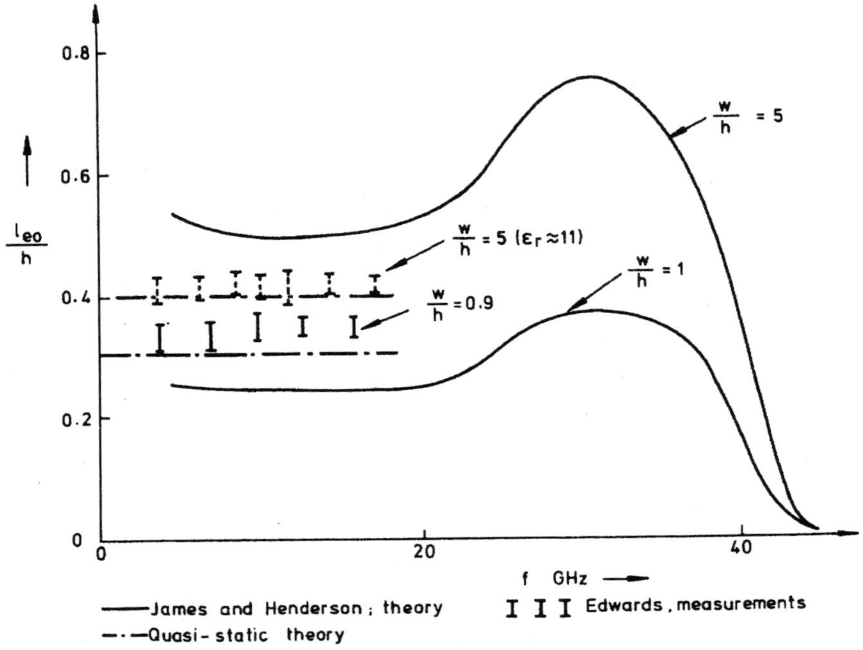

Figure 7.25 Open-circuit normalized equivalent end-effect length (l_{eo}/h) as a function of frequency. The data refer to a 0.65 mm thick alumina substrate with a relative permittivity of 10, except for one case where the values were 0.5 mm and 11, as indicated. (Reproduced by permission of the Institution of Electrical Engineers from James and Henderson [167].)

by the one of the authors (Terry Edwards).

The measured results were obtained by a resonator method that will be described in Chapter 9. They are very tedious to perform and require scrupulous attention to repeatability and high accuracy of manufacture. The 'scatter' in the experimental results is felt to be far from satisfactory, especially where the data are intended for use in critical circuits (e.g. filters). There is about 19% scatter in the results obtained on sapphire substrates and about 15% on alumina. Even so, it is clear that the measured results differ quite markedly from the frequency-dependent theoretical predictions. Results from quasi-static calculations are shown as horizontal chain lines in Figure 7.25. These fit the measurements fairly well up to 18 GHz, although the calculated results are slightly lower in both cases.

It would be useful and interesting to investigate experimentally the behaviour in the 20–40 GHz range for microstrip open-circuits on alumina. Resonator methods, to be described in Chapter 9, might be attempted, but these pose considerable difficulties at such high frequencies because of the presence of higher-order modes. It may even be possible for transverse-resonant modes to become excited. A better approach might be to model the structure for measurement at much lower frequencies.

Other researchers, notably Itoh [275], have theoretically deduced a nonlinear l_{eo}/h variation with frequency. Itoh's results also show up-turns in l_{eo}/h at both lower and higher frequencies, but the absolute values of his end-effect lengths were substantially

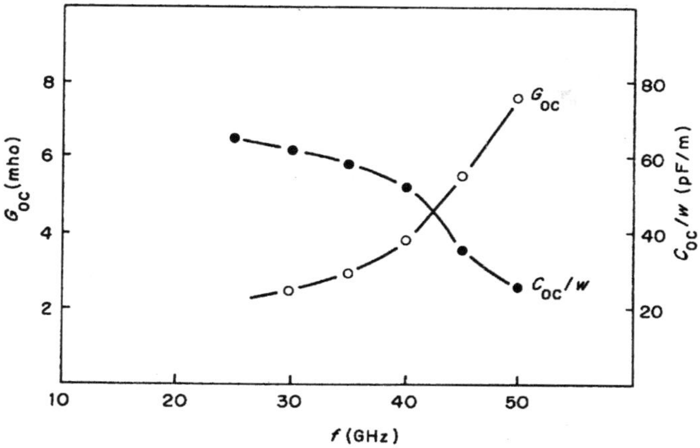

Figure 7.26 Radiation conductance G_{oc} and normalized capacitance C_{oc}/w of an open-circuited microstrip line as functions of frequency ($\varepsilon_r = 9.6$, $w/h = 1$, $h = 0.6$ mm). (© 1985 IEEE. Reprinted, with permission, from Katehi and Alexopoulos [276].)

lower than any other reported data. This is at least partially due to the shielded nature of the microstrip which was investigated, and one of the present writers (Terry Edwards) has performed a number of computations using Itoh's technique with the top shielding cover steadily raised. The results show that the end-effect length then increases. However, a further source of possible error in this technique is probably the necessity for subtracting two large numbers (resonant frequencies) of similar magnitudes.

Katehi and Alexopoulos [276] have described the frequency-dependent characteristics of microstrip discontinuities on a theoretical basis — concentrating largely on open-circuited lines and also on the series gap. They evaluate the current distributions in the vicinity of the discontinuities by means of Pocklington's integral equation, the dyadic Green's function and impedance matrices. They extend their calculations to 50 GHz, and derive results for radiation conductances, equivalent length extensions and gap capacitances.

The results obtained by Katehi and Alexopoulos are shown in Figures 7.26, 7.27, 7.28 and 7.29. The radiation conductances both rise with frequency, as should be anticipated from physical considerations. It is interesting that the equivalent length extension ('excess length') rises slightly, to peak at about 40 GHz, then falling with frequency as predicted elsewhere independently.

A frequency-dependent equivalent circuit for gap discontinuities has been reported by Ozmehmet [277]. This includes the radiation resistance and separates gap capacitances (series) into fringing field and 'plate' components. Therefore, the series structure of the equivalent circuit shown as Figure 7.5(b) is enhanced by these additional elements. Ozmehmet shows that this slightly more complex circuit enables wideband MICs to be designed at frequencies up to at least 15 GHz.

A complete electromagnetic modelling of microstrip open end discontinuities has been reported by McLean *et al.* [278]. They use a spectral domain method of similar general type to that described in Chapter 4, and they include space wave and surface wave radiation by implementing the exact spectral domain Green's function.

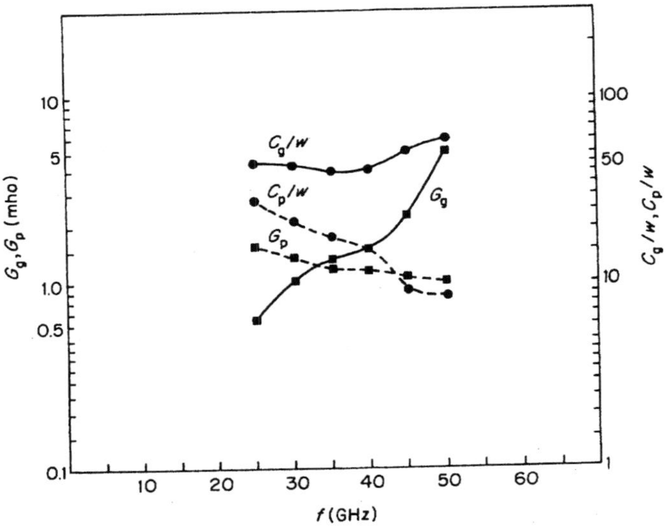

Figure 7.27 Gap-discontinuity radiation conductances G_g and G_p, and normalized capacitances C_g/w and C_p/w as functions of frequency ($\varepsilon_r = 9.6$, $w/h = 1$, $h = 0.6$ mm, $s/h = 0.3762$). (© 1985 IEEE. Reprinted, with permission, from Katehi and Alexopoulos [276]).

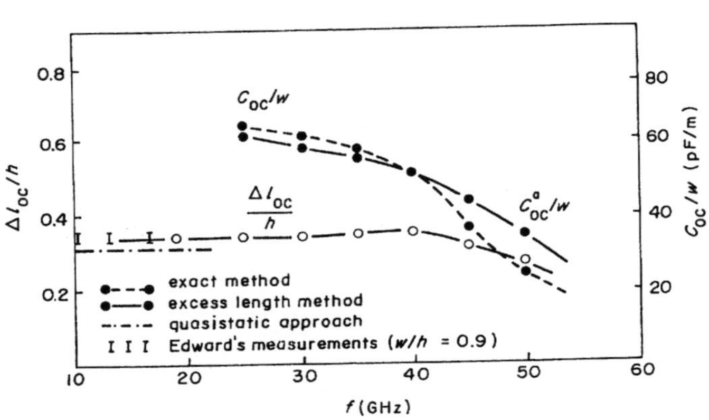

Figure 7.28 Open-circuit normalized excess length and normalized capacitances as functions of frequency ($\varepsilon_r = 9.6$, $w/h = 1$, $h = 0.6$ mm). (© 1985 IEEE. Reprinted, with permission, from Katehi and Alexopoulos [276]).

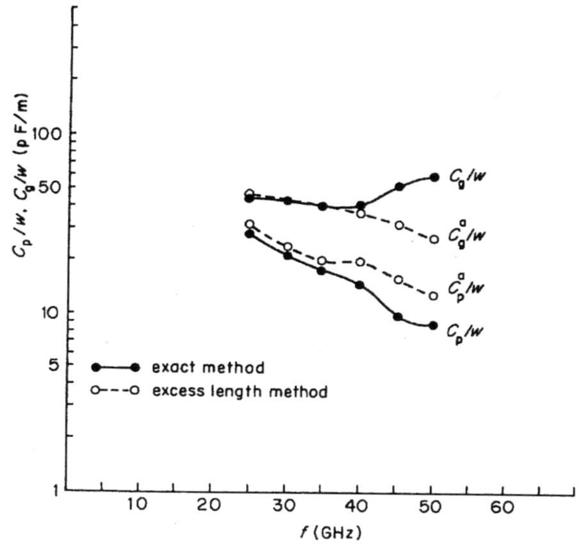

Figure 7.29 Gap-discontinuity normalized capacitances C_p/w and C_g/w as functions of frequency ($\varepsilon_r = 9.6$, $w/h = 1$, $h = 0.6$ mm, $s/h = 0.3762$). (© 1985 IEEE. Reprinted, with permission, from Katehi and Alexopoulos [276]).

In this work, longitudinal and transverse current components are included so that strips ranging from electrically narrow to electrically wide can be analysed. Results are shown for both magnitude and phase of the reflection coefficient as functions of frequency. In the case of an electrically narrow microstrip line on an alumina substrate, both the magnitude and phase of the reflection coefficient decrease with frequency — the magnitude slowly at first, then rapidly for frequencies approaching 30 GHz (about 0.75 at 31 GHz and falling). The phase is approximately 0.05 radians at 5 GHz, 0.25 radians at 25 GHz and then decreasing faster to about 0.5 radians at 31 GHz. These phase calculations are compared with results measured elsewhere, and agreement is very good.

Results are also obtained, in the form of magnitude only as a function of frequency, for a range of line widths applicable to microstrip lines on a substrate having a permittivity of 12.8. Pronounced singularities are now evident for all lines at high frequencies. At first the reflection coefficient continues to decrease rapidly, but this is followed by a further increase, a peak, and finally a resumption of the general fall in value. This behaviour occurs in the vicinity of cut-off for the first TE surface wave, due to the tendency of the transverse current to store as opposed to radiate energy and hence to increase the frequency sensitivity of the reflection coefficient. It is demonstrated that this effect is strengthened as the line width increases, which is consistent with the above argument.

With their significance in, for example, filter realization, some work has been reported by Achkar *et al.* [279] on open end discontinuities of suspended striplines. The analysis uses eigenmodes to determine the dispersion characteristics of suspended substrate striplines, and a modal analysis to find the open end discontinuity characteristics. For this type of structure, fabricated on a substrate of permittivity

2.2, it is shown that the amplitude of the reflection coefficient is essentially zero for all frequencies over the 26–40 GHz range. Its phase, however, increases from about −0.028 to −0.015 radians over this frequency range. Hence the equivalent end-effect length also increases slowly with frequency — although it is still only around one micron at 40 GHz for this structure. The behaviour of the transmission coefficient (S_{12}) was also quantified, but this parameter remains less than 130 dB down on the incident signal power over the entire band.

7.13.2 OTHER DISCONTINUITIES

Easter [268] has performed measurements which indicate that the values of the inductive series equivalent elements for discontinuities can be expected to increase considerably with frequency. Over the frequency range 5–11 GHz, Easter's measurements reveal a 40% increase in the inductance values for lines having $w/h >$ 1.0. For lines with smaller width-to-height ratios, the increase is greatly reduced.

Alternative approaches to the problem of frequency dependence of microstrip discontinuities have been presented by Menzel and Wolff [280], Kompa [281] and Mehran [282]. The complete validity of their magnetic-wall method of modelling microstrip has, however, been questioned by Easter. These researchers present their data in scattering (S) parameter form, which is often difficult to reinterpret as an equivalent circuit.

The work reported by Kompa [281] is, however, particularly interesting, even though most of his computations relate to plastic substrates ($\varepsilon_r = 2.32$). He reports S-parameters for steps in microstrip, using the planar waveguide model to analyse the structure. It is clear that the transmission parameter $|S_{12}|$ is substantially constant up to at least 12 GHz for steps produced on an alumina substrate. A dramatically different behaviour is noted for the plastic substrate, where $|S_{12}|$ decreases very rapidly for moderate-to-high width ratios at high frequencies. For example, the transmission drops to zero at just over 9 GHz for a width ratio of approximately 7.0, whereas when the ratio is only 3.0 there is little change in transmission up to 12 GHz. Kompa describes measurements, a one-section quarter-wavelength transformer (including a field-matching solution to this), and the analysis of a straight-tapered, six-stepped microstrip line.

We now consider analysis and modelling results, in terms of scattering parameters, for a selection of discontinuities.

The use of microstrip circuits fabricated on multilayer substrates is important in many MMIC designs and antenna array feed networks. To meet the circuit design needs discontinuities have been characterized for these types of structures. In particular, Harokopus and Katehi [283] report results for unmitred 90° bends. They employ a space domain method of moments solution of Pocklington's integral equation which is suitable for application to most discontinuities. These researchers present results for various combinations of substrates, generally incorporating high-alumina/low-permittivity Duroid® (i.e. permittivities of 10.2 and 2.2).

Over the frequency range 10–20 GHz, for a 90° bend fabricated on a single layer substrate, it is shown that the magnitude of the reflection coefficient rises slowly from approximately 0.08 to 0.15 (roughly following the prediction using commercially available software). In contrast, the magnitude of the transfer parameter ($|S_{12}|$) falls

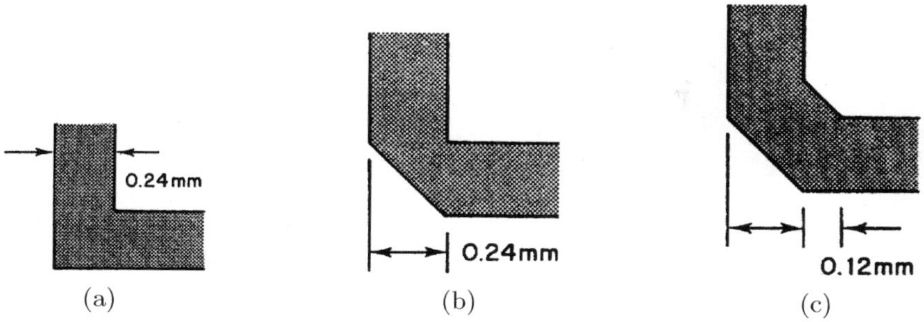

Figure 7.30 Right-angled single bend: (a) direct (i.e. unmitred); (b) mitred; and (c) double-mitred. Substrate parameters: $t = 0.24$ mm, $\varepsilon_r = 9.8$. The phase reference point is chosen at the centre of the straight bend. (© 1990 IEEE. Reprinted, with permission, from Zheng and Chang [284].)

slowly with frequency — from about 0.98 at 10 GHz to approximately 0.88 at 20 GHz. This differs greatly from the predictions of much commercially available software that indicates a value very close to 1.0 over the entire frequency range. For multilayer bends, the general trends of the parameters are the same but the actual values differ considerably. For example, in the case of a low-permittivity Duroid® substrate topped by an alumina substrate (equal thicknesses) the reflection coefficient increases from about 0.2 to 0.4, whilst the transfer parameter decreases from 0.97 to 0.74. This is due to the enhancement of radiation fields where the microstrip itself is held above a ground plane separated by a relatively low-permittivity substrate (Duroid®).

The important case of 90° bends, direct, mitred and double-mitred (see Figure 7.30), has also been investigated by Zheng and Chang [284]. They employed an integral equation approach (with the Galerkin method) to evaluate the performance of these bends.

They found similar trends and values for the coefficients to those reported by Harokopus and Katehi [283] for the unmitred bend. In common with the well known results they also noted that both S_{11} and S_{12} were considerably lower for the optimally mitred bend than for the direct one, over the frequency range 8–40 GHz at least (see also Section 7.7 on page 237). The double-mitred case was, however, particularly interesting since it provided a substantially lower reflection coefficient over the frequency range 22–38 GHz and a lower value of S_{12}, over the entire frequency range (much lower above about 18 GHz). This response is due to interference effects operating between the mitre ends, i.e. a distinctly frequency-dependent phenomenon.

7.13.3 CROSS- AND T-JUNCTIONS

The frequency-dependent behaviour of cross- and T-junctions has been investigated by Wu *et al.* [285]. They use an electromagnetic spectral-domain analysis to develop the scattering parameters of these types of junctions. An N-port junction approach is used together with a concept whereby a 'module' around the junction region contains that section where higher order modes are generated as a consequence of the discontinuity. The current distributions inside the module are expanded and a

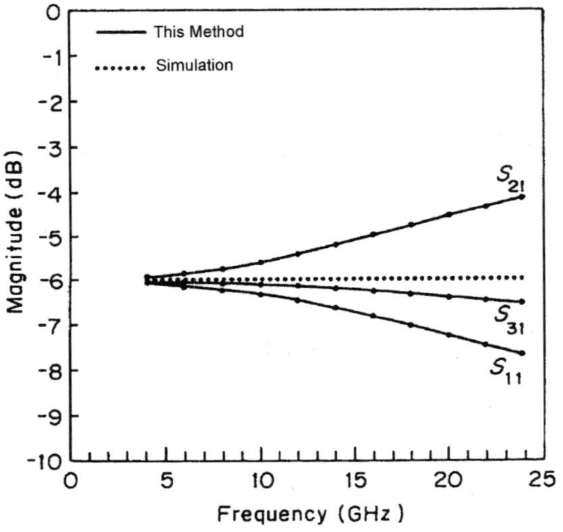

Figure 7.31 Magnitude of S-parameters of a right-angle crossing junction ($\varepsilon_r = 10.2$, $h = 0.635$ mm, $w = 0.61$ mm) (©1990 IEEE. Reprinted, with permission, from Wu, Yang and Alexopoulos [285].)

'nearly Galerkin' method is applied to transform the integral equation into a matrix equation. The scattering parameters are generated from submatrices within the major current, voltage and reflection coefficient interrelated matrices.

Wu *et al.* compare their results, to which experimental data generally fit very well, with quasi-static predictions generated using commercial software over the frequency range 4–24 GHz. For a basic T-junction the magnitude of the transfer parameters S_{21} and S_{31} indicate a greater inequality in the division of power between arms than the software predicted. In general, the disparities amounted to some 10 to 15% — being most marked as frequency increased. Phase performance was more sensitive to the accuracy of the prediction solution and, again for a basic T-junction, the phase of S_{33} is found to differ from the result using this particular modelling software by up to 45° (at 24 GHz). Wu's method, supported by measurements, shows the phase of S_{33} varying only between 170° and 175°.

In the case of a right-angled cross-junction the use of this approach is essential since, for example, some commercial software predicts simple equal power splitting over a wide frequency range. Wu's approach shows that the power separation is distinctly unequal and indeed varies markedly with frequency. Curves applicable to a right-angled cross-junction in microstrip on a high-purity alumina substrate are shown in Figure 7.31.

This method of analysis is also applied to a compensated T-junction — as originally suggested by Dydyk (see Section 7.11 on page 247). In the example analysed by Wu, the top slot is omitted and the intention is to extend the frequency range over which power is equally split between the coupled arms. For such a compensated microstrip T-junction, fabricated on a similar substrate to that described above, it is found that the power is split equally when the structure is driven at any frequency from DC to 16 GHz (compared with DC to 6 GHz for a basic T-junction).

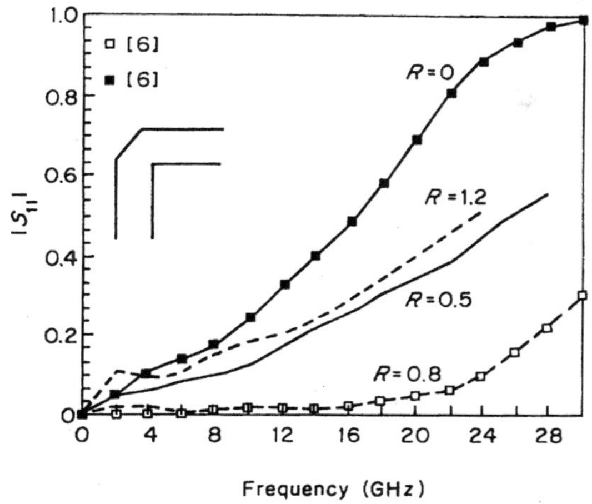

Figure 7.32 Variations in $|S_{11}|$ of a right-angle bend for cut ratio (degree of mitre) R = 0, 0.5, 0.8, 1.2. $W_{1,\text{eff}} = W_{2,\text{eff}} = 1.882$ mm, $d = 0.65$ mm, $\varepsilon_r = 2.53$. (Reprinted, with permission, from Chang, Huang and Itoh [286].)

Chang *et al.* [286] have reported an analytical CAD program applicable to a wide variety of discontinuities. Their method is based on the effective width principle (see Section 5.8.2 on page 139), following which an impedance is derived based on the boundary integral method. The scattering matrix is obtained directly from the impedance matrix.

Chang *et al.* show results for right-angled bends, T-junctions and Y-junctions. In the case of the right-angled bends, a family of curves is computed showing S_{11} as a function of frequency (see Figure 7.33), and these clearly indicate an optimum chamfer or 'cut' ratio of approximately 0.8 for a 0.65 mm thick plastic substrate of permittivity 2.53. It is also clear that this optimum continues for frequencies up to at least 29 GHz. Although the definition contrasts somewhat, the absolute magnitude of this chamfer fraction differs considerably from that given in Section 7.7 (on page 237) here — perhaps because of the much lower permittivity in this case.

Reflection coefficient data are also shown for a T-junction with a vee-cut opposite the leg of the T. With a cut (i.e. mitre) ratio definition similar to that used for the bend, it is clear that when this ratio is 0.4 an approximate optimum exists in that S_{22} is minimized over the frequency range 0 to 16 GHz. This applies to a substrate of the type applicable to the case of the bend above.

An entirely different approach from those described above for the modelling of microstrip cross-junctions has been developed by Giannini *et al.* [287]. They showed how a lumped model can be used to represent the characteristics of cross junctions over a broadband (2 to 12 GHz on alumina substrates). In contrast to most other approaches, Giannini *et al.* derive sophisticated Z-parameter expressions for a set of equivalent elements. The overall configuration is shown in Figure 7.33, where the impedance elements (actually totally reactive) are related to the physical microstrip parameters, permittivity and of course frequency — as well as complicated forms of infinite series. The term $b/2$ is the half-width of the entire cross-junction. The full details will not be set down here, and the interested reader is recommended to follow

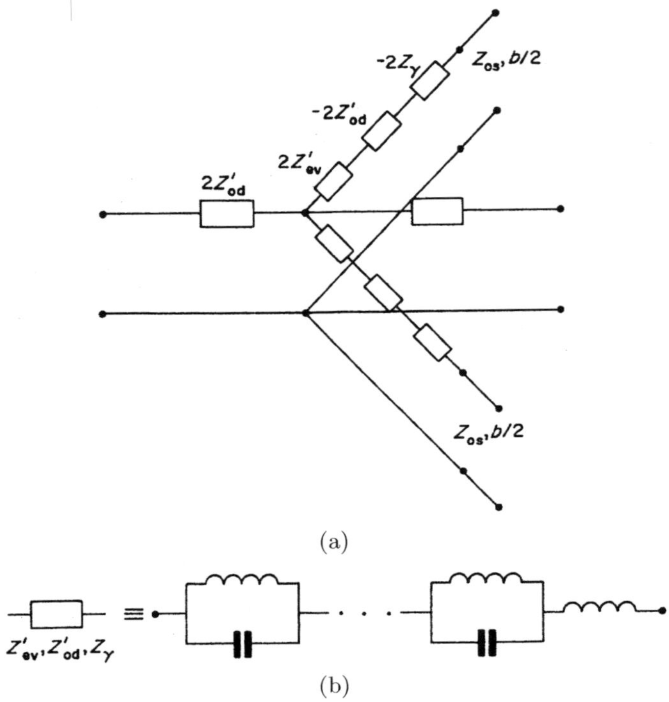

Figure 7.33 Lumped equivalent model for the cross-junction discontinuity. (Reprinted, by permission, from Giannini *et al.* [287].)

the original paper.

Like the approach due to Wu *et al.*, Giannini *et al.* also find that when compared with 'Touchstone'® data obtained using their expressions provide a much better fit to experimental results. They justify this with calculations using a version of Touchstone®, modified ('enhanced') with their families of formulas.

7.13.4 RADIAL BENDS

Most work concerning discontinuities has centred on abrupt changes in the width or direction of the transmission line and relatively little research appears to have been conducted on more continuously changing structures such as radial bends. Reported research by three groups in different locations (the first being headed by Weisshaar at Oregon State University) has, however, redressed the balance somewhat [288]. The nomenclature, equivalent circuit and waveguide model are indicated in Figure 7.34.

In the waveguide model the effective mean radius of curvature R_e is related to the effective microstrip width w_e as shown in Figure 7.34(c). R_e is chosen such that there is always a positive effective inner radius regardless of actual inner and outer radii.

The frequency-dependent behaviour of the radial bend is determined using a second-order perturbation analysis of the equivalent waveguide model together with a mode-matching technique which includes the higher-order modes. Scattering parameters and radiation losses are calculated for alumina (MIC) and GaAs (MMIC) versions of

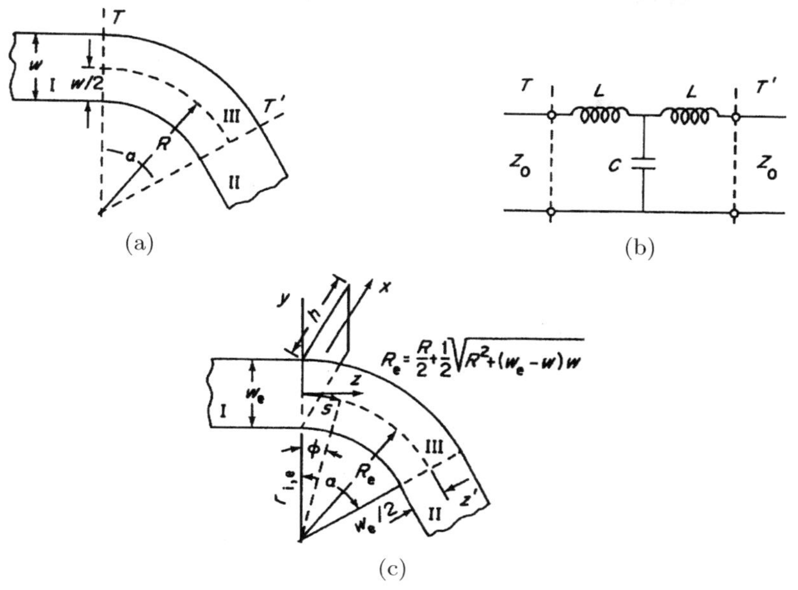

Figure 7.34 The radial microstrip bend: (a) physical, (b) equivalent lumped model (c) waveguide model. (© IEEE 1990. Reprinted, with permission, from Weisshaar *et al.* [288].)

the structure and these are shown, for 50 Ω lines, on Figures 7.35 and 7.36.

Weisshaar *et al.* also report reflection coefficient results for radial bends in 30 Ω lines, and the scattering transfer parameter S_{21}, for a T-junction with an open stub leg having curved bends.

Some researchers have suggested the use of rounding and tapering techniques, as shown in Figure 7.37 to reduce the effects of discontinuities, and these may well be effective and worth considering at least at the lower microwave frequencies. At higher frequencies, however, it would be difficult to accurately quantify the excess conductances and reactances associated with these complex shapes, and it is recommended that the better understood characterization of the conventional discontinuities described in this chapter continue to be used.

7.13.5 FREQUENCY DEPENDENCE OF SHUNT POST PARAMETERS

Finch and Alexopoulos [289] have reported a theoretical study of the frequency dependence of parameters relating to shunt posts in microstrip. They use the planar waveguide model and a multiple expansion method within algorithms for the computation of Schloemilch-type series to analyse the behaviour of these structures. Their approach applies to perfectly conducting, imperfectly conducting (i.e. 'lossy') and composite shunt posts.

These researchers determine scattering parameters (S_{11} and S_{21}) for this type of structure. In the case of a 1.22 mm diameter conducting post in microstrip, having a width of 3 mm, using an alumina substrate of thickness 0.635 mm and permittivity 10, they show that $|S_{11}|$ decreases with frequency. Starting with perfect short-circuit

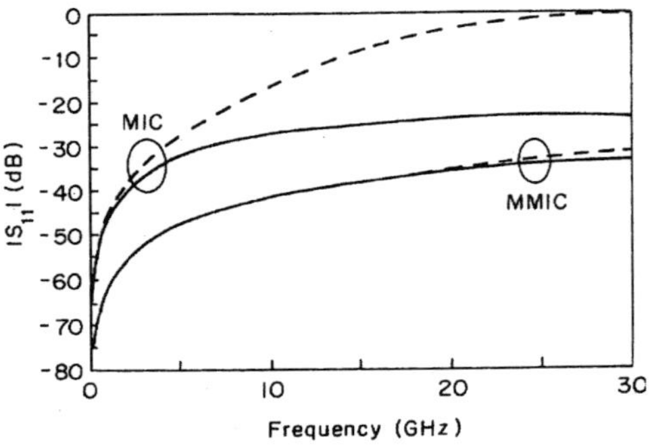

Figure 7.35 S_{11} for 50 Ω, 90° bends with $R = 2w$ on MIC (0.635 mm thick alumina) and MMIC (0.1 mm thick GaAs) substrates (—— waveguide model; - - - quasi-static model). (© IEEE 1990. Reprinted, with permission, from Weisshaar *et al.* [288].)

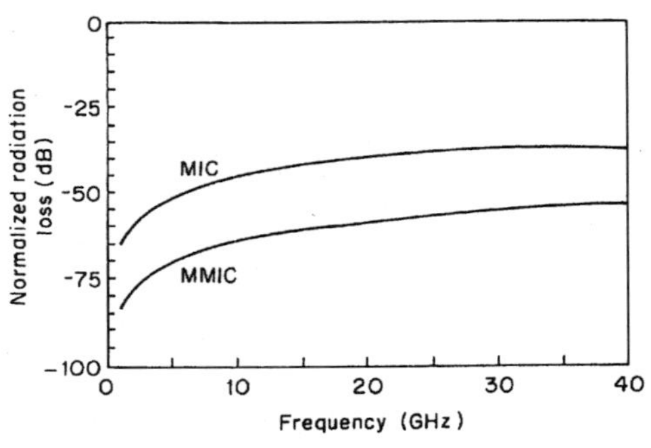

Figure 7.36 Normalized radiation loss for 50 Ω bends with $R = 2w$ on MIC (0.635 mm thick alumina) and MMIC (0.1 mm thick GaAs) substrates. (© IEEE 1990. Reprinted, with permission, from Weisshaar *et al.* [288].)

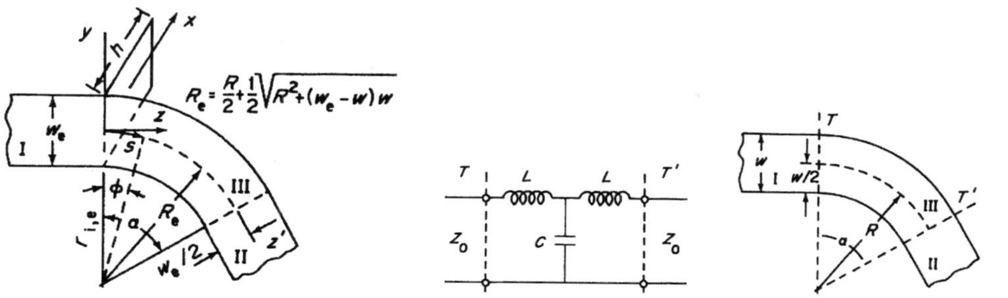

(a) (b) (c)

Figure 7.37 Rounding and tapering techniques that may be applied to reduce the discontinuity effects associated with (a) bends, (b) T-junctions and (c) steps in width.

behaviour at DC, the value of $|S_{11}|$ falls to 0.92 at 12 GHz. Doubtless, the method used by Finch and Alexopoulos could be extended to almost any of these types of structures.

7.14 RECOMMENDATIONS FOR THE CALCULATION OF DISCONTINUITIES

The following recommendations generally apply with reasonable accuracy for frequencies up to about 20 GHz, where the microstrip is on alumina-type substrates. As indicated, some of the formulas also apply at higher frequencies and on quite different substrates, such as various polymers or semiconductors.

7.14.1 FORESHORTENED OPEN-CIRCUITS

Fundamental considerations were presented in Section 7.2.1 on page 228, and (7.7) gives the open-end length extension as a function of the lumped fringing capacitance C_f, and the microstrip line parameters:

$$l_{eo} \approx cZ_0C_f/\sqrt{\varepsilon_{\text{eff}}}. \tag{7.53}$$

It is difficult to specify the accuracy of this expression, but empirical relationships can be set up for C_f, and both Z_0 and ε_{eff} are readily known parameters.

The alternative empirical formula is

$$l_{eo} = 0.412h\left(\frac{\varepsilon_{\text{eff}} + 0.3}{\varepsilon_{\text{eff}} - 0.258}\right)\left(\frac{w/h + 0.262}{w/h + 0.813}\right). \tag{7.54}$$

This expression is useful over a wide range of materials and microstrip lines, but errors can exceed 5% and this can be significant — especially in filter design.

Independent theoretical calculations and measurements verify that either expression can be used satisfactorily at frequencies up to around 18 GHz for alumina-type substrates of less than 1 mm thickness.

See also Section 7.13.1 on page 250 for frequency-dependent effects.

7.14.2 SERIES GAPS

These are usually only of significance in resonator design for measurement or filter realization purposes, where the value of g required for a given order-of-magnitude of coupling (i.e. 'loose', 'tight', ...) can often be determined experimentally. In case such a discontinuity is required in other circuits, some design equations are given in Section 7.3 on page 231, but they will not be repeated here.

7.14.3 SHORT-CIRCUITS

With the present state of technology, short-circuits are unusual and undesirable in microstrip as implemented in hybrid MICs for a variety of reasons, discussed in Section 7.4 on page 233. However, this situation may change and in any event, in the form of vias, short-circuits are important in MMICs. The diameter of the 'optimum metallized hole' [152], applicable at least to alumina-type substrates, is obtainable using

$$\ln\left(\frac{w_{\text{eff}}}{\pi d_e}\right) \approx \left(\frac{\pi d_e}{w_{\text{eff}}}\right)^2 \tag{7.55}$$

where $d_e = 0.03 + 0.44d$, d is the actual hole diameter, and w_{eff} is the effective microstrip width.

This expression yields the diameter of such a hole situated at the end of foreshortened microstrip which provides a broadband short-circuit up to about 18 GHz. It may be noted that the previous expression in w_{eff} and d_e has the form

$$\ln(x) \approx x^{-2} \tag{7.56}$$

in which

$$x = \left(\frac{w_{\text{eff}}}{\pi d_e}\right) \tag{7.57}$$

and this equation in x can be solved iteratively. In fact a total of around six or seven iterations reveal that $x = 1.532$ satisfies the equation quite accurately. The values of d_e and hence d are then easily found (w_{eff} can be found from (5.49), Section 5.8.2). on page 139).

7.14.4 RIGHT-ANGLED BENDS: MITRING

Expressions were given in Section 7.6 on page 235 for the equivalent circuit capacitance and inductance associated with a right-angled bend in microstrip.

The compensating techniques referred to in Section 7.7 on page 237 are, however, probably of more practical significance to the circuit designer, especially the simple device known as the mitred or chamfered bend. It was indicated that the approximately optimum degree, or extent, of chamfer (Figure 7.9(a)), should be

$$b \approx 0.57w \quad \text{for alumina-like substrates} \tag{7.58}$$

(but see also Section 7.13.2 on page 256 and, particularly, Section 7.13.3 on page 257).

7.14.5 STEPS IN WIDTH

Rather approximately, the equivalent length extension of the wider line (width w_2) can be determined as a function of the fully open-circuit value

$$l_{es} \approx l_{eo}\left(1 - \frac{w_1}{w_2}\right). \tag{7.59}$$

More accurate expressions can be found by incorporating the inductance and capacitance for the equivalent circuit of the bend itself; these are taken directly from the work of Garg and Bahl [263]:

$$\frac{L}{h} = 40.5\left(\frac{w_1}{w_2} - 1.0\right) - 75\frac{w_1}{w_2} + 0.2\left(\frac{w_1}{w_2} - 1.0\right)^2 \text{ nH/m}. \tag{7.60}$$

This formula (7.38) is quoted as having an accuracy of better than 5% provided that

$$w_1/w_2 \leq 5.0 \text{ and } w_2/h = 1.0 \ . \tag{7.61}$$

The following expressions for C were derived by curve fitting earlier theoretical results:

For $\varepsilon_r \leq 10$; $1.5 \leq w_2/w_1 \leq 3.5$

$$\frac{C}{\sqrt{w_1 w_2}} = (10.1 \log \varepsilon_r + 2.33)\frac{w_2}{w_1} - 12.6 \log \varepsilon_r - 3.17 \text{ pF/m}. \tag{7.62}$$

For $\varepsilon_r = 9.6$; $3.5 \leq w_2/w_1 \leq 10$:

$$\frac{C}{\sqrt{w_1 w_2}} = 130 \log(w_2/w_1) - 44 \text{ pF/m}. \tag{7.63}$$

The error involved in applying the first equation, which applies to relatively slight steps (w_2/w_1 maximum is 3.5), can be as high as 10%. The error in applying the second equation is claimed to be less than 0.5%. Appropriate extra equivalent lengths of the microstrip line can then be determined in terms of L and C.

7.14.6 TRANSVERSE SLIT

As outlined in Section 7.9 on page 242, this yields the effect of a lumped inductance, ΔL, in the plane of the slit, given by

$$\frac{\Delta L}{h} = \frac{\mu_0 \pi}{2}\left(\frac{a'}{A}\right)^2 \tag{7.64}$$

where

$$\frac{a'}{A} = 1 - \frac{Z_{0(\text{air})}}{Z'_{0(\text{air})}}. \tag{7.65}$$

For design, the procedure is then as follows:

(a) Calculate the characteristic impedance of the air-filled microstrip line, width w — $Z_0(\text{air})$.
(b) Calculate the characteristic impedance of an air-filled microstrip line of width $(w-a)$ — $(Z_0'(\text{air}))$.
(c) Evaluate a'/A with these results, (a) and (b), using the second expression, and hence calculate ΔL using the first expression.

The accuracy of these equations is very difficult to estimate, but the main application appears to be in trimming microstrip stubs, etc., for correct electrical length. That is, the slit could be deliberately introduced at a late manufacturing stage. The expressions tend to break down as the 'slit' becomes too narrow (capacitive) or too wide. In the latter case the 'slit' tends towards a length of (distributed) microstrip line of reduced width.

7.14.7 THE T-JUNCTION

This was fully described in Section 7.10 on page 244 and further in Section 7.13.3 on page 257. Although it is clear that substantial further work is required, especially for frequencies much in excess of 10 GHz, the following expressions hold approximately, provided $2w_{\text{eff1}}/\lambda_{g1} > 0.3$, with

$$w_{\text{eff1.2}} = \frac{h\eta}{Z_{0(1.2)}\sqrt{\varepsilon_{\text{eff}}}}. \tag{7.66}$$

The transformation ratio n is given by

$$n^2 = \left[\frac{\sin\{\pi(w_{\text{eff1}}/\lambda_{g1})(Z_{0(1)}/Z_{0(2)})\}}{\pi(w_{\text{eff1}}/\lambda_{g1})(Z_{0(1)}/Z_{0(2)})}\right]^2 \left[1 - \{\pi(w_{\text{eff1}}/\lambda_{g1})(d_2/w_{\text{eff1}})\}^2\right]. \tag{7.67}$$

The displacement of the reference plane for the primary arm (1) is

$$\frac{d_1}{w_{\text{eff2}}} = 0.05\frac{Z_{0(1)}}{Z_{0(2)}}n^2. \tag{7.68}$$

The displacement of the reference plane for the secondary arm (2) is

$$\begin{aligned}\frac{d_2}{w_{\text{eff1}}} = {} & 0.5 - \left\{0.076 + 0.2\left[\frac{2w_{\text{eff1}}}{\lambda_{g1}}\right]^2\right. \\ & \left. +0.663\exp\left[-1.71\frac{Z_{0(1)}}{Z_{0(2)}}\right] - 0.172\ln\left[\frac{Z_{0(1)}}{Z_{0(2)}}\right]\right\}\frac{Z_{0(1)}}{Z_{0(2)}}.\end{aligned} \tag{7.69}$$

(Replace $Z_{0(1)}/Z_{0(2)}$ by its inverse, i.e. $Z_{0(2)}/Z_{0(1)}$ when $Z_{0(1)}/Z_{0(2)} \geq 2.0$.) Also, the shunt capacitance is determined by the following expressions:

For the condition $Z_{0(1)}/Z_{0(2)} \leq 0.5$:

$$\frac{\omega C\lambda_{g1}}{Y_{0(1)}w_{\text{eff1}}} = \left(\frac{2w_{\text{eff1}}}{\lambda_{g1}} - 1\right)\frac{Z_{0(1)}}{Z_{0(2)}}. \tag{7.70}$$

For the condition $Z_{0(1)}/Z_{0(2)} \geq 0.5$:

$$\frac{\omega C \lambda_{g1}}{Y_{0(1)} w_{\text{eff1}}} = \left(\frac{2w_{\text{eff1}}}{\lambda_{g1}} - 1 \right) \left(2 - 3\frac{Z_{0(1)}}{Z_{0(2)}} \right). \tag{7.71}$$

Also refer to Section 7.13.3 on page 257 for frequency-dependent effects.

7.14.8 THE ASYMMETRIC CROSS-JUNCTION

This structure was considered in Section 7.12 on page 247, and a fairly comprehensive equivalent circuit was shown which is again applicable for frequencies up to about 12 GHz. The capacitance is given simply by

$$C_m = \frac{c}{Z_0 \sqrt{\varepsilon_{\text{eff}}}}. \tag{7.72}$$

Alternatively, and probably more powerfully at somewhat higher frequencies, the cross-junction may be considered as two T-junctions operating from the same plane in the main microstrip line. Some elaboration of the equivalent circuits may be necessary if interactions between the branch arms are to be taken into account.

For virtually all the discontinuities, it is clear that substantial further work is necessary before techniques are available for circuit designers working at frequencies beyond about 1 > 12 GHz. Up to these frequencies, and occasionally beyond, the expressions given may be used in design.

Further design information, including dispersive effects, is provided in Section 7.13.3 on page 257. This should prove useful in support of design at frequencies up to 25 GHz.

7.15 STRIPLINE DISCONTINUITIES

These generally follow the same types of discontinuities as in the case of microstrip. The major differences between the media (mostly the fact that stripline is axially symmetric) result in significant differences in the characterization of the stripline discontinuities. The layouts applicable to each stripline discontinuity are identical to those for microstrip.

7.15.1 BENDS

This has the same layout and equivalent circuit as Figure 6.6. As with microstrip, mitring again significantly reduces the discontinuity effects, and this approach is recommended.

7.15.2 VIAS

The implementation of vias (short-circuits) in stripline is in fact somewhat more difficult than with microstrip or coplanar waveguide. This is because each via must connect through to both the upper and lower ground planes.

Since stripline is not applicable to MMICs, the advantages of vias in MMIC technology are not available.

7.15.3 JUNCTIONS

The most important junctions are the T-junction and the asymmetric cross-junction. For microstrip the layout and equivalent circuit are shown in Figure 7.19, and the reference planes and hypothetical widths are indicated in Figure 7.21. These notations also apply to the stripline case. Again, as with the right-angled bend discussed above, compensation should be considered and a similar approach to that used for microstrip is recommended.

The layout and an equivalent circuit for an asymmetric cross-over junction are shown in Figure 7.22, and this also applies to stripline.

8

Parallel-coupled Lines and Directional Couplers

8.1 STRUCTURE AND APPLICATIONS

The arrangement shown in Figure 8.1 illustrates the basic structure under consideration in this chapter.

It will be assumed that both microstrip lines have the same widths, which is nearly always the case in practical applications. Some work has also been carried out on multiple arrays of such parallel, edge-coupled lines, but we can usefully restrict the considerations to just two lines throughout most of this chapter.

There are two general areas of application for these structures:

(a) Directional couplers — for use in a variety of circuits including balanced mixers, balanced amplifiers, phase shifters, attenuators, modulators, discriminators, and measurement bridges.
(b) Filters, delay lines, and matching networks — often using arrays of parallel-coupled microstrips as resonant elements.

In the first instance, (a), a prescribed amount of the incident power is required to be coupled out of the system. Thus, for example, a '3 dB coupler' is one in which half of the power input is coupled from one microstrip line into another and then on to separate circuitry. We shall see later in this chapter that good coupler performance is achievable if special modifications are made to the basically simple structure, consisting of two parallel microstrips.

Microstrip filters (b) employing the parallel-coupled structure are usually of the bandpass or bandstop types. An example of a design approach employing resonators coupled in this manner is given in Chapter 10.

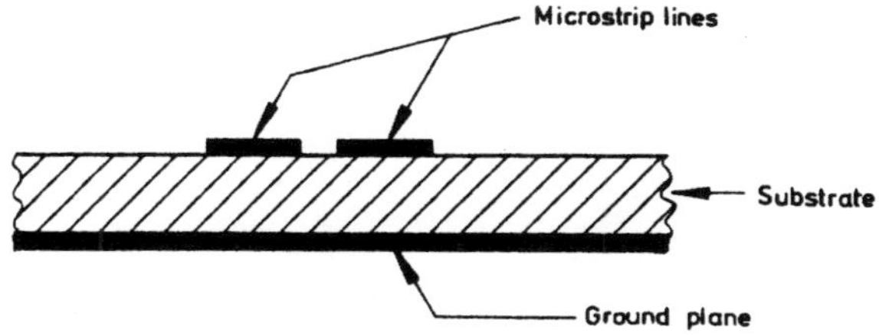

Figure 8.1 A pair of parallel, edge-coupled microstrip lines.

Apart from direct applications, the calculations described here are also useful for estimating the degree of unwanted coupling between runs of microstrip lines on the same substrate. These calculations may also be judiciously applied to the problem of estimating the degree of crosstalk between such microstrip lines when they are used in high-speed pulse and digital systems.

In this chapter we principally consider the following features:

(a) directional coupler specification,
(b) calculation of cross-sectional dimensions (for a desired amount of coupling),
(c) calculation of appropriate coupled-region physical length,
(d) frequency response characteristics,
(e) coupler directivity,
(f) coupler compensation techniques,
(g) 'special' designs (including Lange),
(h) thickness effects, losses, and fabrication tolerances,
(i) crosstalk between microstrips used in high-speed digital systems.

Features (b), (c), (d) and (h) are not only applicable to directional couplers, they are also important in some filter design techniques and in crosstalk calculations (i).

8.2 PARAMETERS AND INITIAL SPECIFICATION

As described in Section 1.10 on page 23, the even- and odd-mode characteristic impedances (Z_{0e} and Z_{0o}) are major design parameters for any parallel-coupled transmission line configuration — whatever its application. These impedances are obtained at an early design stage and they are functions of the degree of coupling (C) and the single-line terminating characteristic impedance (Z_0). The relationships between Z_{0e} and Z_{0o} and the physical dimensions of the coupled structure (including the substrate permittivity) are therefore of prime significance to the designer. As with single microstrip lines, so also in this coupled situation we can determine Z_{0e} and Z_{0o} from known physical dimensions, which amounts to analysis. Alternatively, with greater difficulty, we may synthesize the physical structure from starting values of the impedances. Both of these procedures are useful in practice.

For coupler design the important starting parameters are defined in Section A.5.2 on page 441 of Appendix A. They are coupling factor C, transmission factor T, directivity D and isolation I. Microstrip couplers, unless compensated by one of the methods to be described, exhibit a performance which can be far removed from the ideal in some cases. This is due to the different phase velocities associated with the even and odd modes, the main effect of which is to degrade the directivity. The initial specification for such a coupler will include:

(a) coupling factor C (usually in decibels, dB) at the centre frequency f_0,
(b) permittivity and thickness of the substrate,
(c) terminating characteristic impedance Z_0 (usually 50 Ω),
(d) bandwidth and centre frequency,
(e) coupling factor tolerance over the band (sometimes),
(f) lowest acceptable directivity D in decibels (dB).

From this information the designer must ultimately determine the widths of the

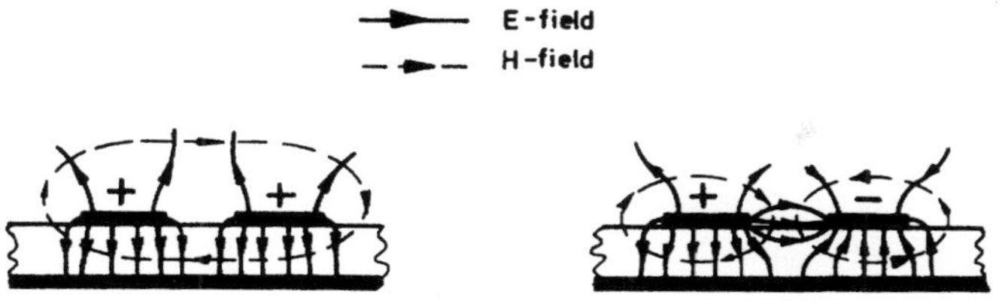

(a) (b)

Figure 8.2 Field distributions resulting from (a) even-mode and (b) odd-mode excitation of parallel-coupled microstrip lines.

microstrip lines, the separation between them, and the length of the coupled region. The final synthesis problem is then similar to that of a coupler. In the design of filters or matching networks using parallel-coupled lines we usually begin with either insertion loss as a function of frequency or VSWR requirements over some band. In either event, we still arrive at desired values of Z_{0e}, and Z_{0o} and the final synthesis problem is then similar to that of a coupler.

8.3 COUPLED MICROSTRIP LINES

In Chapter 1 we considered the basic properties of TEM lines in a parallel-coupled configuration. Just as single microstrip lines are quasi-TEM, so coupled microstrip lines also exhibit a field pattern which is strictly quasi-TEM. We may, however, examine even-mode and odd-mode excitation (Figure 8.2) for the coupled microstrips.

Notice that the relative proportions of fields shared between the dielectric substrate and the air are different. Thus, the phase velocities are also different, a fact which represents an extra design complication.

It is important to note that all parallel-line couplers, whether true TEM or not, have the odd- and even-mode property which always results in Z_{0e} and Z_{0o}. However, true TEM couplers yield equal phase velocities for each mode whereas microstrip (quasi-TEM) and certain other structures yield odd-mode and even-mode phase velocities, v_{po} and v_{pe}.

Referring to Appendix A this means that, strictly speaking, $\theta_e \neq \theta_o$ and Equation (A.39) cannot be evoked to exactly satisfy the equality relationship given (because only a single θ was present there). There is no reason why we should not continue to define the coupling factor as Equation (A.57), but the remaining relationships, for V_2, V_3 and V_4 are all more complicated. We shall return to this problem from a design viewpoint shortly.

The cross-sectional dimensions which we require from a design procedure are w and s, as indicated in Figure 8.3. Both the even and the odd modes exist under the normal circuit excitation conditions associated with a microstrip coupler. However, coupler calculation and design proceeds from consideration of the effects due to each mode taken separately.

Analysis yielding Z_{0e} and Z_{0o}, as functions of the shape ratio of w/h, s/h and ε_r, has been carried out by several groups — notably Bryant and Weiss [99]. Although this computer-orientated work is subject to some debate regarding accuracy

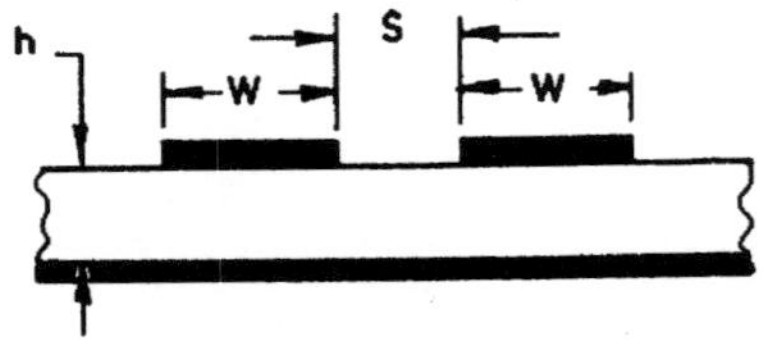

Figure 8.3 Cross-sectional dimensions: nomenclature.

(especially for tight coupling) it is a fundamental approach and graphs of results are very readily available in microwave handbooks. The method of analysis due to Bryant and Weiss [99] amounts to a rigorous solution of the electrostatic problem for coupled microstrip lines. (Incidentally, single strips are also solved.) They determine capacitances on a static TEM basis. Longitudinal components of the fields are neglected and the thickness of the microstrips is also neglected. A formulation based essentially on Laplace's equation is employed with the proper boundary conditions inserted. Iterative numerical methods are used in the computer solution and the impedances are finally obtained as follows (either mode).

We start with the effective permittivity (see Equations (4.10) and (4.11)):

$$\varepsilon_{\text{eff}} = \frac{C}{C_1} \tag{8.1}$$

where C is the microstrip capacitance per strip with the substrate present and C_1 is the microstrip capacitance per strip with the substrate absent (i.e. air spaced). The velocity of propagation v_p is then, from Equation (4.10)

$$v_p = \frac{c}{\sqrt{\varepsilon_{\text{eff}}}} \tag{8.2}$$

and the characteristic impedance is Equation (4.3)

$$Z_0 = \frac{1}{v_p C} = \frac{1}{cC_1\sqrt{\varepsilon_{\text{eff}}}}. \tag{8.3}$$

For odd and even modes we therefore have, respectively,

$$Z_{0o} = \frac{1}{cC_{1o}\sqrt{\varepsilon_{\text{eff},o}}} \tag{8.4}$$

and

$$Z_{0e} = \frac{1}{cC_{1e}\sqrt{\varepsilon_{\text{eff},e}}}. \tag{8.5}$$

The calculations required, leading to C_{1o}, C_{1e} and $\varepsilon_{\text{eff,e}}$, $\varepsilon_{\text{eff,o}}$, are carried out in a manner identical to that described in Section 4.8 on page 97. Some numerical data obtained using this analysis technique are presented graphically here in Figure 8.4. In design, Z_{0e} and Z_{0o} are marked on the vertical axis of a graph for the appropriate substrate material (ε_r). It is then necessary to read these values across horizontally until two points are found lying vertically above one another, and at identical values of s/h. Dropping a vertical ordinate through these points then also yields the required

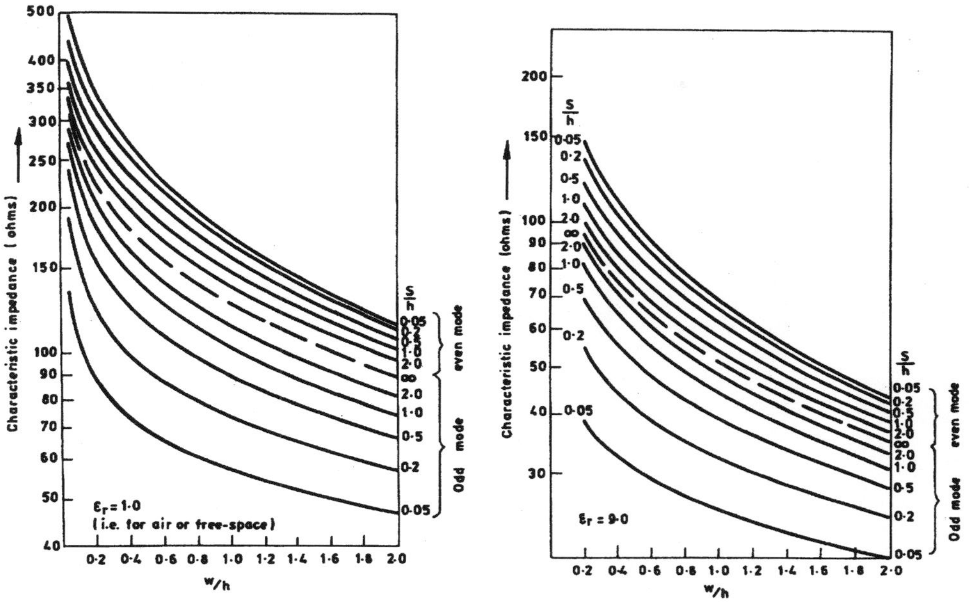

Figure 8.4 Numerical results for the even- and odd-mode characteristic impedances of parallel-coupled microstrip lines: (a) air spaced and (b) with a substrate having relative permittivity ε_r. (© 1968 IEEE. Reprinted, with permission, from Bryant and Weiss [99].)

value of w/h. Interpolation and other difficulties with the graphs makes this an approximate and rather clumsy design technique. However, it may be useful in certain instances and a specific microstrip coupler design, using these curves, will be outlined shortly.

8.4 CHARACTERISTIC IMPEDANCES IN TERMS OF THE COUPLING FACTOR (C)

First, we repeat Equations (A.67), (A.39), (A.68) and (A.69) observing that several of these expressions are approximate in the microstrip case due to the unequal phase velocities. We have:

Coupling Factor:

$$C' = 20 \log \left| \frac{Z_{0e} - Z_{0o}}{Z_{0e} + Z_{0o}} \right| \quad \text{dB} \tag{8.6}$$

Impedance relationship:

$$Z_0^2 \approx Z_{0e} Z_{0o} \tag{8.7}$$

and, from these expressions, the impedances required are

$$Z_{0e} \approx Z_0 \sqrt{\frac{1 + 10^{C'/20}}{1 - 10^{C'/20}}} \tag{8.8}$$

and

$$Z_{0o} \approx Z_0 \sqrt{\frac{1 - 10^{C'/20}}{1 + 10^{C'/20}}}. \tag{8.9}$$

The last two equations are the basic initial design expressions. The designer will know the characteristic impedance of the external lines connecting to the coupler (Z_0), often 50 Ω, and will also know the value of the mid-band coupling factor C' in decibels — or some quantities inherently involving C'.

As the coupling increases the approximation involved in Equation (8.7) worsens steadily. An accurate relationship can be obtained by setting θ_e and θ_o appropriately into Equations (A.24) and (A.25) and adding the matrices to obtain

$$\begin{bmatrix} V_1 \\ I_1 \end{bmatrix} = \begin{bmatrix} \cos\theta_e + \cos\theta_o & j(Z_{0e}\sin\theta_e + Z_{0o}\sin\theta_o) \\ j(Y_{0e}\sin\theta_e + Y_{0o}\sin\theta_o) & \cos\theta_e + \cos\theta_o \end{bmatrix} \begin{bmatrix} V_2 \\ I_2 \end{bmatrix}. \tag{8.10}$$

The condition for a perfect input match and perfect isolation can be written down directly from this because diagonal terms involving impedances must be made equal; thus

$$\frac{Z_{0e}}{Z_0}\sin\theta_e + \frac{Z_{0o}}{Z_0}\sin\theta_o = \frac{Z_0}{Z_{0e}}\sin\theta_e + \frac{Z_0}{Z_{0o}}\sin\theta_o \tag{8.11}$$

or

$$Z_0^2 = Z_{0e}Z_{0o}\frac{Z_{0e}\sin\theta_e + Z_{0o}\sin\theta_o}{Z_{0e}\sin\theta_o + Z_{0o}\sin\theta_e}. \tag{8.12}$$

This is a frequency-dependent result because θ_e and θ_o are both functions of frequency.

Having made a 'first-cut' design θ_e and θ_o may be evaluated and the value of Z_0 can then be obtained from Equation (8.12). If Z_0 is found to differ too greatly from the specified value, then the geometry (w and s) might be varied slightly to correct the result.

In many cases, certainly for 10 dB couplings and looser, the approximation offered by Equation (8.7) is sufficient.

8.5 SEMI-EMPIRICAL ANALYSIS FORMULAS AS A DESIGN AID

Analysis formulas are available for the determination of characteristic impedances to an accuracy of within 3% [290, 291]. These could be useful to the designer who might start with an approximate synthesis and employ the analysis formulas to provide an improvement in accuracy using an iterative technique. Air-line and substrate-line capacitances form the basis of the expressions, and the capacitances are broken down as shown in Figure 8.5. The accuracy of better than 3% holds over the ranges:

$$0.2 \leq w/h \leq 2.0; \quad 0.05 \leq s/h \leq 2.0, \quad \text{and} \quad \varepsilon_r \geq 1. \tag{8.13}$$

The total capacitances for each mode can then be written as

$$C_e = C_p + C_f + C'_f \tag{8.14}$$

and

$$C_o = C_p + C_f + C_{ga} + C_{gd}. \tag{8.15}$$

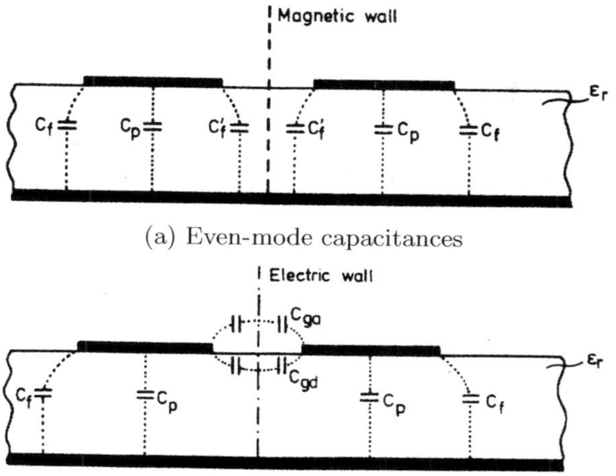

(a) Even-mode capacitances

(b) Odd-mode capacitances

Figure 8.5 Separation of capacitances for use in the analysis. (Reproduced by permission of Artech House Books, Inc., Norwood, Mass., from Gupta *et al.* [88], Figure 8.15.)

The capacitance C_p simply relates to the parallel-plate line value given by

$$C_p = \varepsilon_o \varepsilon_r \frac{w}{h}. \tag{8.16}$$

Also, C_f is simply the fringing capacitance due to each microstrip taken alone, as if for a single strip. This is given by

$$2C_f = \frac{\sqrt{\varepsilon_{\text{eff}}}}{cZ_0} - C_p \tag{8.17}$$

where c is the free-space velocity and ε_{eff}, and Z_0 are obtained by single-strip static-TEM methods.

An empirical expression for C_f' is given as follows:

$$C_f' = \frac{C_f}{1 + (A\,h/s)\tanh(8s/h)} \sqrt{\frac{\varepsilon_r}{\varepsilon_{\text{eff}}}} \tag{8.18}$$

where

$$A = \exp\{-0.1\exp(2.33 - 2.53w/h)\}. \tag{8.19}$$

As can be seen from Figure 8.5, C_{ga} and C_{gd} represent, respectively, odd-mode fringing field capacitances for the air and dielectric regions across the coupling gap. C_{ga} was obtained by using an equivalent coplanar strip geometry calculation, yielding

$$C_{ga} = \varepsilon_0 \frac{K(k')}{K(k)} \tag{8.20}$$

where

$$k = \frac{s/h}{s/h + 2w/h} \tag{8.21}$$

$$k' = \sqrt{1 - k^2} \tag{8.22}$$

and the ratio of the elliptic functions is

For $0 \leq k^2 \leq 0.5$:

$$\frac{K(k')}{K(k)} = \frac{1}{\pi} \ln \left(2 \frac{1+\sqrt{k'}}{1-\sqrt{k'}} \right). \tag{8.23}$$

For $0.5 \leq k^2 \leq 1.0$:

$$\frac{K(k')}{K(k)} = \frac{\pi}{\ln\{2(1+\sqrt{k})/(1-\sqrt{k})\}}. \tag{8.24}$$

C_{gd} was determined differently. The capacitance for coupled striplines (triplate) was modified to give

$$C_{gd} = \frac{\varepsilon_0 \varepsilon_r}{\pi} \ln \left\{ \coth \left(\frac{\pi}{4} \frac{s}{h} \right) \right\} + 0.65 C_f \left(\frac{0.02}{s/h} \sqrt{\varepsilon_r} + 1 - \varepsilon_r^{-2} \right). \tag{8.25}$$

The even-mode and odd-mode characteristic impedances are then

$$Z_{0e} = \left(c\sqrt{C_e C_{e1}} \right)^{-1} \tag{8.26}$$

and

$$Z_{0o} = \left(c\sqrt{C_o C_{o1}} \right)^{-1} \tag{8.27}$$

where c is the velocity in free space and the second subscript 1 refers to a free-space ('air') line.

The effective microstrip permittivities are

$$\varepsilon_{\mathrm{eff},e} = C_e / C_{e1} \tag{8.28}$$

and

$$\varepsilon_{\mathrm{eff},o} = C_o / C_{o1}. \tag{8.29}$$

This concludes the presentation of the semi-empirical analysis. We next consider an approximate synthesis technique.

8.6 AN APPROXIMATE SYNTHESIS TECHNIQUE

In this potentially powerful procedure the shape ratios w/h and s/h are determined from Z_{0e} and Z_{0o} information either using just two universal graphs, or with the aid of a relatively simple computer routine.

In both cases, there are two distinct stages. One of these comprises a determination of equivalent single microstrip shape ratios $(w/h)s$. The second stage relates the required w/h and s/h for the coupled structure to the equivalent single microstrip shape ratios. The final results are made independent of the substrate permittivity for a considerable range and are given by Akhtarzad *et al.* [292], not only in graphical form, but also as closed formulas. A summary of the design stages is thus:

(a) Determine shape ratios for equivalent single microstrip lines.
(b) Obtain the shape ratio w/h and the spacing ratio s/h for the desired coupled microstrip structure using the single-line shape ratios found at stage (a).

For design stage (a) we use the following relationships:

$$Z_{0se} = \frac{Z_{0e}}{2} \quad \text{(for single microstrip shape ratio } (w/h)_{se}\text{)} \tag{8.30}$$

and

$$Z_{0so} = \frac{Z_{0o}}{2} \quad \text{(for single microstrip shape ratio } (w/h)_{so}\text{).} \tag{8.31}$$

The appropriate shape ratios are then determined directly from a graph or from closed-form expressions which are readily available. The graphical technique described in Chapter 4 is recommended or the approximate static-TEM formulas given there may be used for more comprehensive CAD.

Akhtarzad *et al.* [292] give a graph for determining w/h and s/h from the single-line values now known. This graph is suitable for many approximate designs and therefore it is repeated here (Figure 8.6) with some additional information to guide the user.

To perform design using this graph, we start with stage (a) and evaluate $(w/h)_{se}$ and $(w/h)_{so}$ using single-line static-TEM methods. We then select the two curves of Figure 8.6 which have the appropriate w/h parameter for each mode as just evaluated (solid curves for odd mode, broken lines for even mode). The point of intersection of these two selected lines is the 'design point'. Reading across horizontally yields the required w/h for each of the coupled lines, and reading down vertically gives their normalized spacing s/h. A numerical example of design using this graph will be given shortly. Difficulties may be encountered in crowded regions of the graph, such as low w/h combined with low s/h, and if the single-line-derived w/h values do not happen to fit the values 2, 3, 4, 6, 8, or 10, then some fairly awkward interpolation is necessary. To overcome these problems, Akhtarzad *et al.* [292] have given a family of useful expressions the most complete of which are quoted here first:

$$\left(\frac{w}{h}\right)_{se} = \frac{2}{\pi}\cosh^{-1}\left(\frac{2d-g+1}{g+1}\right) \tag{8.32}$$

$$\left(\frac{w}{h}\right)_{so} = \frac{2}{\pi}\cosh^{-1}\left(\frac{2d-g-1}{g-1}\right) + \frac{4}{\pi\left(1+\varepsilon_r/2\right)}\cosh^{-1}\left(1+2\frac{w/h}{s/h}\right), \quad \varepsilon \le 6 \tag{8.33}$$

or

$$\left(\frac{w}{h}\right)_{so} = \frac{2}{\pi}\cosh^{-1}\left(\frac{2d-g-1}{g-1}\right) + \frac{1}{\pi}\cosh^{-1}\left(1+2\frac{w/h}{s/h}\right), \quad \varepsilon \ge 6 \tag{8.34}$$

where

$$g = \cosh\left(\frac{\pi s}{2h}\right) \tag{8.35}$$

and

$$d = \cosh\left(\pi\frac{w}{h} + \frac{\pi s}{2h}\right). \tag{8.36}$$

Equation (8.32) has to be solved simultaneously with either Equation (8.33) or (8.34) as appropriate for the particular permittivity range. The curves in Figure 8.6 were originally obtained by Akhtarzad *et al.* for the median value ($\varepsilon_r = 6$) of

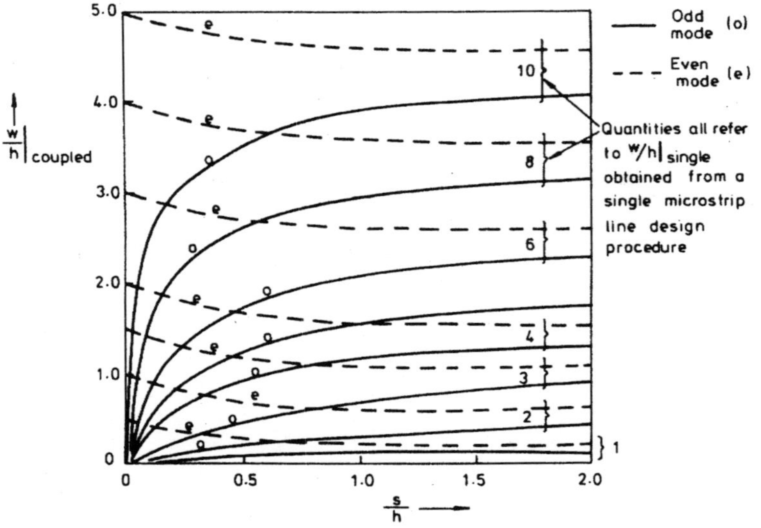

Figure 8.6 'General' curves for use in the synthesis technique. (© 1975 IEEE. Reprinted, with permission, from Akhtarzad *et al.* [292].)

the permittivity. As an approximation the second terms may be neglected in both Equations (8.33) and (8.34). Then an explicit formula for s/h is obtained:

$$\frac{s}{h} = \frac{2}{\pi} \cosh^{-1} \left\{ \frac{\cosh\left[(\pi/2)\ (w/h)_{se}\right] + \cosh\left[(\pi/2)\,(w/h)_{so}\right] - 2}{\cosh\left[(\pi/2)\ (w/h)_{so}\right] - \cosh\left[(\pi/2)\,(w/h)_{se}\right]} \right\}. \tag{8.37}$$

In any case, the functional calculations are quite complicated whether one uses the more accurate expressions or starts with this formula, Equation (8.37).

Also, when compared with the results of Bryant and Weiss [99] an error of the order of 10% has been quoted for this synthesis technique [292]. If this magnitude of error is actually found in practice then the synthesis technique may still be used to obtain an initial design, after which the more accurate analysis of Garg and Bahl (Section 8.5 on page 274) could be employed to correct the design dimensions. The sequence would be:

(a) With the required Z_{0e} and Z_{0o} determine $(w/h)_1$ and $(s/h)_1$ by approximate synthesis.
(b) Using these values, $(w/h)_1$ and $(s/h)_1$, recalculate Z_{0e} and Z_{0o} by means of the semi-empirical analysis formulas given in Section 8.5 on page 274.
(c) Compare the new values thus calculated for Z_{0e} and Z_{0o} with those originally required. A suitable programmable calculator, spreadsheet, or computer routine will then enable iterations to be made, incrementing w/h and s/h until the impedances agree within a specified tolerance.

Although this procedure can yield coupled-line design to approximately 3% accuracy, there is still clearly scope for significant improvement to aid the circuit designer.

In particular, either the synthesis technique due to Akhtarzad *et al.* might be modified to obtain greater accuracy or a new technique might be developed.

8.7 A SPECIFIC EXAMPLE: DESIGN OF A 10 DB MICROSTRIP COUPLER

We will approach this by two methods:

(a) using Bryant and Weiss curves,
(b) using Akhtarzad synthesis curves.

For the moment, in this section, we shall only determine the cross-sectional dimensions. Evaluation of the length of a coupler is considered shortly.

The specification is:

coupling coefficient	C	$= -10$ dB;
single microstrip feedline characteristic impedances	Z_0	$= 50\ \Omega$;
substrate permittivity	ε_r	$= 9.0$;
thickness	(h)	$= 1$ mm;
system centre frequency (mid-band for the coupler)	f_0	$= 5$ GHz.

The frequency is low enough for dispersion effects (dealt with later) to be neglected to a first approximation.

Design calculations:

Z_{0e} and Z_{0o} are obtained using Equations (8.8) and (8.9). This yields

$$Z_{0e} = 69.5\Omega \quad \text{and} \quad Z_{0o} = 36\ \Omega. \tag{8.38}$$

Next, always check that the relation $Z_0 \approx \sqrt{Z_{0e}Z_{0o}}$ holds, although the approximation is not good for tight (around -3 dB or more) coupling. Hence, in this case

$$Z_0 \approx \sqrt{Z_{0e}Z_{0o}} = \sqrt{69.5 \times 36} = 50.02\ \Omega. \tag{8.39}$$

8.7.1 USE OF BRYANT AND WEISS' CURVES

With Bryant and Weiss' method (Section 8.3 on page 271), choose the $\varepsilon_r = 9$ family of curves. Select two points, one on the 'even-mode' set of curves and the other on the 'odd-mode' set of curves. These points are controlled by the necessity for them to lie on the same w/h ordinate value, and on curves with an identical s/h parameter, i.e. they must both have the same w/h and s/h. Then w/h and s/h are read off as 0.87 and 0.25. The procedure is shown in the sketched graph of Figure 8.7.

From the values obtained we can determine w and s easily, since the substrate thickness h is given as 1 mm:

$$w = 0.87 \times 1 = 0.87 \text{ mm} \tag{8.40}$$

and

$$s = 0.25 \times 1 = 0.25 \text{ mm}. \tag{8.41}$$

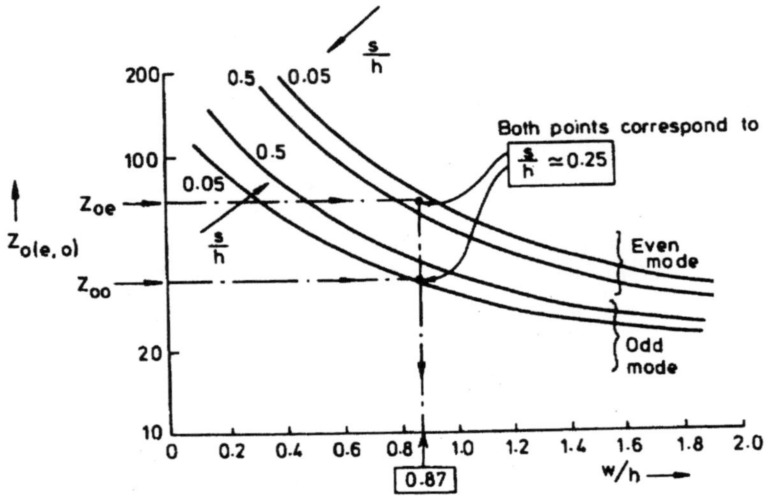

Figure 8.7 Graphical design procedure yielding the shape ratios for a 10 dB microstrip directional coupler.

8.7.2 SYNTHESIS USING AKHTARZAD'S TECHNIQUE

First we must calculate 'single' microstrip characteristic impedances, using Equations (8.30) and (8.31):

$$Z_{0se} = \frac{Z_{0e}}{2} \approx 35\ \Omega \tag{8.42}$$

and

$$Z_{0so} = \frac{Z_{0o}}{2} \approx 18\ \Omega \tag{8.43}$$

Details of the actual calculations will not be given here because they follow the same routine that was given in Chapter 4. Approximate values are

$$(w/h)_{se} \approx 2.0 \quad \text{and} \quad (w/h)_{so} \approx 5.0\ . \tag{8.44}$$

Now, using Figure 8.6, we first select the broken (even-mode) curve which has parameter $(w/h)_{se} = 2$. Next, select the solid (odd-mode) curve which has parameter $(w/h)_{so} = 5$. The point where these two curves intersect is the 'design point' and its coordinates gives the results:

$$w/h \approx 0.85 \quad \text{and} \quad s/h \approx 0.25\ . \tag{8.45}$$

Hence, because $h = 1.0$ mm:

$$w \approx 0.85 \quad \text{mm} \quad \text{and} \quad s \approx 0.25\ \text{mm}. \tag{8.46}$$

8.7.3 COMPARISON OF METHODS

Although only graphical techniques have been used for the preceding calculations it can be seen that the values obtained compared well in this case. Notice that, for the synthesis method, no special set of curves was required to be selected — the curves used (Figure 8.6) are 'universal' in that they apply over a wide range of substrate permittivities.

8.8 COUPLED-REGION LENGTH

The requirement for coupled-region length is determined and given in Appendix A following Equation (A.61). It is seen that the maximum degree of coupling occurs when the length of the coupled region is made $\lambda_{gm}/4$ where λ_{gm} is the mid-band wavelength. At all other wavelengths the coupling decreases. As a rough first approximation, only really applicable to loose amounts of coupling (perhaps less than -10 dB), it is possible to consider a single, i.e. uncoupled, microstrip of width w and to evaluate the mid-band wavelength as described in Chapter 3. However, the error involved in this can rise to 10% or more and the mean of the even-mode and odd-mode wavelengths should be taken for greater accuracy. It is best, for design purposes, to work in terms of characteristic impedances. We first recollect the three general forms in which characteristic impedance may be written, Equations (1.23), (1.24) and (1.25):

$$Z_0 = \sqrt{\frac{L}{C}} = \frac{1}{v_p C} = v_p L. \tag{8.47}$$

The last form of this expression is the most convenient one to use here, since we may assume that the inductance L remains constant independent of whether a substrate is present or absent. (This is really another way of stating that only the electric flux is affected by changing the dielectric material on the substrate, for a given mode.)

Consider the odd mode on microstrip. For the actual situation with a substrate present, Figure 8.8(a), we have

$$Z_{0o} = v_{po} L_o. \tag{8.48}$$

When the substrate is removed, as in Figure 8.8(b), the velocity is simply the free-space limit value c. For that situation the characteristic impedance is given by

$$Z_{01o} = c L_o. \tag{8.49}$$

Equations (8.48) and (8.49) can be simply solved for v_{po}, giving

$$v_{po} = c \frac{Z_{0o}}{Z_{01o}}. \tag{8.50}$$

By a similar consideration for the even mode:

$$v_{pe} = c \frac{Z_{0e}}{Z_{01e}}. \tag{8.51}$$

The odd- and even-mode wavelengths are then easily found at any frequency f by using

$$\lambda_{go} = v_{po}/f \quad \text{and} \quad \lambda_{ge} = v_{pe}/f. \tag{8.52}$$

The characteristic impedances Z_{0e} and Z_{0o} are known early on in the design calculations, as explained in Section 8.4. Then the cross-sectional dimensions w/h

(a) (b)

Figure 8.8 Parallel-coupled microstrips with the substrate (a) present and (b) absent (only the odd mode is indicated).

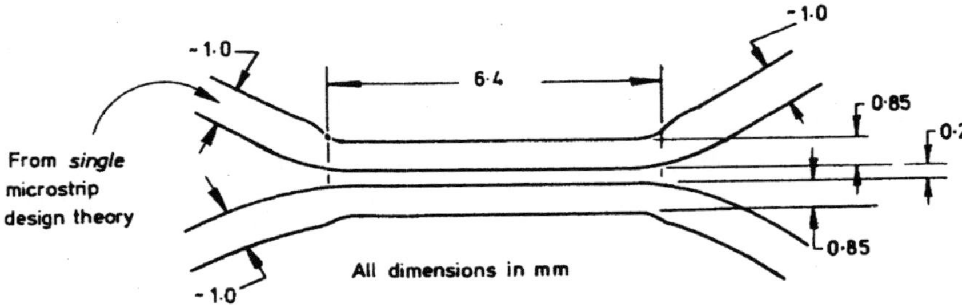

Figure 8.9 Final numerical dimensional values for the −10 dB coupler design shown in the text.

and s/h are obtained. Once these are known the air-spaced characteristic impedance Z_{01e} and Z_{01o} can either be obtained from Bryant and Weiss' curves applicable to $\varepsilon_r = 1$, or by calculations using the analysis given in Section 8.5 on page 274.

We can finally combine Equations (8.49), (8.51), and (8.52), and set frequency F (in gigahertz) to give

$$\lambda_{ge} \approx \frac{300}{F} \frac{Z_{0e}}{Z_{01e}} \quad \text{mm} \tag{8.53}$$

and

$$\lambda_{go} \approx \frac{300}{F} \frac{Z_{0o}}{Z_{01o}} \quad \text{mm.} \tag{8.54}$$

Returning to the numerical design example of Section 8.7 on page 279, we now apply the values of w/h and s/h to the Bryant and Weiss curves ($\varepsilon_r = 1$) and obtain

$$Z_{01e} = 185\ \Omega \quad \text{and} \quad Z_{01o} = 75\ \Omega. \tag{8.55}$$

With the given frequency of 5 GHz, Equations (8.53) and (8.54) then yield

$$\lambda_{go} = 28.8 \text{ mm} \quad \text{and} \quad \lambda_{ge} = 22.54 \text{ mm.} \tag{8.56}$$

The mean of these two wavelengths (λ_{gm}) is 25.67 mm and thus the length of the coupled region ($\lambda_{gm}/4$) is 8.42 mm. With the approximations involved, and also considering that we have neglected dispersion, the value 8.4 mm is sensible. The final complete microstrip coupler design is illustrated, in plan view, in Figure 8.9.

In some cases, especially at higher frequencies, the length may become very small and difficult to define in manufacture. Then the length can be made a convenient odd number of quarter-wavelengths:

$$l = (2n-1)\,\lambda_{gm}/4. \tag{8.57}$$

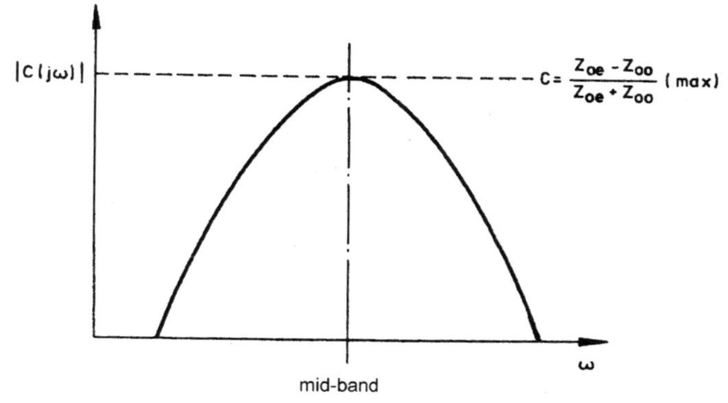

Figure 8.10 The characteristic shape of the frequency response for a loosely coupled ideal matched coupler.

It is, however, important to appreciate that the coupler would then exhibit n responses with frequency.

Such edge-coupled microstrip couplers, based on alumina substrates, have been reported operating at frequencies at least as high as 30 GHz [293]. Tripathi and Hill's 10 dB coupler of this form exhibits two roughly sinusoidal responses: the first centred on 9 GHz and the second on 24 GHz. The reason that the second response is significantly departed from a third harmonic lies in the dispersive nature of these structures. This aspect is covered in the next section concerning frequency response.

Tripathi and Hill's coupler also has deteriorating isolation and directivity — worsening as frequency increases. For example, the isolation is about 20 dB at 9 GHz but has deteriorated to 10 dB at 24 GHz. It is shown that SPICE models accurately predict the performance over the entire 1 to 30 GHz band.

8.9 FREQUENCY RESPONSE

8.9.1 OVERALL EFFECTS AND GETSINGER'S MODEL

So far we have concentrated on 'mid-band' design. The full frequency-dependent (i.e. phase-angle-dependent) coupling coefficient developed in Appendix A as Equation (A.60) is reproduced here, with phase velocity differences neglected:

$$C(\theta) = C(j\omega) = \frac{jC\sin\theta}{\sqrt{1-C^2}\cos\theta + j\sin\theta} \tag{8.58}$$

where C is the mid-band coupling factor given by Equation (A.57) and $\theta = \beta l = 2\pi l/\lambda_g$.

Accepting the validity of the result for the loosely coupled frequency response of the ideal matched coupler, the actual shape would be similar to that given in Figure 8.10.

This general form of response is typical for couplers of this type, since they are difficult to construct for even moderately tight couplings: C is generally smaller than -3 dB in designs with this geometry.

In several applications such a poor frequency response is intolerable, and a number of techniques have been developed to 'flatten' the response. (The phase responses are also significant, and there are also techniques to improve the directivity by near-equalization of the even- and odd-mode phase velocities.) Some of these techniques will be dealt with shortly. As with single microstrip lines, dispersion is also present in the behaviour of these coupled microstrips; this problem is dealt with first.

For calculations on parallel-coupled microstrips, especially at frequencies exceeding a few gigahertz, dispersion is significant and it is essential to have quantitative information on this dispersion. Only then can wavelengths be at all accurately calculated. The problem of dispersion in single microstrip lines was discussed in detail in Chapter 5. Coupled microstrips suffer dispersion for precisely the same physical reasons as single microstrip, but the amount of dispersion depends upon the mode under consideration. The effective microstrip permittivities are now both somewhat different functions of frequency which involve the parameters indicated as follows:

$$\varepsilon_{\text{eff},e}\left(f, h, \varepsilon_r, \varepsilon_{\text{eff},e}, Z_{0e}\right) \tag{8.59}$$

and

$$\varepsilon_{\text{eff},o}\left(f, h, \varepsilon_r, \varepsilon_{\text{eff},o}, Z_{0o}\right). \tag{8.60}$$

Getsinger [294] has shown how his expressions, which were originally derived for single microstrip, can be adapted to suit the coupled microstrip problem (Getsinger's single-line expressions were given in Chapter 5). For use with coupled lines the following approximate substitutions are made where Getsinger's method is adopted:

(a) In the even mode the two strips are at the same potential and the total current is twice that on a single strip. Thus, the total mode impedance is half that of a single strip and the dispersion for even-mode propagation is calculated by substituting

$$Z_0 \approx \frac{1}{2} Z_{0e}. \tag{8.61}$$

(b) In the odd mode the two strips are at opposite potentials and the voltage between strips is twice that of a single strip to ground. Thus, the total mode impedance is twice that of a single strip and the dispersion for odd-mode propagation is calculated by substituting

$$Z_0 \approx 2Z_{0o}. \tag{8.62}$$

in Getsinger's single-line equations.

So, for example, the even-mode dispersion expression runs as follows:

$$\varepsilon_{\text{eff},e}\left(f\right) = \varepsilon_r - \frac{\varepsilon_r - \varepsilon_{\text{eff},e}}{1 + G_e\left(f/f_{pe}\right)^2} \tag{8.63}$$

with

$$f_{pe} = \frac{Z_{0e}}{4\mu_0 h} \tag{8.64}$$

and

$$G_e = 0.6 + 0.0045 Z_{0e}. \tag{8.65}$$

Some caution is necessary regarding the accuracy of using the substitution (b) since Easter and Gupta [295] have indicated that an overestimation of odd-mode dispersion is obtained. This assessment is based on comparisons with the theoretical treatment due to Krage and Haddad, [141] and with the measurements due to Richings and Easter [296] (measurements are dealt with in Chapter 9).

It is suggested that empirical modifications might be made to Getsinger's odd-mode substitution to provide more accurate design formulas but an approach yielding greater accuracy is available and is described in the next section.

8.9.2 MORE ACCURATE DESIGN EXPRESSIONS, INCLUDING DISPERSION

Kirschning and Jansen [297] have given design expressions for the modal dispersion in parallel coupled microstrip lines. These workers develop a coupled model for the structure, in which they use normalized parameters as follows:

$$u = w/h, \quad g = s/h \quad \text{and} \quad f_n = f/h \quad (\text{in GHz/mm units}). \tag{8.66}$$

They quote the ranges of validity for which the design expressions remain accurate:

$$0.1 \leq u \leq 10, \quad 0.1 \leq g \leq 10 \quad \text{and} \quad 1 \leq \varepsilon_r \leq 18. \tag{8.67}$$

The static-TEM formula for the even mode effective microstrip permittivity is provided by Hammerstad and Jansen as follows:

$$\varepsilon_{\text{effe}}(0) = 0.5(\varepsilon_r + 1) + 0.5(\varepsilon_r - 1)(1 + 10/v)^{-[a_e(v)b_e(\varepsilon_r)]} \tag{8.68}$$

where

$$\begin{aligned} v &= u\left(20 + g^2\right)/\left(u + g^2\right) + g\exp(-g) \\ a_e(v) &= 1 + \ln\left\{\left[v^4 + (v/52)^2\right]/\left(v^4 + 0.432\right)\right\}/49 + \ln\left[1 + (v/18.1)^3\right]/18.7 \\ b_e(\varepsilon_r) &= 0.564\left\{(\varepsilon_r - 0.9)/(\varepsilon_r + 3)\right\}^{0.053}. \end{aligned} \tag{8.69}$$

In contrast, the static-TEM formula applicable to the odd mode effective microstrip permittivity was remodelled by Kirschning and Jansen so that an improved accuracy of 0.5% applies over the whole range of parameters. This permittivity is given by

$$\varepsilon_{\text{eff},o}(0) = [0.5(\varepsilon_r + 1) + a_o(u, \varepsilon_r) + \varepsilon_{\text{eff}}(0)]\exp\left(-c_o g^{d_o}\right) + \varepsilon_{\text{eff}}(0) \tag{8.70}$$

where

$$\begin{aligned} a_o(u, \varepsilon_r) &= 0.7287\left[\varepsilon_{\text{eff}}(0) - 0.5(\varepsilon_r + 1)\right]\left[1 - \exp(-0.179u)\right] \\ b_o(\varepsilon_r) &= 0.747\varepsilon_r/(0.15 + \varepsilon_r) \\ c_o &= b_o(\varepsilon_r) - [b_o(\varepsilon_r) - 0.207]\exp(-0.414u) \\ d_o &= 0.593 + 0.694\exp(-0.562u) \end{aligned} \tag{8.71}$$

(where, as elsewhere in this text, $\varepsilon_{\text{eff}}(0)$ refers to zero-thickness single microstrip line).

The functional form in which modal dispersion is introduced for both modes is similar to that shown for single microstrip, described in Chapter 5. Based upon this the general expression then takes the form:

$$\varepsilon_{\text{effe,o}}(f_n) = \varepsilon_r - [\varepsilon_r - \varepsilon_{\text{eff},e,o}(0)]/[1 + F_{e,o}(f_n)]. \tag{8.72}$$

In this expression it is of course F that determines the detailed frequency dependence.

The even-mode result for F, given by Kirschning and Jansen, is

$$F_e(f_n) = P_1 P_2 \{(P_3 P_4 + 0.1844 P_7) f_n\}^{1.5763} \tag{8.73}$$

where

$$\begin{aligned}
P_1 &= 0.27488 + \left[0.6315 + 0.525/(1+0.0157 f_n)^{20}\right] u - 0.065683 \exp(-8.7513u) \\
P_2 &= 0.33622\left[1 - \exp(-0.03442\varepsilon_r)\right] \\
P_3 &= 0.0363 \exp(-4.6u)\left\{1 - \exp\left[-(f_n/38.7)^{4.97}\right]\right\} \\
P_4 &= 1 + 2.751\left[1 - \exp\left\{-(\varepsilon_r/15.916)^8\right\}\right] \\
P_5 &= 0.334 \exp\left[-3.3(\varepsilon_r/15)^3\right] + 0.746 \\
P_6 &= P_5 \exp\left[-(f_n/18)^{0.368}\right]
\end{aligned}$$

and

$$P_7 = 1 + 4.069 P_6 g^{0.479} \exp\left(-1.347 g^{0.595} - 0.17 g^{2.5}\right).$$

Also, the result for F applicable to the odd mode is

$$F_0(f_n) = P_1 P_2 \{(P_3 P_4 + 0.1844) f_n P_{15}\}^{1.5763} \tag{8.74}$$

with

$$\begin{aligned}
P_8 &= 0.7168\left[1 + 1.076/\{1 + 0.0576(\varepsilon_r - 1)\}\right] \\
P_9 &= P_8 - 0.7913\left\{1 - \exp\left[-(f_n/20)^{1.424}\right]\right] \tan^{-1}\left[2.481(\varepsilon_r/8)^{0.946}\right\} \\
P_{10} &= 0.242(\varepsilon_r - 1)^{0.55} \\
P_{11} &= 0.6366\left[\exp(-0.3401 f_n) - 1\right] \tan^{-1}\left[1.263(u/3)^{1.629}\right] \\
P_{12} &= P_9 + (1 - P_9)/\left(1 + 1.183 u^{1.376}\right) \\
P_{13} &= 1.695 P_{10}/(0.414 + 1.605 P_{10}) \\
P_{14} &= 0.8928 + 0.1072\left\{1 - \exp\left[-0.42(f_n/20)^{3.215}\right]\right\} \\
P_{15} &= \text{abs}\left[1 - 0.8928(1 + P_{11}) P_{12} \exp\left(-P_{13} g^{1.092}\right)/P_{14}\right].
\end{aligned}$$

The upper frequency limitation for these expressions depends upon the substrate parameters. The limit for retention of accuracy approaches 40 GHz for 0.635 mm thick substrates, which is fairly typical for alumina. The accuracy is also preserved for thin substrates with a permittivity of 12.9 (GaAs) and a normalized frequency of 30 GHz/mm, which corresponds to high millimetre-wave frequencies for gallium arsenide substrates some tens of microns thick. In all cases, the maximum error bound is 1.4%. It is important to appreciate that these design expressions hold accuracy only where the microstrip circuit is relatively undisturbed, i.e. effectively unshielded and with no affecting adjacent conductors or dielectrics. In situations where such practical disturbances exist, various design corrections can be used (e.g. similar to those described in Chapter 4), but in practice it is generally found that, provided the lines are reasonably tightly coupled, the unshielded results provide a good design basis.

Kirschning and Jansen have also given design expressions for the frequency-dependent characteristic impedances. They showed that earlier formulas given by Hammerstad and Jansen for the static-TEM impedances yielded small errors that would accumulate for these purposes and they reported improved design expressions. For the even mode the static-TEM impedance is given by

$$Z_{0e}(0) = Z_0(0)\left[\varepsilon_{\text{eff}}(0)/\varepsilon_{\text{eff},e}(0)\right]^{0.5} / \left\{1 - \left[Z_0(0)/377\right]\left[\varepsilon_{\text{eff}}(0)\right]^{0.5} Q_4\right\} \tag{8.75}$$

where

$$\begin{aligned} Q_1 &= 0.8695u^{0.194} \\ Q_2 &= 1 + 0.7519g + 0.189g^{2.31} \\ Q_3 &= 0.1975 + \left[16.6 + (8.4/g)^6\right]^{-0.387} + \ln\left\{g^{10}/\left[1 + (g/3.4)^{10}\right]\right\}/241 \\ Q_4 &= (2Q_1/Q_2)\left\{\exp(-g)\,u^{Q_3} + \left[2 - \exp(-g)\right]u^{-Q_3}\right\}^{-1} \end{aligned}$$

and the static-TEM odd mode impedance is given by

$$Z_{0o}(0) = Z_0(0)\left[\varepsilon_{\text{eff}}(0)/\varepsilon_{\text{eff},o}(0)\right]^{0.5} / \left\{1 - \left[Z_0(0)/377\right]\left[\varepsilon_{\text{eff}}(0)\right]^{0.5} Q_{10}\right\} \tag{8.76}$$

where

$$\begin{aligned} Q_5 &= 1.794 + 1.14\ln\left[1 + 0.638/\left(g + 0.517g^{2.43}\right)\right] \\ Q_6 &= 0.2305 + \ln\left\{g^{10}/\left[1 + (g/5.8)^{10}\right]\right\}/281.3 + \ln\left(1 + 0.598g^{1.154}\right)/5.1 \\ Q_7 &= \left(10 + 190g^2\right)/\left(1 + 82.3g^3\right) \\ Q_8 &= \exp\left[-6.5 - 0.95\ln(g) - (g/0.15)^5\right] \\ Q_9 &= \ln(Q_7)(Q_8 + 1/16.5) \\ Q_{10} &= Q_2^{-1}\left\{Q_2Q_4 - Q_5\exp\left[\ln(u)\,Q_6u^{-Q_9}\right]\right\}. \end{aligned}$$

These design formulas are accurate to within 0.6% over the full range of validity given here.

The frequency-dependent expressions then follow:

$$\begin{aligned} Z_{0e}(f_n) &= Z_{0e}(0)\left\{0.9408\left[\varepsilon_{\text{eff}}(f_n)\right]^{C_e} - 0.9603\right\}^{Q_0} \\ &\quad / \left\{(0.9408 - d_e)\left[\varepsilon_{\text{eff}}(f_n)\right]^{C_e} - 0.9603\right\}^{Q_0} \end{aligned} \tag{8.77}$$

where

$$\begin{aligned} C_e &= 1 + 1.275\left\{1 - \exp\left[-0.004625p_e\varepsilon_r^{1.674}\left(f_n/18.365\right)^{2.745}\right]\right\} \\ &\quad - Q_{12} + Q_{16} - Q_{17} + Q_{18} + Q_{20} \\ d_e &= 5.086q_e\left[r_e/\left(0.3838 + 0.386q_e\right)\right] \end{aligned}$$

$$\begin{aligned} & \times \left[\exp\left(-22.2u^{1.92}\right)/\left(1+1.2992r_e\right)\right]\left\{\left(\varepsilon_r-1\right)^6/\left[1+10\left(\varepsilon_r-1\right)^6\right]\right\} \\ p_e &= 4.766\exp\left(-3.228u^{0.641}\right) \\ q_e &= 0.016+\left(0.0514\varepsilon_r Q_{21}\right)^{4.524} \\ r_e &= \left(f_n/28.843\right)^{12} \end{aligned} \tag{8.78}$$

and

$$\begin{aligned} Q_{11} &= 0.893\left[1-0.3/\left\{1+0.7\left(\varepsilon_r-1\right)\right\}\right] \\ Q_{12} &= 2.121\left[\left(f_n/20\right)^{4.91}/\left\{1+Q_{11}\left(f_n/20\right)^{4.91}\right\}\right]\exp\left(-2.87g\right)g^{0.902} \\ Q_{13} &= 1+0.038\left(\varepsilon_r/8\right)^{5.1} \\ Q_{14} &= 1+1.203\left(\varepsilon_r/15\right)^4/\left[1+\left(\varepsilon_r/15\right)^4\right] \\ Q_{15} &= 1.887\exp\left(-1.5g^{0.84}\right)g^{Q_{14}}\left[1+0.41\left(f_n/15\right)^3 u^{2/Q_{13}}\left(0.125+u^{1.626/Q_{13}}\right)\right]^{-1} \\ Q_{16} &= \left\{1+9/\left[1+0.403\left(\varepsilon_r-1\right)^2\right]\right\}Q_{15} \\ Q_{17} &= 0.394\left\{1-\exp\left[-1.47\left(u/7\right)^{0.672}\right]\right\}\left\{1-\exp\left[-4.25\left(f_n/20\right)^{1.87}\right]\right\} \\ Q_{18} &= 0.61\left\{1-\exp\left[-2.13\left(u/8\right)^{1.593}\right]\right\}/\left(1+6.544g^{4.17}\right) \\ Q_{19} &= 0.21g^4\left\{\left(1+0.18g^{4.9}\right)\left(1+0.1u^2\right)\left[1+\left(f_n/24\right)^3\right]\right\}^{-1} \\ Q_{20} &= \text{abs}\left\{1-42.54g^{0.133}\exp\left(-0.812g\right)u^{2.5}/\left(1+0.033u^{2.5}\right)\right\}. \end{aligned}$$

In these expressions $\varepsilon_{\text{eff}}(f_n)$ denotes the effective microstrip permittivity as a function of frequency, given in Chapter 5 here. Similarly, the appropriate design formulas applicable to the frequency-dependent odd mode impedance are

$$\begin{aligned} Z_{0o}(f_n) &= Z_0(f_n)+\left[Z_{0o}(0)\left\{\varepsilon_{\text{effo}}(f_n)/\varepsilon_{\text{effo}}(0)\right\}^{Q_{22}}\right. \\ &\quad \left.-Z_0(f_n)Q_{23}\right]\left\{1+Q_{24}+\left(0.46g\right)^{2.2}Q_{25}\right\}^{-1} \end{aligned} \tag{8.79}$$

where

$$\begin{aligned} Q_{22} &= 0.925\left(f_n/Q_{26}\right)^{1.536}/\left[1+0.3\left(f_n/30\right)^{1.536}\right] \\ Q_{23} &= 1+0.005f_nQ_{27}\left\{\left[1+0.812\left(f_n/15\right)^{1.9}\right]\left(1+0.025u^2\right)\right\}^{-1} \\ Q_{24} &= 2.506Q_{28}u^{0.894}\left[\left(1+1.3u\right)f_n/99.25\right]^{4.29}\left(3.575+u^{0.894}\right)^{-1} \\ Q_{25} &= \left[0.3f_n^2/\left(10+f_n^2\right)\right\}\left\{1+2.333\left(\varepsilon_r-1\right)^2/\left\{5+\left(\varepsilon_r-1\right)^2\right]\right\} \\ Q_{26} &= 30-22.2\left\{\left[\left(\varepsilon_r-1\right)/13\right]^{12}/1+3\left\{\left(\varepsilon_r-1\right)/13\right]^{12}\right\}-Q_{29} \\ Q_{27} &= 0.4g^{0.84}\left\{1+2.5\left(\varepsilon_r-1\right)^{1.5}/\left[5+\left(\varepsilon_r-1\right)^{1.5}\right]\right\} \\ Q_{28} &= 0.149\left(\varepsilon_r-1\right)^3/\left[94.5+0.038\left(\varepsilon_r-1\right)^3\right] \\ Q_{29} &= 15.16/\left[1+0.196\left(\varepsilon_r-1\right)^2\right]. \end{aligned}$$

Here again, the range of accurate applicability of these expressions is the same as that defined earlier here. The maximum error is 2.5% up to $f_n = 30$ GHz/mm but, if the specified maximum substrate permittivity is reduced from 18 to 12.9 (GaAs), then these expressions can be used up to $f_n = 25$ GHz/mm with the accuracy remaining guaranteed. It is noted again that this corresponds to an actual operating frequency of approximately 40 GHz in the case of a 0.635 mm thick substrate.

In design the principal sets of expressions are those for the effective microstrip permittivities since these ultimately form the major quantity in the determination of physical lengths. The frequency-dependent characteristic impedances should be used within dynamically interactive CAD routines to set the requirements for matching the termination line impedance (usually 50 Ω).

Since the determination of physical lengths is of paramount importance we also need to know semi-open end equivalent lengths for the even and odd modes. Kirschning and Jansen give the following expressions for these equivalent lengths, beginning with the length for the even mode

$$\Delta l_e = \left[\Delta l\left(2u, \varepsilon_r\right) - \Delta l\left(u, \varepsilon_r\right) + 0.0198hg^{R_1}\right] \exp\left(-0.328g^{2.244}\right) + \Delta l\left(u, \varepsilon_r\right) \quad (8.80)$$

where

$$R_1 = 1.187\left[1 - \exp\left(-0.069u^{2.1}\right)\right] \quad (8.81)$$

and

$$\Delta l_o = \left\{\Delta l\left(u, \varepsilon_r\right) - hR_3\right\}\left\{1 - \exp\left(-R_4\right)\right\} + hR_3 \quad (8.82)$$

where, in this expression:

$$\begin{aligned}
R_2 &= 0.343u^{0.6187} + \left\{0.45\varepsilon_r / \left(1 + \varepsilon_r\right)\right\} u^{\{1.357+1.65/(1+0.7\varepsilon_r)\}} \\
R_3 &= 0.2974\left\{1 - \exp\left(-R_2\right)\right\} \\
R_4 &= \left(0.271 + 0.0281\varepsilon_r\right) g^{\{1.167\varepsilon_r/(0.66+\varepsilon_r)\}} \\
&\quad + \left[1.025\varepsilon_r / \left(0.687 + \varepsilon_r\right)\right] g^{[0.958\varepsilon_r/(0.706+\varepsilon_r)]}.
\end{aligned}$$

Although the accuracy of these expressions is approximately 5% over the range defined above, with a slight increase also with frequency, the equivalent end-effect lengths usually only amount to a second-order contribution in determining the final physical lengths of the structure. For this reason, it is expected that these expressions will be sufficiently accurate for most requirements of effective parallel-coupled microstrip designs.

8.9.3 COMPLETE COUPLING SECTION RESPONSE

Once $\varepsilon_{\text{effe}}(f)$ and $\varepsilon_{\text{effo}}(f)$ have been evaluated then the appropriate wavelengths can be calculated in a manner similar to that employed with single microstrips. To solve for the full frequency response of the coupler, in microstrip form, it is necessary to return to the full voltage expressions developed in Appendix A. It is also necessary to incorporate the previous dispersion calculations so that the non-linearly varying wavelengths are taken into account. This is a fairly onerous task that is best set up on a computer. To get some idea of the magnitude of the calculations consider the coupling factor, as determined from Equations (A.40) and (A.44), but with θ appropriately replaced by θ_e and θ_o:

$$\begin{aligned} C &= V_{1e} - V_{1o} = \frac{Z_{0e} Z_0 \cos\theta_e + jZ_{0e}^2 \sin\theta_e}{2Z_{0e} Z_0 \cos\theta_e + j\,(Z_{0e}^2 + Z_0^2)\sin\theta_e} \\ &\quad - \frac{Z_{0o} Z_0 \cos\theta_o + jZ_{0o}^2 \sin\theta_o}{2Z_{0o} Z_0 \cos\theta_o + j\,(Z_{0o}^2 + Z_0^2)\sin\theta_o}. \end{aligned} \tag{8.83}$$

8.10 COUPLER DIRECTIVITY

This quantity has already been defined as the undesirable coupling which can occur to port 4 (see Figure A.8). The directivity D was also quoted as

$$D = \left|\frac{V_4}{V_3}\right|. \tag{8.84}$$

Now, for a pure TEM-mode (i.e. ideal) coupler, Equation (A.65) shows that $V_4 = 0$ and the directivity would also be zero because V_3 is finite. With microstrip, the differing field patterns associated with the odd and even modes give rise to different phase velocities. We have already seen that this feature greatly complicates the coupling factor expression and a similar complication arises for the voltage appearing at port 4 (see Figure A.8. on page 442). In fact, whereas $V4$ cancels to zero ideally, this is now far from true.

Expressions for V_3 and V_4 are obtained by using Equations (A.26) and (A.30), and also Equations (A.28) and (A.32), as follows:

$$\begin{aligned} V_3 = V_{1e} - V_{1o} &= \frac{Z_{0e} Z_0 \cos\theta_e + jZ_{0e}^2 \sin\theta_e}{2Z_{0e} Z_0 \cos\theta_e + j\,(Z_{0e}^2 + Z_0^2)\sin\theta_e} \\ &\quad - \frac{Z_{0o} Z_0 \cos\theta_o + jZ_{0o}^2 \sin\theta_o}{2Z_{0o} Z_0 \cos\theta_o + j\,(Z_{0o}^2 + Z_0^2)\sin\theta_o} \end{aligned} \tag{8.85}$$

$$\begin{aligned} V_4 = V_{2e} - V_{2o} &= \frac{Z_0 Z_{0e}}{2Z_{0e} Z_0 \cos\theta_e + j\,(Z_{0e}^2 + Z_0^2)\sin\theta_e} \\ &\quad - \frac{Z_0 Z_{0o}}{2Z_{0o} Z_0 \cos\theta_o + j\,(Z_{0o}^2 + Z_0^2)\sin\theta_o}. \end{aligned} \tag{8.86}$$

Dividing Equation (8.86) by Equation (8.85) and taking the modulus yields

$$D = \left|\frac{\chi Z_0 Z_{0e} - \psi Z_0 Z_{0o}}{\chi\,(Z_{0e} Z_0 \cos\theta_e + jZ_{0e}^2 \sin\theta_e) - \psi\,(Z_{0o} Z_0 \cos\theta_o + jZ_{0o}^2 \sin\theta_o)}\right| \tag{8.87}$$

where

$$\chi = 2Z_{0o} Z_0 \cos\theta_o + j\left(Z_{0o}^2 + Z_0^2\right)\sin\theta_o \tag{8.88}$$

and

$$\psi = 2Z_{0e} Z_0 \cos\theta_e + j\left(Z_{0e}^2 + Z_0^2\right)\sin\theta_e. \tag{8.89}$$

Calculation using these expressions is lengthy and a simpler alternative expression has been given for the directivity [298]

$$D = \left[\frac{4\,|\zeta|}{\pi\Delta \left(1 - |\zeta|^2\right)} \right]^{-2} \tag{8.90}$$

in which

$$\begin{aligned} \zeta &= \left(\frac{p_e}{1+p_e^2}\right) - \left(\frac{p_o}{1+p_o^2}\right) \\ p_e &= \frac{Z_{0e} - Z_0}{Z_{0e} + Z_0} \\ p_o &= \frac{Z_{0o} - Z_0}{Z_{0o} + Z_0} \end{aligned}$$

and

$$\Delta = \frac{\lambda_{go}}{\lambda_{ge}} - 1. \tag{8.91}$$

Equation (8.90) and (8.92) clearly demonstrate that, when $\lambda_{go} = \lambda_{ge}$ the directivity goes to infinity.

Values much better than 12 or 14 dB are difficult to achieve for directivity with this basic form of microstrip coupler. (The basic structure is still very useful in filter design and as a starting point for couplers of improved performance.) We shall now examine some techniques beginning at:

(a) equalizing, or approximately equalizing, the phase velocities so that the directivity (at least) is improved;
(b) modifying the overall coupler structure so that the frequency response is made much flatter, i.e. the bandwidth is greatly increased; and, finally,
(c) completely different structures where, for example, coupling factors much greater than −3 dB can be achieved (−3 dB is approaching the limit with parallel-coupled microstrips).

8.11 SPECIAL COUPLER DESIGNS WITH IMPROVED PERFORMANCE

8.11.1 THE 'LANGE' COUPLER

This structure owes its name to its inventor Julius Lange [299] who first reported the strictly empirical concepts and development in 1969. Lange's aim was to produce a nearly octave-bandwidth coupler having a coupling factor of around −3 dB. In his design, true quadrature coupling over an octave is realized as a consequence of the interdigital coupling section which compensates for even- and odd-mode phase velocity dispersion over the wide frequency range. It is necessary to bond directly, via very short wires, between transversely interposed 'fingers' of the coupler. In this way, power is coupled optimally to the desired port, as well as the remaining power (after losses) still being fed to a direct port. The arrangement is illustrated in Figure 8.11.

The bonding wires should look, electrically, as close as possible to short-circuits — or at least very small lumped inductances. This means that their lengths l_s must be

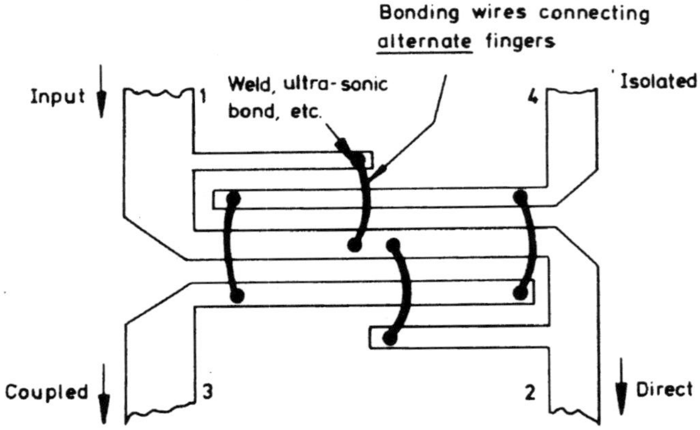

Figure 8.11 Basic form of the Lange- coupler. (©1969 IEEE. Reprinted, with permission, from Lange [329].)

kept as short as possible: $l_s \ll \lambda_{gm}/4$ where λ_{gm} is the mid-band wavelength. Notice that the input-to-direct output link meanders through the structure.

The physically short finger elements are approximately $\lambda_{gh}/4$ at the highest frequency in the band desired, whereas the entire length within the central region of the coupler is approximately $\lambda_{gl}/4$ at the lowest frequency of the band desired. These features, together with properly phased travelling-wave reinforcement due to the bonding wires, give the excellent performance. One really significant (and often important) characteristic of this device is the very good relative phase shift over the band. This is generally better than $\pm 2°$ on the nominal $90°$ phase shift. Also, the VSWR is generally less than 1.15 over an octave bandwidth.

In practice, especially at higher frequencies, several bonding wires are required in parallel connection in place of each single one shown in Figure 8.11 in order to reduce inductance, and therefore enhance performance. This results in extra manufacturing expense, since the wires must be judiciously bonded in by hand or by precision computer-controlled machines. The spacings are about 75 μm, compared with only 10 μm for the simple, parallel, microstrip -3 dB coupler (on alumina).

An early design procedure was reported by Presser [300]. In Presser's design a coupling factor design repeatability of ± 0.05 dB was generally obtained and a gap-variation sensitivity of approximately 0.03 dB/μm was observed. This means that a ± 2.5 μm tolerance on the coupling gap would yield approximately ± 0.1 dB coupling factor variation. (Parallel-coupled microstrip tolerances are discussed in more detail in Section 8.12.3 on page 305.) The couplers designed and measured by Presser showed a minimum directivity of approximately -24 dB, corresponding to an isolation of about -30 dB.

A design procedure, together with recommendations, is now given for Lange couplers. This follows the work of Ou [301] and also Osmani [302] whose approach has previously been consolidated and described by Fusco [303].

Ou developed the following elegant result:

$$Z_0^2 = \frac{Z_{0e} Z_{0o} \left(Z_{0e} + Z_{0o}\right)^2}{\left\{Z_{0e} + (k-1)\, Z_{0o}\right\} \left\{Z_{0o} + (k-1)\, Z_{0e}\right\}} \tag{8.92}$$

where k, the number of lines within the coupler, has to be an even integer.

A further important result derived by Ou is that the voltage coupling ratio may be expressed as

$$\begin{aligned} C &= 10^{-[(\text{voltage coupling ratio in dB})/20]} \\ &= \frac{(k-1)\,Z_{0e}^2 - (k-1)\,Z_{0o}^2}{(k-1)\,(Z_{0e}^2 + Z_{0o}^2) + 2Z_{0e}Z_{0o}}. \end{aligned} \tag{8.93}$$

As described previously, here it is essential to know the even- and odd-mode characteristic impedances in order to synthesize the coupled microstrip cross-section. Osmani showed that, by combining the simple expression in terms of C: $(1+C)/(1-C)$, with the above expression for C and multiplying the result by the expression for Z_0^2, Z_{0e} and Z_{0o} can be determined as follows:

$$Z_{0e} = Z_{0o}\frac{C+q}{(k-1)\,(1-C)} \tag{8.94}$$

and

$$Z_{0o} = Z_0\left(\frac{1-C}{1+C}\right)^{\frac{1}{2}}\frac{(k-1)\,(1+q)}{(C+q)+(k-1)\,(1-C)} \tag{8.95}$$

where

$$q = \left\{C^2 + \left(1-C^2\right)(k-1)^2\right\}. \tag{8.96}$$

In a Lange coupler the even-mode and odd-mode phase velocities are ideally equal, and this is approximately the case in practice. Therefore the overall length, which must be close to a quarter of a wavelength, can be based upon either phase velocity. The inherently broadband performance of this type of coupler means that this overall length must actually be a quarter of a wavelength calculated at the lowest frequency in the band under consideration (e.g. 10 GHz for a 10–40 GHz coupler).

By contrast, the finger length must, for the same reason of maintaining broadband operation, be set to one quarter wavelength calculated at the highest frequency in the band under consideration (e.g. 40 GHz for a 10–40 GHz coupler).

However, largely capacitive end-effect discontinuities tend to slightly reduce the actual physical lengths in each case and the expressions given in Section 8.9.2 on page 285 should be used to estimate the corrections required as a result. The bond wires or other direct connection links also have an effect on the design values but, as explained earlier, by paralleling several bonds the overall inductance is greatly reduced.

Two examples will now be given showing the results of using these expressions as first steps in the design of Lange couplers.

Example 1

Assume that we are required to design a four-finger Lange coupler to provide 10 dB of coupling in a 50 Ω system operating at 10 GHz. In this case, a soft plastic substrate is to be used with a thickness of 0.234 mm and a permittivity of 2.23.

We first note that, since $C = 10^{-(10/20)}$, $C = 0.3162$. This, together with the specification of 4 fingers, leads to q:

$$q^2 = \left[0.3162^2 + \left(1-0.3162^2\right)(4-1)^2\right] \tag{8.97}$$

which yields $q = 2.864$. Then, calculating Z_{0o} next:

$$Z_{0o} = 50\left(\frac{1-0.3162}{1+0.3162}\right)^{\frac{1}{2}}\left[\frac{3\,(3.864)}{(0.3162+2.864)+3\,(1-0.3162)}\right] \tag{8.98}$$

yielding $Z_{0o} = 80\ \Omega$. We now have sufficient quantities to determine Z_{0e}:

$$Z_{0e} = 80\left(\frac{0.3162+2.864}{3\,(1-0.3162)}\right) \tag{8.99}$$

yielding $Z_{0e} = 124\ \Omega$.

Applying the coupled line synthesis approach as described in Section 8.6 on page 276, here we find that these impedances are satisfied by a structure whose cross-sectional parameters are:

$$w/h = 1.103 \quad \text{and} \quad s/h = 0.39\ .$$

Since the substrate thickness is 0.254 mm, we require coupled microstrips having widths of 0.28 mm and separations (gaps) of approximately 0.1 mm.

We now consider briefly the contrasting situation if we had, instead of choosing a Lange coupler, attempted to design this coupler as a single, simple parallel-coupled element. In this case, using the design procedure described earlier in this chapter, we would require the following cross-section parameters:

$$w/h = 2.335 \quad \text{and} \quad s/h = 0.095\ .$$

These parameters yield actual width and gap values as follows:

$$w = 0.593\ \text{mm} \quad \text{and} \quad s = 0.024\ \text{mm}.$$

The width is just over twice and the gap is approximately one quarter the value found for the Lange coupler. The difference in width requirement is not generally serious although it does have implications for 'real estate' and infers ultimate operating frequency limitations. However, the much smaller gap results in very significant manufacturing and operational problems. In particular, the small gap (only 24 μm) leads to difficulties in terms of manufacturing tolerances, repeatability and yield — on a hybrid MIC. Also, even if these difficulties are largely overcome, there remains the limitation of voltage handling capability.

The final design requirements, centring on the calculation of the coupled microstrip lengths, must follow the use of expressions such as those given earlier here (Kirschning and Jansen [297]). In this way the physical lengths are determined but this is left as an exercise for the reader.

Example 2

For this example only the main specification and the resulting design values are given here. In Example 1 we considered a 10 dB coupler, but 3 dB couplers are commonly required for use in balanced amplifiers, attenuators and phase shifters. Balanced amplifiers are considered in more detail in Chapter 11. Let us assume we require an octave-bandwidth Lange coupler to be built on an alumina substrate

0.5 mm thick. Experience in using the above design expressions shows that a good first guess for the design iteration is to set the width of the coupled microstrips 38 μm and the separating gap 25 μm. Many practical amplifier applications demand a bandwidth much greater than one octave — often nearer two octaves or more. A significant increase in bandwidth can be obtained by slightly overcoupling the coupler. In this example we can transform the design into one covering two octaves by reducing the width and separation by about 30% and 40%, respectively (again as first approximations at the start of a design and simulation routine). This of course results in $w = 25\ \mu$m and $s = 15\ \mu$m.

Waugh [304] has reported a sensitivity analysis of the Lange coupler. He (Waugh) points out that, conventionally, designers impose $\pm 0.25\ \mu$m tolerances on their designs to ensure that specifications are met. This, however, leads to a corresponding deterioration in yield of the hybrid MICs. Waugh shows that for a 2.7 dB octave-bandwidth coupler designed on 0.635 mm alumina dimensional errors of 6 μm in both gap and line widths can still enable the coupler to function according to specification.

Most of the major microwave CAE packages embody Lange coupler models.

Designs using multiconductor configurations yielding coupling factors as tight as 1.5 dB have also been described [305]. Quadrature-coupling arrangements using these principles are particularly significant in some microwave transistor amplifier circuits and 2–18 GHz bandwidths are obtainable.

8.11.2 THE 'UNFOLDED LANGE' COUPLER

This is a development of the original Lange coupler just described and was reported by Waugh and Lacombe in 1972 [306]. In Figure 8.12, which shows this arrangement, it can be seen that some line elements and two straps are saved. The two straps are saved essentially by avoiding the strapped feed across to the direct port and, instead, retaining a 'straight-through' direct port connection. Slightly less coupling is obtained, but the most serious trade-off appears to be the phase-shift variation over the band. This now has a worst value of $-8°$ departure away from the nominal 90°. The fact of this phase departure and the lack of a DC path between input and opposite port have probably represented the main reasons why this version of the Lange coupler seems to have been largely avoided.

Both forms of 'Lange' coupler require the connection of short wire straps, and this can pose repeatability problems in production (although there may, of course, be other short wires associated with active devices). For this reason, among others, alternative techniques have been developed.

8.11.3 SHIELDED PARALLEL-COUPLED MICROSTRIPS

This type of structure also tends to nearly equalize even- and odd-mode phase velocities. The geometry is shown in the drawing of Figure 8.13.

In this arrangement there is a considerable proportion of the total field in the air region above the coupler, whatever the mode. Thus the proportional redistribution of electric field into air space for the odd mode can be made much less than that for the unshielded structure. As a consequence the phase velocities can easily be within approximately 1% of one another, as shown by Weiss [307]. In order to obtain these

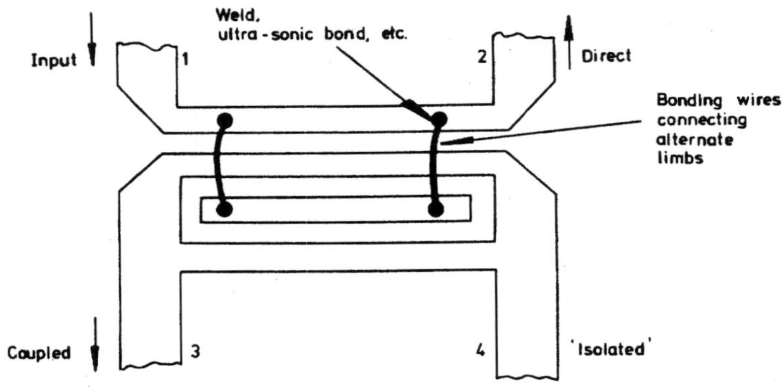

Figure 8.12 The 'unfolded' Lange coupler. (©1972 IEEE. Reprinted, with permission, from Waugh and Lacombe [306].)

advantages, it is necessary for the shield-spacing d to be around or less than the substrate height h:

$$d \leq h.$$

Although closely shielded circuits impose some manufacturing and other difficulties, they do yield certain specific design advantages, because the structures are electrically well defined.

Several researchers have reported studies on parallel-coupled microstrips including shielded structures and multiconductor geometries. For example, Kouki *et al.* [308] describe the finite element modelling of such structures analysed on a two-dimensional basis using an asymptotic boundary condition. Their method can accept conductors with arbitrary cross-sections in addition to arbitrarily inhomogeneous dielectrics (e.g. multilayer substrates). They demonstrate capacitance matrix results for shielded coupled pairs as well as six-conductor arrays of microstrips.

8.11.4 THE USE OF A DIELECTRIC OVERLAY

In this approach a layer of substrate-type dielectric material is fixed in the local region above the coupled (conventional) microstrip lines. The fixing is usually carried out with stycast epoxy, and the result is shown in Figure 8.14. Some very effective results can be achieved when the thickness of the overlay is only about 50 μm on an alumina substrate of 0.5–0.65 mm thickness.

The main principle is that electric field in the region immediately above the coupling gap experiences a permittivity similar to that of the substrate (instead of air when no overlay is present). This is most effective for the odd mode, when the field in the gap region is greatest. Thus the odd-mode phase velocity is reduced to approach, within a few per cent, the value of the even-mode phase velocity. If the overlay thickness is made approximately the same as that of the substrate and it overlaps the outer edges of the microstrips by at least half a strip width, then the phase velocity disparity is less than 1%.

It is likely that production repeatability problems would be experienced with this

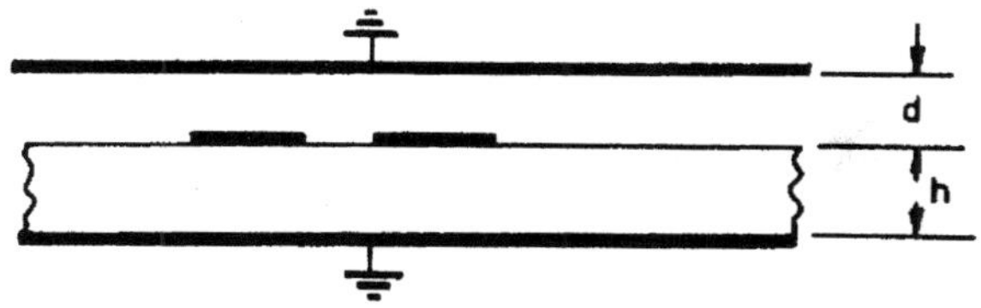

Figure 8.13 Parallel-coupled microstrips with a grounded shielding top plate.

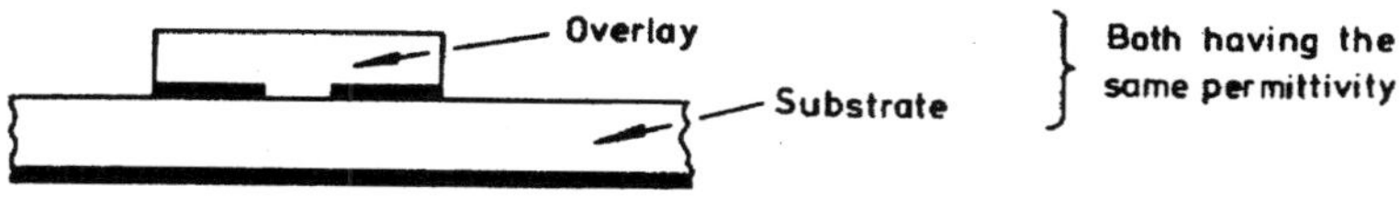

Figure 8.14 Parallel-coupled microstrips with overlaid dielectric compensation. (Reproduced by permission of the Institution of Electrical Engineers from Shelad and Spielman [309].)

technique, and epoxy bonding of the overlay is difficult to keep free of air bubbles. However, some excellent results have been reported for a −10 dB coupler.

8.11.5 THE INCORPORATION OF LUMPED CAPACITORS

It has been shown that the connection of lumped capacitors across a pair of coupled microstrip lines, as indicated schematically in Figure 8.15, will equalize the phase velocities over quite a broadband [71].

An accurate method for the determination of these compensating capacitances, applicable over a wide range of coupling, has been derived by Dydyk [310]. In his approach, Dydyk uses the standard ABCD matrix to represent both the even and odd modes for this structure, noting that the capacitance is only effective for the odd mode. The matrices therefore take the forms

$$\begin{bmatrix} A & B \\ C & D \end{bmatrix}_e = \begin{bmatrix} \cos\theta_e & jZ_{0e}\sin\theta_e \\ jY_{0e}\sin\theta_e & \cos\theta_e \end{bmatrix} \tag{8.100}$$

and

$$\begin{bmatrix} A & B \\ C & D \end{bmatrix}_o = \begin{bmatrix} (\cos\theta_o - Z_{0o}2\omega C\sin\theta_o) & jZ_{0o}\sin\theta_o \\ j\left\{4\omega C\cos\theta_o + \left(Y_{0o} - Z_{0o}(2\omega C)^2\sin\theta_o\right)\right\} & (\cos\theta_o - 2\omega C Z_{0o}\sin\theta_o) \end{bmatrix}. \tag{8.101}$$

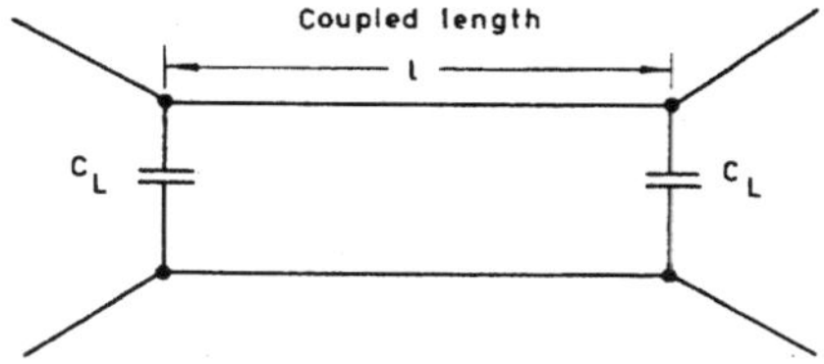

Figure 8.15 Schematic illustration of parallel-coupled lines with lumped capacitors bridging the ends to provide compensation.

Note is then taken of the fact that, in the ideal case of equal even and odd mode electrical lengths, the ABCD matrix for this case becomes

$$\begin{bmatrix} A & B \\ C & D \end{bmatrix}_{oi} = \begin{bmatrix} \cos\theta_e & jZ_{0oi}\sin\theta_e \\ jY_{0oi}\sin\theta_e & \cos\theta_e \end{bmatrix}. \tag{8.102}$$

where Z_{0oi} and Y_{0oi} are respectively the odd-mode characteristic impedance and admittance in this ideal case. This leads to the required capacitance by first equating the A and B element entries in the odd-mode matrices, giving

$$\cos\theta_e = \cos\theta_o - 2\omega C_L Z_{0o}\sin\theta_o \tag{8.103}$$

$$Z_{0o}\sin\theta_o = Z_{0oi}\sin\theta_e. \tag{8.104}$$

Combining these expressions and recognizing that the even mode electrical length is 90° at the centre frequency leads simply to the desired formula

$$C_L = \cos\theta_o/2\omega Z_{0oi}. \tag{8.105}$$

The electrical length can also be written in terms of the effective microstrip permittivities:

$$\theta_o = \frac{2\pi l}{\lambda_{go}} = \frac{\pi}{2}\sqrt{\frac{\varepsilon_{\text{eff},o}}{\varepsilon_{\text{eff},ei}}} \tag{8.106}$$

so that the compensating capacitance, as a function of frequency, is given by

$$C_L = \frac{\cos\left(\frac{\pi}{2}\sqrt{\varepsilon_{\text{eff},o}/\varepsilon_{\text{eff},ei}}\right)}{2Z_{0oi}\omega}. \tag{8.107}$$

From the foregoing considerations it also follows that the actual odd mode characteristic impedance may be expressed as

$$Z_{0oa} = \frac{Z_{0oi}}{\sqrt{1 - \left[\cos\left(\frac{\pi}{2}\sqrt{\frac{\varepsilon_{\text{eff},o}}{\varepsilon_{\text{eff},e}}}\right)\right]^2}} \tag{8.108}$$

indicating that Z_{0oa} must in general be greater than Z_{0oi}. This in turn implies that the conductors must be set narrower than initially calculated (before compensation), and the separating gap must be increased. Both requirements are completely acceptable in practice. Dydyk also shows that at the coupler's centre frequency the ABCD matrices reduce to

$$\begin{bmatrix} A & B \\ C & D \end{bmatrix}_{e} = \begin{bmatrix} 0 & jZ_{0e} \\ jY_{0e} & 0 \end{bmatrix} \tag{8.109}$$

and

$$\begin{bmatrix} A & B \\ C & D \end{bmatrix}_{o} = \begin{bmatrix} 0 & jZ_{0oi} \\ jY_{0oi} & 0 \end{bmatrix} \tag{8.110}$$

which represents a situation identical to that of an ideal directional coupler having the following scattering matrix:

$$[S] = \begin{bmatrix} 0 & C & -j\sqrt{1-C^2} & 0 \\ C & 0 & 0 & -j\sqrt{1-C^2} \\ -j\sqrt{1-C^2} & 0 & 0 & C \\ 0 & -j\sqrt{1-C^2} & C & 0 \end{bmatrix} \tag{8.111}$$

Table 8.1 Parameters for compensated and uncompensated coupled sections. The data in this table are due to Dydyk [310].

Parameter	Uncompensated	Compensated
Centre frequency (GHz)	12.0	12.0
Coupling C (dB)		
Desired	7.0	7.0
Achieved	7.25	7.0
Z_0 (Ω)	50	50
Z_{0e} (Ω)	80.85	80.85
Z_{0o} (Ω)	30.92	31.71
$\varepsilon_{\text{eff},e}$	7.20	7.20
$\varepsilon_{\text{eff},o}$	5.31	5.32
w (mm)	0.444	0.440
s (mm)	0.081	0.089
l (mm)	2.505	2.327
Directivity		
Desired	Infinite	Infinite
Achievable	12.32	Essentially infinite

where C is related to Z_{0e}, Z_{0o} and Z_0, as defined earlier in this chapter. From these results it should be clear that the isolation is ideally zero and the directivity infinite in this case. In practice, due to losses and with the frequency departed from the design centre, the actual directivity will degrade. As an example of the application of this theory to a practical design of a coupled section on an alumina substrate, it is found that the requirements are: a compensating capacitance of 0.047 pF, an actual odd mode impedance of 31.7 Ω and a coupled region length of just over 2.3 mm. Dydyk applied Agilent Technologies 'Linecalc' software to both uncompensated and compensated coupled sections (both having nominally 7 dB coupling) to generate the data repeated here in Table 8.1. Rickard [71] has shown that a coupling factor variation of only ±0.3 dB (on −10 dB) is achievable over a two-and-a-half-octave bandwidth: 8–20 GHz. The directivity was 20 dB over this band, compared with 5–13 dB uncompensated, and the VSWR was always better than 1.4. No information was reported on the phase-shift variation over the band. This performance considerably exceeds that of a comparable coupler using a dielectric overlay. However, it may be difficult to implement this approach with fairly large coupling factors (e.g. −3 dB or greater), because of the already very narrow coupling gap. At low frequencies the capacitors might be formed either using a thin-film add-on technique or conceivably as chip components. At high frequencies the capacitors could simply be interdigital structures at the coupler ends.

8.11.6 THE EFFECT OF A DIELECTRICALLY ANISOTROPIC SUBSTRATE

Provided the crystal axes of a dielectrically anisotropic substrate are correctly aligned with respect to the microstrip then the net electric displacements are similar for both even and odd modes. Thus, the phase velocities will approach equality, depending upon the actual values of the permittivity tensor members.

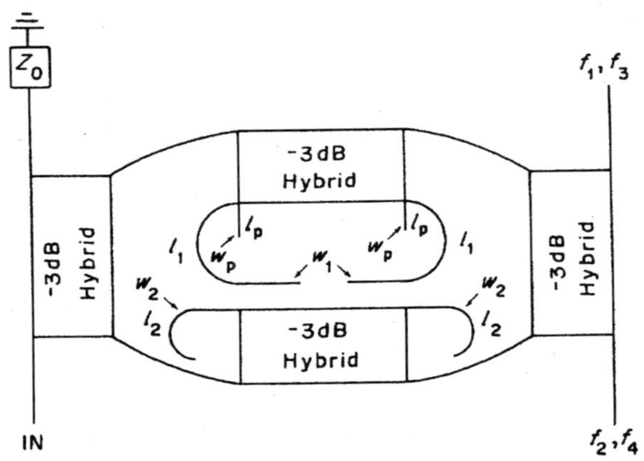

Figure 8.16 Schematic diagram of a quadruplexer using ×4, 3 dB hybrids. (Reprinted, with permission, from Uysal, Aghrami and Mohamed [312].)

Szentkuti [311] has shown that the degree of anisotropy of sapphire (see Chapter 3) and of some other readily available substrate materials is not sufficient to produce a really significant improvement. If a material having permittivity values of 25 and 10 were available then Szentkuti shows that the velocities differ by less than 1% over the range:

$$0.05 \leq \frac{w}{h} \leq 1.2 \tag{8.112}$$

where $s/h = 0.05$. Also, unfortunately, in the case of sapphire at least, the crystal axis alignment would mean that the wave would experience a very complicated permittivity environment, changing in the direction of propagation and dependent on the actual location on the substrate.

8.11.7 MICROSTRIP MULTIPLEXERS

Various planar or quasi-planar technologies have been adopted to facilitate the function of multiplexing — the separation of a broadband input signal into a number of contiguous channels. Technological approaches to this problem at microwave frequencies include both microstrip and also suspended-substrate stripline. An example of a microstrip quadruplexer has been reported by Uysal *et al.* [312]. Their device comprises four 3 dB directional couplers and four phase shifting networks, as indicated in Figure 8.16. The design is critically dependent upon differing electrical lengths for the microstrip lines connected to the two intermediate 3 dB couplers as shown in Figure 8.16. Uysal described an experimental quadruplexer designed to split a 2–18 GHz signal into four contiguous output channels each 4 GHz wide. The design was realized in MIC form using a standard 25 mm × 50 mm alumina substrate. Some empirical adjustment of the phase lengths is indicated, and it appears clear that there is room for further advances in this design approach.

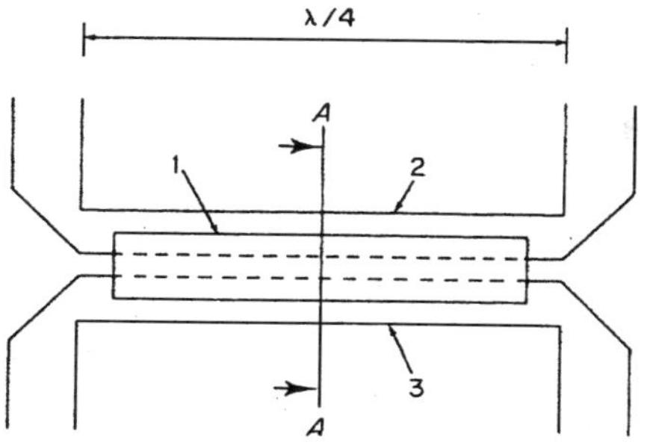

Figure 8.17 A plan schematic view of a single section semi-re-entrant coupler. (© 1990 IEEE. Reprinted, with permission, from Nakajima, Yamashita and Asa [313].)

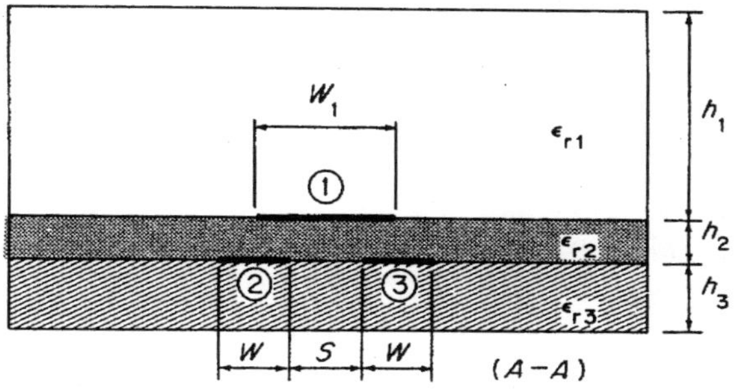

Figure 8.18 A cross-sectional view of a semi-re-entrant coupled section. (© 1990 IEEE. Reprinted, with permission, from Nakajima, Yamashita and Asa [313].)

8.11.8 MULTISECTION COUPLERS

As they tend to occupy considerable substrate area ('real estate'), particularly in terms of length, multisection couplers have generally not found much favour amongst solid-state microwave designers. However, at least one novel approach to the design of this type of coupler has been reported by Nakajima [313]. This team described a 2–18 GHz multisection coupler providing 6 dB of coupling using five semi-re-entrant tight-coupling sections.

A single section of such a coupler structure is shown in plan-schematic form in Figure 8.17 and in cross-section in Figure 8.18. A plan-schematic diagram of the entire five-section coupler is shown in Figure 8.19. Nakajima *et al.* use a capacitance matrix approach and provide a table showing all the design parameters for the three semi-re-entrant coupled sections and the two parallel (edge) coupled sections. All the physical dimensions are fully realizable — the first and narrowest coupling gap is 0.08 mm and the narrowest width is 0.23 mm for section three.

The coupler was fabricated on a Duroid® 5880 substrate of nominally 0.508 mm thickness. Relative through-arm power transfer remained about 2 dB whilst the coupled arm delivered power around 6 dB down on the incident power (with approximately 2 dB of ripple).

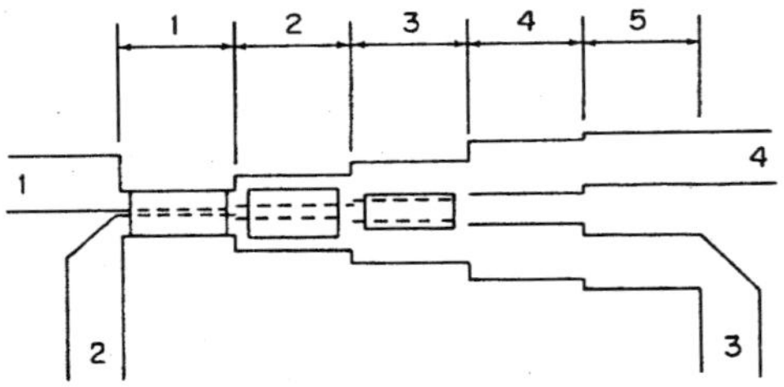

Figure 8.19 The plan schematic view of a 5-section directional coupler. (© 1990 IEEE. Reprinted, with permission, from Nakajima, Yamashita and Asa [313].)

8.11.9 RE-ENTRANT MODE COUPLERS

The types of fine-line geometries and interdigital structures described in earlier sections of this chapter lead to design and fabrication difficulties where tight coupling is required. One method of overcoming some of these difficulties is to consider combline approaches (dealt with in Section 8.13 on page 306), but very fine interdigital geometries, with their attendant production difficulties, remain.

The re-entrant concept is not new since Seymour Cohn originally described it for realization in coaxial form and later Lavendol showed how it could be adapted for stripline technology. More recently Pavio and Sutton [314] have described a quadrature coupler using this approach. Pavio and Sutton's structure, shown in cross-section in Figure 8.20, comprises a subsidiary parallel-coupled line arrangement with a 'floating under-plane' of restricted width w_1. This subsidiary arrangement is layered on top of a conventional alumina or gallium arsenide substrate with a ground plane.

Several types of circuits are fabricated by Pavio and Sutton using this technology. In one case a 3 dB coupler was fabricated on a 0.381 mm thick alumina substrate and a polyimide film, 0.0075 mm thick and of permittivity 3.7, was used for the subsidiary arrangement. Capacitive compensation was incorporated, as described in Section 8.11.5 (on page 297) of this chapter, and was realized in the form of interdigitated capacitors. Although transmission loss was similar to that of a conventional coupler, the VSWR was superior. The phase response varied in the range 82°–87° over the 5–19 GHz band, indicating that further work is required to more closely phase-linearize these types of devices. These researchers also described a 180° balun designed according to these principles (baluns are described in Chapter 10).

8.11.10 PATCH COUPLERS

At the higher millimetre-wave frequencies (high tens into hundreds of GHz) none of the previous coupler technologies are particularly practical, on account of the extremely small overall dimensions. Many designers therefore adopt technologies other than microwave (e.g. finline, imageline).

It is, however, possible to consider special forms of microstrip patch coupler at these high frequencies, and Fusco and Meruga [315] have described such a device, shown in schematic plan view in Figure 8.21. They describe a three-dimensional wave analysis of the coupler using a TLM numerical simulation to develop amplitude and phase responses over the frequency range 80–94 GHz. With this technique field distribution

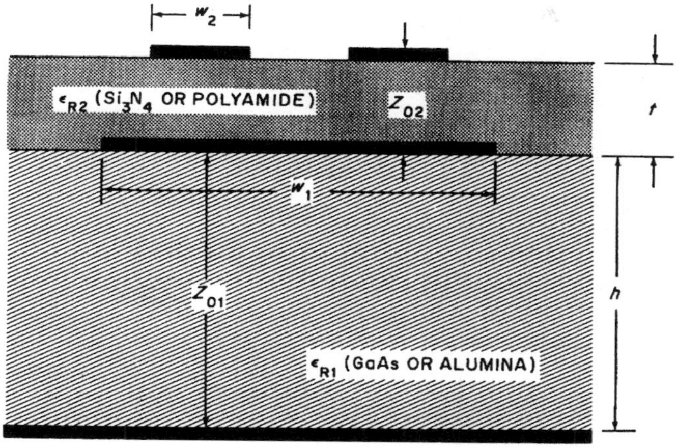

Figure 8.20 Microstrip re-entrant mode coupler cross-section. (© 1990 IEEE. Reprinted, with permission, from Pavio and Sutton [314].)

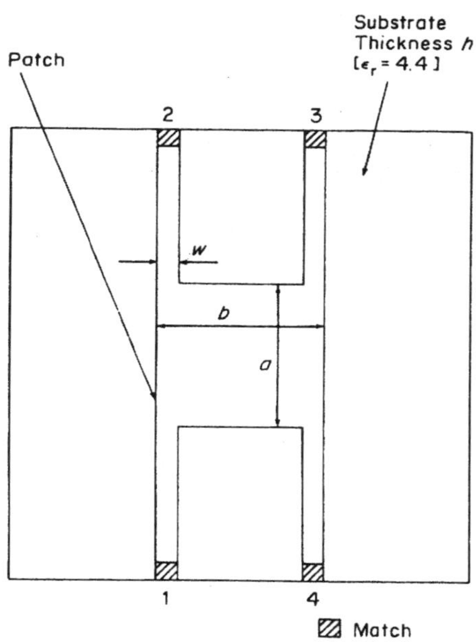

Figure 8.21 Geometry of a patch coupler. (© 1990 IEEE. Reprinted, with permission, from Fusco and Merugu [315].)

and radiation characteristics can also be identified.

8.12 THICKNESS EFFECTS, POWER LOSSES AND FABRICATION TOLERANCES

8.12.1 THICKNESS EFFECTS

So far in this chapter, we have neglected the effects due to the finite thickness of the microstrip lines.

Thickness effects play only a small part in the accurate design of coupled lines, provided the lines are of thick- or thin-film manufacture and the coupling is around −10 dB. Rizzoli [316] has found that thick-film strips taper at the edges, thus allowing 'thin-strip' calculations to be effective.

The nominal spacing for a -3 dB coupler on alumina is about 10 μm. Therefore, the effect of, say, 5 μm thickness 'side walls' must be very significant in such a case, especially where the side walls approach a vertical disposition. This is now a considerably complicated structure to analyse theoretically.

For design corrections it seems sensible to introduce modifications taking into account strip thickness and enabling the expressions of Section 8.5 on page 274 to be used. With the thickness t we require a modified form of Equation (8.17) for $C_f(t)$:

$$C_f(t) = 0.5\left\{\frac{\sqrt{\varepsilon_{\text{eff}}(t)}}{cZ_0(t)} - C_p\right\} \tag{8.113}$$

where $\varepsilon_{\text{eff}}(t)$ and $Z_0(t)$ are for single microstrip lines and are given in Chapter 4 Equations (4.29) to (4.35) and Equations (4.80) to (4.83). An extra term is also required in Equation (8.15) to account for the increased gap capacitance, which is entirely due to electric field in air. This is

$$C_{gt} = 2\varepsilon_0\frac{t}{s}. \tag{8.114}$$

These expressions, Equations (8.113) and (8.114), were originally given by Gupta *et al.* [88].

It is found that, for 'thin' microstrip on alumina, the odd-mode characteristic impedance decreases by 2 or 3% due to the thickness effect. The change is negligible for the even mode.

8.12.2 POWER LOSSES

In coupled microstrip lines there are two principal sources of power loss: one due to the conductors and the other due to the dielectric. Garg and Bahl [290] have reported expressions for both the conductor and dielectric losses and these expressions are now given here, together with extra information which enables the effects of surface roughness to be determined (for conductor loss, see Chapter 5 for single-line details):

$$\alpha_{c,i} = \left\{\frac{8.686R_s}{120\pi Z_{0i}hc\left[C_{ai}(t)\right]^2}\right\}$$

$$\times\left[\frac{\partial C_{ai}(t)}{\partial(w/h)}\left(1+\frac{w}{h}\right)-\frac{\partial C_{ai}(t)}{\partial(s/h)}\left(1-\frac{s}{h}\right)+\frac{\partial C_{ai}(t)}{\partial(t/h)}\left(1+\frac{t}{h}\right)\right]$$
$$\text{dB/unit length} \tag{8.115}$$

where this expression holds for the strips and ground plane combined, and $C_{ai}(t)$ = odd- or even-mode line capacitance with air as the dielectric, R_s = sheet resistivity (ohms per square metre) of the metallization, $i= o, e$ — for the odd or even mode.

The attenuation α_{co} due to the odd mode is always greater than that due to the even mode, and the sensitivity of α_{co} to changes in the spacing s is also greatest. This is consistent with a physical interpretation, where it must be remembered that interruptions are then (for the odd mode) being forced upon opposing-edge current components.

The loss for each mode is first calculated separately; then the total loss is determined as the average of these two results.

It should be noted that Equation (8.115) yields a rather low estimate of the conductor loss in practical situations. Compare, for example, the information supplied by Jansen [144]. With most plastics the discrepancy is very significant (several decibels) and there is a noticeable difference even when a polished sapphire substrate is used. Surface roughness is the dominant effect, although the precise nature of the conductor material is also very important. We previously considered these effects in Chapter 5, and a modifying expression Equation (4.86) was given to improve the prediction of conductor loss. Exactly the same expression, with identical precautions, should be used to modify (8.114) for coupled-line conductor losses.

Dielectric loss is almost always much less than conductor loss, due to the low-loss substrates that are generally used. The dielectric loss is given by [88]

$$\alpha_{d,i} = 27.3\frac{\varepsilon_r\left(\varepsilon_{\text{eff},i}-1\right)\tan\delta}{\sqrt{\varepsilon_{\text{eff},i}}\left(\varepsilon_r-1\right)\lambda_0}\ \text{dB/unit length} \tag{8.116}$$

where $\tan\delta$ is the loss tangent of the dielectric material and λ_0, is the free-space wavelength.

In X-band (8–12 GHz) we typically achieve $\alpha_D \cong 1.2$ dB/m using an alumina substrate with a dielectric loss tangent of 0.0005. On the other hand, the conductor loss is approximately 8 dB/m for moderate line spacings, increasing to 20 dB/m or more for close spacing.

8.12.3 EFFECTS OF FABRICATION TOLERANCES

This has been investigated by several workers, including Shamasundra and Gupta [317]. We are primarily interested in the maximum change in the coupling factor:

$$\frac{|\Delta C|_{\max}}{C} = \left|\frac{\Delta w}{w}S_{C(w)}\right| + \left|\frac{\Delta h}{h}S_{C(h)}\right| + \left|\frac{\Delta s}{s}S_{C(s)}\right| + \left|\frac{\Delta\varepsilon_r}{\varepsilon_r}S_{C(\varepsilon_r)}\right| \tag{8.117}$$

where C has been previously defined as

$$C = \frac{Z_{0e}-Z_{0o}}{Z_{0e}+Z_{0o}}. \tag{8.118}$$

The sensitivity terms are defined by

$$S_{C(p)} = \frac{p}{C}\frac{\partial C}{\partial p} \tag{8.119}$$

where C is the quantity whose sensitivity is to be determined, due to a change in a controlling parameter p.

A complete calculation of any sensitivity is actually quite a lengthy procedure. Complete functions giving the coupling factor have to be written down, as far as possible, using expressions given in Section 8.5 on page 274. The derivative then has to be determined, as required, for Equation (8.119). For coupled microstrip lines on alumina a 4% change in coupling factor will generally result from the following tolerance excursions:

$$\Delta s = \pm 2.5 \ \mu\text{m} \tag{8.120}$$
$$\Delta w = \pm 2.5 \ \mu\text{m} \tag{8.121}$$
$$\Delta h = \pm 25 \ \mu m \ (\text{on } h = 0.65 \ \text{mm}) \tag{8.122}$$
$$\Delta \varepsilon_r = \pm 0.25 \ (\text{on } \varepsilon_r = 9.7)\,. \tag{8.123}$$

Comparatively large variations tend to occur in the substrate thickness h, and this usually influences the coupling factor C to a greater extent than other parameter variations.

For relatively tight coupling, certainly with $s/h < 0.02$, tolerance excursions in s influence C to an increasing extent until these excursions begin to actually dominate.

It is useful and interesting to note some figures which apply to a nominally -6 dB coupler manufactured on alumina. A 2.54% variation in coupling has been quoted due to tolerances on various parameters [317]. The principal contributions to this variation are Δh (1.39%) and Δs (1.01%). This weighting remains approximately the same for somewhat lower coupling factors, i.e. as s is increased, but the total percentage change rises to just over 6% [317].

Compared with the coupling factor, the VSWR is much less sensitive to variations in the physical parameters; the main controlling parameters are ε_r and h. Over a wide range of coupling factors the VSWR remains close to a constant value, which generally lies between 1.2 and 1.4. The sensitivity of the VSWR only increases significantly when the coupling becomes very tight — around -3 dB or more. As previously discussed, different types of coupling structure are usually employed for such values.

8.13 PLANAR COMBLINE DIRECTIONAL COUPLERS

All of the coupling arrangements so far described here have three identifiable characteristics:

(a) They are 'backward-wave' devices, i.e. the coupled wave travels in the opposite direction to the input wave.
(b) They are all basically quadrature couplers, i.e. the coupled wave is 90° phase displaced, or close to that, with respect to the input wave.
(c) The maximum achievable coupling factor is in the region of -1.5 dB.

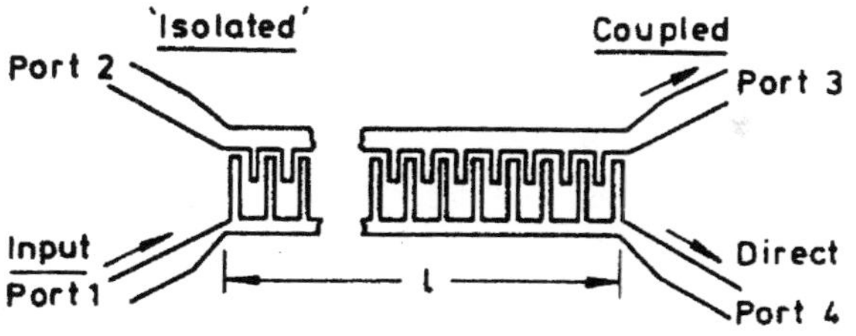

Figure 8.22 The general layout of a planar combline directional coupler. (Reproduced from [330], by permission of the Controller, HMSO.)

Even this requires special multiconductor configurations. It is imposed by close tolerance requirements leading to poor reproducibility.

These backward-wave microstrip couplers suit a wide variety of circuit applications ranging from parallel-coupled filters to balanced transistor amplifiers. However, there are some applications, such as mixer circuits, where a 180° coupler phase shift is required.

Although not strictly a microstrip arrangement, planar combline couplers [318] are briefly considered here because they do yield 180° phase shifts and possess especially tight coupling. Planar combline directional couplers offer forward coupling of very large amounts; in fact approaching 0 dB can be achieved at frequencies around and exceeding 12 GHz without any serious tolerance problems. A diagram of such a coupler is shown in Figure 8.22.

This arrangement may be regarded as a system of two interpenetrating comb transmission lines. As a coupler it derives its properties from the predominantly capacitive coupling between the two lines which arises from the interpenetration of the transverse 'finger' elements. For fuller details of the device properties, design, and performance the references given should be consulted.

Rigg and Carroll [319] have given a detailed design procedure for a general n-line 'herringbone' coupler, the fundamental structure of which is similar to the combline. They applied the design procedure to successfully realize a three-line equal power-splitting balun with a 3 dB bandwidth of three octaves centred on 1.8 GHz, a 180° phase relationship, and a linear phase variation of 100°/GHz across the band.

An analysis and design technique for combline couplers has been reported by Islam [320].

8.14 CROSSTALK AND SIGNAL DISTORTION BETWEEN MICROSTRIP LINES USED IN DIGITAL SYSTEMS

As remarked in Chapters 1 and 4, microstrip lines are increasingly being used as the interconnect medium in electronic logic circuitry such as that involved in moderate-to-high-speed digital systems. High-density packaging and short rise-time pulses are common in such systems and thus the problem of crosstalk 'noise' can be considerable. Inevitably there will be microstrip lines which are intended to operate, ideally, as if they were isolated from other microstrips on the same substrate. In practice, significant coupling ('crosstalk') can exist even when the lines are separated by several strip widths. We are therefore still concerned with a structure having the same general

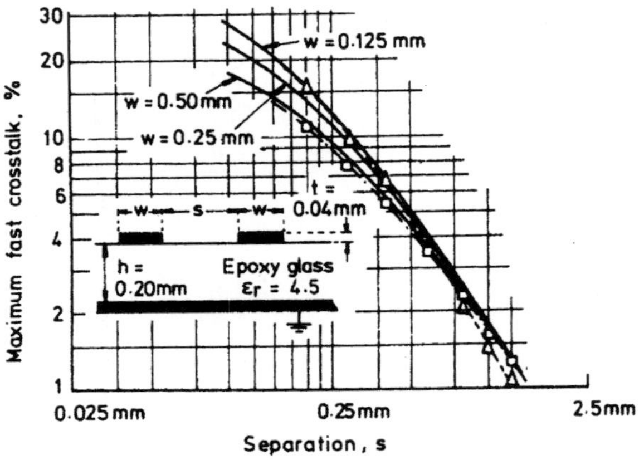

Figure 8.23 Maximum fast crosstalk (per cent) as a function of separation s for a structure having the dimensions shown on the inset diagram. (Reproduced by permission of Dr I. Catt [321], his Figure 25.)

geometry shown in Figure 8.1, except that the separation s is now usually much greater than we have so far encountered. All the relationships developed for this structure still hold here. The nomenclature is the same as that used in Appendix A and the lines are terminated in the characteristic impedance:

$$Z_0^2 \approx Z_{0e} Z_{0o}. \tag{8.124}$$

It is necessary to evaluate the maximum fast crosstalk, as a percentage, in order to design effective digital systems using microstrip transmission lines. In most digital subsystems, an epoxy-glass reinforced circuit board (such as FR-4) is used which will form the microstrip substrate; it has a permittivity in the region 4.5-4.8. Some careful measurements have been performed by Catt [321] with adjacent microstrip lines on such a substrate, and a family of curves resulting from those measurements are given here as Figure 8.23. A few points calculated using Equation (A.57) are also given on the graph. Equation (A.57) is repeated here for convenience:

$$C = \frac{Z_{0e} - Z_{0o}}{Z_{0e} + Z_{0o}}. \tag{8.125}$$

The data points on Figure 8.23 indicate values computed using a program based upon the work in Bryant and Weiss [99] where the triangles refer to $w/h = 2.5$ and the squares refer to $w/h = 0.625$.

All the points fit very closely to the measured curve. Thus, at least on a 'thin' epoxy-glass substrate, this expression for the coupling factor C may be reliably used to calculate the maximum fast crosstalk between microstrip lines. There is one difficulty in the calculations here: data for Z_{0e} and Z_{0o} are difficult to obtain. Where $s/h < 2.0$ these impedances may be determined using the analysis formulas given in Section 8.5. However, in most crosstalk problems we have $s/h > 2.0$ (this corresponds, for example, to a separation of 2.0 mm on a 1.0 mm thick substrate). In cases where approximate values are acceptable, the available curves may be used and extrapolated. If greater accuracy is required, then Z_{0e} and Z_{0o} must be computed accurately because, as s/h

increases, Z_{0e} approaches Z_{0o}, and Equation (A.57) involves the subtraction of two quantities of similar magnitudes. The full computer analysis due to Bryant and Weiss, outlined in Section 8.3 on page 271, may well have to be used.

It should be mentioned that several researchers have advocated the use of an expression for crosstalk which involves the square roots of the characteristic impedances [322]. When compared with the well established equation, Equation (A.57), and with measurements, this expression yields results that are approximately 50% too low.

Coekin [323] has pointed out that the crosstalk can be halved if a grounded conductor can possibly be interposed between the two signal-carrying runs of microstrip. Also, if this interposed conductor can actually be connected through to the ground plane via close-spaced plated-through holes, then a further halving of the crosstalk can be achieved. Thus, if we started with 2% crosstalk between adjacent microstrips we could reduce this to 0.5% by this method — if sufficient space is physically available between the lines. With MMICs the availability of such space could of course be highly problematical.

For ultra-high density integration, such as MCMs with flip-chip logic ICs, alumina (or similar) would almost certainly be used as substrate material. The temptation to implement close line spacings must be resisted, because of the severe crosstalk problem that can arise.

An important transient analysis of distortion and coupling in lossy parallel-coupled microstrips is reported by Gilb and Balanis [324]. These researchers use a spectral domain approach together with complex permittivity to characterize the frequency dependent complex propagation coefficient over a wide frequency range. Original (undistorted) Gaussian-shaped pulses approximately 50 ps in duration are considered and it is shown that both the signal and, more noticeably, the sense (coupled) line responses are significantly distorted. The principal reason for this distortion is the presence of dispersion. An example of the predicted sense line distortion is given in Figure 8.24.

If the microstrip lines amount to interconnects on a densely packed substrate then these results should be used as a guide for the designer. Gilb and Balanis only report results for a line separation of 1 mm on one type of substrate (permittivity 12.2). It is otherwise presumed that the distortion decreases as the separation increases, but quantification for a variety of substrates and geometries would be useful.

Another important example of published work concerns a study of the step response of multiparallel-coupled lines. This work was performed by Orhanovic, Tripathi and Wang [325] who used a numerical solution to a set of partial differential equations describing the structures.

These researchers provide results applying to as many as three coupled lines of unequal widths. Most of the published results refer to structures fabricated on 0.1 mm thick alumina substrates, and they show the step responses as terminal voltage waveforms.

Discontinuities associated with coupled lines have been investigated by Sabban and Gupta [326]. They use a planar-lumped model to evaluate performance on the basis of equivalent planar waveguides and they apply the analysis to a coupled microstrip section with chamfered right-angled bends connecting to feeding single lines.

Sabban and Gupta report results up to 18 GHz which show that their model

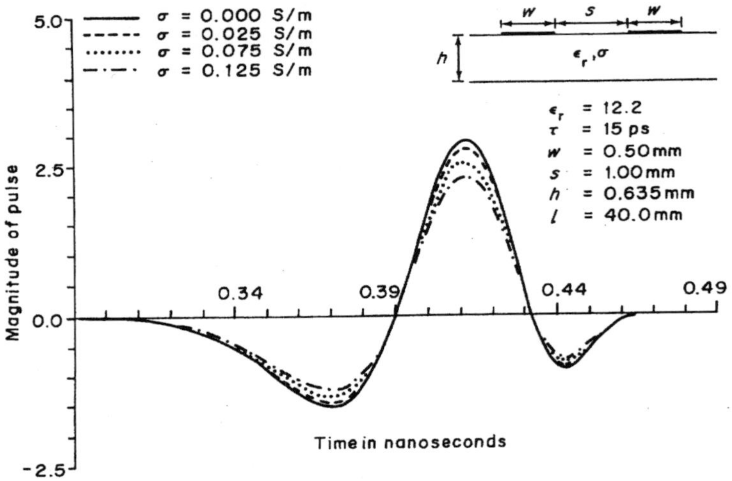

Figure 8.24 Pulse distortion on coupled, lossy microstrip lines, sense line response. (© 1990 IEEE. Reprinted, with permissions from Gilb and Balanis [324].)

provides an accurate prediction of performance, based on a 0.1 mm thick GaAs substrate.

Amongst other microstrip structures (including meander lines), Wertgen and Jansen [327] report theoretical results, using a novel Green's function database technique, for a coupled right-angled bend fabricated on a 0.2 mm thick GaAs substrate. The results extend to a maximum frequency of 26 GHz.

8.15 CHOICE OF STRUCTURE AND DESIGN RECOMMENDATIONS

In this chapter, most of the attention has been devoted to the basic parallel-edge-coupled microstrip line structure. This is because of the fundamental nature of that structure, with its importance in relatively simple couplers and its significance in circuit applications such as parallel-coupled bandpass filters. The design of this structure is therefore summarized first here with reference to coupler realization.

8.15.1 DESIGN PROCEDURE FOR COUPLED MICROSTRIPS, WHERE THE MID-BAND COUPLING FACTOR $C < -3$ dB

The design procedure has the following steps:

(a) Provided the coupling is loose enough for the relationship $Z_0^2 \approx Z_{0e}Z_{0o}$ to hold with sufficient accuracy, then Equations (8.8) and Equation (8.9) give the required characteristic impedance:

Even mode:

$$Z_{0e} = Z_0\sqrt{\frac{H}{Q}} \qquad (8.126)$$

Odd mode:

$$Z_{0o} = Z_0 \sqrt{\frac{Q}{H}} \tag{8.127}$$

where $H = 1 + 10^{(C'/20)}$ and $Q = 1 - 10^{(C'/20)}$, and the coupling factor C' is directly substituted as '$-C'$ dB'. We note also that this is the mid-band coupling factor.

(b) An approximate synthesis is then performed using the method due to Akhtarzad *et al.* (see Section 8.6 on page 276). A simple CAD algorithm may be devised using the formulas provided. It is likely that discrepancies in the region of 10% will arise in this process, which yields initial values of the 'shape ratios' w/h and s/h.

(c) These initial values can then be used to determine the corresponding impedances, Z_{0e} and Z_{0o}, employing the quite accurate (within 3%) analysis formulas given in Section 8.5. If higher accuracy is required, and the increased time and cost are acceptable, the much more accurate computational technique developed by Bryant and Weiss may be used. This was explained in Section 8.3. Some adjustment for finite conductor thickness may sometimes be necessary — particularly with hybrid MICs.

(d) These new (stage (c)) values for Z_{0e} and Z_{0o} are compared with the original requirements, determined at stage (a), and discrepancies are noted. Shape ratios w/h and s/h can now be adjusted slightly to correct the impedances, and hence the coupling factor C', to the desired values. It will usually be best if s alone is slightly altered, since the resulting mismatch to the feed lines will be small.

For guidance, we observe that s affects the odd mode more than the even mode. Increasing s means that more field is within the (air) coupling gap and Z_{0o} increases. As a rough approximation x% change in s results in $x/10$% change in Z_{0o}. For example, a 5% increase in s yields about 0.5% increase in Z_{0o}.

Finally the value of the mid-band coupling factor must be checked, which is given by Equation (8.6):

$$C' = 20 \log \left| \frac{Z_{0e} - Z_{0o}}{Z_{0e} + Z_{0o}} \right| . \tag{8.128}$$

8.15.2 *RELATIVELY LARGE COUPLING FACTORS (TYPICALLY $C \geq -3dB$)*

The problems of unequal even- and odd- mode wave velocities worsen rapidly as the coupling increases, e.g. the approximation $Z_0^2 \approx Z_{0e}Z_{0o}$ becomes poor when $C' \geq -3$ dB. An exact relationship for Z_0 was given as Equation (8.12), but this involves the even and odd mode electrical lengths θ_e and θ_o. In order to find these lengths we need, at the outset, to know the geometry (i.e. the shape ratios) for the structure, which is just what we were trying to find in the first place. One approach could be:

(a) Carry through a rough, first-order synthesis using the approximation $Z_0^2 \approx Z_{0e}Z_{0o}$, and hence an initial w/h and s/h.

(b) Calculate approximate values of θ_e and θ_o using the values obtained at step (a). The analysis procedure given in Section 8.5 on page 274 is well suited to this calculation, Equations (8.28) and (8.29) yielding the effective microstrip permittivities. The electrical lengths are

$$\theta_{e,o} = \frac{2\pi l}{\lambda_g}\sqrt{\varepsilon_{\text{effe},o}}. \tag{8.129}$$

An approximate value of Z_0 is then obtained using Equation (8.12). If Z_0 differs from that of the feed lines by an unacceptable amount (VSWR specification), the width w of the microstrip lines should be altered to compensate. The discussion of Section 4.12.3 on page 112 contains guidance on this type of adjustment. Altering w will alter Z_{0e}, Z_{0o}, θ_e and θ_o; thus the procedure must be iterated to find the optimum value of w.

(c) The evaluation of w must also ensure that the required mid-band coupling factor is obtained, from Equation (8.6):

$$C' = 20\log\left|\frac{Z_{0e} - Z_{0o}}{Z_{0e} + Z_{0o}}\right| \quad \text{dB}. \tag{8.130}$$

(d) When this procedure has been used and the parameter values have been met as closely as possible, the final values of Z_{0e} and Z_{0o} are used to synthesize, as described in Section 8.15.1 on page 310.

(e) For close coupling, which is the case here, the effect of finite microstrip thickness is often significant, and the analysis procedure should account for this by incorporating Equations (8.113) and (8.114) of Section 8.12. on page 304.

8.15.3 LENGTH OF THE COUPLED REGION

In a directional coupler, the coupled region extends for one-quarter of a wavelength at mid-band, and the designer requires the physical length of this region. Since, in microstrip, we are faced with the problem of different phase velocities applicable to the even and odd modes, we must also consider the different wavelengths associated with these modes. It is necessary to approximate the design to suit the average value of wavelength. This problem was treated in detail in Section 8.8 on page 281 and the expressions derived, Equations (8.53) and (8.54) are repeated here:

$$\lambda_{ge} \approx \frac{300}{F}\frac{Z_{0e}}{Z_{01e}} \quad \text{mm} \tag{8.131}$$

$$\lambda_{go} \approx \frac{300}{F}\frac{Z_{0o}}{Z_{01o}} \quad \text{mm} \tag{8.132}$$

where F is directly substituted in gigahertz, $Z_{0e,o}$ are the even and odd mode characteristic impedances for the coupled microstrips with a substrate, and $Z_{01e,o}$ are the same impedances, but for the air-spaced microstrips. The average value

$$\lambda_{gm} \approx \frac{1}{2}\left(\lambda_{ge} + \lambda_{go}\right) \tag{8.133}$$

is taken for approximate design purposes. For further improved accuracy, a weighted-average calculation has been suggested [328].

8.15.4 FREQUENCY RESPONSE

There are two fundamental effects governing the frequency-dependent behaviour of coupled microstrips:

(a) the basic frequency response of any pair of parallel-coupled lines (neglecting any dispersion),
(b) the effects of microstrip dispersion.

These features, (a) and (b), were discussed in detail in Section 8.9 on page 283.

Great care is needed when using the fairly complicated expressions involved — equations of the form of Equations (8.63) incorporated into (8.65). Although it appears that the even-mode electrical length θ_e may be evaluated 'fairly accurately' using Equation (8.63) and $\theta_e = 2\pi l/\lambda_{ge}$, there is evidence that the odd mode calculation overestimates changes in $\varepsilon_{\mathrm{eff},o}(f)$ [295].

For greater accuracy, particularly for frequencies into the millimetre-wave ranges, the expressions due to Kirschning and Jansen [297] should be used (see Section 8.9.2 on page 285).

8.15.5 COUPLED STRUCTURES WITH IMPROVED PERFORMANCE

Although several alternative methods have been reported, the technique invented by Lange [299] still yields an electrical performance which is generally superior to that of its 'closest rivals'. The main disadvantage is the requirement for short wire connections. A design procedure now exists for the 'Lange coupler' (see [300–303]), and combinations of Lange couplers have been built which cover the wide frequency range 2–18 GHz — and higher. The design repeatability of such couplers is generally ± 0.05 dB, on the coupling factor, that is ± 2.5 μm excursions in coupling gaps result in ± 0.1 dB changes in coupling factor.

Of the other techniques which have been developed the end-connected compensating lumped capacitors (Section 8.11.5 on page 297) is possibly the next best — from an electrical standpoint. This is a particularly elegant technique, because it can usually be realized as an entirely planar structure. (We recollect that the Lange coupler requires arrays of bond wires in its implementation.) However, when tight or fairly tight coupling is required ($C' >\approx -3$ dB) the already narrow coupling gap makes it difficult to incorporate the lumped capacitive structures in this capacitor-compensated coupler.

The two equal capacitances predominantly affect the odd-mode and (8.105) gave the required value:

$$C_L = \frac{\cos\theta_o}{4\pi f Z_{0oi}}. \tag{8.134}$$

Typically, for a -10 dB coupler, $\pm$ 0.3 dB is achievable, with a directivity of 20 dB and VSWR$<$ 1.4 over the frequency range $8.5 \leq f \leq 20$ GHz. Closely shielded couplers may be used in some instances but, to achieve good compensation, the close proximity of the shield to the circuit can cause difficulties. Dielectric overlays are a further possible alternative but, as with close shielding, they can make it difficult to design and control under production conditions.

8.15.6 *EFFECTS OF CONDUCTOR THICKNESS, POWER LOSSES AND PRODUCTION TOLERANCES*

It has already been noted that the principal effect of finite conductor thickness is to reduce Z_{0o} slightly. This reduction amounts to approximately 2–3% for a 5 μm conductor on alumina, where the coupling gap is in the approximate range $10 \leq s \leq 500\mu$m.

Equations (8.116) and (8.117) enable one to calculate the sensitivity of coupling factor to fabrication tolerances. The expressions will not be reproduced here, but the following tolerance excursions typically give a worst-case change of 4% in coupling factor (for alumina substrates):

$$\Delta s = \pm 2.5\ \mu\text{m} \tag{8.135}$$

$$\Delta w = \pm 2.5\ \mu\text{m} \tag{8.136}$$

$$\Delta h = \pm 25\ \mu\text{m (on } h = 0.65 \text{ mm)} \tag{8.137}$$

$$\Delta \varepsilon_r = \pm 0.25\ (\text{on } \varepsilon_r = 9.7)\,. \tag{8.138}$$

8.15.7 *CROSSTALK BETWEEN MICROSTRIP LINES USED IN DIGITAL SYSTEMS*

This has been discussed in Section 8.14. With the provisions given in the main discussion the following expression should be used to evaluate the crosstalk:

$$C = \frac{Z_{0e} - Z_{0o}}{Z_{0e} + Z_{0o}} \tag{8.139}$$

(which is identical to that used for the coupling factor). Now, because the spacing is usually very substantial, and often $s \gg 2w$ here, Z_{0e} is very close to the value of Z_{0o} and we have to evaluate a small difference. Thus, accurate computer-based techniques are recommended, or at least empirically curve-fitted formulas from curves using such computer-based techniques.

8.15.8 *POST-MANUFACTURE CIRCUIT ADJUSTMENT*

After testing it may well be found that a completed coupled-line circuit requires adjustment. Either the actual circuit may be etched, laser trimmed, etc., or the artwork used in preparing the circuit may be judiciously modified to correct errors.

Changing s or w slightly will alter the impedances and hence the coupling factor. It will generally be necessary to alter both w and s at the same time so that the correct impedance relationships are preserved. For example, if s is increased by 1% then Z_{0o} (mainly) will also increase, by approximately 0.1% — see Section 8.15.1 on page 310. Therefore, w/h should be increased very slightly, by less than 1% to reduce the impedance again.

The above is only an outline of the procedure: in practice, the precise trends for a particular case would soon become clear.

9

Power Capabilities, Transitions and Measurement Techniques

9.1 POWER-HANDLING CAPABILITIES

There are two main aspects to any consideration of power-handling in microstrip: Continuously Working (CW) and pulse operation. Under CW conditions the major problems and limitations are thermal, whereas under pulse conditions the principal limitation is dielectric breakdown.

9.1.1 MAXIMUM AVERAGE POWER P_{ma} UNDER CW CONDITIONS

This is controlled by heat flow and the associated temperature rise due to conductor and dielectric losses (principally conductor losses in most cases). The problem is very well treated in detail by Gupta *et al.* [88], and only the salient features will be presented here. Gupta calculates the power absorbed by a unit length of microstrip line, due to conductor losses. The density of heat flow, resulting from both conductor and dielectric losses, is then determined and the consequent temperature rise is found. The final expression for the temperature rise above ambient is

$$\Delta T = \frac{0.2303h}{K}\left[\frac{\alpha_C}{w_{\text{eff}}} + \frac{\alpha_D}{2w_{\text{eff}}(f)}\right] \ {}^{o}C/W \tag{9.1}$$

where α_C and α_D are the conductor and dielectric losses, respectively, in decibels per metre (length), w_{eff} and $w_{\text{eff}}(f)$ are effective microstrip widths (note that $w_{\text{eff}}(f)$ rises slowly with increasing frequency, see Section 5.8.2 on page 139), and K is the thermal conductivity of the substrate. Thus, as expected, the substrate dominates the temperature rise due to continuous power transmitted along the line. Apart from beryllium oxide (BeO), alumina and sapphire give the best (lowest) temperature rise characteristics when compared with other common substrates. For a wide range of microwave frequencies, 4–20 GHz, 50 Ω lines on alumina and sapphire substrates yield a figure of approximately 0.02°C/W. Progressively worse materials are, for the same conditions:

silicon	0.04°C/W
gallium arsenide (GaAs)	0.06°C/W
quartz	0.2°C/W
polystyrene	1.0°C/W.

If we consider a temperature rise (θ) of 95°C above ambient, a 50 Ω microstrip line on an alumina substrate could then carry a maximum average power of

$$P_{ma} = \frac{\theta}{\Delta T} = \frac{75}{0.02} = 3750 \text{ W}. \tag{9.2}$$

In practice, this is somewhat (not greatly) frequency-dependent, but such microstrip lines can be expected to reliably carry a few kilowatts of average power. In contrast, lines on polystyrene or other plastic substrates are restricted to only around 100 W. Beryllium oxide provides a thermally excellent substrate, with $\Delta T \approx 1 \times 10^{-3}$ °C/W, and an average power of about 95 kW can be handled by a 50 Ω line on beryllium oxide at 10 GHz. However, beryllium oxide is rarely used as a substrate because its dust is toxic. This ceramic is encountered as a pressing attached firmly to power semiconductor devices, e.g. the collector of a transistor, the drain of a power FET, or the anode of an IMPATT. In microstrip, this portion of the device would be bonded to the ground plane using high thermal-conductivity materials. The power limitations on the actual microstrip then remain as previously outlined.

In most cases, therefore, alumina or sapphire substrates are likely to be encountered. Some improvement can be obtained when wider microstrip lines are used, that is lines of lower characteristic impedances. Since both the effective widths w_{eff} and $w_{\text{eff}}(f)$ in Equation (9.1) are strongly dependent upon the physical width w, then ΔT will decrease as the width increases. If the previously calculated 50 Ω line on alumina is, instead, 20 Ω, then the power capacity will be increased by a factor of approximately four and up to about 15 kW may be transmitted. Unfortunately, most practical circuits will require narrow lines, of relatively high impedance, and the power capacity will therefore be decreased. Referring to the same example as before, a 100 Ω line will be restricted to approximately 490 W maximum CW power.

9.1.2 PEAK (PULSE) POWER-HANDLING CAPABILITY

In a microwave system employing MICs (or MMICs), the coaxial connectors, rather than the microstrip, are likely to set the ultimate limit to the peak power. To maintain a low VSWR, i.e. to achieve a good transition, the final portion of the coaxial connector, which actually contacts the microstrip, is almost invariably of a 3 mm subminiature form. This is usually the case even where N-type or APC coaxial lines are used in the main system external to the MIC or MMIC. The final portion of the coaxial connector sets a limit because of air breakdown. With any transmission line having a characteristic impedance Z_0 and maximum (breakdown) voltage V_{mb}, the peak power allowable is given by

$$P_{ma} = \frac{V_{mb}^2}{2Z_0}. \tag{9.3}$$

Now the breakdown electric field strength of dry air is 3 MV/m, and the internal difference in radii of the 50 Ω connector is approximately 1.5 mm — as can be seen from Figure 9.1 shown in the next section. Thus the breakdown voltage is

$$3 \times 10^6 \times 1.5 \times 10^{-3} = 4.5 \text{ kV} \tag{9.4}$$

and the peak power is, from Equation (9.3):

$$P_{ma} \approx 200 \text{ kW}. \tag{9.5}$$

In practice it would probably be unwise to greatly exceed 100 kW. Voltage antinodes due to internal mismatches reduce the allowable power still further.

Gupta *et al.* [88] mention two useful improvements which can be made to the microstrip itself in order to improve peak power handling:

(a) Since sharp edges intensify the electric field and thus reduce the allowable breakdown voltage, thick, rounded conductors are preferred.
(b) To avoid air breakdown near the edges, the strip conductor can be painted with a dielectric paint having a permittivity identical to that of the substrate.

A careful technique of preferential etching should enable (a) to be achieved and (b) is relatively straightforward to implement. However, the introduction of such dielectric paint along the edges of the microstrip conductor will significantly alter the line parameters (ε_{eff} and Z_0) and it is not clear how one might calculate appropriate design adjustments. An empirical cut-and-try approach thus seems inevitable here, including tests for the final microstrip-limited peak power when this, and not the transitions, forms the limit.

9.2 COAXIAL-TO-MICROSTRIP TRANSITIONS

These are available in a variety of forms, the choice mainly depending on the overall electrical quality required. For non-critical applications, commercially produced transitions, operating from semi-rigid (3.58 mm diameter) coaxial lines, are available from a number of companies. The final contact to the microstrip is variously formed as a wedge or a tab. In most applications the wedge will probably be preferred as it is more mechanically rugged. Some connections to a microstrip circuit can often be made with comparatively low-grade (and therefore relatively cheap) 'transitions' — for DC bias ports inexpensive BNC connectors can simply be used. Where only temporary connection is required, a pressure contact is quite satisfactory. A permanent transition will usually be completed with some form of epoxy bond surrounding the tab-microstrip metals.

It will be appreciated that the transition region, from the 'cylindrical' coaxial structure to the planar rectangular microstrip, presents quite a considerable discontinuity in the system. When used in conjunction with a microstrip line having 50 Ω characteristic impedance the commercial transitions typically yield a VSWR equal to 1.06 at moderate (e.g. X-band) frequencies. This is adequate for a wide variety of operating circuit and system applications.

In critical situations, such as measurement arrangements, a far superior VSWR performance is often desired and various compensation techniques have been developed. For a microwave transistor characterization jig a form of screw compensation has been developed [331]. A metal screw is introduced through a hole in the jig beneath the ground plane, in the vicinity of the transition. There is a further hole in line with this and actually in the ground plane, thus exposing a small region of the substrate underside surface (this arrangement can be seen in Figure

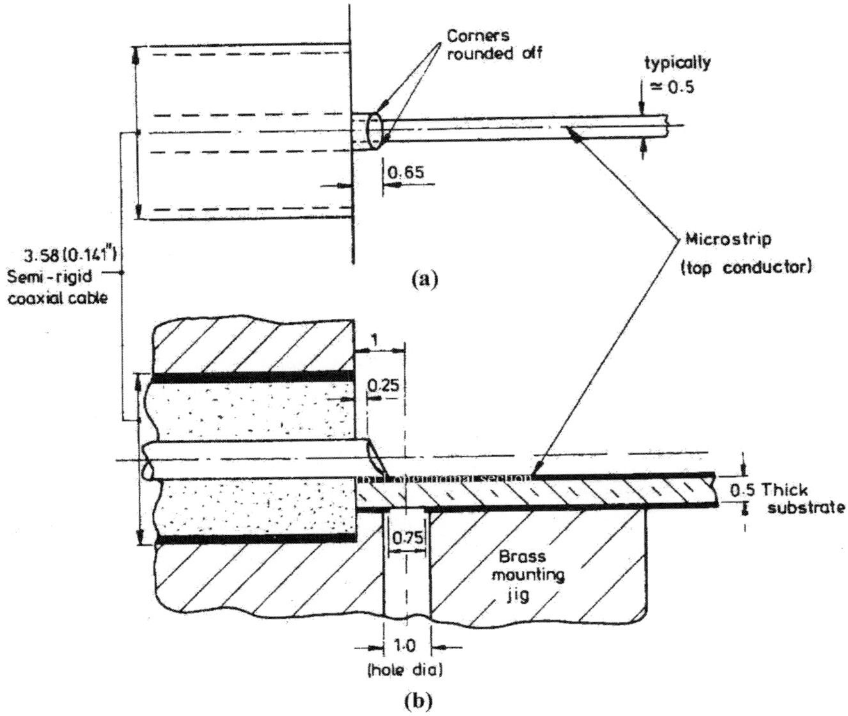

Figure 9.1 Compensated coaxial-microstrip transition (all dimensions in millimetres): (a) plan view; and (b) longitudinal section. (© 1996 IEEE. Reprinted, with permission, from England [332].)

9.1(b). The exact field situation is complex, but it seems clear that the screw mainly serves to reduce the microstrip line capacitance very locally so as to compensate (reduce) the transition capacitance. Extremely careful manufacture and time-domain reflectometry measurements have resulted in the transition shown in Figure 9.1 [332]. The dimensions shown apply to a transition launching into a 50 Ω microstrip line on a 0.5 mm thick sapphire substrate (C axis perpendicular to the ground plane), and a reflection coefficient less than 0.01 was obtained. With some extra dimensional and shape refinements, described in the paper, the achievement amounts to a reflection coefficient less than 0.005 (VSWR ≈ 1.01) over the range $1 \leq f \leq 18$ GHz. Over the same frequency range, the insertion loss was less than 0.3 dB.

The use of alumina substrates demands some modifications. For a similar performance a 1.0 mm diameter ground-plane hole was found to be necessary, where the microstrip characteristic impedance is approximately 50 Ω. For further details the reference should be consulted.

In the case of a different substrate (thickness and/or permittivity), it is necessary to repeat these careful measurements with a cut-and-try approach all over again.

If it could be developed, a general design procedure for such transitions would be most useful.

9.3 WAVEGUIDE-TO-MICROSTRIP TRANSITIONS

External to an MIC or MMIC, the principal transmission medium for microwaves at millimetric wavelengths consists of rectangular waveguide. Where microstrip is used at such wavelengths transitions from waveguide-to-microstrip are therefore required. Since the waveguide itself is relatively narrowband the transitions are also only required to operate over typically 10 to 50% bandwidths.

One essentially very narrowband solution is to first introduce a waveguide-to-coaxial line transition (the T-transformer, for example) and then to follow this with a coaxial line-to-microstrip transition as already described. This may be convenient in some restricted cases.

Transitions of improved bandwidth and other advantages are, however, conceivable. The desirable features of any waveguide-to-microstrip transitions are:

(a) Low return loss. This means that inherent reflections will be low and reflection measurements may be made on the microstrip circuit.
(b) Low insertion loss.
(c) Sufficient bandwidth for the application.
(d) Fairly easy and well reproducible re-connection to the microstrip.
(e) An in-line design for ease of connecting test equipment.
(f) Straightforward mechanical reproduction.

Several different solutions to the problem have been presented in the literature and some contrasting techniques will be presented here.

9.3.1 RIDGELINE TRANSFORMER INSERT

In the design approach of Schneider *et al.* [333], the waveguide impedance is transformed to the microstrip impedance by use of a fairly broadband stepped ridgeline transformer which is mechanically connected to the microstrip by a tab and a single pressure screw. A four-step ridgeline transformer designed to work over the 29.5–31.3 GHz band gave more than 30 dB return loss (reflection coefficient < 0.03) and less than 0.1 dB insertion loss over the band. A quartz substrate of 0.96 mm thickness was used for the microstrip circuit and the arrangement is shown in Figure 9.2.

For a required bandwidth and return loss, the necessary impedance at each step in the transformer can be calculated using Cohn's design method [334]. The stepped single ridgeline selected is quite compatible with the asymmetrical microstrip line. It is also easily machined and connected to the waveguide. The mechanical dimensions required of ridgeline structures, to realize a specified impedance, can be calculated from information given by Hopfer [335]. So that the substrate will stop against ridgeline, the last (deepest) step of this ridgeline is made sufficiently deep — actually 2.84 mm in this case. By tapering the edge of the transformer at the junction, the local reactance there is decreased, hence improving the performance.

Good contact is maintained both to the microstrip and its ground plane by applying pressure via a single insulated screw operating on the contact RG-96/U waveguide tab (this screw is not shown in Figure 9.2). For further details the paper by Schneider *et al.* [333] should be consulted.

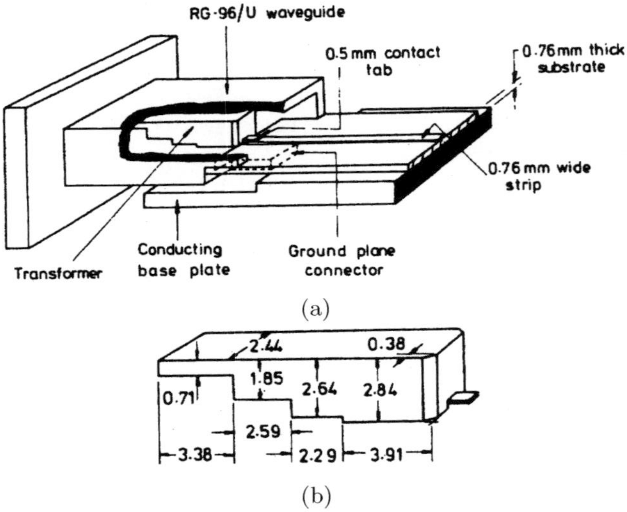

Figure 9.2 Rectangular waveguide-to-microstrip transition using a ridgeline transformer for the 27.5–31.3 GHz band: (a) complete mechanical structure and (b) ridgeline transformer dimensions (all in millimetres). (Reprinted with permission from *The Bell System Technical Journal*, ©1969, the American Telephone and Telegraph Company, from Schneider *et al.* [333], Figure 5.)

Menzel and Klassen [336] have reported developments of the ridged waveguide to microstrip transition, including a 30 GHz design based on alumina-substrated microstrip, and a 60 GHz design in which waveguide 15 was connected to microstrip on a GaAs substrate. These researchers use mode matching techniques based upon an LSE and LSM field analysis. The structures are shown in Figure 9.3. Using an unloaded ridged waveguide in transition with 0.056 mm wide microstrip on a 0.2 mm thick GaAs substrate the minimum reflection coefficient S_{11} was found to be about 0.15 for an optimum ridge width of 0.5 mm.

9.3.2 MODE CHANGER AND BALUN

A different approach was adopted by van Heuven [337]. This transition begins, in the waveguide portion, with a form of mode-changing section which rotates and concentrates the fields until they principally exist in a parallel line. This symmetrical line is then matched to the asymmetrical microstrip line by the use of a 'balun' portion. The arrangement is shown in some detail in Figure 9.4; the substrate is usually quartz.

The conductors on the substrate are electrically connected to the top and bottom walls of the waveguide, in the region marked (A) in Figure 9.4(b). Each conductor is then gradually tapered until, towards the end of the region marked (B), a balanced-conductor or 'symmetrical' line is formed. The characteristic impedance of this symmetrical line should be approximately the same as that of the final microstrip. In the region marked (C) a balun is formed; notice that one side of this has to be the microstrip ground plane, and therefore this side is electrically connected back to the

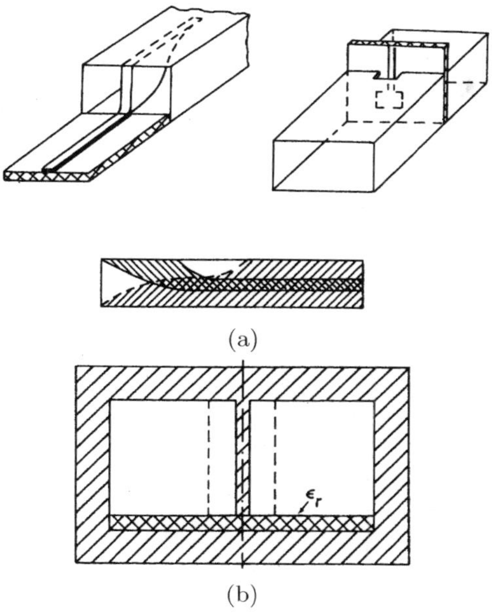

Figure 9.3 Microstrip transitions: (a) transitions from metal waveguide to microstrip; and (b) waveguide-to-microstrip transitions using ridged waveguide. (Reprinted, with permission, from Menzel and Klassen [336].)

(ground) top and bottom waveguide walls.

Serrated chokes may be used to produce good RF connections to the waveguide. Since the ground plane divides the waveguide into two separated guides, each having a cut-off frequency well above the band in use, further waveguide-mode propagation beyond the transition is prevented. The height of the guide is increased above normal to compensate for the loading due to the quartz substrate. Van Heuven's transition gave the following performance:

Bandwidth	18–26 GHz, and, over this,
VSWR	1.11
insertion loss	0.2 dB

Over a significant, but narrower, band (17.7–19.7 GHz) the VSWR remained less than 1.1. This corresponds to a reflection coefficient less than 0.05, which is only slightly worse than for Schneider's transition [333], although this was operating at a higher frequency.

Further details, including mechanical dimensions, can be found in van Heuven's paper [337]. Final design adjustments are largely empirically determined.

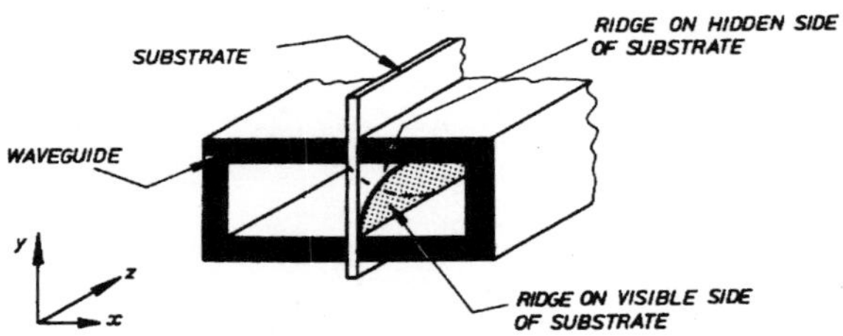

(a) Waveguide-end view

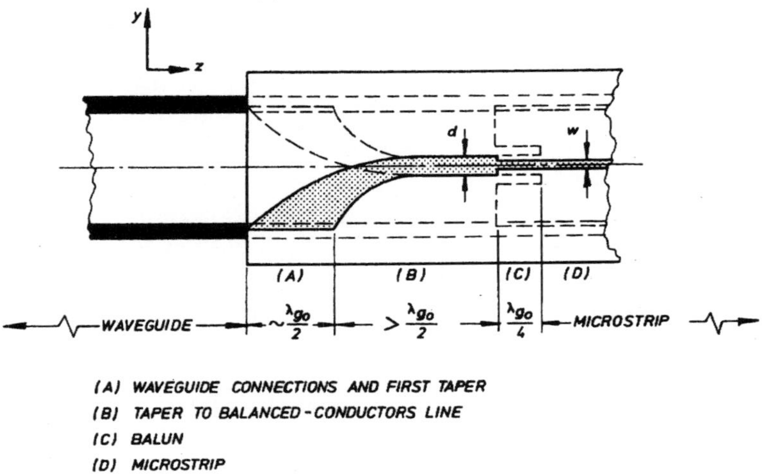

(b) Longitudinal View

Figure 9.4 Waveguide-to-microstrip transition using tapers and a balun. (© 1996 IEEE. Reprinted, with permission, from van Heuven [337].)

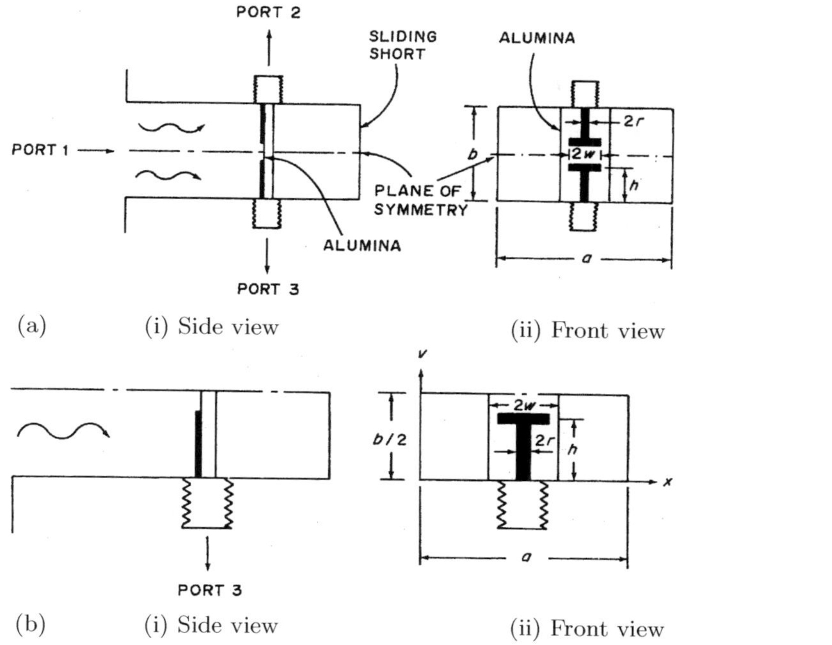

Figure 9.5 Waveguide-to-microstrip power splitter: (a) Structure of the waveguide to microstrip power divider. The side view illustrates that the microwave signal is coupled to two T antennas on an alumina substrate. The front view of the power divider consists of two T-shaped coupling antennas. (b) Simplified model using symmetry of the waveguide to microstrip power divider. The plane of symmetry in (a) is an electric wall where a metal plate can be inserted without changing the field distribution. Thus, the power divider problem is changed to the waveguide-to-microstrip transition design. (© 1990. Reprinted, with permission, from Wu, Schneider and Trambarulo [338].)

9.3.3 A WAVEGUIDE-TO-MICROSTRIP POWER SPLITTER

A type of waveguide-to-microstrip power splitter that is suitable for at least the lower microwave frequencies has been developed by Wu *et al.* [338]. This device uses a T-bar waveguide to microstrip transition, the theory for which is based upon current distributions and the dyadic Green's function techniques — treating the T-bar as an internal antenna. From this analysis an expression for the T-bar radiation resistance is obtained. The physical structure and simplified model are shown in Figure 9.5.

The design goal is to select a T-bar structure whose radiation resistance is matched to 50 Ω. The design described by these researchers exhibits an output resistance which is within 2 Ω (4%) of the requisite 50 Ω value. A return loss below 20 dB is obtained over the 3.3–4.6 GHz band and the power difference between output ports is below 0.1 dB.

9.3.4 SLOT-COUPLED ANTENNA WAVEGUIDE-TO-MICROSTRIP TRANSITION

With the advancement into high-millimetre-wave frequencies, having important applications such as automotive radars and seekers, the need for inexpensive, reliable and effective waveguide-to-microstrip transitions operating in the 77 GHz and 85 GHz bands is highly significant. Recognizing that the problem is essentially one associated with antenna-based transitions Grabherr *et al.* [339] developed a slot-coupled antenna-based waveguide-to-microstrip transition.

In Grabherr's transition the microwave energy is coupled into the microstrip line through a slot in the ground that connects to a patch antenna element located on an additional, small substrate inserted into the waveguide. The metallic waveguide is situated only on the back of the microstrip circuit and, because of the geometry, almost all of the microwave energy is coupled through the slot, to the patch antenna and this energy is radiated into the waveguide.

To suit millimetre-wave conditions, a step is implemented in the waveguide widths so that the additional substrate with the radiating patch can be automatically positioned across the ledges of the step, and fixed in place by soldering or similar. Grabherr *et al.* solve the calculation problem for this transition by introducing a shielding enclosure for the structure and using a Green's function approach with computer optimization.

The practical implementations of Grabherr's transition comprise an alumina-based microstrip and a fused-quartz-based patch element. Radiation from the top side of the substrates was found to be extremely low in all instances and, at 76.5 GHz and also at 85 GHz, the net insertion loss of each transition was approximately 0.3 dB.

9.4 TRANSITIONS BETWEEN OTHER MEDIA AND MICROSTRIP

As discussed in Chapter 3, a variety of other media exists which serve some useful functions in certain special applications and operate effectively at approximately the same frequencies as microstrip. Techniques have been developed which provide transitions between some of these structures and microstrip.

The first example we take is the triplicate-to-microstrip transition. Although not fully analysed in detail as an uncompensated transition, this type of transition has been simply treated by Hall and James [340], who also report some measurements of power loss due to radiation. On a plastic ($\varepsilon_r = 2.32$) substrate this loss was found to be approximately 11%. Such an uncompensated transition may well yield an acceptable performance when the microstrip substrate is of plastic, but worse results are likely with alumina and some form of compensation would be desirable. This might be of the type previously described in Section 9.2 on page 317 for coaxial-to-microstrip transitions.

A microstrip-to-slotline transition was first described by Robinson and Allen [341] and the details have been thoroughly discussed by Gupta *et al.* [88]. The simple basic structure yields a VSWR$<$ 1.2 over the frequency range 2–4 GHz. With some care in design, and the microstrip orientated at right angles to the slotline, a VSWR$<$ 1.1 was achieved over a higher range of frequencies, namely 8–10 GHz. It is apparent that a six-port junction will provide further improvements, and the structure is useful up

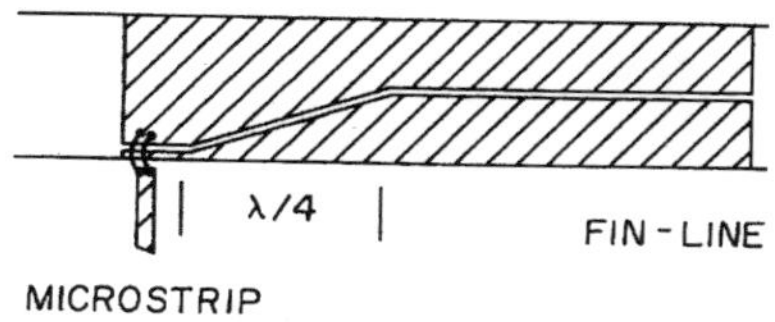

Figure 9.6 Microstrip-to-finline transition (Reprinted, with permission, from Climer [344].)

to a frequency of at least 12 GHz [342].

The design of dielectric-loaded slotline to microstrip transitions has been reported by Qin and Wang [343]. In their design approach, they have the microstrip coupled transversely across the slotline which is constructed on a relatively high permittivity substrate (permittivity greater than 9.6) with a lower permittivity (2.5) polymer dielectric overlay. Qin and Wang provide design equations covering the 3–4 GHz band.

A fairly simple finline to microstrip transition, designed as one component in a single-sideband mixer operating over the 26–38 GHz frequency range, is described by Climer [344]. This component, shown in outline in Figure 9.6, is built on a quartz substrate and the entire mixer design was based on microstrip. The approach of applying the local oscillator input to the finline port ensures that this oscillator is isolated from both the signal input and its output, as well as being rejected by the diplexer. Disadvantages include higher signal losses and greater difficulty in designing the signal filter.

Horng *et al.* [170] have presented a full-wave analysis of shielded microstrip line-to-line transitions (edge-coupled and overlay structures). The derivation of their results depends upon a method of moments, the use of dyadic Green's function and a Galerkin procedure. They provide a check on results using a power conservation approach and show that all results are within 1% accuracy. Data are presented for high-grade alumina-type substrates operated over the frequency range 0.5–11 GHz. Transitions between microstrip and coplanar waveguide are briefly considered by Gupta *et al.* [88] and more extensively by Viitanen and Kuismann [345] who report an analysis of this type of transition at frequencies up to 20 GHz. Further information on microstrip-coplanar waveguide transitions is provided in Chapter 6.

9.5 INSTRUMENTATION SYSTEMS FOR MICROSTRIP MEASUREMENTS

In general we have to choose between frequency-domain and time-domain measurements. The great majority of microwave measurements are best performed in the frequency domain because, with present systems technology, a high acuracy and a large amount of information are then potentially available. However, there are some requirements where time-domain measurements are valuable, and high-resolution Time-Domain Reflectometer (TDR) systems should then be used. Pulse-rise times less than 28 ps are common with such systems; more details will be given in a later section.

In this section we are concerned with frequency-domain instrumentation. There are

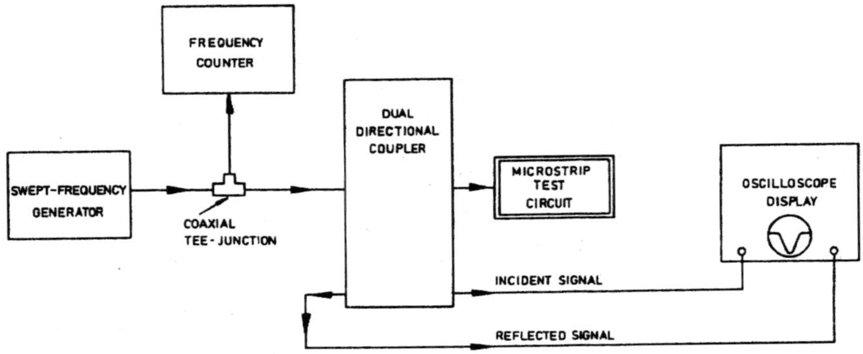

Figure 9.7 Example of a one-port instrumentation system (typically having all-coaxial interconnections).

several questions that should be considered, for example:

(a) Are single-frequency techniques suitable?
(b) Are swept-frequency techniques suitable?
(c) Are single-port or two-port (or more) measurements desired?
(d) What information is actually required about the circuit?
(e) How important is the speed and ease of measurement?
(f) What accuracy is required?

Single-frequency techniques were used in early microwave measurements, but these have been largely superseded by swept techniques. In some instances, Standing-Wave Indicators (SWI) are useful, and these are generally operated in conjunction with a swept-frequency test signal.

At the more sophisticated level, computer-augmented fully automatic measurement systems are available; the cost is naturally relatively high. Such systems employ sweep (frequency) generators, frequency synthesizers, down-converters, detectors, signal-processing electronics, and a computer to control the system. Full amplitude and phase information is determined for any circuit under study. One-port or two-port circuits are readily handled; and systems have been developed to cope with multiport studies [342].

For many microstrip measurements such comprehensive systems are not necessary and, indeed, often only the magnitude of a one-port network refection coefficient is required. A suitable specific system is indicated in Figure 9.7. This arrangement may be used for substrate measurements and many measurements on microstrip resonators, including dispersion, discontinuities and various Q-factors.

The microstrip circuit under test is connected to the test port of a dual directional coupler via a coaxial-to-microstrip transition. The dual directional coupler has high directivity and good, uniform coupling throughout the band. It also exhibits an effective source impedance close to 50 Ω, i.e. a good match throughout the band. This directional coupler enables the ratio:

$$\frac{\text{Reflected signal (from the microstrip circuit)}}{\text{Incident signal (direct from the swept-signal generator)}} \tag{9.6}$$

to be determined and displayed on an adapted CRO. There remain three further vital items:

(a) a Swept-Signal Generator (SSG)
(b) a frequency counter
(c) a special oscilloscope-display indicator.

Item (a), the SSG, continuously varies or 'sweeps' the RF test signal, generated by a suitable source, so that a frequency is obtained that varies linearly with time. A variety of sweep rates are available ranging from some megahertz per second to several gigahertz per millisecond. The slow rates are particularly useful when driving an 'X-Y' pen recorder read-out, and the faster rates are generally used when displaying information on the oscilloscope screen. All solid-state sources cover, individually, approximate octave bands from around 0.5 to 18 GHz. Typical RF output power is 50–100 mW and the entire system will operate continuously over the very wide frequency range. (The directional coupler may have to be changed for lower frequencies, in the hundreds of megahertz bands.) With some adaptation, the same system configuration may be used, in coaxial technology, up to about 50 GHz in special cases. Beyond this some waveguide systems are available to well over 100 GHz. At these higher frequencies the RF sources are usually backward-wave oscillator tubes and the available output power is often lower than that obtainable when the frequency is below 18 GHz. For some measurements, such as Q-factor, the flatness of the RF power as a function of frequency is important and the 'raw' generators cannot directly provide a sufficiently flat output. In such cases, an external feedback 'levelling loop' must be connected. This consists of a directional coupler and a detector which feeds a correcting DC signal to levelling PIN diodes that are usually already provided within the SSG (but this should always be checked).

Item (b), the frequency counter, ensures that any required single frequency can be accurately recorded directly. Most commercially available counters operate up to at least 18 GHz on direct input, the first stage of frequency down-conversion is often implemented by means of a phase-locked converter. Higher frequencies can be accommodated with extra conversion and suitable scale calibration. Accuracies of ± 2 MHz on 10 GHz are generally achieved.

Item (c), the 'CRO' indicator, displays the relative signal amplitude as a function of frequency. The ratio of the reflected signal amplitude to that of the incident signal (reference) is continuously displayed. These signals are actually introduced to the main CRO signal-processing electronics after down-conversion to lower frequencies, and the down-converter has been omitted from Figure 9.7, since it can be incorporated within the CRO unit. Amplifiers and precision attenuators enable accurate calibration of the vertical scale to be achieved and typical scalings are 0.25 dB/div., 1 dB/div., 5 dB/div. and 10 dB/div.

Good synchronization is essential between the sweep-generating circuits of the SSG and the horizontal sweep within the 'CRO'. In practice, careful adjustment of the SSG sweep rate is necessary. Slow manual sweep of the SSG output is most useful for finally selecting identified points on the characteristic; the frequency and relative amplitude of these points can then be recorded.

9.6 MEASUREMENT OF SUBSTRATE PROPERTIES

Unlike many coaxial lines and waveguides, microstrip depends inherently on the properties of its dielectric substrate. Suitable substrates were described in Chapter 3. The most significant property is the permittivity and its variations.

In the early stages of preparing microwave integrated circuits (hybrid MICs), substrates are easily obtained that are metallized all over their broad faces. This structure, shown in Figure 9.8(a), can be treated as a parallel-plate resonator to which an RF signal may be coupled (Figure 9.9). The permittivity of the substrate material is then deduced from the expression

$$\varepsilon_r = \left(\frac{c}{2lf_{n,m}}\right)^2 (n^2 + m^2) \tag{9.7}$$

where ε_r is strictly the value in a direction normal to the broad faces (important — especially in the case of anisotropic substrates), l is the length of the substrate sides, and n, m are mode numbers.

It is possible to form an almost totally enclosed rectangular resonator by metallizing not only the broad faces, but also the narrow edges of a substrate. An aperture is then needed for coupling; the arrangement is illustrated in Figure 9.10(a) with extra details of the field in the coupling region given in Figure 9.10(b). Errors involved with such resonator measurements have been described by various researchers. Howell [346] concentrated on the effects of losses and fringing, Ladbrooke *et al.* [347] gave details of coupling errors, and Owens *et al.* [348] pursued a reexamination of the results given by both Howell and Ladbrooke. The deductions due to Owens *et al.* are significant, because the permittivities of sapphire were under study, and these values have already been accurately determined by independent methods. Although it is very difficult to accurately account for a correction due to edge fringing, some adjustment must be made for this in order to achieve accurate results.

9.7 MICROSTRIP RESONATOR METHODS

Resonator methods are generally preferred for microstrip measurements because other conceivable methods suffer from significant difficulties that will first be briefly described.

Microstrip short-circuit terminations that have definable characteristics are very hard to implement (see Chapter 7), and matched loads having good quality with low VSWR are difficult to achieve. Any transmission measurements in microstrip therefore have to be made through transitions to other lines (usually coaxial) in which matched loads, etc., are available. Errors are then introduced due to the microstrip-to-coaxial transitions, and the influence of these errors on required parameters can be very difficult to establish, although substantial improvements in transition design have recently been achieved. Standing-wave measurement techniques have also been developed, to a limited extent, but the precision obtainable does not yet approach coaxial or waveguide standards.

If a microstrip line is formed as some closed loop on a substrate, or left open- or short-circuited at both ends, it represents a resonator at certain microwave frequencies. By rather loosely coupling an RF test signal to such a resonator, so

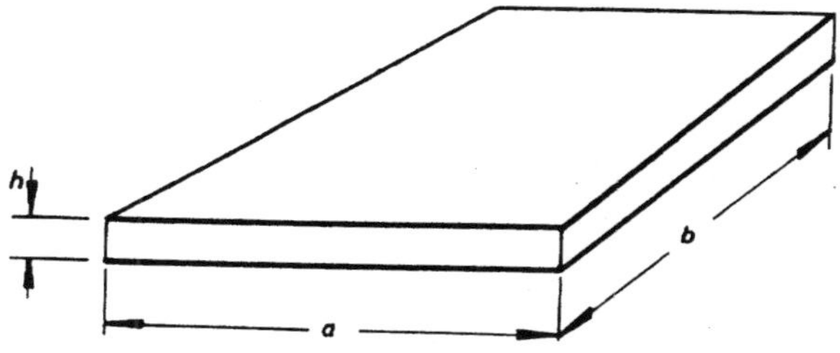

(a) Parallel-plate resonator

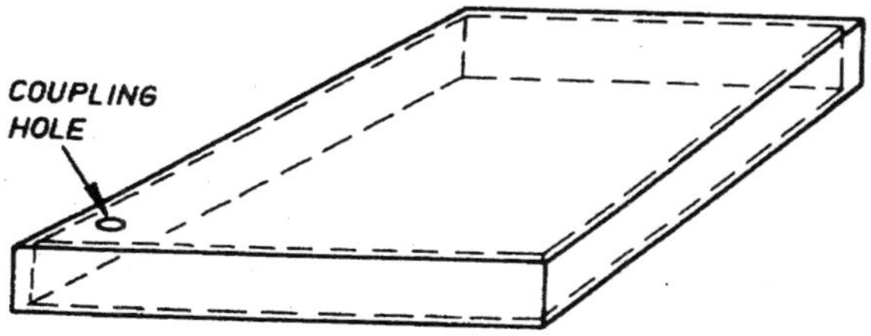

(b) Dielectric-filled, almost wholly metallized resonator

Figure 9.8 Metallized substrates forming rectangular microwave resonators.

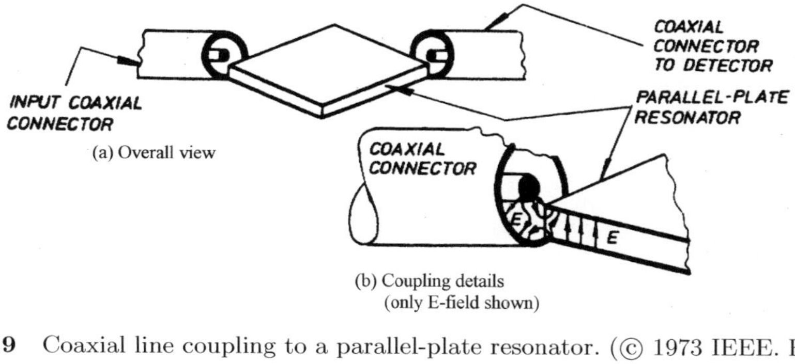

Figure 9.9 Coaxial line coupling to a parallel-plate resonator. (© 1973 IEEE. Reprinted, with permission, from Ladbrooke *et al.* [347].)

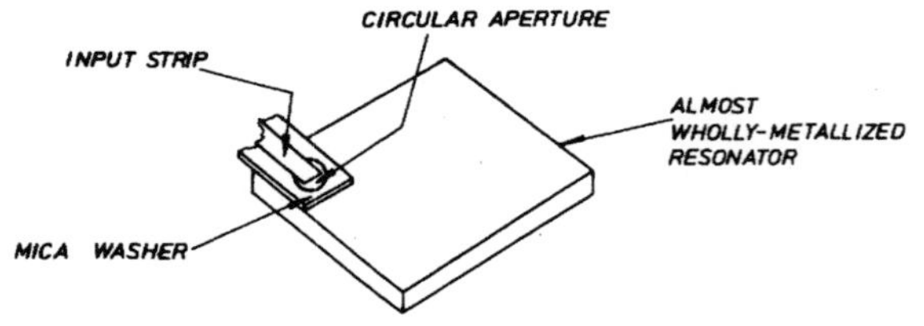

(a) Overall view

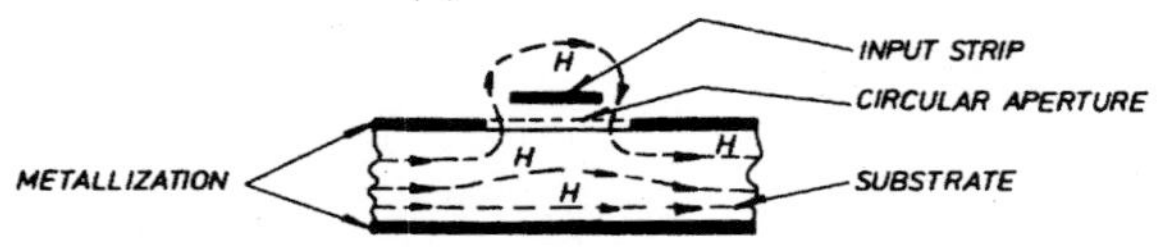

(b) Fields

Figure 9.10 Coupling via an aperture to an almost wholly metallized resonator. In (b) only the H-field is shown and the mica washer omitted for clarity. (© 1993 IEEE. Reprinted, with permission, from Ladbrooke *et al.* [347].)

that frequency-pulling effects are avoided, the reflected energy can be studied and resonator characteristics may then be deduced.

This section deals with a variety of resonator configurations useful for determining quantities such as effective microstrip permittivity (ε_{eff}) and various discontinuity parameters. Q-factor measurements are described in Section 9.8 on page 340 and some measurement techniques for parallel-coupled microstrips are outlined in Section 9.9 on page 341.

9.7.1 THE RING RESONATOR

This arrangement seems to be due to Troughton [349, 350], principally for dispersion measurements, and results obtained using ring resonators have been quoted in support (or otherwise) of certain dispersion theories and other measured results. Although other shapes have been tried, the most popular design is the simple circular ring shown in Figure 9.11.

Since the ring has no open ends it is almost free of radiative losses and may be used to determine line losses alone. For rings in which $w/h < 1$ and l (mean circumference) $\gg w$ we have, at resonance,

$$l = \frac{n\lambda_g}{2} \tag{9.8}$$

where n is the integer order of resonance. If dispersion measurements are to be performed, then $\varepsilon_{\text{eff}}(f)$ is required and, since $c/f\sqrt{\varepsilon_{\text{eff}}(f)} = \lambda_g$, Equation (9.8) becomes

$$\frac{nc}{2f\sqrt{\varepsilon_{\text{eff}}(f)}} = l. \tag{9.9}$$

With f in gigahertz and l in millimetres, this gives

$$\varepsilon_{\text{eff}}(f) = \left(\frac{150n}{fl}\right)^2. \tag{9.10}$$

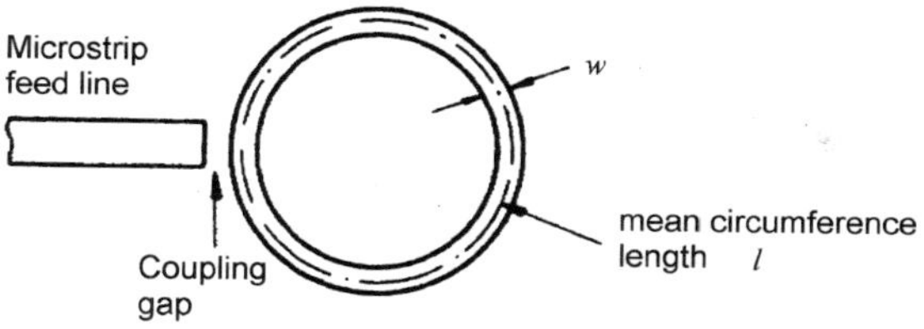

Figure 9.11 Ring resonator.

For n resonances with a particular ring, n values of $\varepsilon_{\text{eff}}(f)$ are therefore obtainable.

The main advantage usually claimed for ring resonators is the freedom from open-end effects. However, while this appears to be an important advantage, it by no means signifies that the electrical length of the ring can be accurately established, for the following drawbacks apply to ring resonators:

(a) curvature effects, which result in the mean physical circumference of the ring not being the most accurate value to use in calculating resonance wavelengths;
(b) local radial distortion of the field in the vicinity of the coupling gap;
(c) field interactions across the ring;
(d) the possibility of resonance splitting due to width non-uniformities around the ring.

The problem of curvature effects, (a), appears to have been satisfactorily resolved by Owens [351] who demonstrated that a curvature correction can be applied to greatly improve the accuracy of ring measurements of dispersion. Both the measurements and the computations were very carefully performed and were compared independently. The cited paper should be consulted for further details.

It is clear that the curvature problem is greatest for wide ($w/h \gg 1$) rings of relatively small diameter. The drawback of field interactions, (c), is also greater for $w/h \gg 1$ and for rings of relatively small radii. Therefore, if rings of sufficiently large radius are used, problems (a) and (c) are diminished and the overall effect of (b) is reduced, since the coupling region occupies a smaller fraction of the total ring length. However, it must be remembered that a larger ring increases the chance of variability not only in width, but also in the substrate permittivity and thickness under the line, and the possibility of resonance splitting, (d), still remains.

9.7.2 THE SIDE-COUPLED, OPEN-CIRCUIT-TERMINATED, STRAIGHT RESONATOR [349, 350]

The microstrip configuration for this structure is shown in Figure 9.12. This microstrip resonator, in common with several other structures which are used, has abruptly open-circuited ends. The behaviour of this 'discontinuity' is such that a considerable proportion of the field fringes beyond the physical ends of the line, and the effect of this is best accounted for by considering the line to be somewhat longer electrically.

This situation has already been dealt with (see Section 7.2.1 on page 228) and the same notation will be used here. There are two open-end-effect extensions to be considered, as indicated in Figure 9.13. It is apparent that a total *effective* length of $(l + 2l_{eo})$ exists, where l_{eo} is a hypothetical extension of the line to make up for the

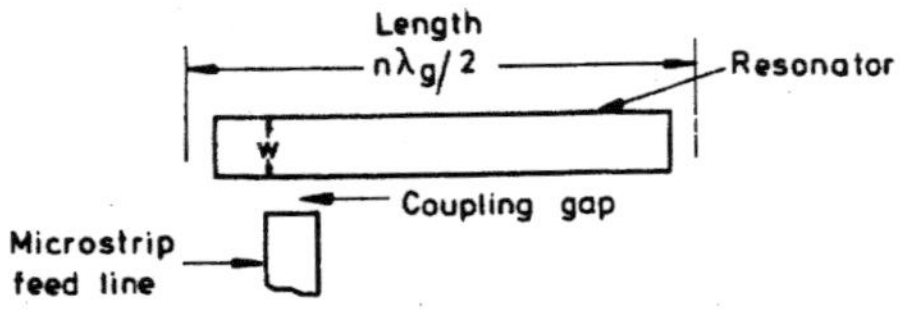

Figure 9.12 Side-coupled open-ended straight resonator.

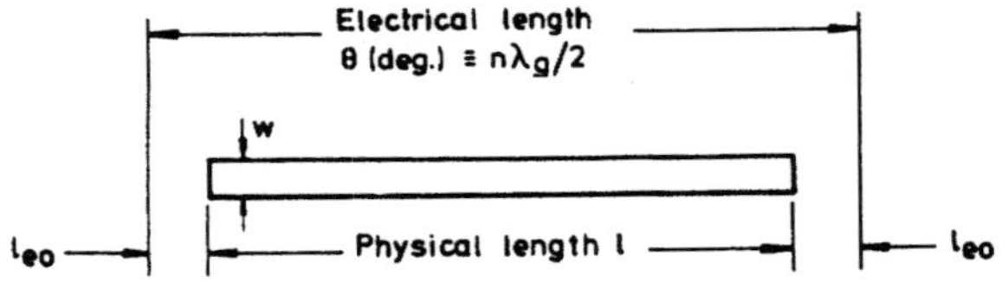

Figure 9.13 Illustrating the effective length of an open-ended straight resonator.

end-fringing. Also, the electrical length θ can be written as

$$\theta = \beta(l + 2l_{eo}). \tag{9.11}$$

With similar considerations to those used for the ring resonator, the dispersion expression for this straight resonator becomes

$$\varepsilon_{\text{eff}}(f) = \left\{ \frac{150n}{f(l + 2l_{eo})} \right\}^2. \tag{9.12}$$

This particular straight resonator technique may occasionally be useful, but it possesses the following disadvantages:

(a) There is the local transverse distortion of the field in the region of the side coupling gap. This is an identical problem to (b), mentioned under ring resonators, and it may be expected to lead to even larger errors with straight resonators, because the coupling region can occupy a relatively greater fraction of the total line length.
(b) The position of the feed line, relative to either open end of the resonator, is fairly important. A symmetrical location at the centre of the resonant line may often be best. In any event, other positions lead to limitations over the number of resonances which may be excited.

Positioning the feed line at right angles to the resonator and in the vicinity of the open-circuit end considerably complicates the fringing fields there, and is thus undesirable.

9.7.3 SERIES-GAP COUPLING OF MICROSTRIPS

The properties and parameters of a symmetrical series-gap are covered in Section 7.3 on page 231. A considerable amount of success has been enjoyed in several areas of microstrip measurements and in some filter designs by the use of end-to-end series-gap coupling (these types of filters are considered in Chapter 10).

There are, in general, two possibilities for feeding with series-gap structures where the feed line may have a characteristic impedance Z_{02} differing considerably from that

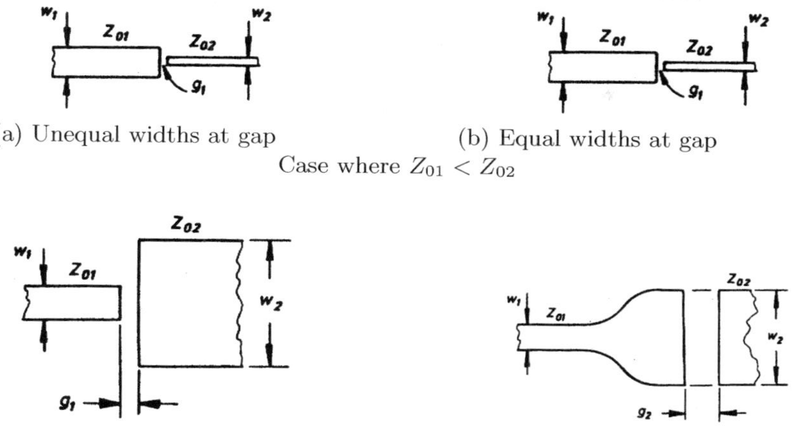

(a) Unequal widths at gap (b) Equal widths at gap

Case where $Z_{01} < Z_{02}$

(c) Unequal widths at gap (d) Equal widths at gap

Case where $Z_{01} > Z_{02}$

Figure 9.14 Practical microstrip series-gap-coupled arrangements.

of the feed line Z_{01}. The types of geometry which are possible are indicated in Figure 9.14.

Coupling configurations (a) and (c) ('asymmetrical longitudinally') are often disadvantageous for at least two reasons:

(1) It is hard to predict the degree of coupling, let alone assess the effects of adjusted gaps.
(2) These configurations have not, as yet, been analysed theoretically. Equivalent circuits and parameters are therefore unavailable.

Coupling configurations (b) and (d) overcome these objections.

Practical coupling gap designs for microstrip measurement resonators are constrained by a prime requirement that the degree of coupling be kept as loose as possible, while still offering a tolerable signal-to-noise ratio for measurement. Even slight loading of the resonator due to the external circuit will affect the resonant frequency. Furthermore, all the techniques about to be outlined employ two coupling gaps. These must be manufactured as identically as possible in each case, because otherwise the gap end-effect lengths must definitely differ due to the differing capacitance values. Errors will then result, since the expressions applicable to each measurement method assume equal coupling gaps, and hence identical end-lengths. Different end-effect lengths at the gaps result in too many unknowns, thus prohibiting accurate solution of the resonance expressions.

These circuits do not have the disadvantages which apply to the arrangements discussed in the previous paragraphs. They do, however, all involve the end-effect lengths, both at the gaps and at the abruptly open-circuited ends.

Richings [352] performed experiments to ascertain suitable coupling gap separations (g) for resonators on alumina, and we have carried out other investigations where the substrate material is sapphire. For resonators on 0.5 mm thick monocrystalline sapphire, the coupling gap separation may be calculated approximately from the empirical expression

$$g = \frac{8,000 \log Z_0}{Z_0} - 148 \quad \mu\text{m} \tag{9.13}$$

where Z_0 is the characteristic impedance of the microstrip line forming the resonator.

Use of this formula ensures that fairly loose coupling is maintained. When resonators are to be constructed on alumina the gap needs to be somewhat greater for similar conditions to apply. In the case of 0.65 mm thick alumina substrates, with permittivity approximately 10.0, gap separations are approximately 30% greater than those calculated for sapphire. This is because the thicker material and lower permittivity both serve to increase the influence of the series capacitance, and hence to increase the coupling.

9.7.4 SERIES-GAP-COUPLED STRAIGHT RESONATOR PAIRS [353]

This arrangement is particularly useful for dispersion measurements, and Figure 9.15 shows both the physical layout and effective lengths involved. In this method, the combined end-effect length l_e is eliminated for the purpose of calculating the effective microstrip permittivity $\varepsilon_{\text{eff}}(f)$ using measured resonant frequencies and lengths of each resonator. For this technique to operate successfully, we make $l_2 \approx 2l_1$. (Incidentally, this means that l_2 resonates at many more frequencies than l_1 which has the extra advantage of providing further measurement data.) Refer to Figure 9.15, and assume that:

(a) l_{eo} and l_{eg} are both independent of resonator length so that $l_{eo1} = l_{eo2} = l_{eo}$,
(b) $l_e = l_{eo} + l_{eg}$ (combined end-effect length).

Then

$$l_1 + l_e = \frac{\lambda_{g1}}{2} \tag{9.14}$$

and

$$l_2 + l_e = n\lambda_{g2} \tag{9.15}$$

where the unknowns are λ_{g1}, λ_{g2}, and l_e — and n is the (integral) order of resonance.

From these expressions, Equations (9.14) and (9.15), eliminating l_e:

$$l_2 - l_1 = n\lambda_{g2} - \frac{n\lambda_{g1}}{2}. \tag{9.16}$$

Now there is the general relationship between $\varepsilon_{\text{eff}}(f)$ and wavelengths:

$$\varepsilon_{\text{eff}}(f) = \left(\frac{\lambda_0}{\lambda_g}\right)^2. \tag{9.17}$$

Therefore, we may write

$$\varepsilon_{\text{eff}}(f) = \left(\frac{c}{\lambda_{g1} f_1}\right)^2 = \left(\frac{c}{\lambda_{g2} f_2}\right)^2 \tag{9.18}$$

where f_1 applies to λ_{g1}, and f_2 applies to λ_{g2}. We may allow the same $\varepsilon_{\text{eff}}(f)$ value for each result since f_1 and f_2 have deliberately been forced to be close values in

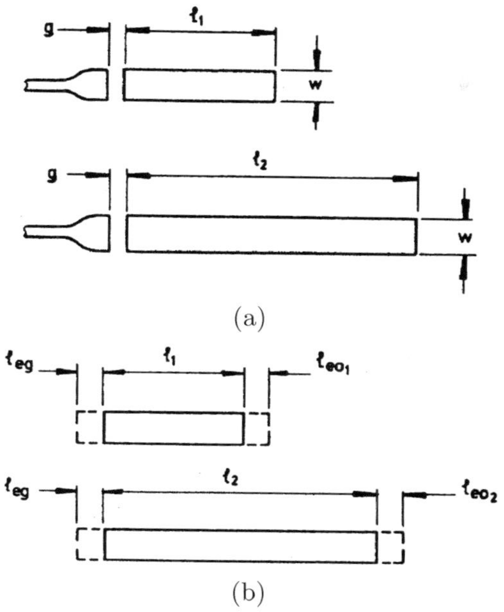

Figure 9.15 Open-circuited pairs of straight resonators: (a) practical arrangement; and (b) effective lengths of the resonators.

designing the resonators, and $\varepsilon_{\text{eff}}(f)$ changes slowly with frequency. Equation (9.16) therefore becomes

$$l_2 - l_1 = \frac{nc}{f_2\sqrt{\varepsilon_{\text{eff}}(f)}} - \frac{nc}{2f_1\sqrt{\varepsilon_{\text{eff}}(f)}} \tag{9.19}$$

and solving this for $\varepsilon_{\text{eff}}(f)$ finally gives

$$\varepsilon_{\text{eff}}(f) = \left\{\frac{nc(2f_1 - f_2)}{2f_1 f_2(l_2 - l_1)}\right\}^2 . \tag{9.20}$$

This expression may be used to obtain dispersion results in cases where both resonators, l_1 and l_2, are resonant at close frequencies. It is also necessary to evaluate the combined end-effect length l_e. Equations (9.14) and (9.15) can readily be solved for l_e. Initially, dividing the equations eliminates several common quantities (again assuming $\varepsilon_{\text{eff}}(f)$ to be constant):

$$\frac{l_1 + l_e}{l_2 + l_e} = \frac{\lambda_{g1}/2}{\lambda_{g2}} = \frac{c/2f_1\sqrt{\varepsilon_{\text{eff}}(f)}}{c/f_2\sqrt{\varepsilon_{\text{eff}}(f)}} = \frac{f_2}{2f_1} \tag{9.21}$$

and after a little manipulation, this yields the required end-effect length:

$$l_e = \frac{f_2 l_2 - 2f_1 l_1}{2f_1 - f_2}. \tag{9.22}$$

An approximate *a priori* knowledge of l_e is required so that $l_2 = (2l_1 + l_e)$, which ensures that f_1 and f_2 are close frequencies.

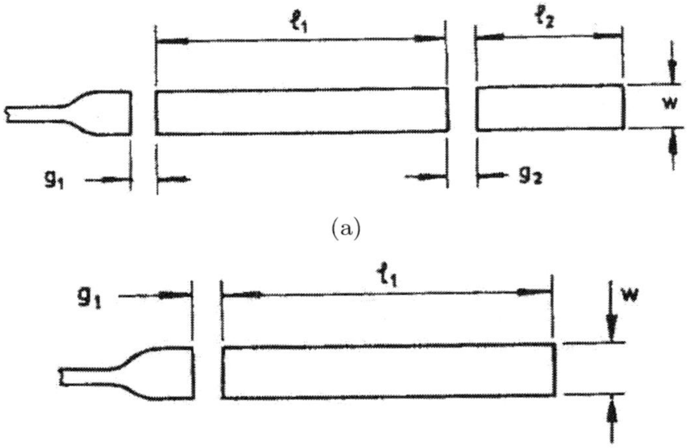

(b)

Figure 9.16 The two-stage resonator technique originally devised by Richings and Easter: (a) stage 1; and (b) stage 2 with the section of length l_2 removed.

The special requirements arising from this method are that, for each set of results, a pair of resonators must be made with identical widths and coupling gaps. Both resonators may lie on the same substrate, or each may occupy one of a pair of substrates with very closely matched thicknesses, in order to accommodate a long resonator across a diagonal or to avoid resonator interaction.

Additional values of $\varepsilon_{\text{eff}}(f)$ can be obtained from the odd-mode resonances of the longer resonators, where

$$\begin{aligned} l_2 + l_e &= (2n-1)\frac{\lambda_{g2}}{2} \\ &= (2n-1)\frac{c}{2f_2\sqrt{\varepsilon_{\text{eff}}(f)}} \end{aligned} \tag{9.23}$$

or

$$\varepsilon_{\text{eff}}(f) = \left\{ \frac{(2n-1)c}{2f_2(l_2+l_e)} \right\}^2 \tag{9.24}$$

where the appropriate value of l_e is interpolated between values calculated from the adjacent frequencies, using Equation (9.22).

9.7.5 THE RESONANT TECHNIQUE DUE TO RICHINGS AND EASTER

Richings and Easter [352,356]) appear to have been the first to devise an experimental technique enabling separation of the end-effects due to the series-gap and the foreshortened open-circuit. The new feature which is introduced is a second series-gap-coupled quarter-wavelength line resonator. As in the previous, paired resonators method, at least a very approximate *a priori* knowledge of the end-effect lengths is required. The physical structure is shown in Figure 9.16.

The quarter-wavelength line, of physical length l_2, transforms the open-circuit at the foreshortened end of l_2 to an effective short-circuit in the vicinity of the gap g_2.

This is the crucial feature of the method, because it means that with two gaps the initial circuit can be solved (with length l_1) involving gap end-effect extensions. For 'stage 2' of the measurements the quarter-wavelength line l_2 is removed by chemical etching. If we assume that the foreshortened open-circuit end-effect lengths are all identical and equal to l_{eo} and that the gap end-effect lengths are equal to l_{eg} we may adopt the same notation as in Section 9.7.4 on page 334 (preceding Equation (9.14). Thus, we obtain the following expressions for 'stage 1' shown in Figure 9.16, resonator l_1:

$$n\lambda'_g/2 = l_1 + 2l_{eg} \tag{9.25}$$

and, with l_2 removed,

$$n\frac{\lambda''_g}{2} = l_1 + l_{eg} + l_{eo} \tag{9.26}$$

where n is the (integer) order of resonance and the frequencies must differ slightly, and hence the different wavelengths λ'_g and λ''_g.

These wavelengths are evaluated by measuring the two resonant frequencies f' and f'', knowing $\varepsilon_{\text{eff}}(f)$, and applying the procedure previously described following Equation (9.17) in Section 9.7.4 on page 334. The effective microstrip permittivity $\varepsilon_{\text{eff}}(f)$ can either be determined independently (Section 9.7.4 on page 334), or by constructing a further circuit under this method and eliminating l_{eo} and l_{eg} from the resonance expressions.

We can now evaluate l_{eo} and l_{eg}. Setting the experimentally determined quantities $n\lambda'_g/2 = x$ and $n\lambda''_g/2 = y$ we obtain, from Equation (9.25),

$$l_{eg} = \frac{x - l_1}{2} \tag{9.27}$$

and, from Equation (9.26),

$$l_{eo} = y - l_{eg} - l_1. \tag{9.28}$$

Using this technique, and also using sets of varied-length resonators to accommodate different frequencies, Richings [352] obtained several useful and consistent results.

For adequate accuracy in measurement, the gaps must be made identical to within a few micrometres for resonators manufactured on alumina-type substrates.

9.7.6 THE SYMMETRICAL STRAIGHT RESONATOR [354]

All previously described straight resonator measurement methods have suffered from the drawback of requiring some *a priori* knowledge of the end-effect equivalent lengths. The 'symmetrical resonator' method does not require this rather uncertain item of design information. This arrangement also avoids the critically placed quarter-wave transforming line associated with Richings' method. The basic structure is shown in Figure 9.17.

Electrically, taking a plane situated in the physical centre of the microstrip resonator, a nearly identical circuit is seen whether one looks towards the source (coupled in through g_1) or towards the matched load. In the network analyser measuring system, the source impedance is generally very close to 50 Ω, so a matched

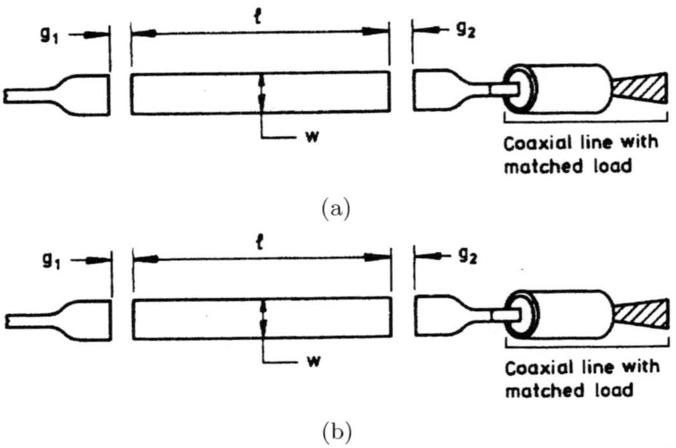

Figure 9.17 Symmetrically loaded resonator technique: (a) stage 1; and (b) stage 2 with the load and load-section removed.

50 Ω load is also used. The gaps g_1 and g_2 are kept very nearly equal and set for loose coupling.

For 'stage 1', referring to Figure 9.17(a) and using the notation previously employed, the resonance equation is

$$\frac{1}{2} n\lambda_{g1} = l + 2l_{eg} \tag{9.29}$$

where the microstrip wavelength λ_{g1} is known from the effective microstrip permittivity $\varepsilon_{\text{eff}}(f)$ obtained by another method, such as the paired resonators. The resonant frequency f, is measured, and l_{eg} becomes

$$l_{eg} = \frac{nc}{4f_1\sqrt{\varepsilon_{\text{eff}}(f)}} - l. \tag{9.30}$$

With the coaxial-matched load removed and the second section of line chemically etched away from the substrate, we have 'stage 2' of this method. The resonance equation in this case is, referring to Figure 9.17(b), and using the previously employed notation,

$$\frac{1}{2} n\lambda_{g2} = l + l_{eg} + l_{eo} \tag{9.31}$$

whence

$$l_{eo} = \frac{nc}{2f_2\sqrt{\varepsilon_{\text{eff}}(f)}} - l - l_{eg} \tag{9.32}$$

where all quantities on the right-hand side are known, including the new measured resonant frequency f_2.

This general type of resonator layout has also been used to calibrate Through-Reflect-Lines (TRL) for MMIC on-wafer microstrip lines [355]. Calibrated versus resonator-derived loss measurements were reported over the frequency range 1–22 GHz for magnitude and from 1 to 26 GHz for phase. Both reflection coefficient $S_{11}$and forward transmission coefficient S_{21}, were measured for microstrip and coplanar lines. It was concluded that microstrip provides a superior TRL capability, given the relatively good agreement between calibrated and independent measurements.

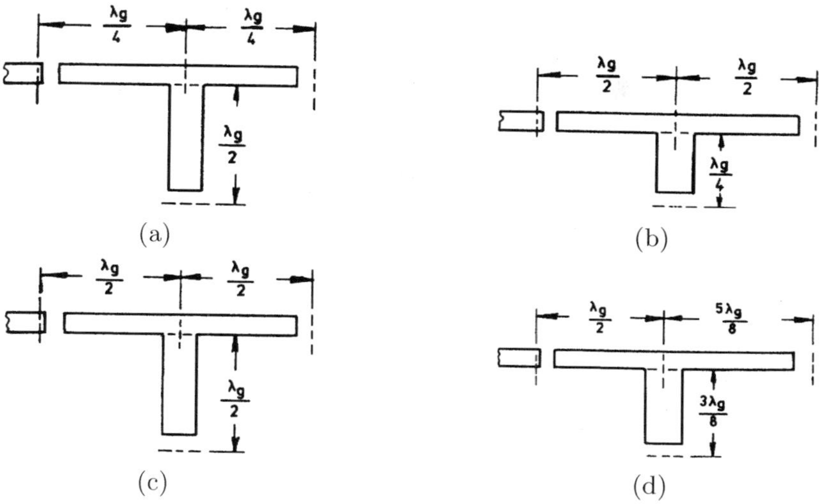

Figure 9.18 T-junction measurement arrangements: (a) L_1 measurement; (b) L_2 measurement; (c) measurement of C; and (d) measurement of n. (© 1975 IEEE. Reprinted, with permission, from Easter [356].)

9.7.7 RESONANCE METHODS FOR THE DETERMINATION OF DISCONTINUITIES OTHER THAN OPEN CIRCUITS [268]

In order to measure quantities other than equivalent discontinuity capacitance, it is necessary to set up resonator arrangements which ensure effective cancellation of the capacitive effect in the vicinity of the discontinuity. For the determination of inductance, a voltage node must be placed in the discontinuity region and use is made of the fact that such a node will occur $\lambda_g/4$ away from a voltage maximum (at a true open-circuit plane). Thus, it is important to accurately know the wavelength and open-end effect for the line concerned. The arrangement shown in Figure 9.18(a) [356] indicates how through-arm inductances L_1 may be measured. The measurement of L_3, and of capacitance C, is indicated in Figure 9.18(b) and (c), respectively.

In all cases, the parameters are first determined as equivalent extra lengths of the resonator, and then finally as lumped equivalent elements. For the measurement of the impedance ratio n^2 it is necessary that the transformed impedance (in the vicinity of the junction) be finite and also well defined. To obtain this, the intermediate eighth-wavelength sections of line are used, since these yield input reactances equal to $-jZ_0$ Ω.

In order to complete the measurements, the secondary arm of the T-junction has to be removed by etching-off, and the circuits have to be remeasured. The demands on the accuracy of initial circuit fabrication and the etching-off operation are generally high. For further details, the cited reference should be consulted.

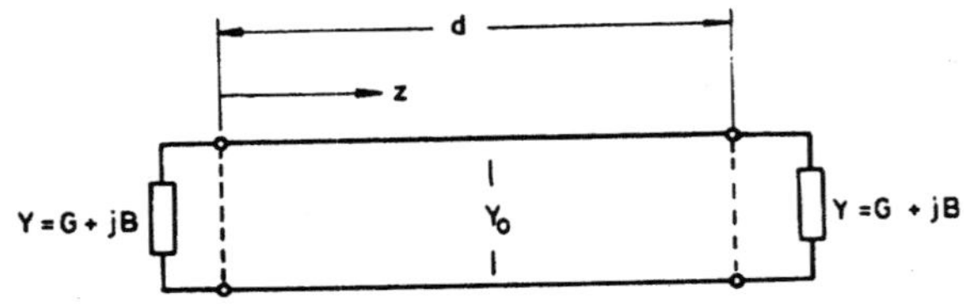

Figure 9.19 Equivalent equal loads ($G + jB$) at both ends of a straight, symmetrical, open-ended resonator.

9.8 Q-FACTOR MEASUREMENTS

Reliable techniques for the determination of the Q-factor of a microwave resonator are well known. Network-analyser-compatible swept-frequency techniques for measuring the Q-factor have been well described in a comprehensive paper by Aitken [357]. These techniques are well suited to microstrip resonator measurements where, for example, the instrumentation system shown in Figure 9.7 is in use.

Both the Q-factor and the coupling coefficient are obtained using the techniques described by Aitken. It is indicated that errors in measured Q-factors can be kept below 10% provided the coupling coefficient η (Aitken's β) is within the range

$$0.15 \leq \eta \leq 0.75 \ . \tag{9.33}$$

When η is below 0.15 the signal-to-noise ratio becomes very poor and the measurement accuracy suffers, although the accuracy can be improved if signal-averaging techniques are used. For straight microstrip resonators, the coupling gap separation given by Equation (9.13) should ensure that η lies within the range given, but some adjustment may occasionally be required. Coupling gaps for ring resonators require experimental optimization.

Measurements of the Q-factors of ring resonators yield the losses due, predominantly, to conductor resistivity and substrate dielectric losses. There is very little loss due to radiation.

On the other hand, measurements conducted using open-circuit-ended straight resonators are influenced by the 'end-effects'. The Q-factor due to these end-effects alone may be found in terms of the reflection coefficients ρ at each end. This is useful since, once Q is measured, ρ is known and the parameters which describe the behaviour of the open-circuit can be obtained. To simplify the analysis we start by assuming that the microstrip resonator is isolated [358], i.e. the coupling does not influence the Q-factor significantly. The situation is shown in Figure 9.19.

It is shown in Appendix B on page 449 that the external Q-factor applicable to this type of resonator can be expressed in terms of the end reflection coefficient ρ as:

$$Q_e = \frac{\pi l \rho}{1 - \rho^2}. \tag{9.34}$$

(This is given in Appendix B as Equation (B.39) on page 454.)

By simple substitution in Equation (9.34) we can obtain the results for end reflection coefficients, given measured external Q-factors.

For example, an external Q-factor of 29.8 leads to an end reflection coefficient of 0.9. This would be a poor Q-factor if it applied to a resonator for most circuit design purposes and should more likely be the attribute of some resonant structure designed for a microstrip antenna. In contrast, a measured external Q-factor of 312 (which is fairly typical of microstrip in circuits) corresponds to an end reflection coefficient of 0.99.

It should be noted that the actual measured Q-factor will also include both conductor and dielectric losses, which are accounted for by an unloaded Q-factor, Q_u. Therefore, the total measured Q for the resonator is given by

$$(1/Q) = (1/Q_u) + (1/Q_e). \tag{9.35}$$

Since Q_u can easily be measured, using the simple arrangements described earlier in this chapter (Sections 9.5 and 9.7.1 on pages 325 and 330 respectively), Q_e becomes the only unknown in Equation (9.35) and can readily be found.

9.9 MEASUREMENTS ON PARALLEL-COUPLED MICROSTRIPS

The basic principles of parallel-coupled microstrip lines, and design procedures using them, were discussed in Chapter 8.

In order to check any theory and develop sound designs, it is highly desirable that adequate measurements can be performed to support the theoretically calculated results. Gupta *et al.* [88] have fully described some test arrangements for determining coupled microstrip line parameters. Particular parameters required are:

- even- and odd-mode characteristic impedances
- even- and odd-mode phase velocities
- coupling factor (principally the mid-band, i.e. maximum value).

Gupta describes the measurement technique, originally given by Napoli and Hughes [298], in which the even and odd modes are separately excited, and outputs from the coupled lines are fed into the 'reference' and 'test' channels of a microwave network analyser. Resonant, coupled, half-wavelength lines are used, and the wavelengths applicable to each mode are found from the resonant frequencies. It is suggested that a slotted line be inserted in the test channel input to determine the VSWR values, and hence the characteristic impedances from

$$Z_{0e} = Z_0\sqrt{r_e} \tag{9.36}$$

and

$$Z_{0o} = \frac{Z_0}{\sqrt{r_o}} \tag{9.37}$$

where r_e and r_o are VSWR values for the even and odd excitations respectively.

However, it would seem that the VSWR could equally well be obtained from reflection coefficient measurements using the network analyser. (The required interrelationship is given as Equation (1.32) in Chapter 1.) Since, especially with 'loose' coupling, Z_{0e} and Z_{0o} do not differ greatly, the measurements of r_e and r_o must be accurate. The value of Z_0 must also be known with confidence.

Richings [352] points out that Napoli and Hughes' measurement technique is fairly complex, requiring four identical coupling gaps and four transitions. Richings has

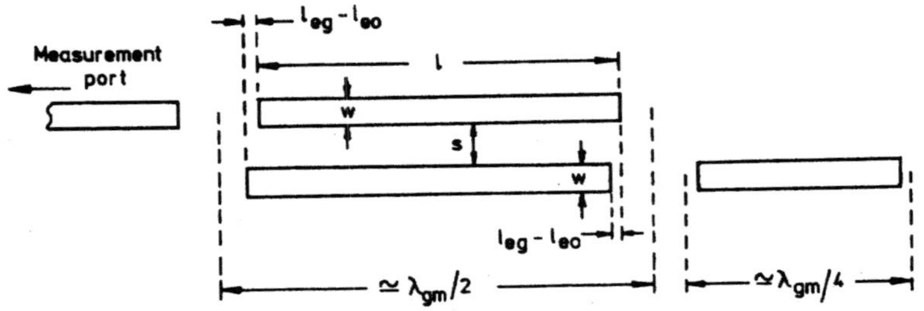

Figure 9.20 Resonant method for determining the phase velocities in two parallel-coupled microstrips.

given details of a single-port measurement technique, the basic circuit for which is shown in Figure 9.20.

This technique has been used for open microstrip circuits and also for shielded circuits. Careful measurements account for the end-effects, and the resonance conditions are

$$\frac{1}{2}n\lambda_{ge} = l + 2l_{ege} + l_{eoe} \tag{9.38}$$

and

$$\frac{1}{2}n\lambda_{go} = l + l_{ego} + l_{eoo} \tag{9.39}$$

where the subscripts, and the equivalent end-effect lengths, have the meanings previously given. Two distinct resonances are observable, leading to the measurement of two frequencies f_{oe} and f_{oo}. Effective permittivities and phase velocities can then be evaluated from the expressions

$$\varepsilon_{\text{effe}} = \left(\frac{c}{f_{oe}\lambda_{ge}}\right)^2 \tag{9.40}$$

$$\varepsilon_{\text{effo}} = \left(\frac{c}{f_{oo}\lambda_{go}}\right)^2 \tag{9.41}$$

$$v_{pe} = \frac{c}{\sqrt{\varepsilon_{effe}}} \tag{9.42}$$

and

$$v_{po} = \frac{c}{\sqrt{\varepsilon_{effo}}}. \tag{9.43}$$

The accuracy of this technique is subject to the limitations discussed for the measurement of single line parameters (Section 9.7.5 on page 336). Further details are given in Richings' thesis [352], where power-loss measurements are also described.

Richings has given a different circuit for the determination of the coupling factor (Figure 9.21). As described in Appendix A (the comments following Equation (A.61)) the maximum coupling occurs when the coupled region extends for a quarter-wavelength. This is provided by the overlapping length, $l = \lambda_{gm}/4$ in Figure 9.21. The resonance condition for this circuit is

$$\frac{Z_0}{\tan\theta} + \frac{1}{2}\left(\frac{Z_{0e}}{\tan\theta_e} + \frac{Z_{0o}}{\tan\theta_o}\right) \pm \frac{1}{2}\left(\frac{Z_{0e}}{\sin\theta_e} - \frac{Z_{0o}}{\sin\theta_o}\right) = 0. \tag{9.44}$$

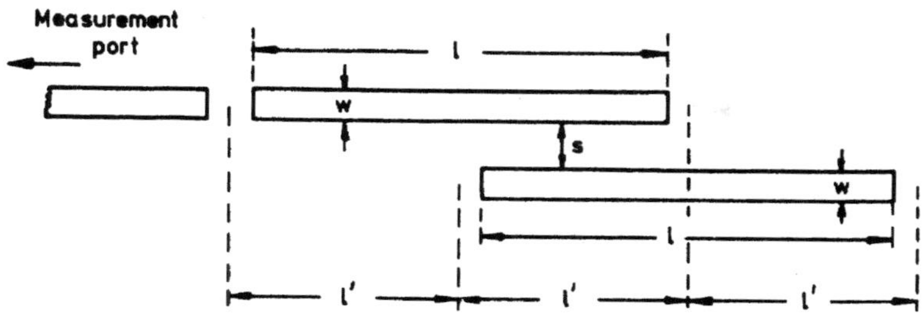

Figure 9.21 Resonant method for determining the coupling factor between two parallel-coupled microstrips, $l' = \lambda_{gm}/4$.

The appropriate values of Z_{0e} and Z_{0o} are obtained by theoretical calculation, and the electrical lengths $(\theta_{e,o})$ are determined by an iterative method using a computer. Thus, the coupling factor can also be evaluated.

Rizzoli [359] has demonstrated a method using properly terminated parallel-coupled microstrip to ensure separate excitation of even and odd modes. Rizzoli's method requires short-circuits to ground, which can be effectively implemented in microstrip at high microwave frequencies (see Section 5.4 on page 122), but which become difficult at millimetre-wave frequencies.

Babu and Banmali [360] have described a technique for measuring coupled microstrip line parameters without the use of a network analyser. Only a signal generator, a conventional standing-wave indicator, a variable short-circuit stub, and two coaxial-to-microstrip transitions are required. These researchers show how scattering-parameter measurements can be performed, with corrections for the transitions, to obtain the characteristic impedances, phase velocities and losses. Circle diagram techniques are employed, and the cited reference should be read for further details.

It is difficult to assess the accuracy of this method from the given information because so few final results are actually quoted.

9.10 STANDING-WAVE INDICATORS IN MICROSTRIP

All the measurements described so far in this chapter have required transitions from the measuring system transmission medium (often coaxial line) to the microstrip circuit under study. Even the best, carefully compensated transitions must interfere to some extent with reflections from the microstrip circuit. This fact prompted the development of some Standing-Wave Indicators (SWI) in microstrip.

Possibly the most thoroughly designed microstrip SWI is reported by Ladbrooke [361]. In this indicator the circuit, whose characteristics are to be studied, is made to move in relation to a fixed probe. Instead of a slot, as used with conventional coaxial or waveguide SWI, a thin dielectric leaf has the circuit pattern printed on it and this leaf is added to the basic structure. Ladbrooke employed a mica slice for the thin leaf and pressed this down uniformly onto the main substrate by using a backing of expanded polystyrene (ε_r= 1.04). The reference cited gives further details.

It may be considered unlikely that a microstrip SWI would be useful in routine microstrip measurements. This is because computer-augmented automatic correction is readily achievable, effectively removing the parameters of the transitions from the 'raw' measurements made in coaxial line or waveguide systems. Desktop or laptop of ever-increasing power-to-cost ratios tend to make this approach very attractive.

However, a particular virtue of a microstrip SWI is the facility for probing the fields around, and even substantially away from, the circuit structure. In his paper, Ladbrooke [361], for example, describes the measurement of a wave launched from the foreshortened open-circuit end of a microstrip line. This is predominantly due to the surface-wave fields discussed in Section 5.10 on page 147.

9.11 TIME-DOMAIN REFLECTOMETRY (TDR) TECHNIQUES

The basic principles and operating practices associated with TDR techniques, in which a pulse is transmitted through the circuit, are well known and have been published elsewhere. Although frequency-domain studies generally yield accurate information regarding the performance of microwave circuits and systems, TDR possesses the following possible advantages:

(a) immediate recognition of the nature of various elements, by waveform examination,
(b) approximate location of specific elements or discontinuities in the circuit or system,
(c) fairly simple instrumentation.

Since the operation of TDR depends inherently on pulse reflections from the elements under study, we return to the reflection coefficient Γ caused by any load Z_L

$$\Gamma = \frac{Z_L - Z_0}{Z_L + Z_0}. \tag{9.45}$$

In several simple but significant cases, this is easily interpreted, e.g.

Matched load:

$Z_L = Z_0$, and hence, by using Equation (9.45), $\Gamma = 0$.

There is no reflection, and an incident pulse transmitted will not suffer any distortion due to this load.

Short circuit:

$Z_L = 0$, and hence, again by using Equation (9.45), $\Gamma = -1$.

This means that an incident pulse transmitted will suffer complete cancellation in the round-trip time required for the pulse to reach and return from the short-circuit location.

Thus, TDR can be effectively used for checking (broadband) matched circuits, short circuits, and a wide variety of other elements and discontinuities — provided that the system resolution is sufficiently precise. The time resolution is particularly critical.

We recall the delay time calculated for 50 Ω microstrip on an alumina substrate which was given in Section 4.11 on page 109. This was 8.6 ns/m, or 8.6 ps/mm. Using this value, we find that a TDR with a system pulse-rise time of about 30 ps can only resolve distance separations of approximately 2 mm, at best, along 50 Ω microstrip lines. In practice, the effects of multiple reflections and noise complicate the displayed waveform considerably and limit the resolution. Companies such as Picosecond Pulse Labs provide systems that have input pulse rise times down to several picoseconds — therefore enabing substantially finer resolution to be achieved.

Piller [362] has presented a very interesting extension of both TDR and TDT (Time-Domain Transmission) measurement techniques. Substantial information is given, and Piller claims that time constants down to approximately 0.1 ps are within the capabilities of his methods. The principal features of this technique are the careful interpretation of maximum pulse amplitude and the separate employment of a ramp waveform. Using a 30 ps TDR system, especially adapted, where the minimum voltage is 2 mV and the maximum voltage is 1 V, it is claimed that the following minimum circuit parameter values can be resolved:

With TDR:

$$C_g \approx 15 \text{ fF}$$
$$Ll \approx 35 \text{ pH}$$

With TDT:

$$C_s \approx 1 \text{ fF}$$
$$L_g \approx 1 \text{ pH}$$

where

subscript 'g' means 'to ground'
subscript 'l' means 'lead'
subscript 's' means 'series'.

Measurement precautions mentioned by Piller include:

(a) a calibration measurement with a very narrow gap (so that $C_g \rightarrow 0$)
(b) the use of a good reference point
(c) high noise rejection and jitter rejection
(d) expansion of the waveform plots or traces
(e) use of an averaging interpretation of the measurements.

Piller's paper [362] should be consulted for further details.

10

Interconnects and Filters in Passive RFICs and MICs

10.1 RADIO FREQUENCY INTEGRATED CIRCUITS (RFICs)

The expression radio frequency integrated circuits, or RFICs, refers to circuits that operate from about 1 GHz up to 5 GHz. This frequency range is sure to change as new markets open up. One of the factors that distinguishes RFICs from other RF and microwave integrated circuit technology is the use of lumped elements rather than distributed elements. Distributed elements are based on transmission-line structures having linear dimensions which are a significant part of a wavelength — with a quarter wavelength being a special length as resonant effects are then obtained. At 1 GHz, a quarter-wavelength long line is 3.8 cm in silicon dioxide, the medium in which a line is fabricated on a silicon chip. This reduces to 7.7 mm at 5 GHz. With GaAs, GaAs itself can be used as the transmission line medium, and these dimensions become 1.97 cm and 3.9 mm respectively. These dimensions are too large for chips at the low end of the microwave spectrum (i.e 5 GHz and below), and so lumped elements must be used. RF circuits require many passive components for matching networks, RF chokes for bias (i.e. inductors that block RF but provide a lossless DC connection), harmonic tuning, and to ensure stability at frequencies below the frequencies of operation. The primary lumped elements on RFICs are resistors, capacitors and inductors.

Before commencing on a discussion of lumped elements for RFIC we should point out one of the other distinguishing features of RFICs. This is the use of differential signalling for nearly all of the analogue signals on-chip and not only for the RF paths. This is necessary to overcome the limitations of silicon MOSFET-based circuitry including significant substrate coupling. This also utilizes the availability of complementary devices and differential circuits to remove many of the nonlinear effects that are manifest with single-sided circuit technologies using, for example, MESFETs which are only available as n-type. Transmission line structures are rarely used on-chip at low microwave frequencies as there is insufficient room to realize distributed components which typically involve quarter-wavelength long lines. However, from an interconnect perspective, the differential signalling requires that the RF connections to the chip also be differential and preferably use differential transmission lines. These were discussed extensively in Section 6.11 on page 218.

Table 10.1 On-chip resistor properties. (Material from Lee [363] and other sources.)

Type	Resistivity (Ω per square)	Tolerance	Temp. Coeff. (ppm/°C)	Comment
Metal interconnect	0.01	35%	3900	Aluminium, small resistances up to 10 Ω.
Polysilicon, silicided	5–10	50%	1000	Small resistances.
Polysilicon, unsilicided	50–100	35%	variable	Poor tolerancing.
Source drain diffusion	25–200	80%	2000	Generally ion implantation used; large parasitic capacitance from semconductor junction; voltage dependence; limited to non-critical functions.
Wells	1000–10000	80%	3000–5000	
MOS transistor	1000–10000	35%	very high	Voltage dependence.

10.1.1 ON-CHIP RESISTORS

On-chip resistors can be obtained using doped-semiconductor regions, thin metal lines or, in the case of Si RFICs, polysilicon. The desired properties include good linearity, reasonable tolerance control, low parasitic capacitance, and low temperature coefficient. The operating temperature range required for RF products, such as cellular telephones, is broad from −30 °C to +85 °C, with the end result being that resistor values are poorly controlled. All practical materials have significant resistivity variations with temperature over this range. However, resistor ratios are generally well controlled so there is a design preference for circuits whose functionality is determined by ratios. The properties of on-chip resistor technologies are summarized in Table 10.1. In addition, specialized processes have been developed so that resistors can be made using nichrome (NiCr) or sichrome (SiCr), which have low temperature coefficients (about 100 ppm/°C) and good tolerancing achieved using laser trimming. However, these are not available in general.

10.1.2 ON-CHIP CAPACITORS

There are three primary forms of on-chip capacitor:

- Metal-Oxide-Metal (MOM) capacitor — using the interconnect metallization.
- Metal-Oxide-Semiconductor (MOS) capacitor — essentially a MOS transistor.
- semiconductor junction capacitor — either the capacitance of reverse biased *pn* junction or Schottky barrier.

The properties of these capacitors are presented in Table 10.2.

MOM capacitance can be realized as a parallel plate capacitance, but multiple levels of metallization can be used to increase the capacitance density. Relatively low

Table 10.2 On-chip capacitor properties. (Material from Lee [363] and other sources [364–368].)

Type	C ($fF/\mu m^2$)	Toler-ance	Temp. Coeff. (ppm/°C)	Comment
Metal-Oxide-Metal	0.05	35%	50	Capacitance determined by geometry, low loss, good Q.
MOS	1–5	≥20%	30	High loss, low Q, voltage dependent, strong temperature dependence.
Junction	0.1–1.0	≥20%	200–1,000	Strong voltage and temperature dependence. Range is for low leakage (yielding high Q capacitance and currently used doping levels.

capacitance values of up to 500 $fF/\mu m^2$ are available because of the large dielectric thickness of 0.5–1 μm between metal layers. This thickness is required in normal interconnect circumstances to minimize interconnect-to-interconnect coupling so it is unlikely to change much. The distance between the bottom metallization layer to the substrate is comparable to the metal layer separation, and so the capacitance between the bottom plate of a capacitor and the substrate leads to significant capacitance — in the range of 10–30% of the metal-to-metal capacitance [363]. This series capacitor connection must be considered in design. Regions of high dielectric constant material, such as ferro-electrics, are used in memory chips to obtain high capacitance but this is unlikely to be available in RFIC processes. This also results in a strongly nonlinear capacitance.

An alternative MOM capacitance is available using lateral arrangments of interconnects on the same layer. That is, adjacent metal structures are separated by a small horizontal gap. Again, there are two distinct metal connections and a smaller metal separation can be obtained using photolithography than available using dielectric thickness. However, the capacitance density is only increased by a factor of around 3.

Both types of MOM capacitance, parallel-plate and lateral, are geometrically defined, are voltage independent, have very low temperature coefficients, and have initial fabrication tolerances of 20–30%. At intermediate frequencies, and potentially at RF frequencies, many analog designs (e.g. active filters) use transconductance tuning to achieve frequency-response precision. In these cases the capacitance tolerancing can be compensated for.

MOS capacitors use a MOS transistor with a parallel-plate capacitance between the gate of a MOS transistor and a heavily inverted channel. The drain and source are connected in this configuration and the separation between the 'conductors' is thin being the gate oxide thickness. As technology progresses this will continue to thin. This leads to high values of capacitance although with weak voltage dependence. Junction capacitance is realized as the capacitance of a reverse-biased semiconductor junction. This capacitance can be quite large, but has a strong voltage dependence.

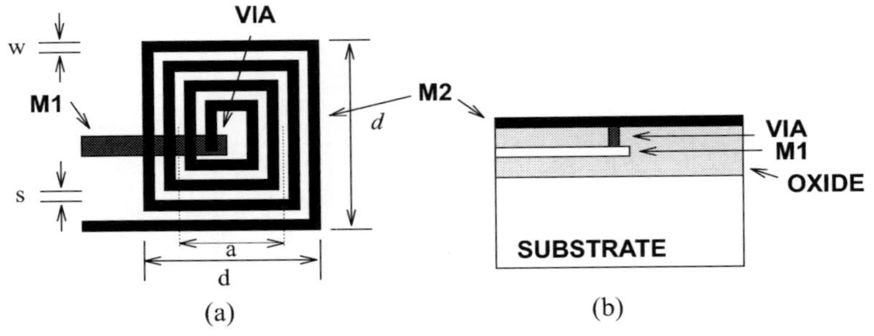

Figure 10.1 An on-chip spiral inductor: (a) plan view; and (b) side view.

This voltage dependence can be utilized to realize tunable circuits, for example, a voltage controlled oscillator.

Lengths of low impedance transmission line can also be used to realize narrowband capacitances. These are discussed in Section 10.3.3 on page 361 with reference to Figure 10.17 on page 365. The line lengths required are generally too long for integrated circuits, except at millimetre-wave frequencies (above 30 GHz), and so are most commonly used off-chip.

10.1.3 PLANAR INDUCTORS

Inductors are important components in RF and microwave circuits. In addition to their role in matching networks, they are used to provide bias to active devices while effectively blocking RF signals from the bias circuitry. Inductors of up to 10 nH can be fabricated on-chip. Values above this consume too much die area and either the entire inductance, or the majority of a required inductance, are fabricated off-chip, more often on an LTCC substrate or as part of the RFIC package. Bond wires can also be used to realize small inductances in the 0.5–1 nH range. The advantage of having a portion of a large inductance on-chip is reduced sensitivity to die attach (bond-wire, etc.) connections.

An on-chip spiral inductor, the most common type, is shown in Figure 10.1 in both plan and side views. An expression for the inductance of this structure was developed by Wheeler [369]:

$$L \approx \frac{9.4\mu_o n^2 a^2}{11d - 7a} \tag{10.1}$$

where a is the mean radius of the spiral, and n is the number of turns. This formula was derived for circular coils, but its accuracy for square spirals has been determined by Lee [363] to be within 5% of values derived using electromagnetic simulation. It is therefore a very useful formula in the early stages of design, but electromagnetic analysis is required to obtain the necessary accuracy in design.

Fields produced by the spiral inductor penetrate the substrate, and as a ground plane is located at a relatively short distance, the eddy currents on the ground plane reduce the inductance that would otherwise be obtained. The eddy current in the ground conductor rotates in a direction opposite to that of the spiral itself (i.e. counterclockwise opposing clockwise). As a result, the inductance of the image inductor in the ground is in the opposite direction to that produced by the spiral itself, with the consequent effect that the effective total inductance is reduced. By creating a broken, or perforated, conductor pattern the ground inductance is largely eliminated [370]. Thus, with care, reasonably good inductors, with high Q, can be realized on GaAs chips as the GaAs substrate is semi-insulating. The situation is different with silicon because of the finite conductivity of silicon substrates, which therefore supports eddy currents.

One of the major sources of loss for inductors on silicon is substrate loss due to the finite conductivity of the substrate and the resulting current flow. These induced currents follow a path under the conductors of the spiral and, just as with ground plane eddy currents, lower the inductance achieved, Mernyei *et al.* [371] proposed a scheme to reduce substrate eddy currents. Most silicon substrates are at least slightly doped, usually of p-type. With heavily doped n-type strips arranged radially from the centre axis of the spiral, the eddy currents are blocked. They achieved an increase of the Q from 5.3 to 6.0 at 3.5 GHz and from 4.3 to 4.5 at 2.0 GHz. In the spiral inductor model, the effect was also to reduce the shunt resistance and capacitance to the substrate.

Parasitic capacitance results in resonance of the on-chip inductance structure and hence the frequency of operation. With GaAs ($\varepsilon_r = 12.85$) the effective permittivity of the medium can be reduced by adding a polyimide layer ($\varepsilon_r = 3.2$) and using metallization on top of this layer [372]. Thus, the capacitance is substantially reduced. This is at the cost of poorer thermal management as the thermal conductivity of polyimide is substantially lower (about 100 times) than that of GaAs, and so the power handling capability is compromised. Coupling this with thicker metallization to reduce resistance can result in a Q that is 50% larger and a self-resonant frequency that is 25% higher [372].

The properties of on-chip inductor realizations are summarized in Table 10.3.

We now return to the issue of the finite conductivity of the silicon substrate. The inducement of charges in the silicon and the insignificant skin depth of the silicon substrate has the effect of increasing the capacitance of an interconnect line over silicon as the electric field lines are terminated on the substrate charges. (This effect is in addition to the induction of eddy currents in the substrate as discussed earlier.) Now the magnetic field lines penetrate some distance into the substrate so that the LC product is greater than if the substrate was insulating (as with GaAs). The effect is that the velocity of propagation along the interconnect ($= 1/\sqrt{LC}$) is reduced, leading to what is called the slow wave effect. This means that very small inductances can be realized using short lengths of interconnect arranged so that fields, particularly the magnetic field, penetrates the substrate. This effect can be adequately simulated using planar electromagnetic simulators that allow the conductivity of media to be specified. Simulation is necessary as the effect is a complex function of geometry and substrate conductivity, and generalizations available for use in design are not available.

The lowest loss inductors are obtained by etching away the underlying substrate or by using insulating or very high resistivity bulk material. Loss is also reduced by separating the planar inductor from the ground plane and additional dielectric layers have been deposited on a chip to achieve this [390]. When all steps have been taken then the dominant loss mechanism is current crowding [393]. This is a particular problem with multi-turn inductors which are required to realize high inductance

Table 10.3 Inductance and Q of planar inductors: d is the side-dimension shown in Figure 10.1(a), n is the number of turns, L and Q_{max} are at f_0, f_0 is the operating frequency, and f_{SR} is the self-resonant frequency. Mn indicates the metal layer used. Adapted in part from Burghartz [373].

Type	Ref. year	d (μm)	n	L (nH)	Q_{max} @f_0	f_0 (GHz)	f_{SR} (GHz)
M2 on Si	[370] 1997	280	6	8	7	1.8	3.5
M2 on Si	[371] 1998	–	2	1.8	6.0	3.5	–
M2 on Si	[371] 1998	–	3	2.8	4.8	2.0	–
M2 on GaAs	[372] 2000	210	11	1	44	11.5	21
MLS on GaAs	[372] 2000	440	23	6.7	33	6.4	7
M3/M2/M1[1] on p^--Si	[374] 1995	226	3	1.45	24	2.3	> 20
MLS[2] on p^--Si	[375] 1996	226	8	6.8	7.8	1.3	5.6
M3 on p^--Si	[376] 1996	400	6	10.2	7.5	1	8
M3/M2/M1 on n-well Si	[377] 1997	145	$9\frac{1}{4}$	6	2.6	1.5	5.8
MLS[2] on p^--Si	[378] 1995	150	3	15.5	2.64	0.42	–
M2 on p^--Si	[379] 1990	230	9	9.7	3	0.9	2.5
M3/M2/M1 on p^--Si	[380] 1995	226	4	2.1	9.3	2.4	20
6 μm Au on HR Si[3]	[381] 1994	300	2	2.88	12	3.2	> 10
2.5 μm on sapphire	[382] 1996	226	3	1.45	40	5.8	> 20
4 μm Cu on quartz	[383] 1998	226	3	1.5	60	5	> 20
M2 on suspended substrate	[384] 1983	440	20	100	–	–	2.9
3.5 μm Al on glass	[385] 1997	600	2	2.6	39	4.3	10.5
M2 on SOS[4]	[386] 1996	150	5	4	11.9	3.0	13.9
4 μm Al on HR Si[3]	[387] 1997	290	8	12.0	15.3	3.25	9.0
M2 on suspended substrate	[388] 1996	200	4	1.1	28	18	84
1.5 μm Cu suspended above cavity on Si	[389] 2000	265	7	10.4	30	8.0	10.1
4 μm AlCu on polyimide on Si	[390] 2000	–	$3\frac{1}{2}$	1.1	18	10	–
3 μm Al on ferro-magnetic thin-film on Si	[391] 2000	337	$4\frac{1}{4}$	7.2	7	1	1.3
20 μm Al on Si	[392] 2000	500	$2\frac{1}{2}$	4.2	21	2.47	8.1
20 μm Al on Si	[392] 2000	170	10	12.4	9	1.82	6.84

[1]Metallization levels of the multi-level interconnects.
[2]MLS: A MultiLayer Stacked coil inductor made using multi-level interconnects.
[3]HR Si: High Resistivity silicon.
[4]SOS: Silicon On Sapphire.

values. Current crowding results when the magnetic field of one turn penetrates an adjacent trace creating eddy currents so that current peaks on the inside edge of the victim trace (towards the centre of the spiral) and reduces on the outside edge. This constricts current and results in higher resistance than would be predicted from skin effect and DC resistance alone [393].

Planar inductors are often fabricated in close proximity to each other. The coupling of adjacent planar inductors depends on the separation of the inductors, shielding, geometry and the resistivity of the underlying substrate. An effective measure of shielding is to use a discontinuous guard ring [394]. This however reduces the originally designed inductance values because of the image currents induced in the ring [395].

Lengths of high impedance transmission line can also be used to realize narrowband inductance. These are discussed in Section 10.3.3 on page 361 with reference to Figure 10.15 on page 363. The line lengths required are generally too long for integrated circuits, except at millimetre-wave frequencies (above 30 GHz), and so are most commonly used off-chip.

10.2 TERMINATIONS AND ATTENUATORS IN MIC TECHNOLOGY

In waveguides or coaxial lines it is usually preferable for short-circuits to be realized rather than open-circuits, since relatively high-quality short-circuits can be manufactured, and the effective location of an open-circuit is hard to define and moves with frequency. In microstrip the opposite situation applies, open-circuits being in far greater use than short-circuits, although neither form of termination can be made with coaxial or waveguide quality due to the hybrid fields associated with microstrip. These terminations are discussed in greater detail in Sections 7.2 and 7.4. (on pages 227 and 233 respectively). The application of open-circuit terminations will be seen in conjunction with filters and other passive circuits, which are dealt with later on in this chapter and active circuits are considered in Chapter 11.

With some care in fairly simple circuit designs quite broadband matched terminations can be realized. For frequencies up to about 12 GHz a simple rectangular resistor deposition is suitable, and may be expected to yield a VSWR below 1.1. The actual deposition may be thick or thin film. The resistor is kept in-line with the microstrip, and the far end is shorted to ground via 'wrap-around' metallization — an approach that is suited to MICs, although not for MMICs.

At frequencies above about 12 GHz series-inductive reactance becomes an increasingly severe problem, and the resistor element has to be viewed as a series LR network — at least to a first approximation. Lacombe [396] devised a fairly simple broadband matching arrangement for a 50 Ω resistor to form a 50 Ω microstrip termination with VSWR < 1.5 over the range $1 < f < 18$ GHz. The details will not be outlined here, because considerably improved designs for matched terminations have been developed more recently. These improved designs rely upon attenuator structures, which are therefore considered first.

Rickard [71] has described a π-type microwave attenuator, manufactured in thick-film microstrip. The electrical circuit configuration and the microstrip layout are both indicated in Figure 10.2. Resistance values are $R_s = 17\ \Omega$ and $R_p = 290\ \Omega$, both with $\pm$ 15% tolerance. To maintain lumped behaviour, the resistances are made

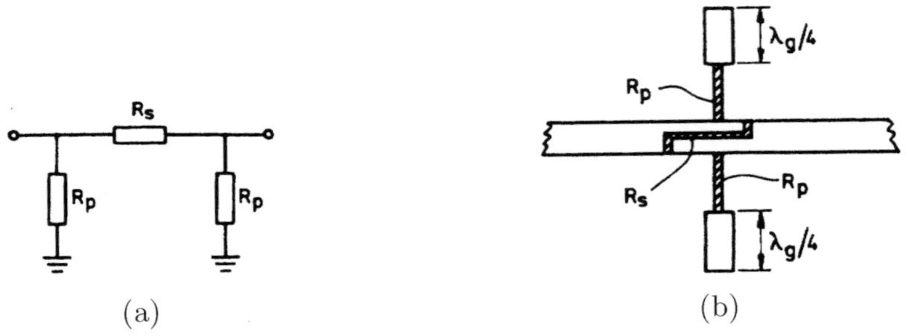

Figure 10.2 Network and microstrip circuit layout for a π-type attenuator: (a) circuit topology; and (b) microstrip circuit.

less than an eighth-wavelength long at the highest frequency (i.e. less than 1 mm). The interconnecting resistor, R_s is formed as a 'sideways S' shape within the width of the microstrip. The resistors R_p are effectively grounded by using quarter-wave open-circuit transforming stubs, each having a characteristic impedance of 30 Ω. This attenuator had a final performance, over the frequency range $8 \leq f \leq 20$ GHz, as follows:

Attenuation	3 dB ±0.2 dB
VSWR	< 1.1

Such attenuators have several important applications. For example, they are used in broadband power-levelling loops where the microwave output power is maintained constant over a wide range of frequencies.

Finlay *et al.* [397] have described a precision microstrip matched load using two 3 dB attenuator pads of a slightly different basic design to that described here (actually 'pseudo-tee'). These two pads are followed by a 50 Ω resistor which is grounded at its far end. The length of microstrip line between the pads is approximately $\lambda_g/4$ at 15 GHz and between the final pad and the resistor is approximately $\lambda_g/16$ at 15 GHz. An excellent performance is obtained, load VSWR being less than 1.1 for all frequencies up to 18 GHz. This is somewhat degraded by the SMA transition to the microstrip line, which results in a VSWR below 1.2 up to 13 GHz, but also a peak of 1.22 at about 15 GHz. Such performance is compatible with most requirements likely to be encountered in microstrip circuits.

10.3 FURTHER THICK AND THIN FILM MIC PASSIVE COMPONENTS

Thick film passive components are used in hybrid MICs and LTCC-based packages. Terminations and attenuators are actually very basic forms of MIC passive circuit components. In the following sections of this chapter we examine other, more complex, types of passive structures developed in hybrid MIC format that may use either thick or thin film realization.

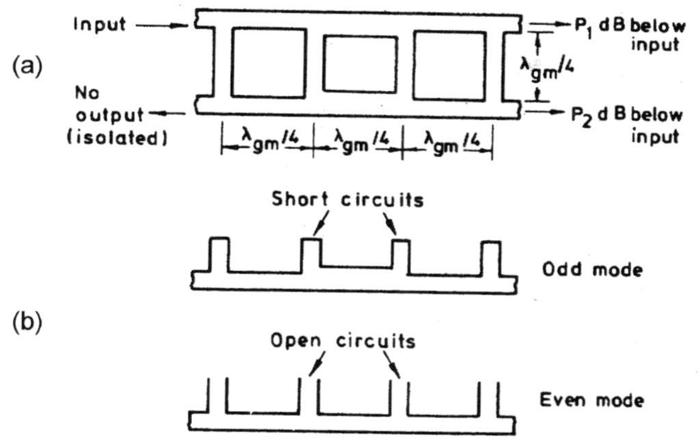

Figure 10.3 Branch-line directional coupler in MIC form: (a) physical layout; and (b) odd- and even-mode equivalent circuits. (Reproduced by permission of Wiley Eastern Limited, from Gupta and Singh [398].)

10.3.1 BRANCH-TYPE COUPLERS AND POWER DIVIDERS

Branch-type couplers have DC as well as microwave coupling which is a significant difference compared to the characteristics of the couplers considered earlier. Although the inherent bandwidths of branch-type couplers are not particularly large (typically up to about 50%), they can be used for comparatively high microwave power transmission.

A branch-line directional coupler is shown in Figure 10.3(a), with the 'even' and 'odd' mode equivalent 'circuits' given in Figure 10.3(b). The terms 'even' mode and 'odd' mode refer to the overall effects of having voltage nodes or voltage antinodes present at the branch- or main-line junctions, respectively. In Figure 10.3(a), λ_{gm} is the mid-band wavelength in the microstrip line.

Particularly significant properties of these couplers are:

(a) Special-purpose chokes or filters may be connected into the joining branches, especially if these are made (say) $3\lambda_{gm}/4$ in length rather than only $\lambda_{gm}/4$.
(b) High microwave power may be handled with very little danger of breakover.
(c) They are well suited to tight coupling requirements: 0 dB is feasible over a 50% bandwidth for a six-branch coupler, provided that a VSWR of about 1.2 is tolerable. (A fourteen-branch arrangement gives a VSWR of 1.05, but naturally occupies much more substrate area.)
(d) They may be used for the purpose of impedance transformation [208].

'Standard' Designs

Design techniques appropriate for stripline have been described by Howe [400]. These may be translated into the microstrip form by following the procedures given in Chapters 4 and 5, and Sections 7.10, 7.11 and 7.14.7 (on pages 244, 247 and 266) respectively) for the T-junctions.

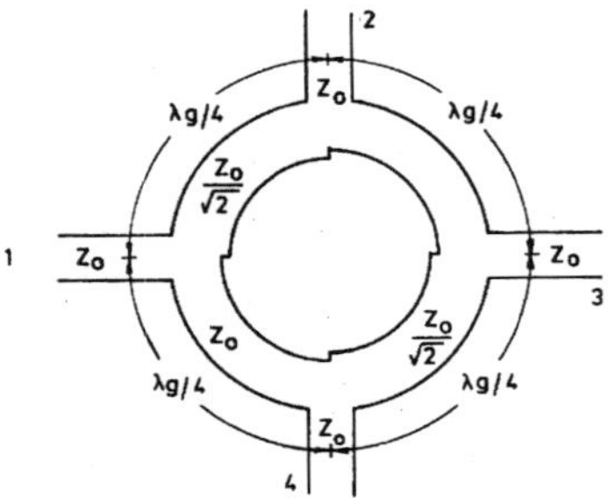

Figure 10.4 Ring form of 3-dB branch-line directional coupler in MIC form. (Reproduced by permission of Wiley Eastern Limited, from Gupta and Singh [398].)

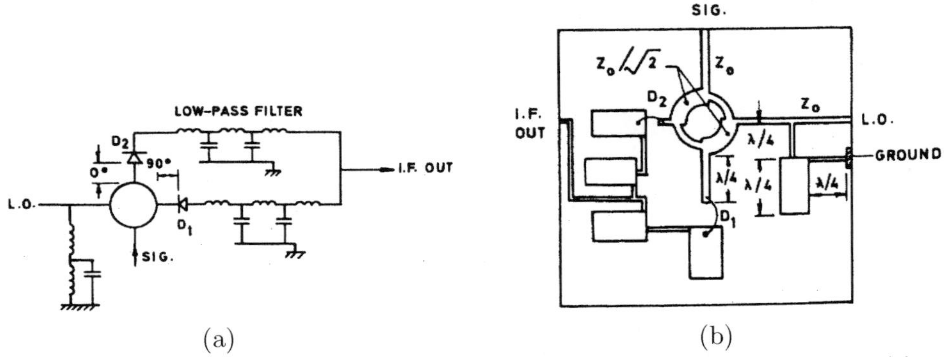

Figure 10.5 Balanced mixer circuit based upon the structure of Figure 10.4: (a) schematic layout; and (b) microstrip circuit. (Reprinted by permission of Wiley Eastern Limited, from Gupta and Singh [398].)

When only four branching sections are required, this branch-line coupler may be reformed into a ring shape; this is shown in Figure 10.4. One important application for this arrangement is offered by the balanced mixer (and also phase detector) illustrated in Figure 10.5. It can be readily verified, by counting up the total number of quarter-wavelength phase shifts in each path, that the signal and local oscillator waves arrive at each Schottky-barrier diode in relative antiphase. This is a design requirement for this type of mixer. Unwanted waves are filtered out by the various low-pass filters shown on the circuit. The design of such filters is described in Section 10.3.3. on page 361. The use of this 3 dB ring coupler enables both the signal and the local oscillator powers to be coupled to the mixer diodes with very little loss, resulting in high sensitivity.

It is also worth considering the so-called 'rat-race' circuit, or 'hybrid-ring', which is shown in Figure 10.6. Output signals from ports 2 and 4 differ in phase by 180° (in contrast to the branch-line coupler, where the phase difference is 90°). An interesting and important design feature arises by considering the quarter-wave transformer action of this coupler. Only ports 2 and 4 exhibit this action, because port 3 is half-wave separated from the input feeding port 1. Thus, the net effective load on the inner ring lines feeding ports 2 and 4 amounts to $2Z_0$ (two Z_0 loads appearing,

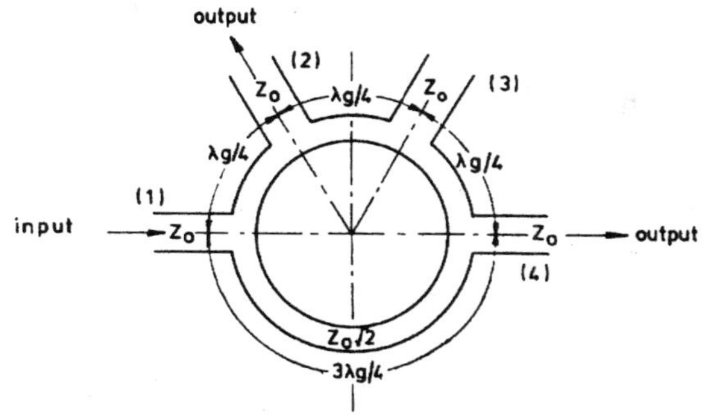

Figure 10.6 'Rat-race' or 'hybrid-ring' coupler. (Reproduced by permission of Wiley Eastern Limited, from Gupta and Singh [398].)

equivalently, in series). Now, the characteristic impedance Z_0 of any quarter-wave-transforming line between two impedances Z_{01} and Z_{02} is known to equal $\sqrt{Z_{01}Z_{02}}$; this result was developed in Section 1.7 on page 15. In this case, the two impedances are Z_0 and $2Z_0$ respectively, so the impedance of the intervening quarter-wave line (i.e. the ring) must be

$$Z_0' = \sqrt{Z_0 \cdot 2Z_0} \tag{10.2}$$

or

$$Z_0' = \sqrt{2}Z_0. \tag{10.3}$$

Thus, the characteristic impedance of the line forming the ring itself must be $\sqrt{2}$ times that of the feeder line impedances. Where the impedance of all feeders is 50 Ω, the ring characteristic impedance is therefore 70.7 Ω.

Broadband and Millimetre-Wave Branchline Couplers

Considerable effort continues to be expended towards improving the design flexibility, expanding the bandwidth and extending the upper frequency capability of branchline and similar couplers. For example, Wright and Judah [401] describe a particularly broadband hybrid ring with flat coupling, and Ashworth [402] developed design expressions for a two-branch broadband hybrid ring (or a two-branch guide) coupler of any coupling coefficient.

Mayer and Knochel [403] describe branchline couplers with improved design flexibility and broad bandwidth. They show that, for a simple four-port branchline coupler, the general impedances Z_x of the 'inner coupler' are given by:

$$Z_x/Z_0 = \{1 + S_{11}(f_o)\}/[1 - S_{11}(f_o)] \tag{10.4}$$

and hence the specific impedances are given by

$$Z_2/Z_0 = \sqrt{-\frac{3}{2}\left(1 - \frac{Z_1}{Z_0}\right) + \sqrt{\frac{9}{4}\left(1 - \frac{Z_1}{Z_0}\right)^2 + \frac{Z_1}{Z_0}}} \tag{10.5}$$

and

$$Z_1/Z_0 = Z_2\Big/(Z_o\sqrt{2}). \tag{10.6}$$

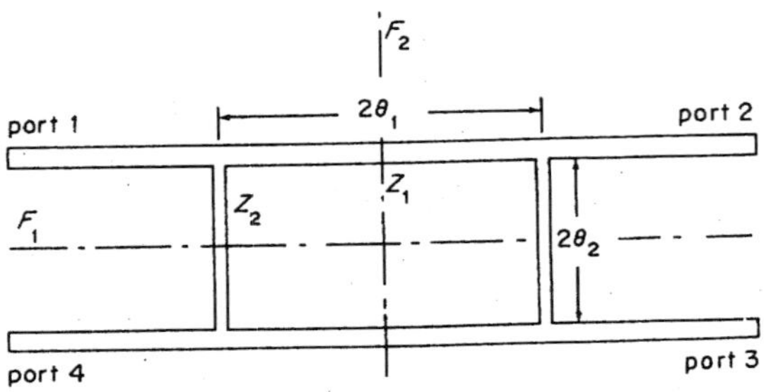

Figure 10.7 Two-branch coupler with two-fold symmetry ($\theta_{1,2}$= line lengths, $Z_{1,2}$ = characteristic impedances). (© 1990 IEEE. Reprinted, with permission, from Mayer and Knochel [403].)

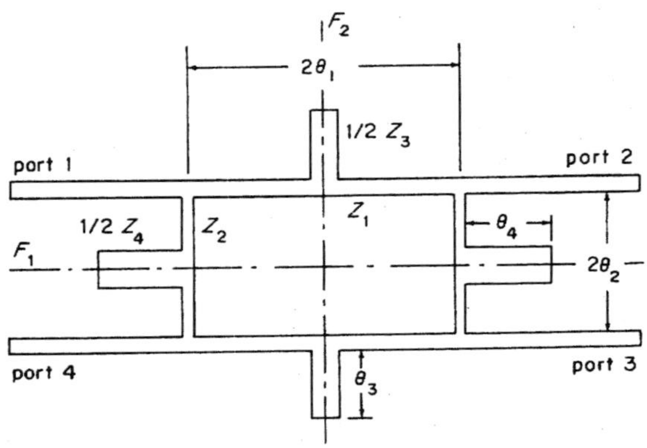

Figure 10.8 Extended two-branch coupler with open stubs at the symmetry-planes. (© 1990 IEEE. Reprinted, with permission, from Mayer and Knochel [403].)

These impedances refer to a branchline coupler of similar layout to one section of the device shown in Figure 10.3(a). The inner coupler is that bound by the first series and parallel arms, respectively, and this is also shown in Figure 10.7. Mayer and Knochel [403] also describe a two-branch coupler modified by the addition of half-wavelength open-circuited stubs centrally located on each inner arm as shown in Figure 10.8. The way this approach serves to broadband the coupler amounts to relocating, each side of the original centre frequency, two frequencies at which S_{11} is a maximum. Further details are described in the reference.

Mayer and Knochel show that it is possible to increase the bandwidth from 10% with an unmodified coupler to more than 40% with the stub-modified device, whilst maintaining very flat amplitude characteristics.

These researchers also describe some more sophisticated broadband couplers and high-power in-phase power dividers [404]. An example of a broadbanded 0°/180° coupler is shown in Figure 10.9, where the stubs are clearly visible. A folded stub is used (Z_6) for one of the inner arms of this design. The high-power dividers described use the all-reactive broadband principle already outlined and therefore avoid the problem of the resistive element required in the Wilkinson divider, and the difficulty of restricted bandwidth associated with the Gysel circuit. Bandwidths around 50% are

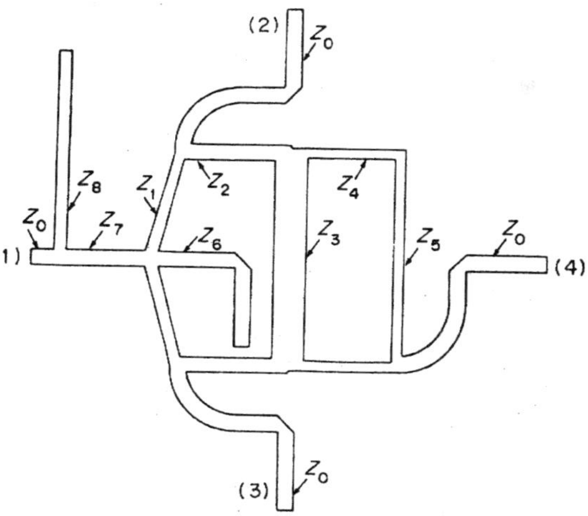

Figure 10.9 Broadbanded 0°/180°-coupler. (© 1990 IEEE. Reprinted, with permission, from Knochel and Mayer [404]).

readily achievable. Others have also described various types of hybrid power dividers, but usually with resistive elements.

The branchline coupler principle is by no means restricted to microwave (below 30 GHz) applications and, for example, Meaney [405] has described a novel design for millimetre-wave systems. Meaney points out that there are some problems with the direct use of microstrip in these types of application. In particular, there is the restricted range of realizable characteristic impedances, mentioned several times earlier in this text. There is also the fact that as the frequency increases, so the aspect ratios of quarter-wavelength lines decrease, to the point where the physical lengths become shorter than the widths. This effect causes degradation of input match, bandwidth and isolation. There are also problems associated with mutual coupling and discontinuities.

By using curved inner series lines, Meaney's design enables all input ports to be mutually orthogonally disposed and the discontinuities are also mostly eliminated. This coupler is based upon a three-branch structure to take advantage of the broadbanding effect of this topology, and it uses three quarter-wavelength series lines together with quarter-wavelength parallel lines. Since the relatively long series lines are those with the lowest characteristic impedance, and therefore are the widest, so their aspect ratios are maintained at reasonable values. The coupler provides a nominal 3 dB insertion loss (variations between 2.7 and 4.7 dB) with good return loss and isolation performance over the 26–40 GHz band.

When using curved microstrip lines it is important to account accurately for the effect of line width. For example, Roy *et al.* [406] have shown that the design of rat-race hybrids is influenced markedly by the line width, and they develop an expression for a suitable modification of this width allowing accurate design. Branchline couplers and other similar circuit structures are readily amenable to CAD.

Other recent developments centre on MMIC realization of couplers and power dividers (the topic of MMICs is considered in Chapter 11).

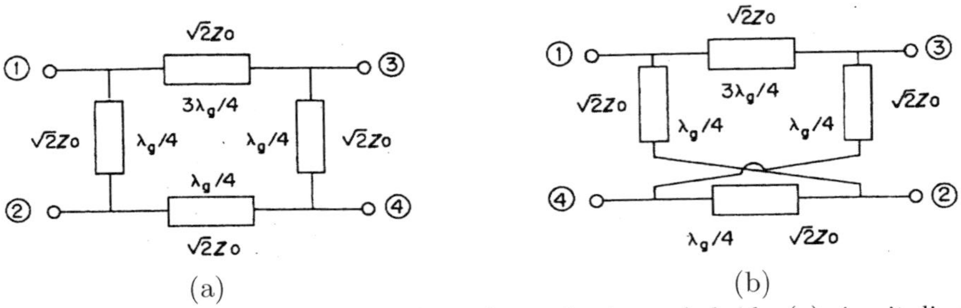

Figure 10.10 Conventional and port-interchanged rat-race hybrids: (a) circuit diagram of a conventional rat-race hybrid; (b) circuit diagram of a port-interchanged rat-race hybrid. (Reprinted, with permission, from Nakamoto, Tokumitsu and Aidawa [408].)

Tokumitsu *et al.* [407] described a novel MMIC realization approach using three micron/three layer dielectric films. They describe layered microstrip (thin film microstrip, or TFMS), inverted microstrip and also both normal and inverted 'quasi-coplanar' transmission structures. They designed and built 12 GHz four-port Wilkinson power divider circuits using vertical connections, 15 GHz branchline couplers and 20 GHz port-interchanged rat-race hybrids (using a line crossover) — all based on this technology. A more detailed exposition of the port-interchanged rat-race hybrid, shown with comparison with the conventional topology in Figure 10.10, is presented in a separate paper by this group [408], where it is pointed out that the new version provides greater MMIC design flexibility. In addition, Tokumitsu *et al.* describe a 6-stage distributed amplifier using TFMS.

10.3.2 MICROSTRIP BALUNS

The balun, a balanced-to-unbalanced transmission converter, was invented by Marchand in 1944 and has been available commercially in (mainly) coaxial form for several decades. MIC realizations of this type of circuit element have only emerged in comparatively recent times. Interest is stimulated primarily by the specific circuit requirements of mixers, multipliers, Class B push-pull amplifiers, differential RF paths on RFICs, and antennas where balanced transmission line structures are needed to feed parts of the circuit, whereas unbalanced lines must then form the remainder of the circuit.

Baluns have been designed for both MIC and MMIC implementation. For example, Pavio and Kikel [409] report a compensated balun which provides an amplitude response that is flat within 1 dB over the 6–18 GHz bandwidth (both in hybrid and monolithic forms). Phase matching between the output arms is very good, being $180° \pm 2°$. The circuit construction is shown in Figure 10.11, in which it can be seen that suspended substrate technology is avoided and the transformation is achieved by means of a multilayer conductor structure. Tight coupling is obtained by using a very thin second dielectric layer t_n and, for the MMIC balun, this is realized with a 2-micron thick layer of silicon nitride on the GaAs surface.

Barber [410] has described enhanced coupled, even mode terminated, baluns and also mixers using this technology. Barber's approach starts with the creation of a planar balun circuit cell which comprises a pair of coupled transmission lines, bridged

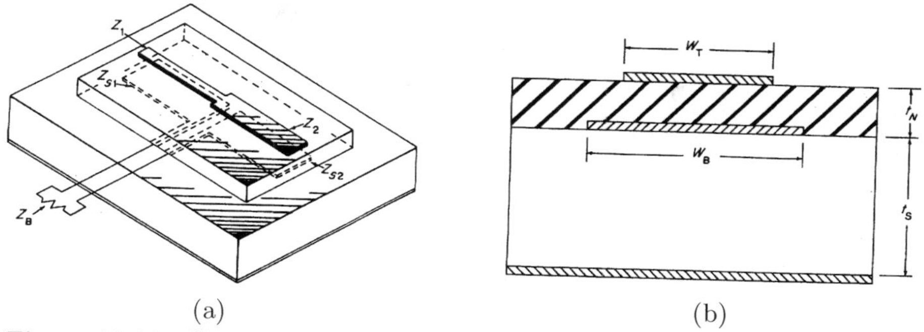

Figure 10.11 Circuit construction and cross-sectional views of baluns: (a) circuit construction of monolithic or hybrid balun; and (b) typical balun structure cross-section. (© 1990 IEEE. Reprinted, with permission, from Pavio and Kikel [409].)

at their outputs by a series-connected pair of capacitors. A final transmission line is connected from the centre point between the capacitors to a load resistor which is grounded. This overall structure forms one balun cell, the circuit and physical cross-section of which is shown in Figure 10.12. To form a complete broadband balun typically eight of these cells are cascaded. One of the two input lines is fed whilst the opposite line is grounded (i.e. unbalanced), and the two outputs are left floating in balanced format. The capacitors are formed in overlay technology and the appropriate line interconnections are made with vias.

These baluns are designed to operate over the 2–26 GHz band, and Barber describes mixers (restricted to the 2–18 GHz band) using this design approach. For the mixers the IF port is formed by diplexing the local oscillator and IF frequencies using reactive filtering. A Schottky monolithic quad of diodes forms the mixing function and the mixers exhibit conversion loss varying between 7 and 10 dB over the band. The entire mixer, incorporating a total of 16 balun cells, occupies a 6 mm square package.

10.3.3 A STRATEGY FOR LOW-PASS MICROWAVE FILTER DESIGN

We shall first review the overall design strategy that may be followed for any microstrip filter. Figure 10.13 shows the typical sequence that starts with a prototype specification (by the insertion-loss approach), and finishes with a practical, fully dimensioned, microstrip circuit.

Low-Pass Filters Formed with Cascaded Microstrips

For this design example we shall employ cascaded sections of microstrip, as indicated against 3(1) of Figure 10.13, although this technique is only really useful at frequencies up to several gigahertz. It is first necessary to analyse the direct substitution of lumped filter elements by such microstrip lengths.

Short ($< \lambda g/4$) lengths of relatively high impedance line will behave predominantly as series inductances. Also, a very short ($\ll \lambda g/4$) length of relatively low impedance line will act predominantly as a shunt capacitance. So a π-network of lumped elements can quite readily be realized with alternate sections of low- and high-impedance

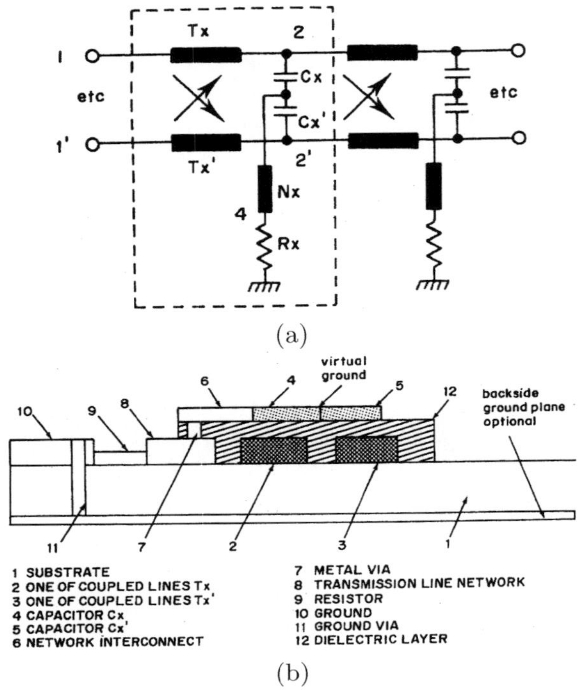

Figure 10.12 Planar baluns: circuit cell and cross-sections: (a) planar balun circuit cell; and (b) cross section of a circuit cell. (© 1990 IEEE. Reprinted, with permission, from Barber [410].)

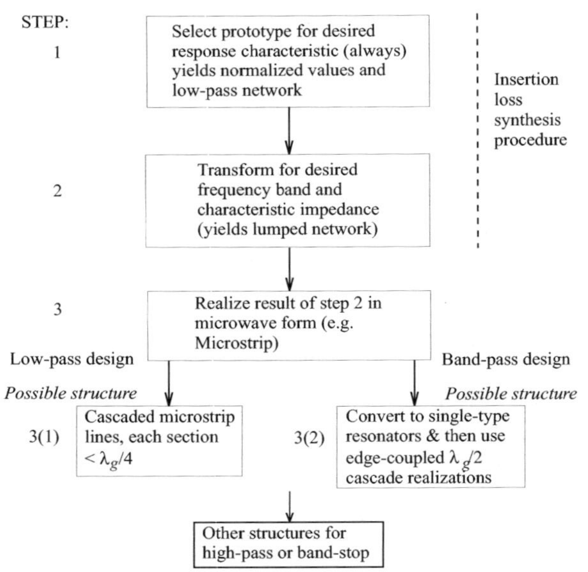

Figure 10.13 Procedure for filter design using cascaded microstrips.

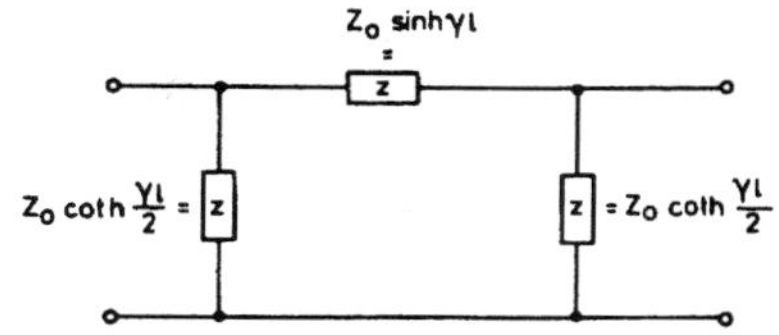

Figure 10.14 Equivalent π-network of impedances representing a length l of transmission line having a propagation coefficient γ.

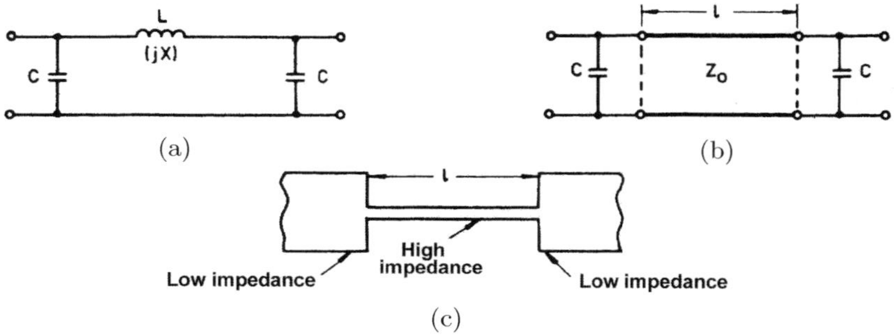

Figure 10.15 Inductive length of line with adjacent capacitive lines: (a) lumped circuit; lumped-distributed equivalent; and (c) microstrip form.

microstrip lines, except that great care is necessary to fully calculate the equivalent line lengths. In particular, consider the π-equivalent arrangement applicable to a section of any transmission line. This was given in Section 1.8 on page 17, and is repeated here as Figure 10.14.

In order to determine the characteristics of the microstrip elements to be implemented we need to evaluate both inductive and capacitive lines. Let us first consider a predominantly inductive length of line as in Figure 10.15. For short ($< \lambda_g/4$) and loss-free lines this reduces to the two equivalent circuits shown in Figures 10.15(a) and 10.15(b). Basic transmission line theory gives the input reactance of the line of length l:

$$X_L = Z_0 \sin(2\pi l/\lambda_g) \tag{10.7}$$

so that the length of this predominantly inductive line is

$$l = \frac{\lambda_g}{2\pi} \sin^{-1}\left(\frac{\omega L}{Z_0}\right). \tag{10.8}$$

Occasionally, the small-angle approximation can be used such that $2\pi l/\lambda_g \ll \pi/4$ and Equation (10.8) simplifies to

$$l \approx f\lambda_g L/Z_0. \tag{10.9}$$

It must be emphasized that the small-angle inequality will frequently not be met, and therefore the full Equation (10.8) will often be necessary in practice. The end susceptances B_L for the equivalent circuit, yielding values of C_L, are also given by

$$B_L = \frac{1}{Z_0} \tan\left(\frac{\pi l}{\lambda_g}\right). \tag{10.10}$$

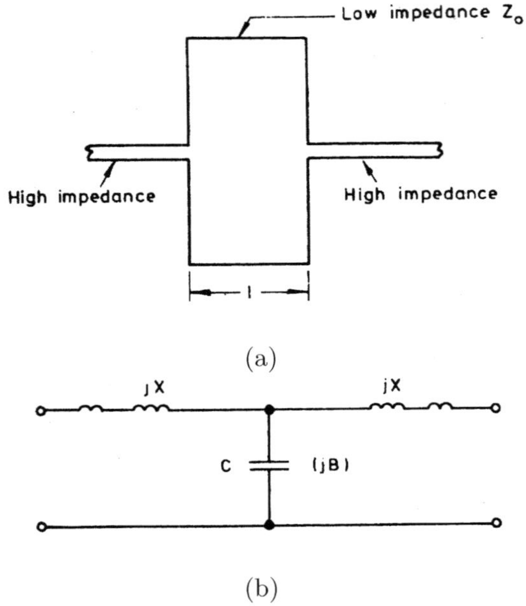

Figure 10.16 Capacitive length of line with adjacent inductive lines: (a) microstrip; and (b) lumped equivalent.

Then, approximately, for short lengths:

$$C_L \approx l(2fZ_0\lambda_g). \tag{10.11}$$

The inductive length of line l will also be presented with adjacent lengths of low impedance and, therefore, wide microstrip lines. This situation results in a significant step discontinuity; the properties of such discontinuities were discussed in Section 7.8 on page 240. The expressions for step capacitance, which were given in that section, should be used to evaluate the discontinuity capacitances here.

We have previously remarked that a short length of line having a relatively low characteristic impedance yields a capacitive element and this is shown, together with its equivalent circuit, in Figure 10.16. The predominating shunt capacitance is determined by first considering the susceptance

$$B = \frac{1}{Z_0} \sin\left(\frac{2\pi l}{\lambda_g}\right) \tag{10.12}$$

so that

$$l = \frac{\lambda_g}{2\pi} \sin^{-1}(\omega C Z_0) \tag{10.13}$$

or, usefully, since these line lengths are almost always very short:

$$l \approx f\lambda_g Z_0 C \tag{10.14}$$

Since this low-impedance length of line has the equivalent T-network shown in Figure 10.16 it must incorporate 'stray' series inductances. These inductance values

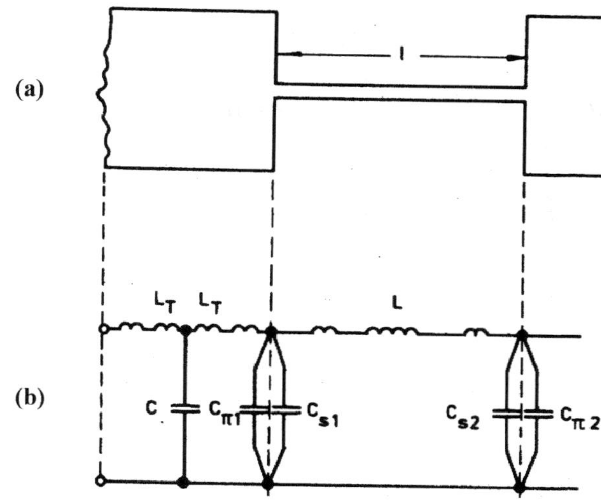

Figure 10.17 Inductive length of microstrip line with adjacent capacitive lines: (a) microstrip line; and (b) lumped equivalent circuit including discontinuity (step) capacitances and inductances.

will now be shown to be usually negligible. We have

$$X_L \approx \pi l Z_0 / \lambda_g. \tag{10.15}$$

Now both Z_0 and l are relatively small, so this complete reactance will only be a very small correction to the main inductive reactance of the adjacent inductive length of line.

The capacitance actually calculated for the capacitive length of line must in fact be less than the required total lumped value C_T because the inductive sections of line and the steps in width already contribute their capacitances, C_L and C_s respectively. Since these capacitances will generally differ due to the differing inductive lengths we finally have, for the capacitive line,

$$C_{\text{line}} = C_T - C_{L1} - C_{L2} - C_{s1} - C_{s2} \tag{10.16}$$

where C_{L1} and C_{L2} are obtained from Equations (10.10) or (10.11), applied to each adjacent inductive line, and C_{s1} and C_{s2} are determined using the expressions given in Section 7.8 of Chapter 7 on page 240. A more complete equivalent circuit for a C-L-C portion of a filter is given in Figure 10.17. At frequencies where these structures are useful, generally only up to a few gigahertz, inductive effects due to the step discontinuities can usually be neglected.

Table 10.4 presents a summary chart showing the results applicable to inductive and capacitive elements realized in the form of microstrip lines.

10.3.4 BANDPASS FILTERS

This section deals mainly with an approach to bandpass microstrip filter design by the use of a cascade of half-wave resonators — each quarter-wave parallel coupled to

Table 10.4 Summary chart for inductive and capacitive elements formed using microstrip lines.

Mainly inductive element		Mainly capacitive element
From Equation (10.7): $X_L = Z_{0L} \sin(2\pi l L/\lambda_{gL})$	and	From Equation (10.7): $B_C = \frac{1}{Z_{0C}} \sin(2\pi l C/\lambda_{gC})$
End-capacitances: $C_L \approx \frac{l_L}{2fZ_{0L}\lambda_{gL}}$, via Equation (10.11)	and	End-inductances† $L_C \approx \frac{l_C Z_{0C}}{2f\lambda_{gC}}$ via Equation (10.15)
Main inductive length: $l_L = \frac{\lambda_{gL}}{2\pi} \sin^{-1}(2\pi f L/Z_{0L})$ via Equation (10.8)		Main capacitive length: $l_C = \frac{\lambda_{gC}}{2\pi} \sin^{-1}(2\pi f C Z_{0C})$ via Equation (10.13)
For short lengths only $l_L \approx f\lambda_{gL} L/Z_{0L}$ (via Equation (10.9))		Since lengths are usually short $l_C \approx f\lambda_{gC} Z_{0C} C$ (via Equation (10.14))

†Usually negligible.

its neighbours. This type of filter yields a narrow-to-moderate bandwidth; structures providing a broader bandwith are considered towards the end of the discussion.

Before proceeding with this we must study some preliminaries, beginning with the transformed (lumped-element) bandpass filter (BPF) network shown in Figure 10.18. The designer would arrive at this at the end of step 2 calculations, as indicated in Figure 10.13. Unfortunately, a serious difficulty is encountered with this network in its present form.

It is evident that direct realization of this filter requires the design and interconnection of two *distinctly different kinds* of resonant LC circuits (hereafter referred to as 'resonators'). Both series and parallel types of resonators would be required in any attempt to directly realize this filter. Any microstrip resonators which could be used and electromagnetically coupled together in some way would individually exhibit one type of resonance — series or parallel — but not both simultaneously from similar structures.

There are two principal approaches to planar coupling between adjacent resonators: end coupling or edge (or 'parallel') coupling. We shall consider both approaches here, but more space is given to edge-coupled BPFs, because these are usually more important.

End-Coupled Bandpass Filters

The general layout of an end-coupled microstrip filter is shown in Figure 10.19. In practice the right-hand side of this filter would either connect to the load (the next stage) or would comprise further filter sections, dependent upon the performance requirements. A microstrip version of this type of filter has been described by

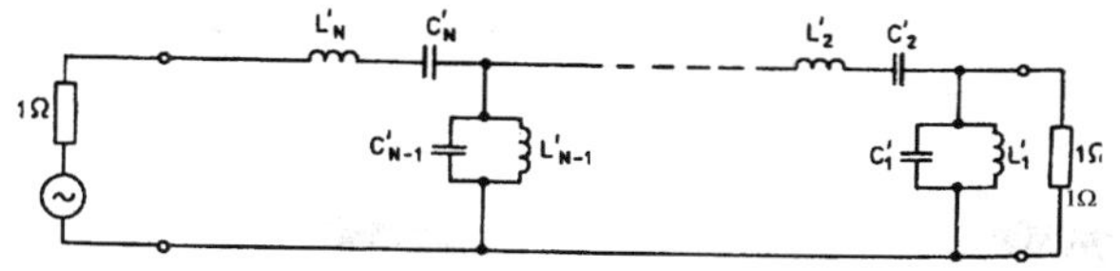

Figure 10.18 Transformed, lumped-element, bandpass filter network.

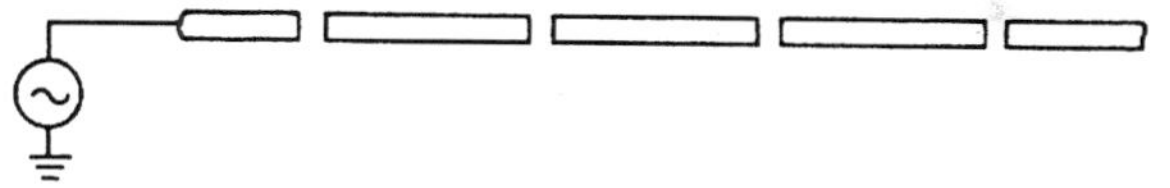

Figure 10.19 General microstrip layout for an end-coupled bandpass filter (series coupling gaps between cascaded straight resonator elements).

Cunningham *et al.* [411], designed to operate at a centre frequency of 28.76 GHz with a 2.78% bandwidth. The application is within a front-end filter in a down-converter for a Microwave Video Distribution System (MVDS).

Further specification parameters for this filter included a local oscillator rejection of at least 10 dB (at 27.6 GHz) and an image rejection (at 26.6 GHz) of at least 35 dB. The filter's insertion loss was simply to be as low as possible. This design approach was chosen because end-coupled filters are known to be relatively narrowband, which should suit the 2.78% bandwidth demanded here. The technology was chosen to be microstrip throughout for reasons of low cost.

In all types of filters based upon transmission line resonators, the coupling gaps between resonators are of prime importance, and the resonator lengths obviously also represent a particularly significant design result. For these end-coupled resonator filters the coupling needs to be as tight as possible, implying gaps that are usually much smaller than the substrate height. The resonator lengths depend not only upon the microstrip wavelength, but also on coupling reactances and the effect of shunt capacitances due to the gaps. Cunningham shows that this length, excluding the effective gap lengths, is given by the following expression:

$$l_j = \frac{\lambda_g}{2\pi}(\pi - X_{j-1,j} - X_{j,j+1}) - b_{j-1,j} - b_{j,j+1} \tag{10.17}$$

where $b_{j-1,j}$ are lengths due to gap shunt capacitances and

$$X_{j-1,j} = 0.5[\tan^{-1}(2B_{j-1,j}/Y_0)]. \tag{10.18}$$

A three-resonator filter was designed to meet the specification with a final bandwidth of 750 MHz and a centre frequency of 28.78 GHz. The out-of-band response (critical in this application) was found to be superior to that of a parallel-coupled filter.

Other types of end-coupled filters have been developed, some only partially based upon microstrip, but usually involving mainly microstrip-type elements. For example, Tzuang *et al.* [416] have reported what they term a quasi-planar broadside filter of this general type. This design, in contrast to conventional end-coupled filters, is relatively broadband (a 30% bandwidth is specified). The filter structure includes a ground plane milled to provide a slot beneath the microstrip section to form suspended microstrip. The reader is referred to the paper for further details, but the amplitude response is shown in Figure 10.20.

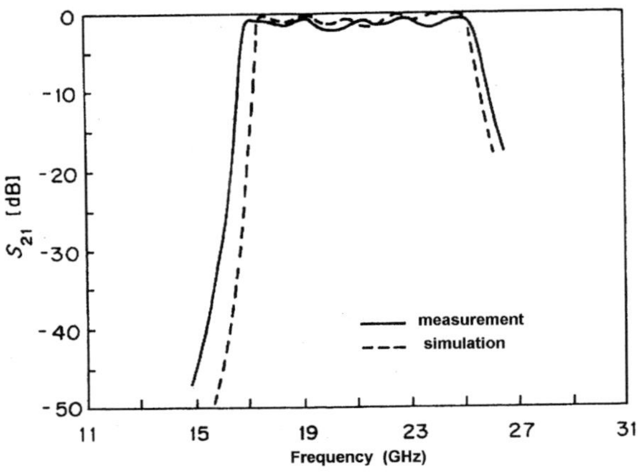

Figure 10.20 Performance of quasi-planar broadside end-coupled bandpass filter. (© 1990 IEEE. Reprinted, with permission, from Tzuang, Chiang and Su [416].)

Parallel-Coupled (or Edge-Coupled) Bandpass Filters

As discussed in Chapter 8 for microstrip couplers, maximum coupling is obtained between physically parallel microstrips when the length of the coupled region is $\lambda_g/4$, or some odd multiple thereof. To achieve resonance, each resonator element has to be $\lambda_g/2$ in length, or any multiple thereof. Therefore, the microstrip circuit must have the general layout shown in Figure 10.21, where

$$l_1, \ldots, l_4 \approx \lambda_g/4. \tag{10.19}$$

We will assume, for the moment, that this fairly straightforward cascade of parallel-coupled (or 'edge-coupled') microstrip resonators can be designed on a basis of all-parallel resonator networks, together with intervening circuits known as inverters. Outlining the four main design steps:

(a) Determine the one-type resonator network, to realize the specification, from the original prototype.
(b) From the network parameters, evaluate the even- and odd-ordered characteristic impedances, Z_{0e} and Z_{0o}, applicable to the parallel-coupled microstrip. (This will be shown in detail presently.)
(c) Relate the values of Z_{0e} and Z_{0o} to microstrip widths and separations (w, s). Procedures for this step are given in detail in Chapter 8.
(d) Calculate the whole resonator length $2l$, slightly less than $\lambda_g/2$ because of the semi-open end-effects, and therefore of the coupled-section length l, which is slightly less than $\lambda_g/4$ for the end-effect reason again.

Here λ_g is the mid-band and average microstrip wavelength. Allowance must be made for the semi-open-circuit microstrip end-effects which exist for all elements in this circuit.

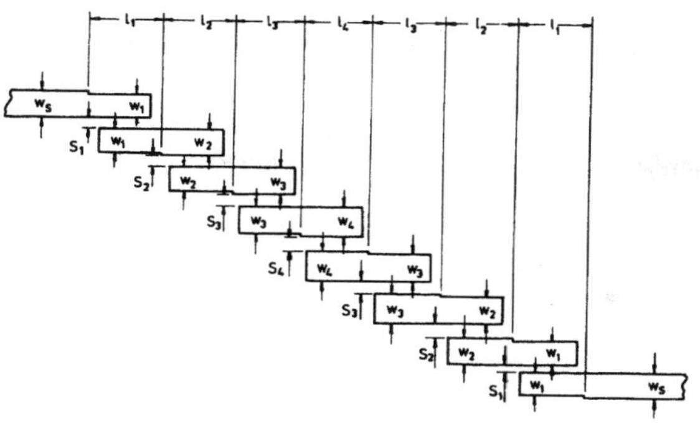

Figure 10.21 General microstrip configuration for a seven-section, parallel-coupled bandpass filter (BPF). (© 1990 IEEE. Reprinted, with permission, from Nakamura and Hibuno *et al.* [419].)

Impedance and Admittance Inverters; Basic Design Principles

The simplest form of an impedance inverter is the quarter-wave transformer. It should be recollected from Section 1.7 on page 15 that such circuit components transform any load impedance by the quantity Z_0^2, where Z_0 is the characteristic impedance of the quarter-wave transforming section of line. As a consequence of this characteristic impedance, squared terms will arise for impedance and admittance inverters.

It has already been stated that a network consisting of admittance inverters, usually called J-inverters, and uniform-type resonant circuits is to be made equivalent to the original network consisting of two types of resonant circuit — series and parallel. While the impedance level will vary through the system, two features must be maintained for equivalence:

(a) The resonant frequency must remain constant: $\omega_0^2 = 1/LC$.
(b) Impedances at similar planes in each network must be equal.

The details of impedance or admittance inverter relationships will not be derived here; the excellent book by Matthaei *et al.* [412] is recommended in this respect.

In the following expressions the 'first coupling structure' is that formed by w_s, s_1, w_1 (Figure 10.21) and the 'final coupling structure' consists also of w_s, s_1, w_1 at the opposite end of the filter. The quantities g refer to the prototype element values, for example $g_0 = 1$ (the normalized termination value) and $g_0 = 0.781$ — from the table for a sixth-order Chebyshev prototype, which is standard data available in many references [412]. The admittance inverter parameters (J) are then given by:

For the first coupling structure:

$$\frac{J_{01}}{Y_0} = \sqrt{\frac{\pi\delta}{2g_0g_1}}. \tag{10.20}$$

For the intermediate coupling structures:

$$\left.\frac{J_{j,j+1}}{Y_0}\right|_{j=1}^{n-1} = \frac{\pi\delta}{2\omega_c'\sqrt{g_jg_{j+1}}}. \tag{10.21}$$

For the final coupling structure:

$$\frac{J_{n,n+1}}{Y_0} = \sqrt{\frac{\pi\delta}{2g_n g_{n+1}}}. \tag{10.22}$$

In these three equations, δ is the fractional bandwidth

$$\delta = (f_2 - f_1)/f_0. \tag{10.23}$$

The frequency transformation from the lowpass prototype filter to the bandpass microwave filter is then

$$\frac{\omega_i'}{\omega_c} = \frac{2}{\delta}\left(\frac{f_i - f_0}{f_0}\right) \tag{10.24}$$

where ω_c' in Equation (10.21) is the prototype cut-off frequency (it equals 1.0) and ω_i' has to be defined in the filter specification. (The full design procedure will presently be shown by means of a specific numerical example, where the meanings of ω_c and ω_i' etc., will become clear.)

To proceed with the microstrip design, we now only require the odd- and even-mode coupled-line impedances Z_{0o} and Z_{0e} These are given by

$$(Z_{0e})_{j,j+1} = Z_0(1 + aZ_0 + a^2 Z_0^2) \tag{10.25}$$

and

$$(Z_{0o})_{j,j+1} = Z_0(1 - aZ_0 + a^2 Z_0^2) \tag{10.26}$$

where $a = J_{j,j+1}$ and Z_0 is the 'system' characteristic impedance (i.e. that of the lines feeding the filter).

Having obtained the values of Z_{0o} and Z_{0e} section by section, the microstrip widths and separations are determined. The synthesis technique due to Akhtarzad *et al.* [292] is recommended for this, with refinements as described in Section 8.6 on page 276. Calculations and a typical design are best shown by a specific example. A 0.635 mm thick alumina substrate with a permittivity of 10.0 is used.

10.3.5 A WORKED NUMERICAL EXAMPLE OF A PARALLEL-COUPLED BANDPASS FILTER

Assume that the specification shown in Figure 10.22 is to be realized, where L is the insertion loss in dB. The passband insertion loss (exactly related to the 'tolerance') has not been specified on the characteristic; it would usually be expected to lie below about 0.1 or 0.2 dB.

Step 1 Using Equations (10.23) and (10.24) a fractional bandwidth of 0.1 and transformation ratio of −1.619 are obtained.

Step 2 Examine filter prototype specifications which meet the insertion-loss requirements. These are available in many handbooks (see, for example, Matthaei *et al.* [412]).
The family of curves shown in Figure 10.23 apply to Chebyshev prototypes all giving a 0.01 dB passband ripple.

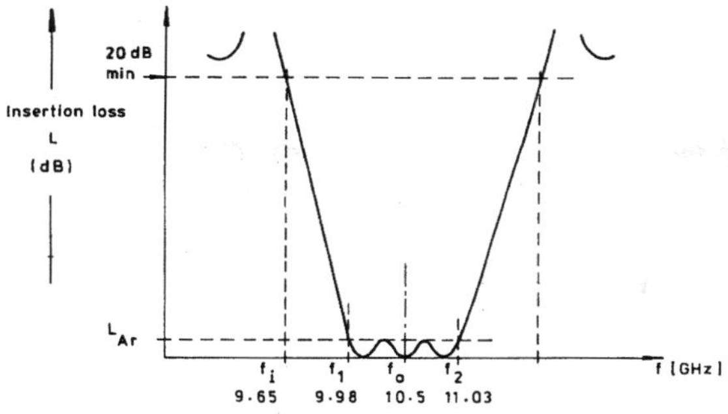

Figure 10.22 Insertion-loss characteristic for the bandpass filter of the example.

Table 10.5 Coefficients for a 6th order Chebyshev filter.

g_0	g_1	g_2	g_3	g_4	g_5	g_6	g_7
1.000	0.781	1.360	1.690	1.535	1.497	0.710	1.101

Notice that the independent variable has been adjusted for convenience to

$$|\omega_i'/\omega_c| - 1. \tag{10.27}$$

Considerable care must be exercised regarding the value of ω_i' — it is the value before transformation and is not equal to ω_i. It must be obtained from Equation (10.24). In this case the value of $(|\omega_i'/\omega_c| - 1)$ is found to be 0.619, which is indicated on Figure 10.23.

A sixth order ($n = 6$) design therefore gives a 23-dB insertion loss at ω_i'. (This is equivalent to the 9.65 GHz point in the final requirement, which specified > 20 dB. For ripple amplitudes > 0.01 dB, only odd order designs are permissible, ensuring that $g_0 = g_n = 1.0$ accurately.) The table of values for a Chebyshev filter with $n = 6$ is given in Table 10.5.

Step 3 Calculate the inverter admittances (normalized for a 50 Ω system) and hence coupled-line impedances using Equation (10.20) to Equation (10.22), and also Equation (10.25) and Equation (10.26). The results are shown in Table 10.6.

Step 4 The microstrip line dimensions are finally obtained as follows (substrate permittivity $\varepsilon_r = 10$):

Table 10.6 Step 3 results.

j	$J_{i,j+1}/Y0$	$(Z_{0e})_{j,j+1}$ (Ω)	$(Z_{0o})_{j,j+1}$ (Ω)
0	0.449	82.5	37.6
1	0.1529	58.8	43.5
2	0.1038	55.7	45.3
3	0.0976	55.4	45.6

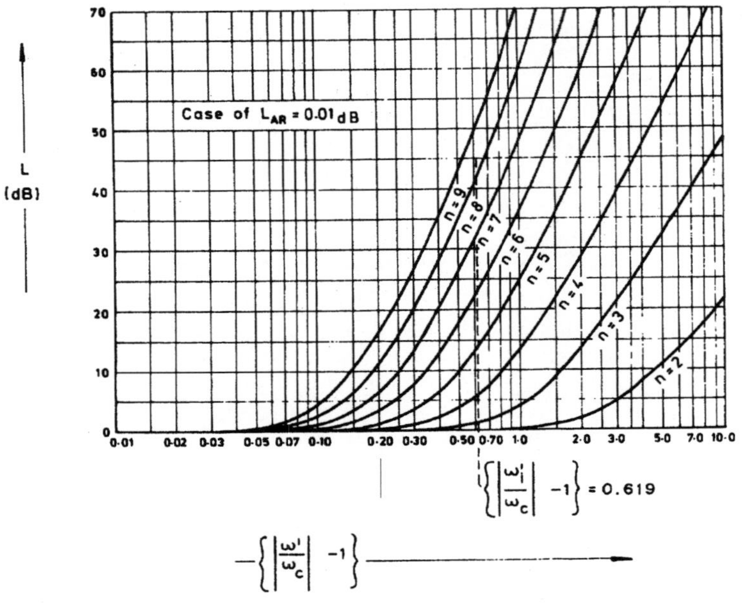

Figure 10.23 Family of Chebyshev prototype filter insertion-loss characteristics for a passband ripple of 0.01 dB. (Reproduced by permission of McGraw-Hill Book Company from Matthaei *et al.* [412] — Figure 4.03–4.)

Using Akhtarzad *et al.* [292] synthesis:

$$j = 0 \begin{cases} Z_{0se} = Z_{0e}/2 = 41.25\ \Omega \\ Z_{0so} = Z_{0o}/2 = 18.8\Omega \\ \text{Equivalent single-line synthesis yields:} \\ \quad (w/h)_{se} = 1.5 \\ \quad (w/h)_{so} = 4.8 \\ \quad \text{Then coupled-line synthesis gives the required parameters:} \\ \quad (w/h) = 0.65 \quad (s/h) = 0.2 \end{cases}$$

$$j = 1 \begin{cases} Z_{0se} = 29.4\Omega \quad Z_{0so} = 21.75\Omega \\ (w/h)_{se} = 2.5\,(w/h)_{so} = 4.0 \\ w/h = 1.0 \quad s/h = 0.68 \end{cases}$$

$$j = 2 \begin{cases} Z_{0se} = 27.85\Omega \quad Z_{0so} = 22.65\Omega \\ (w/h)_{se} = 2.7 \quad (w/h)_{so} = 3.7 \\ w/h \approx 1.0 \quad (s/h) \approx 1.0 \end{cases}$$

$$j = 3 \begin{cases} \text{Since impedances are only } 0.3\ \Omega \\ \text{different from the } j = 2 \text{ case approximately} \\ \text{the same results are obtained here.} \end{cases}$$

Figure 10.24 shows the final dimensions of this filter, assuming the substrate to have a thickness of 0.635 mm. 'Open' end-effects may be determined experimentally, as described in Section 9.7 on page 328. Then the physical lengths of the resonators and the coupled regions are half-wavelengths and quarter-wavelengths respectively,

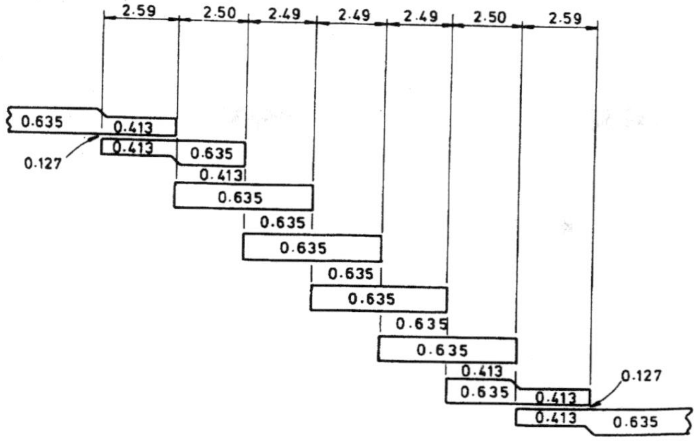

Figure 10.24 Final design dimensions of the bandpass filter calculated in the worked example.

with the equivalent end-effect lengths subtracted. (A wavelength is specified at the centre frequency f_0 and calculated as described in Sections 8.8 and 8.9, on pages 281 and 283 respectively.) The measured amplitude response of this filter is shown in Figure 10.25.

This filter was manufactured using copper-conductor thin-film technology, with the dimensions given on Figure 10.24; no post-manufacture trimming was undertaken. We note that the following results apply:

Parameter	Desired (Figure 10.22)	Measured (Figure 10.25)
3 dB bandwidth (%)	10.5%	12.5%
Centre frequency (GHz)	10.5 GHz	10.7 GHz

Judicious modification (trimming/tuning) is necessary to adjust these values; e.g. making the resonators just 2% longer should result in a closer centre frequency. The bandwidth is particularly sensitive to the separations of the first and last coupled sections, and these separations should be increased slightly so that the bandwidth is reduced.

10.3.6 CAD OF PARALLEL-COUPLED BANDPASS FILTERS

A number of approaches have been developed for the computer-aided design of microstrip bandpass filters. The method described in outline follows largely the design of procedure developed above (i.e. a first-cut design, excluding losses).

The aim is to generate the artwork for a filter having the general layout shown in Figure 10.22, given an insertion loss specification with the format shown in Figure 10.26. This indicates the following input requirements:

1. centre frequency f_0 (GHz),
2. 'ripple' or 'band-edge' insertion loss L_R (dB),
3. insertion loss at the 'attenuation' points L_A (dB),

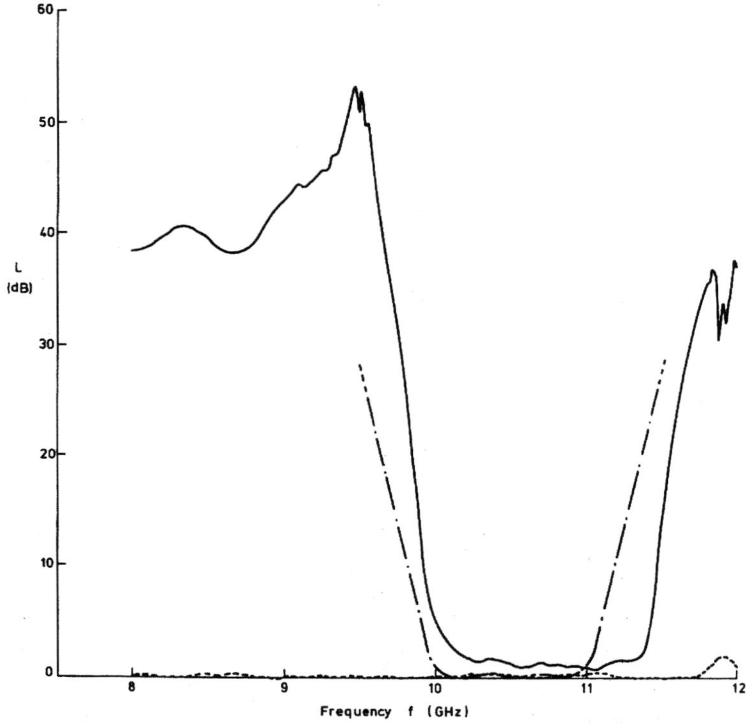

Figure 10.25 Measured (——) compared with specified (— · —) insertion-loss characteristics for the 10–11 GHz parallel-coupled bandpass filter.

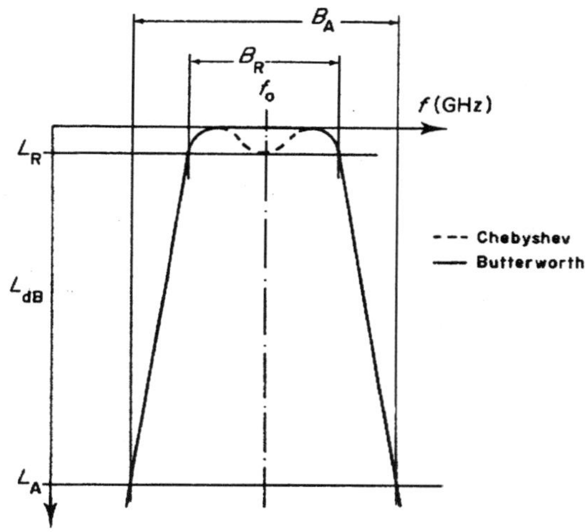

Figure 10.26 General bandpass filter insertion-loss characteristic.

4. 'ripple' bandwidth B_R (GHz), and
5. 'attenuation' bandwidth B_A (GHz).

Four other items of input information, not evident from the insertion loss specification, are also required:

6. the termination characteristic impedance Z_0 (Ω),
7. substrate relative permittivity ε_r, and
8. substrate thickness h (mm).
9. Finally, the appropriate function must be chosen by which the insertion loss is to be modelled. In this case a choice between Chebychev or Butterworth is given.

The main flowchart for the algorithm applicable to this CAD approach is shown in Figure 10.27 and the sequence of events now proceeds as follows:

1. Initially the required number of sections is calculated, and for each section the even and odd mode characteristic impedances Z_{0e} and Z_{0o} are determined, as given by Matthaei *et al.* [412].
2. These impedances (A) are applicable to any TEM-type parallel-coupled bandpass filter, and the next step is for the algorithm to follow either stripline (sometimes referred to as triplate as in the original work by Matthaei *et al.*) or microstrip technology design routes, as directed.
3. The stripline design procedure is relatively straightforward and simple, being direct, and it will not be described here.
4. For microstrip design, the most challenging process is the accurate computation of the final widths (w) and separations (s). Firstly, auxiliary single microstrip parameters are calculated, leading to approximate, starting-point, values of the aspect ratios of w/h and s/h as determined by the synthesis technique due to Akhtarzad *et al.* and provided in Chapter 8.
5. These aspect ratios are then used in an accurate analysis routine, also shown in Chapter 8, to find new values for the section impedances (B).

6. The new impedances (B) are compared with the cardinal values (A) and adjustments are made to w/h and s/h in order to correct the new values.
7. The latest w/h and s/h quantities are only output when the impedance differences (A−B) are acceptably small.
8. Microstrip lengths are now required. Modified dispersion formulas are used to calculate the even-and-odd-mode phase velocities v_e and v_o at the centre frequency f_0 and the weighted-mean phase velocity v_n is then evaluated using

$$v_n = \frac{Z_{0e} + Z_{0o}}{(Z_{0e}/v_e) + (Z_{0o}/v_o)}. \tag{10.28}$$

   Then the microstrip wavelength λ_{gn} is given by

$$\lambda_{gn} = v_n/f_o. \tag{10.29}$$

9. Open-end-effect lengths must be calculated as derived from the single-line model and v_n for the parallel-coupled lines. These are also a function of frequency $l_{e,\text{coupled}}(f)$.
10. Each corrected, physical, 'quarter-wave' length is then

$$l = (\lambda_{gn}/4) - [l_{e,\text{coupled}}(f)] \tag{10.30}$$

and this completes the design algorithm.

10.3.7 IMPROVEMENTS TO THE BASIC EDGE-COUPLED FILTER RESPONSE

Parallel (or edge)-coupled microstrip bandpass filters of the types designed according to principles described above display somewhat unsymmetrical frequency responses in addition to a significant second passband unless compensated. These highly undesirable effects are due to the field characteristics of the odd mode, as described in Chapter 8.

Various techniques are available for compensation of this odd-mode effect and some of these were also outlined in Chapter 8. Bahl [413] has shown that, with capacitive compensation introduced at the ends of each quarter-wave section (see Section 8.11.5 on page 297), the stopband performance of these types of filters is greatly improved. The 'sharpness' or selectivity characteristics of both skirts are improved most markedly in the case of the upper skirt, where the improvement is typically 5–10 dB. Also, the second passband is then more than 40 dB below the level of the fundamental passband — and, to further advantage, slightly shifted in frequency. Compensating capacitances typically lie in the range 0.002–0.055 pF, depending upon the specific filter section and the substrate. Bahl gives examples of GaAs and quartz substrates.

With MMIC realizations of this type of filter these small capacitances could be introduced as a process step — with vias providing the ground connections.

10.3.8 FILTER ANALYSIS AND DESIGN INCLUDING ALL LOSSES

As mentioned earlier, the preceding filter CAD approaches neglect the effects of power losses through the structure. It is extremely difficult, if not impossible, to anticipate

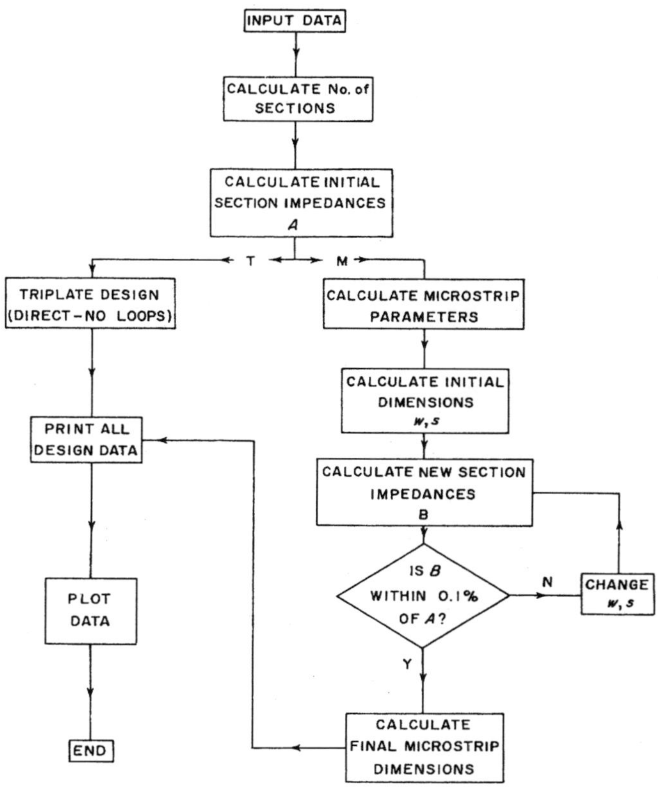

Figure 10.27 Flowchart applicable to the CAD of bandpass filters.

these in quantitative detail and then to actually synthesize a design including all losses. Instead, it is suggested that the design be 'completed' to a first-cut level and that the resulting filter structure is then simulated on the computer to determine its insertion loss response, including power losses. At one extreme the design philosophy could then be aborted and a different, less lossy, approach adopted (e.g. hairpin line resonators might be used if straight resonators were implemented at first). Alternatively, controlled variations on the exact design could be studied for their impact on loss-driven aspects of their characteristics.

The insertion loss of each section of the filter is best characterized in terms of the forward transfer scattering parameter S_{21}. This, in turn, involves section and mode impedances as follows [412]:

$$S_{21zI} = \left. \frac{2Z_{12n}/Z_0}{(1+Z_{11n}/Z_0)^2 - (Z_{12n}/Z_0)^2} \right|_{I(n=\text{mean})} \tag{10.31}$$

where

$$\left. \frac{Z_{11e,o}}{Z_0} \right|_I = -j\left(\frac{Z_{0e}+Z_{0o}}{2Z_0}\right) \cot \theta_{e,o}|_I \tag{10.32}$$

and

$$\left. \frac{Z_{12e,o}}{Z_0} \right|_I = -j\left(\frac{Z_{0e}-Z_{0o}}{2Z_0}\right) \csc \theta_{e,o}|_I \,. \tag{10.33}$$

The appropriate phase angles for use in these expressions should be determined using the formulas given above for the CAD routine, i.e. the weighted-mean phase velocity and microstrip wavelength. Further forward scattering parameter functions are readily available to account for the loaded and unloaded Q-factors of the resonators (accounting for the conductor and dielectric losses) and the radiation loss from the semi-open ends of these resonators.

Dissipative losses:

For the nth resonator we can write the dissipative loss S-parameter as:

$$S_{21_{\text{ln}}} = Q_u/(Q_u - Q_L) \tag{10.34}$$

where the loaded-Q (Q_L) is

$$Q_L = (f_o/\Delta f)\sin[(2n-1)\pi/2N]. \tag{10.35}$$

In this expression, Δf is the 3 dB bandwidth of the filter which is determined from the insertion loss function (i.e. Butterworth or Chebyshev) and N is the total number of resonators.

For each resonator Q_u is found from:

$$1/Q_u = 1/Q_c + 1/Q_d + 1/Q_r. \tag{10.36}$$

The Q-factors due to conductor and dielectric losses are found by using expressions given in Chapter 5 and the radiation conductance G_r is also provided in Chapter 5.

We require an expression for Q_r in terms of G_r, and this is the external Q-factor for a resonator that is equally end-loaded. Such an expression is given in Appendix C and is:

$$Q_r = \frac{\pi\rho}{1-\rho^2} \tag{10.37}$$

where

$$\rho = (G_r Z_o - 1)/(G_r Z_o + 1). \tag{10.38}$$

This expression is also equal to the input scattering parameter $S_{11,rad}$ and therefore, using standard S-parameter theory (Appendix D), the contribution due to this radiation mechanism is given by:

$$|\Delta S_{21}|^2 = 1 - |S_{11rad}|^2 . \tag{10.39}$$

Once all these features are quantified it should be possible to fully simulate the behaviour of the filter, having first converted the S-parameters to an appropriate cascade family of parameters such as $ABCD$.

10.3.9 BANDPASS FILTERS WITH INCREASED BANDWIDTH (>15%)

Where larger bandwidths are desired, one of several alternative configurations may be considered. It has long been appreciated that a series of stubs, all $\lambda_g/4$ in length if open-circuited ($\lambda_g/2$ if short-circuited) and spaced $\lambda_g/4$ apart, will form a bandpass filter with a realizable bandwidth of the order of an octave. The stubs are placed alternately on each side of the microstrip. Kompa and Mehran [414] have reported a frequency-dependent design procedure for this type of filter in microstrip. Tzuang and Lo [415] have also described a tapped combline bandpass filter with a 4 GHz passband.

Straight stubs may in many instances be advantageously replaced by radial stubs as described in Section 10.3.11. on page 380.

Pang [417] has shown that the first and last coupling structures of the parallel-coupled filter, described in the previous section, can be replaced by continuous strip designs involving a quarter-wave transformer. In stripline, Pang demonstrates a 1.6 GHz filter with a 41% 3 dB bandwidth based upon this approach. Apart from the bandwidth improvement, there is an advantage regarding tolerancing, since the regions which originally involved the narrowest coupling gaps are now avoided altogether. The characteristic impedance of the transforming line length is not particularly straightforward to evaluate — it requires the use of the sineplane approach in the design. Microstrip realizations of this approach should be explored.

Mara and Schappacher [418] have given details of an interesting parallel-coupled filter design involving multiline couplings instead of the more usual two lines. Several filters are described. In one case a 6 GHz octave-bandwidth bandpass filter is described, based upon a 0.1 dB ripple five-section Chebyshev prototype. The ends of all the resonators were wire-bonded together (like the Lange coupler described in Section 8.11). Unfortunately, the 6 GHz filter exhibits a spurious passband beginning at about 11.6 GHz, preceded by a characteristic in which the insertion loss does not improve beyond 18 dB — after the main passband. Doubtless some judicious tuning of the filter would overcome these problems.

10.3.10 FURTHER DEVELOPMENTS IN BANDPASS FILTER DESIGN

With the extensive knowledge available for microwave filter design, including the approaches described above, it has been proposed that an 'expert system' could be developed for the rapid and reliable availability of such design processes. Nakamura and Hibino [419] have reported such an expert system. Starting with the specification their procedure uses a microwave database, a knowledge database, a microwave circuit simulator and what they term 'inference' (the effect of interactions between the different requirements).

The knowledge database includes sufficient information to allow for design margins and to choose the appropriate substrate on the basis of frequency of operation, temperature, etc. It also has the ability to ensure that the filter will fit the available substrate space (the real estate).

The microwave circuit simulator is designed to call up one of several commercially available CAE software suites. This is arranged to be available automatically, by operating the system using IP. Final MIC/MMIC pattern drawing is facilitated by an appropriate subsidiary package. Practically any orientation can be accommodated.

Parallel-coupled bandpass filters may be provided as 'stand-alone' components or may be realized as part of a more complex circuit.

The use of hairpin resonators is also important and parallel-coupled bandpass filters implementing these structures, rather than simple straight resonators, are frequently encountered. With these forms of resonators radiation can be reduced by suitably spacing the hairpin arms, and the bends are mitred (see Section 7.14.4 on page 264).

Filters implementing these hairpin-line resonators generally occupy more real estate than their straight resonator counterparts, and clearly this must be taken into account before opting for this approach. Implementation on high-permittivity MMICs and at relatively high (e.g. millimetre-wave) frequencies is often appropriate.

As mentioned earlier in this text, work is also progressing in microwave applications of the important field concerning high-temperature superconductors. Although not strictly high-temperature, one example of work regarding the design of superconductive microstrip filters is that reported by Bonetti and Williams [420]. They use substrate materials such as lanthanum aluminate and lanthanum galate. These materials, for this particular type of application, are relatively recently developed, and Bonetti and Williams report substrate measurements at about 4 GHz. The loss tangents are very low and the permittivities vary between 23.1 and 24.3.

The resulting bandpass filters, operated at 77 K, exhibit symmetrical insertion loss characteristics and a bandwidth of less than 4% (150 MHz). Hairpin line resonators are employed throughout the filters.

10.3.11 MICROSTRIP RADIAL STUBS

For over two decades it has been appreciated that the implementation of radial stubs, as opposed to the straight stubs traditionally used, would improve the performance of many microstrip circuits. Examples of various options concerning these circuit elements are shown in Figure 10.28.

Such stubs may, like their straight counterparts, be either series or shunt-connected as indicated in Figures 10.28(a) and (b). When two shunt-connected radial stubs are

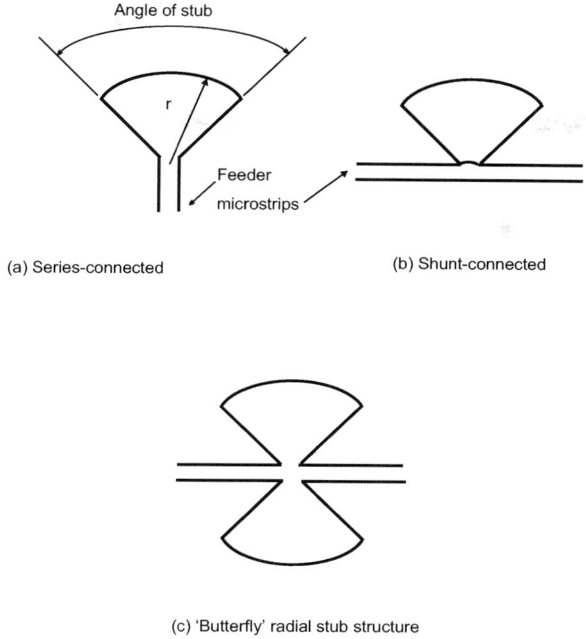

Figure 10.28 Microstrip radial stubs.

introduced in parallel — i.e. one on each side of the microstrip feeder line — the resulting configuration is termed a 'butterfly' structure. This is indicated in Figure 10.28(c).

Critical design parameters include the radius r and the inclusive angle.

For a design on a 0.635 mm thick alumina substrate, Giannini *et al.* [421] show that a stub radius of 5 mm and inclusive angle 60° provide a useful circuit element at approximately 12 GHz. At higher frequencies, towards and through millimetre-wave, the radius must be shortened and/or the angle increased proportionately. If out-of-band restrictions permit, the radius may be increased in odd multiples of quarter-wavelengths so that a physically realizable structure is obtained.

A major advantage of these radial stubs is the input impedance offered over generally broad bands. An input impedance lower than 20 Ω often applies here — compared with the near-infinite impedance associated with straight stubs. This allows comparatively accurate identification of the impedance reference plane, which is highly significant for design.

In practice these types of stubs are lossy, and CAD models accounting for this situation have been reported [422]. Giannini *et al.* [422] also report an interesting application to a microstrip termination. They show that the combination of a 'butterfly' structure and a series-connected radial stub provides a termination having S_{11} better than 20 dB over the 2.5 to 8 GHz frequency range. The termination structure comprises a $3 \times 60°$ array of radial stubs at the end of a microstrip line.

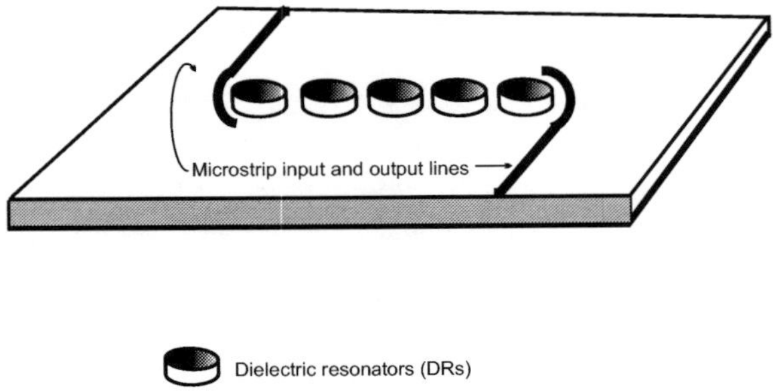

Figure 10.29 A microstrip-connected dielectric resonator bandpass filter configration.

10.3.12 DIELECTRIC RESONATORS AND FILTERS USING THEM

Although not specifically microstrip elements, dielectric resonators are of such significance in several forms of filters and oscillators that a brief consideration of these devices is essential here.

Any cylindrically-shaped dielectric structure can be caused to resonate at some specific frequencies and these depend upon both the permittivity and the physical dimensions. The general theoretical expression for such resonant frequencies, applying to a dielectric resonator of radius a and height d, is

$$f_{mnl} = \frac{c}{2\sqrt{\varepsilon_r}}\sqrt{\left(\frac{\chi_{mn}}{\pi a}\right)^2 + \left(\frac{l}{d}\right)^2} \tag{10.40}$$

where the integer l denotes the number of half-wavelengths in the vertical direction and χ_{mn} is the mth extremum of the Bessel function J_n for a TM mode (or alternatively the mth zero for a TE mode). This is developed as the dual of the well-known metallic cavity analysis [423].

In practice fringing fields, microstrip (or other) line coupling and the presence of the substrate dielectric render the analysis complex and more sophisticated, computer-based, procedures are necessary to evaluate the applicable parameters.

Low-loss ceramic materials with permittivities in the range 21.6 to 152 are generally available that can be supplied in dielectric resonator pill form. Unloaded Q-factors are in the 5000 to 10000 range and physical dimensions generally depend upon the resonant frequencies desired. However, loaded Q-factors depend strongly upon the degree of coupling to the microstrip line, i.e. the proximity of the two elements. When the coupling distance is less than 1 mm the loaded Q-factor is usually below 500 and this has significant consequences in filter and oscillator design [424]. At 4 GHz a 14 mm diameter pill is typical (with $\varepsilon_r = 35$), whereas at higher frequencies dielectric pills having diameters of only a few mm are implemented.

Filter design using conventional resonators (such as the edge-coupled microstrips considered earlier in this chapter) requires several optimally coupled resonators and this is also true with Dielectric-Resonator (DR) based filters. A general arrangement of a DR-based filter is shown in Figure 10.29.

To increase the coupling whilst avoiding excessive proximity of the microstrip lines, curved coupling sections are used. Each DR is also optimally coupled and, as with all filters, a major design aim is to identify appropriate DRs and to establish the degrees

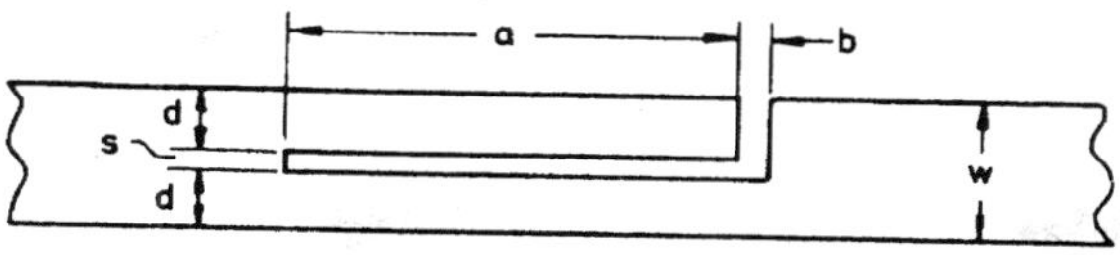

Figure 10.30 One section of a spur-line bandstop filter. (Reproduced by permission of the Institution of Electrical Engineers from Bates [425].)

of coupling applicable to each section. Bahl and Bhartia [424] cover this problem extensively and show an example of a 900 MHz bandwidth bandpass filter design, centred on 9 GHz. The impact of the performance of loaded Q-factors, ranging as low as 100, are shown. With a loaded Q-factor of 500 the insertion loss deteriorates to around 2 dB through the passband, which is probably acceptable in many instances. However, with loaded Q-factor of 100 this insertion loss is around 4 dB, which is almost always unacceptable.

10.3.13 SPURLINE BANDSTOP FILTERS

Bates and Pearson [425, 426] evolved a simple and clever technique for designing a bandstop filter using a resonant structure (Figure 10.30) fabricated within the width of a microstrip line.

It was demonstrated that this filter is almost completely non-dispersive, and the structure is so configured that radiation loss is very low compared with coupled-line or shunt-stub filters. The filter structure was shown to be equivalent to an open-circuit shunt stub of characteristic impedance Z_1, followed by a continuous section of microstrip having a characteristic impedance Z_{12}. These impedances are given by

$$Z_1 = \frac{Z_{0e}}{Z_{0o}}\left(\frac{Z_{0o} + Z_{0e}}{2}\right) \tag{10.41}$$

and

$$Z_{12} = Z_{0o} + Z_{0e}/2 \tag{10.42}$$

where Z_{0e} and Z_{0o} have the meanings defined previously for coupled lines.

This filter can be designed according to the procedure described by Matthaei, Young and Jones [412], and computer-aided optimization can then be employed. Bates [425] gives graphs of Z_1 and Z_{12} as functions of the parameters s/h and w/h, although a designer with access to computer facilities would probably need to develop suitable expressions for these impedances. The length a (see Figure 10.30) is electrically $\lambda_g/4$ at mid-band (f_0), but the gap equivalent length l_{eg} will affect the physical value of a such that

$$a = \frac{3 \times 10^8}{4f_0\sqrt{\varepsilon_{\text{effo}}}} - l_{eg} \tag{10.43}$$

where $\varepsilon_{\text{effo}}$ is the odd-mode effective microstrip permittivity at the centre frequency (see Section 8.9 on page 283) and l_{eg} is the gap end-effect extension (see Section 7.3 on page 231).

The value of the series-gap separation b has to be determined empirically and a value of approximately 50 μm appears to be suitable for designs on alumina, and similar substrates [425]. Once b is chosen, the end-effect length may be found by using

$$l_{eg} = C_{\text{odd}} v_{po} Z_{0o}. \tag{10.44}$$

C_{odd} is calculated from expressions and data given by Benedek and Silvester [265], and Bates [425] gives results which indicate that a very good agreement between calculation and experiment can be achieved for gap separations of about 100 μm or less. The filter design procedure is thus:

(a) From the specified response desired use the technique given by Matthaei *et al.* [412] to find Z_1 and Z_{12}.
(b) Calculate Z_{0e} and Z_{0o} from Z_1 and Z_{12} using Equations (10.41) and (10.42).
(c) Employ computer-aided optimization to ensure that the network so far described meets the desired specification within acceptable limits.
(d) From Z_{0e} and Z_{0o} synthesize the desired s/h and w/h.
(e) Evaluate the length a, approximately.
(f) Choose the series gap separation b ($\approx$ 50μm, on alumina), and hence also determine l_{eg}.
(g) Iterate (e) and (f) until a is found more accurately and the centre frequency f_0 is close to that desired.

The bandstop filter performance is particularly good when two of these spurline structures are arranged back to back in cascade, with a separation slightly less than a quarter-wavelength. For example, where a = 3.2 mm the distance separating the structures should be about 3.0 mm [426]. With such a filter a rejection (insertion loss) exceeding 35 dB was achieved over the band $9.3 < f < 9.7$ GHz.

Several variations of the structure are conceivable. A 'double-sided' arrangement, i.e. two spur-line structures side by side across the microstrip width, has been manufactured for a band centred on 30.5 GHz. The rejection exceeded 29 dB over a 2 GHz band [426]. This general concept, leading to unusual and very compact filters, must be capable of extension into many new designs.

Microstrip lends itself to entirely new structural shapes to perform circuit/system functions which were previously realized using stripline, or coaxial, or waveguide technology.

10.3.14 FILTERS USING SYNTHETIC PERIODIC SUBSTRATES (ELECTROMAGNETIC BANDGAP CRYSTALS)

Artificial electromagnetic bandgap (EBG) materials represent a new class of periodically structured dielectrics which exhibit transmission behaviours analogous to carrier transport in semiconductors. When electromagnetic waves travel in PBGs they behave in a similar manner to that of electrons in semiconductors. Such materials are artificially manufactured to form precisely orientated crystals which control the propagation of the electromagnetic waves.

Such periodic structures are characterized by three specific parameters: the spatial period defining the lattice constant, the fractional volume and the ratio of the permittivities of the individual constituent materials. By making the composite materials periodic in three dimensions, it is possible to control the electromagnetic propagation accurately. A Bragg reflector is a one dimensional form of a PBG.

The crystal opal, comprising stacks of 200 nm diameter spheres of silica, represents a well known form of PBG. Using 'woodpile' layer-by-layer structures of dielectrics, microwave, millimetre-wave and sub-millimetre-wave antennas and filters have been developed [427]. A new metallic PBG has also been developed, and Maagt *et al.* show an example of the construction and insertion loss performance. This microstrip-based bandstop filter rejects signal frequencies in the band 3.4 to 4.8 GHz with an insertion loss better than −62 dB over this frequency range. At frequencies below 2.8 GHz the insertion loss is −2 dB to −3 dB and at higher frequencies around −10 dB is obtained. It is anticipated that many high rejection capability filters will use PBGs in future realizations of mobile communications terminals and related technologies.

10.3.15 PASSIVE MICS WITH SWITCHING ELEMENTS

MICs having switching functions, normally implemented with PIN diodes but increasingly using transistors, and MEMS (for MicroElectro-Mechanical System) switches have been realizable for many years. The basic design concepts are simple, utilizing the fact that specific line lengths (e.g. quarter wave) are switched into or out of the circuit depending upon whether the PIN, transistor switch or MEMS switch is activated or not.

Although strictly a form of active circuit, we mention this application here since it is not accommodated by any of the analogue active functions covered in Chapter 11. Broadband cross point switch matrices have been reported, for example by Powell *et al.* [428] using GaAsFETs (in some cases dual-gate GaAsFETs) and microstrip crossovers. Such switch matrices are readily capable of being implemented in MMIC form and, as 2×2 matrices operating over the 4–10 GHz band, can achieve over 40 dB isolation with less than 4 dB insertion loss. The circuits can be cascaded to form higher order matrices.

Various forms of HEMTs (e.g. PHEMTs), in selected semiconductor technologies (including SiGe), may also be used as the switching device.

10.3.16 ISOLATORS AND CIRCULATORS

Various configurations have been studied to exploit ferrites in hybrid MIC technology and hence to realize isolators and circulators. The characteristics of ferrite in this context are described in Chapters 4 and 5.

Kane and Wong [429] report an isolator design based upon edge-guide mode principles in conjunction with a transverse slot or slit discontinuity (the transverse slit is described in Chapter 7, Section 7.14.6). This configuration is shown in Figure 10.31, where the entire substrate is ferrite, the isolator section occupies an area approximately 10 mm × 8 mm and the slot measures approximately 6 mm × 0.8 mm.

At the centre frequency of 11 GHz the forward insertion loss is about 2 dB, whereas the isolation is better than 18 dB.

A circulator designed for operation over the 31–37 GHz band has been described by Pan *et al.* [430]. Their approach uses a new nickel-zinc ferrite substrate material, which has superior performance to previous materials and which is available in 0.127 and 0.254 mm thicknesses. The adhesion between the ferrite and the metallization layer was poor in previous designs and this is improved in the new design by sputtering

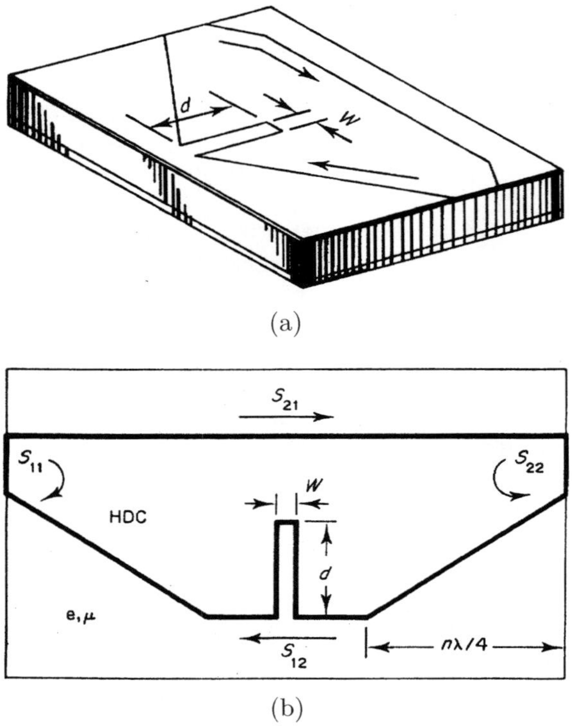

Figure 10.31 An edge-guide mode microstrip isolator with transverse slot discontinuity. (© 1990 IEEE. Reprinted, with permission, from Kane and Wong [429].)

a seed layer of titanium tungsten (TiW) onto the ferrite surface prior to the final metallization. The thicknesses of the TiW and gold or copper layers are 500 and 2000 angstroms respectively.

The overall layout of the stub-tuned circulator is shown in Figure 10.32. It is possible to make circulators of particularly small dimensions using this technology and Pan *et al.* [430] have demonstrated such a device having a diameter of 3.8 mm and thickness 0.127 mm. In the case of a 7.6 mm diameter circulator, a calibrated insertion loss of 0.6 dB and a return loss of better than 17 dB were measured.

This type of circulator is designed primarily for deep space communications but the design principles should be extendable to other applications.

How *et al.* [431] have reported an extensive theoretical and experimental investigation into the performance of an X-band thin-ferrite-film junction circulator. For theoretical formulation, they implement a new effective-field approach assuming TEM-like propagation. How *et al.* show that circulators with insertion losses better than 0.5 dB at X-band can be fabricated providing the film thickness exceeds 100 μm. They also indicate that the quality of the conductor plane is important in terms of reducing the insertion loss. Their (How *et al.*) circulator is fed by three, three-section, impedance-matching transformers.

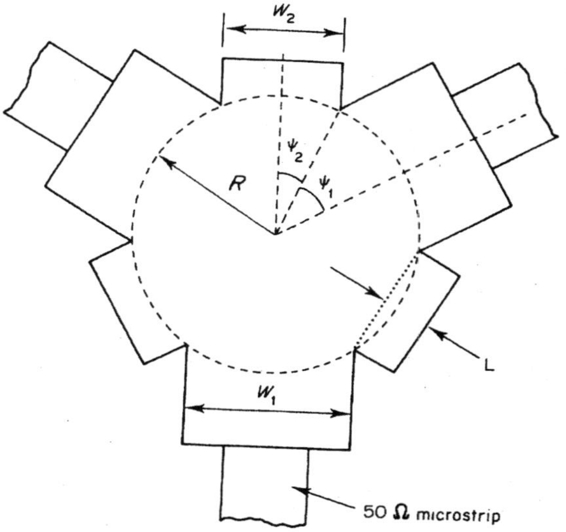

Figure 10.32 Wideband stub tuned microstrip circulator. (© 1990 IEEE. Reprinted, with permission, from Pan, Shih and Riley [430].)

11

Active Digital and Analogue ICs

11.1 INTRODUCTION

In this chapter we will look at the design of high-speed digital and active RF and microwave (i.e. analog) integrated circuits. The emphasis is on application of the transmission line technologies covered in this book. The first section outlines interconnect design for high-speed digital circuits. The clock distribution net requires the most stringent transmission line design approach and will be reviewed in Section 11.2. Section 11.3 introduces a novel high-speed clock distribution architecture which was developed using sound transmission line theory and concepts. Subsequent sections are devoted to the design of RF and microwave analog integrated circuits.

11.1.1 HIGH-SPEED DIGITAL CIRCUITS

Microprocessors can now be clocked at gigahertz rates. Interconnect issues (delay, path loss and signal integrity, including coupling between interconnects) now dominate the performance of many high-speed digital circuits. CMOS is the device technology of choice for digital circuits. In this technology, transistors are fabricated epitaxially on a typically lightly doped silicon substrate. Alternating dielectric and metal layers are grown on top of this. The lowest metal layer is designated M1 (for Metal One), and there can be six or more metal layers. Traditionally, the metal has been aluminium (Al) and the dielectric silicon dioxide (SiO_2). Interconnects on the lowest metal layers have small cross-sections, and are used for local connections, e.g. to from gates by interconnecting transistors. Higher metal layers are used to form longer connections and have larger dimensions to reduce loss. The highest metal layers are used for long haul connections.

On long haul interconnects first incidence switching is required, and not an *RC*-charging-like voltage rise which occurs with multiple reflections and a low-current, high-impedance driver. Typical examples are clock distribution nets and data buses extending across a chip. Here loss and signal integrity (e.g. coupling and reflections) are especially important. These connections have larger dimensions and are designed as transmission lines using the principles presented throughout this book. The most important principle is the provision of a signal return path, and avoiding structures that interrupt the electric and magnetic field. (These principles were discussed in Chapter 1.) Common transmission line technologies include stripline with a signal strip between two metal layers[1]; and coplanar lines both having well defined signal return paths. The stripline uses considerably more metal than other coplanar technologies to provide a signal return path, although design is considerably simplified. The most common coplanar lines are coplanar waveguide (CPW) and differential line. These technologies and principles are migrating to intermediate metal layers as signal speeds increase.

Passivated copper (Cu) conductors (having lower resistivity than Al) have been introduced to keep losses as low as possible in many instances. Low permittivity materials (replacing SiO_2) have been introduced to reduce capacitance and interconnect delay. (These developments were described in Sections 3.7.3 and 3.7.4 on pages 69 and 70, respectively.) Other factors include:

- The extremely small geometry.
- The thin SiO_2 dielectric layer.
- The metal which is sufficiently thin, in fact comparable with a skin depth of the highest frequency component, which complicates the loss mechanisms.
- The fact that the SiO_2 (the type most commonly used) dielectric layer is itself lossy. However, this loss can be reduced dramatically be what is known as 'patterning' — a process also increasingly applied to the conductor-backing metallization in many planar structures.
- Since the SiO_2 dielectric layer is also thin and the conductor backing is present, fields are drawn down to this grounded backing metal so that field lines extend into the silicon which has finite conductivity. Loss also increases because of this effect.

Open CPS has losses of around 3 dB/mm but, because of the above effects, the losses in this case where the CPS has the specific CMOS-silicon structure indicated, may well increase to approximately 4 or 5 dB/mm. In practice, this means that for a 6 mm length, all this leads to a total worst case loss of around 13.5 dB. This implies a power reduction factor exceeding 22 or, more relevantly, a voltage reduction factor of 4.73. This means that a pulse travelling along the 3 mm line will suffer a reduction in level to approximately one-fifth of the sending value. These data apply to a line width of 5 μm. By increasing the width to 30 μm the voltage reduction can be reduced to a factor of three rather than five. It is necessary in both cases to use buffering to maintain voltage levels.

11.2 CLOCK DISTRIBUTION

Of all the nets on a digital integrated circuit, the clock distribution net benefits the most from a transmission line synthesis approach as the timing characteristics of the clock system is a major market differentiator. Low skew, low jitter and fast edge rates are the major consdierations. Also conventional clock distribution systems consume 20–40% of the total chip power. Consequently, the manufacturers of high performance chips devote significant design resources to the design of the clock distribution network. In this and the next section, we will consider the design of high speed clock distribution architectures.

---

[1] These two metal layers can be ground and power distribution planes as well, as they just need to provide signal return paths for the long haul interconnects.

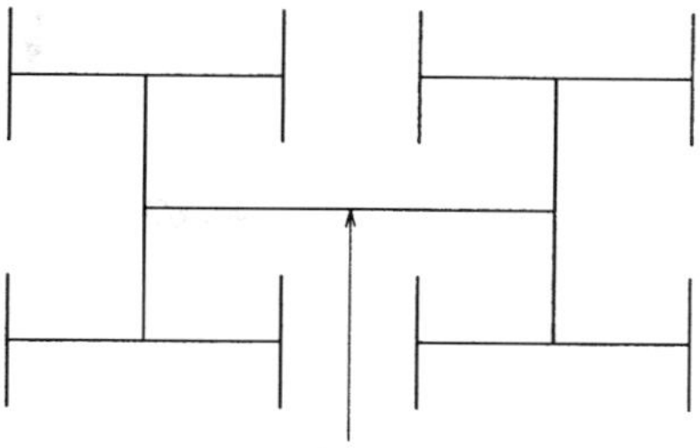

Figure 11.1 A conventional H tree clock distribution architecture.

Clock distribution is one of the most difficult challenges facing the designer of high-speed digital chips. Clock skew[2] and jitter targets become more difficult to meet when clocking at higher frequencies. Designs need to be partitioned if a high speed clock with low skew cannot be distributed across the chip. The partitioning of the design necessarily compromises performance, and this complicates design methodologies. One of the suggestions for coping with the difficulty of distributing a high-speed low-skew clock across a chip is to use a relatively slow global clock and local fast clocks. By 2014 it is estimated that the local clock rates will be 3.5 GHz and the local clock rates will be 10 GHz or more [434]. Each local clock would be synchronized to the global clock using phased locked loops. However, in nearly every situation it would be preferable to distribute the high-speed clock across the whole chip. One of the major concerns with conventional clock distribution architectures and synchronous design is that the majority of a circuit switches at nearly the same time, resulting in very large current drain from the supply. It is estimated [434] that the switching currents will peak at 1000 A unless tightly controlled global clock skews are utilized to avoid simultaneous switching.

There are several different schemes that enable the leading or falling edge to reach circuit elements at about the same time. The most common distribution scheme is shown in Figure 11.1, and is known as the H tree clock distribution net. In this scheme, clock edges propagate outwards from a centralized source that is usually phase locked to an external clock. The final clock loads, mainly the inputs of flip-flop, are at the leaf nodes of the tree. Intermediate buffers are inserted to keep the clock edge fast, maintain voltage levels, and to reduce skew. Wires are carefully sized so as to minimize skew due to differential RC delays. Jitter is mainly a function of the power and ground noise present at the phase lock loop and buffers, and is thus controlled mainly by careful power supply decoupling.

A traditional microwave perspective, although not strictly applicable here, is as follows. Each of the centrally driven interconnects branches into more interconnects, and so on. The design objective is to avoid generating a backward travelling wave at the interconnect junctions. One approach is to match the incoming interconnect to the (possibly) multiple outward going interconnects and another is to use buffering throughout the net. Matching can be done through appropriate design of the

[2] Clock skew is the difference in the time at which the leading or falling edges of a clock reach different nodes on chip. Ideally the leading and falling edges reach every part of the circuit at the same time so that there is zero skew.

characteristic impedances of the interconnects. It is difficult to carry this process beyond one junction, as the widths of the subsequent lines become unreasonably narrow to achieve the required high characteristic impedance. In narrowband amplifier design this is not difficult as reactive elements can be used to achieve the matched condition, but this is not appropriate to the square clock pulses as the matching must be achieved over a very broad bandwidth that includes the many harmonics of the clock frequency. An alternative, as already mentioned, is to use buffers, especially at the interconnect junctions, so that reflections are not an issue. In any event this is required to maintain sharp clock edges and also voltage levels.

Ignoring for the moment the time-of-flight distance from the clock feeds (at the extremities of the tree) to the gates, the clock skew for a perfectly designed tree would be zero. Even then, clock skew is introduced by coupling of noise into the clock distribution net, and careful design of the clock interconnect and its environment is required.

In the traditional 'forward travelling wave approach' just discussed, a large amount of energy is dissipated principally in two mechanisms. First, energy is used by the central clock driver and subsequent buffers to charge and discharge the clock lines. Secondly, the end terminations of the clock lines must totally absorb (i.e. dissipate) the power of the incoming signals so that there are no reflections. The power dissipated in the clock distribution net and the power dissipated by the terminating nodes are about equal. The total clock-related power in high-speed chips is typically 20–40% of the total power dissipated. Note that all of the signal power on the line is required to define the leading and falling edges of the clock, even though this power is largely dissipated to prevent reflections. As well, any signal coupled into the clock distribution line will lead to jitter on the clock signal and the need to tolerate clock skew. Variable loading on the clock distribution net also leads to clock skew.

The comments above lead to two important trains of thought. The first is that the overall power consumed in the clock tree would be reduced by avoiding the necessity to dissipate power (to eliminate reflections). The second is that there is no resilience in the clock distribution net to the effects of clocking and variable loading. Resilience would come from having energy that is committed to maintaining the integrity of the clock signal so that variable coupling to other circuitry has a relatively small effect. Differential transmission line design would also reduce the undesired coupling.

One clock distribution scheme that addresses the energy consumption issue is the salphasic™ clock system [436,437]. In this system a sinusoidal standing wave provides the clock reference. In the simplest implementation, a backplane connecting PCBs, a single transmission line is driven near its centre and terminated at its ends in a reactive load. Buffers tapping signals from this resonant line then form clock pulses that are in sync. A similar scheme as been attempted on-chip. The central problem is that the clock reference has a sinusoidal variation, and hence low slope 'edges.' Thus there are problems in establishing low-skew leading and falling edges. The central feature of the approach is that energy in the standing wave establishes a clock reference, and there are no timing variations along the line.

Another clocking scheme is the rotary clock™ distribution system, described in the next section, which has both of the desired attributes: low power clock distribution and accurately controlled clock skew.

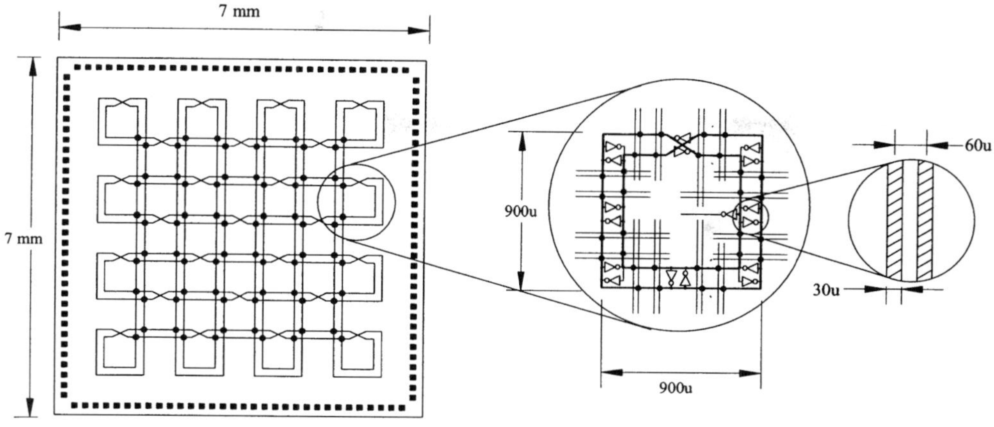

(a) (b) (c)

Figure 11.2 A rotary clock^{TM}: (a) layout of a 2.5 GHz oscillator with 25 interconnected rings of differential lines; (b) details of one ring showing anti-parallel inverter pairs; and (c) details of the differential line dimensions. (© 2000 Multigig, used with permission.)

11.3 ROTARY CLOCK^{TM} DISTRIBUTION

The rotary clock^{TM} distribution architecture [435] shown in Figure 11.2 achieves lower skew gigahertz-speed clocking at significantly lower power levels than does a conventional clock distribution scheme. Figure 11.2(a) is a layout for a 2.5 GHz rotary clock^{TM} having 25 interconnected rings. In Figure 11.2(a) there are four rings on the top row, then three, and the pattern is repeated. Each ring is essentially identical with the details of a ring shown in Figure 11.2(b). The rings consist of differential lines, as shown in Figure 11.2(c), each conductor of which is joined to each other periodically by anti-parallel inverters. (The clock signal would normally be taken from the ring as a differential signal, as indicated in Figure 11.2(b).) This produces a rotating clock wave that rotates around the ring. It is easy at first to associate this with an oscillator comprised of a ring of gates. However, in a ring oscillator the oscillating signal is input to one inverter and the output of that is input to a subsequent inverter and so on. In this structure (the ring oscillator) the clock power is dissipated at each inverter input and regenerated at each output. In the rotary clock^{TM} the anti-parallel inverter pair serves to top-up the clock edge as it passes.

Differential lines were discussed in Section 6.11 on page 218. The essential characteristic of these is that the signal is carried on the conductors comprising the differential line and the two conductors and the signals on them (except for being opposite in sign) are identical. This arrangement makes this type of line relatively immune to noise. The signal on the differential line is not floating, and because of this the outputs of the inverter pair driving the signals on the lines are either at the supply, say V_{DD} or at ground 0 V. Thus a voltage that changes from V_{DD} to 0 V on one line, say line 1, is input to one of the anti-parallel diodes which ensures that there is a transition from 0 V to V_{DD} on the other conductor, line 2. Similarly, the 0 V to V_{DD} transition on line 2 reinforces the V_{DD} to 0 V transition on line 1. The clock signal so generated rotates around the ring, and the clock frequency is set by the electrical length of the ring.

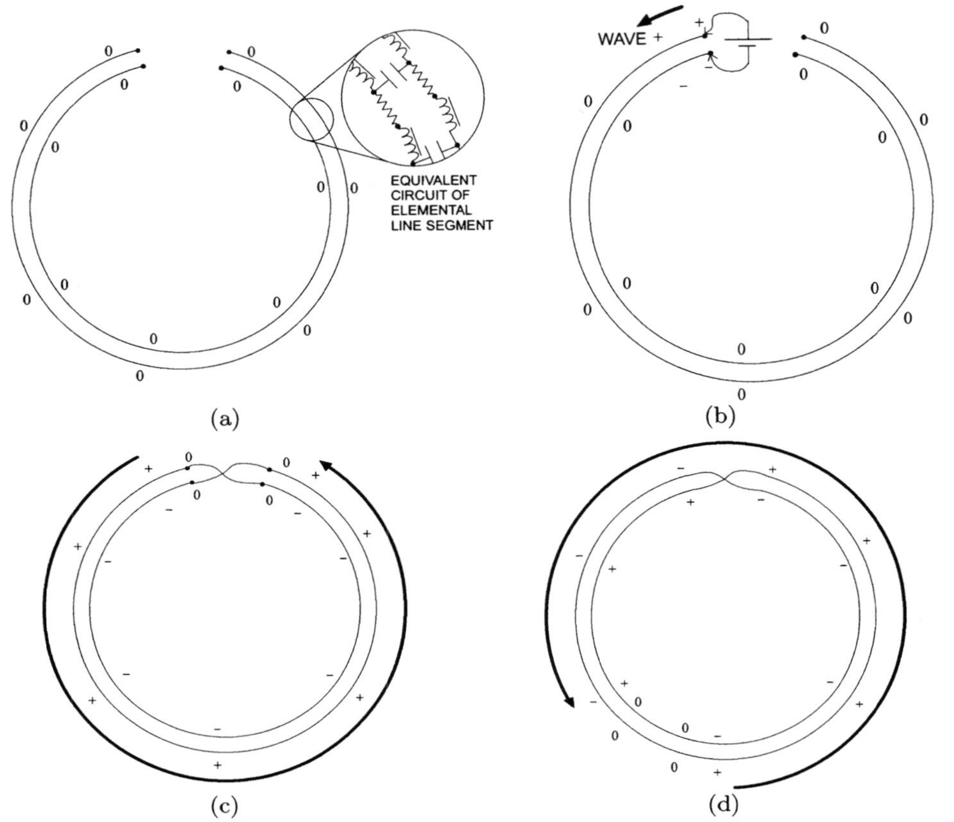

Figure 11.3 The origin of a perpetually rotating square wave: (a) differential transmission line loop; (b) application of a step voltage source; (c) a counter-clockwise rotating square wave; and (d) continuation of the rotating wave through a cross-over. (© 2000 Multigig, used with permission.)

11.3.1 CONCEPTUAL BASIS

The concept behind the rotary clock^{TM} architecture is the establishment of a perpetually oscillating square wave. Illustration of how this is achieved begins by considering the open loop shown in Figure 11.3(a). This loop, or ring, is a transmission line and is generally realized as a differential line. Ignoring losses, the differential line can be modelled by lumped inductors and capacitors, as shown in the inset. With the application of a step voltage, shown in Figure 11.3(b) a differential square wave begins travelling down the transmission line in a counter-clockwise direction. With the voltage source removed and the ends of the differential line cross-connected as shown in Figure 11.3(c), the rotating square wave contines around the ring through the cross-over, Figure 11.3(d). Thus the wave requires two roundtrips to complete a full cycle. Note that the cross-over, rather than a direct connection, is essential to obtain the rotating square wave.

The rotating square wave would continue indefinitely if there were no losses. In

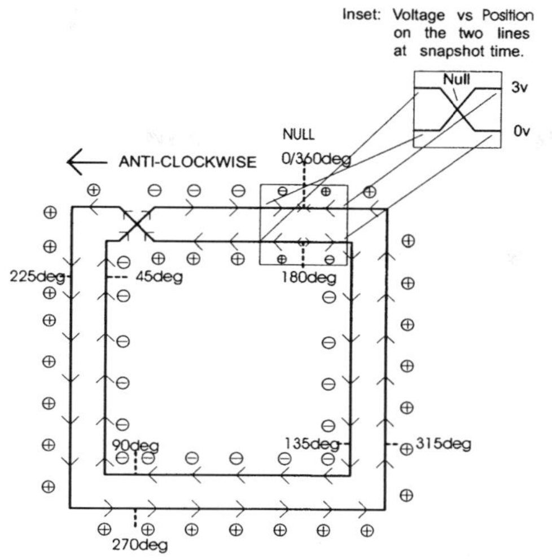

Figure 11.4 Clock phases around a single rotary clock^{TM} ring. Each rotation around the ring results in 180° phase shift. (© 2000 Multigig, used with permission.)

practice an anti-parallel inverter pair is required to compensate for losses, and also to provide a signal to intiate the rotating signal, see Figure 11.2(b). Start-up of the rotating wave begins with power supply ramp-up, or other noise event, and the lateral displacement of the anti-parallel inverter connections results in the preferred direction of rotation. Once the wave is established, very little power is required to maintain the wave, except for that required to compensate for losses of the differential line, and the the small amount of power consumed by the inverter inputs.

The rotating wave has a well defined skew related to the 360° rotation for two complete trips around the ring. The clock skew, in degrees, is shown in Figure 11.4. This can be used in controlled skew design, as is being proposed for high-speed systems [438].

Interconnecting the rings, as shown in Figure 11.2, ensures that the same signal appears on each ring. The process is akin to locking, and provides an opportunity to lock the rotary clock^{TM} to an external source. The locking range so obtained is around ±20%. Frequency dithering over a range of 5% to 10% is required for EMI control.

11.3.2 CIRCUIT MODEL OF A ROTARY CLOCK^{TM}

The rotary clock^{TM} is modelled as short lengths of transmission line between the inverter pairs at which there is substantial capacitive loading. A standard cell for the anti-parallel inverter pair is shown in Figure 11.5, and this is repeated to create the rotary clock^{TM} shown in Figure 11.6. The development of a simple equivalent transmission line model of an inverter pair and a section of transmission line is shown in Figure 11.7.

The per unit length capacitance, C, of the transmission line segments is dwarfed by the capacitive loading C_{INV} of an inverter pair. This is just as well or else the clock would rotate around the ring at the speed of light (reduced by the permittivity of the

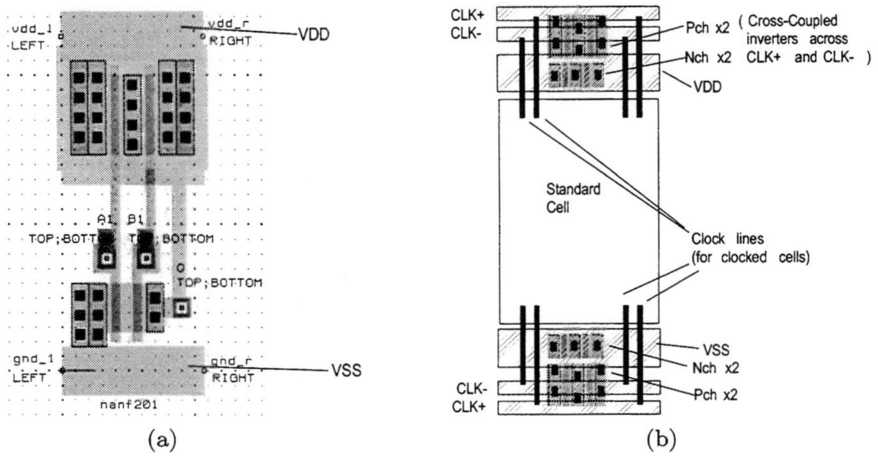

Figure 11.5 Anti-parallel inverter cells: (a) standard cell; and (b) modified standard cell with clock lines at top and bottom using the standard cell in (a). (© 2000 Multigig, used with permission.)

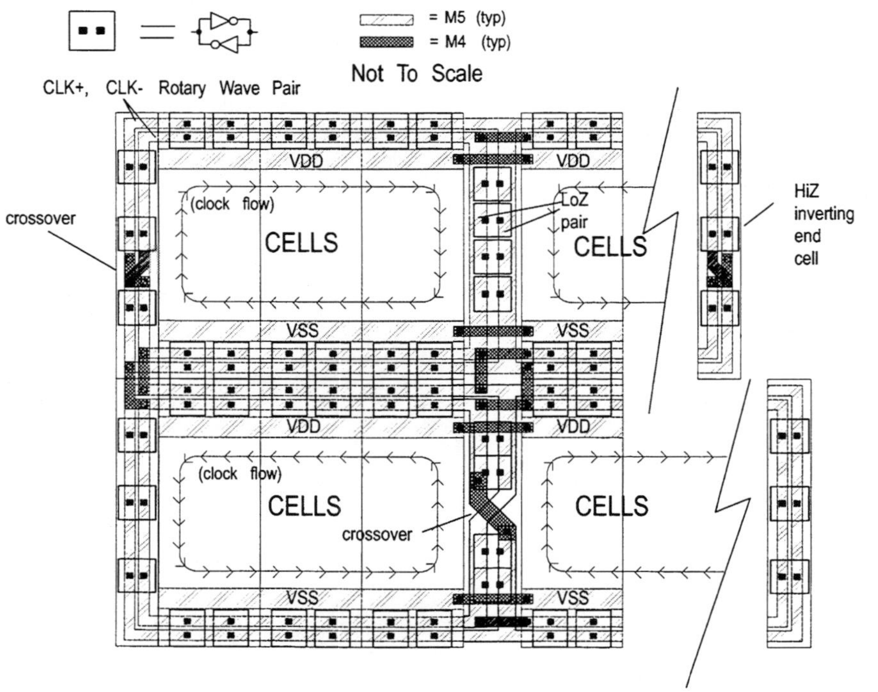

Figure 11.6 Two rings of a rotary clock^{TM} distribution net comprising the modified standard cells of Figure 11.5. (© 2000 Multigig, used with permission.)

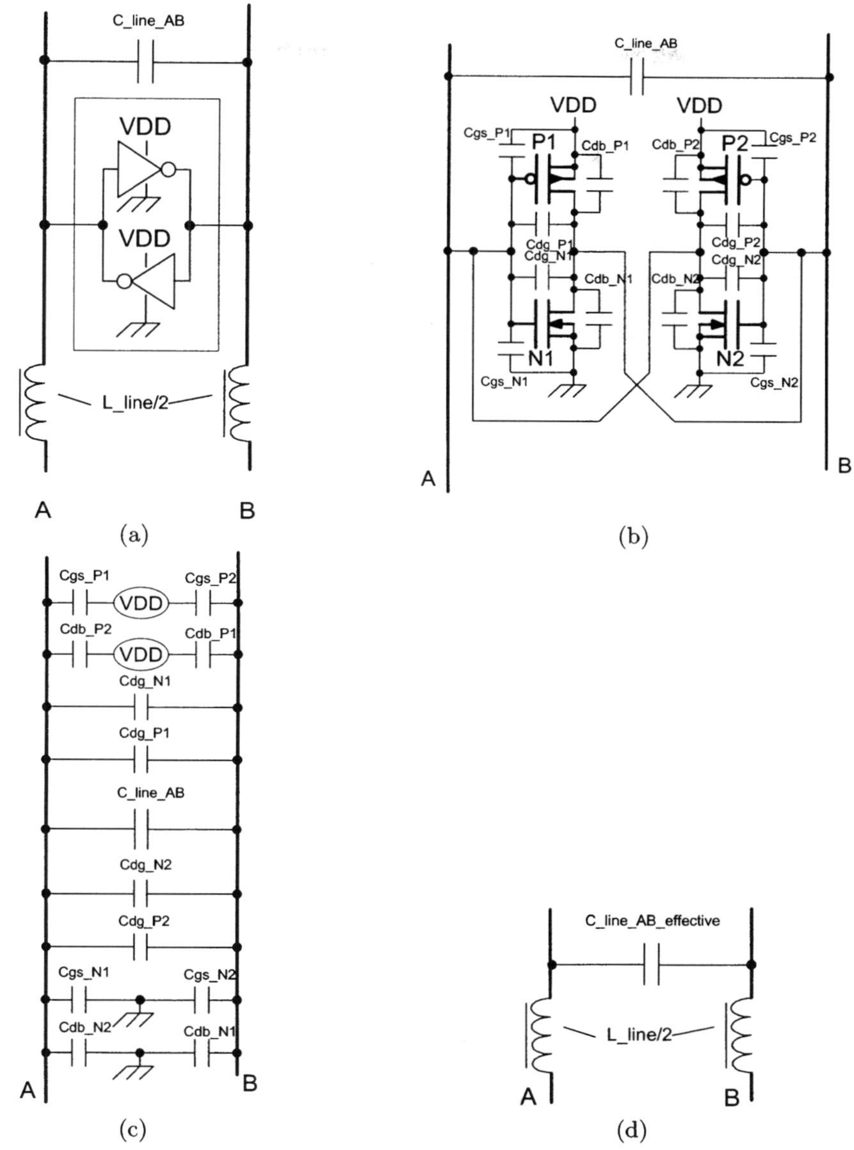

Figure 11.7 Evolution of an equivalent transmission line model: (a) schematic of an anti-parallel inverter section where L_{line} (L_line) is the total line inductance for the inverter-to-inverter length of interconnect; (b) detailed schematic showing transistor capacitances; (c) lumped element representation; and (d) simple equivalent transmission line segment model. (© 2000 Multigig, used with permission.)

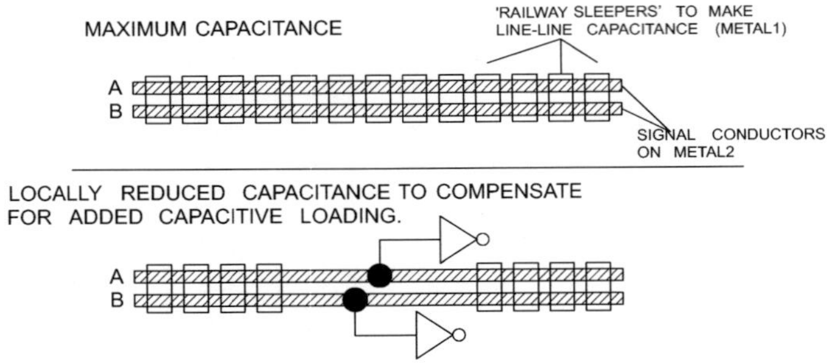

Figure 11.8 A passive capacitive loading scheme for reducing clock frequency. (© 2000 Multigig, used with permission.)

medium). If L is the per unit length inductance of the differential line, and there are $N_{\rm INV}$ anti-parallel inverter pairs per unit length around the ring, then the effective parameters describing the loaded line are:

$$\begin{aligned} L_{\rm EFF} &= L \\ C_{\rm EFF} &= C + 0.5C_{\rm INV}N_{\rm INV} \\ Z_{0,\rm EFF} &= \sqrt{L_{\rm EFF}/C_{\rm EFF}} \\ v_P &= 1/\sqrt{L_{\rm EFF}C_{\rm EFF}}\,. \end{aligned} \tag{11.1}$$

Here $Z_{0,\rm EFF}$ is the effective characteristic impedance of the line and v_P is the phase velocity which is considerably slowed down by the loading of the line by the inverters. Note that for differential signals the input capacitors of each inverter of a pair are in series, so that the total differential capacitance of an inverter pair is $0.5C_{\rm INV}$. (This capacitance is shown as C_line_AB_effective in Figure 11.7(c).)

The effect of buffers for extracting the clock reference is ignored for now, as the simplest implementation would be to have just one buffer per ring, whereas there would be many inverter pairs. The clock frequency f_C is then determined by considering the length l_R of a ring. Thus

$$f_C = v_P/(2l_R) \tag{11.2}$$

as the clock signal must go around the loop twice for a full cycle. Thus the centre frequency of the clock is determined by the length of a ring.

Coarse selection of the clock frequency is obtained by varying the effective capacitance and inductance of the line. This enables parts to be run at different speeds. Differential transmission lines with increased inductance can be obtained by meandering the lines. Additional capacitance is can be inserted using gate-channel capacitance or passive coupling structures, see Figure 11.8.

11.3.3 CASE STUDY: A 3 GHZ ROTARY CLOCK™

A four ring rotary clock™, a high level view of which is presented in Figure 11.9, was implemented in a 0.25 μm, 2.5 V CMOS process. More processing details are

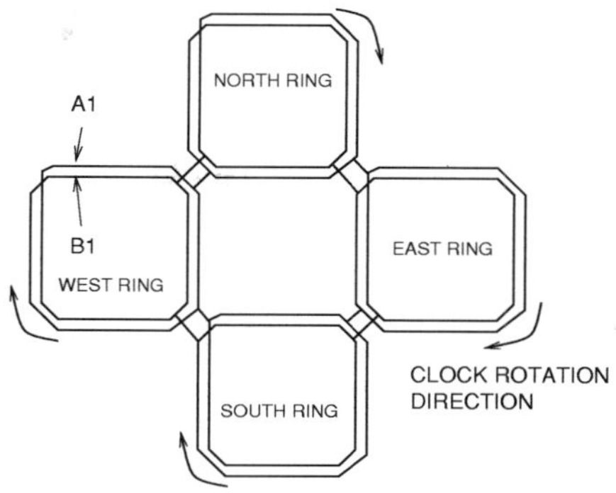

Figure 11.9 A four ring rotary clock™. The arrows indicate the direction of rotation of the clock signal. (© 2000 Multigig, used with permission.)

Table 11.1 Details of the 0.25 μm process.

Parameter	Value
Perimeter	2.4 mm (600 μm per side) 20 μm wide x 1.9 μm thick strips
Pitch	40 μm
Loop DC resistance	7.2 Ω
No. of active nodes (inverter pairs) per ring	40
Spacing between inverter pairs	60 μm
Inverter details:	
n channel devices	w = 37.5 μm, l=0.25 μm
p channel devices	w = 93.7 μm, l=0.25 μm
Transistor model	BSIM3v3 (including substrate, overlap and diffusion capacitaces)

given in Table 11.1. The process has 1.9 μm thick aluminium interconnects[3], and the clock frequency is 3.42 GHz with rise and fall times of less than 20 ps. The interconnect segments were modelled using a 20 pole equivalent LR matrix generated using FASTHENRY simulations. This model is accurate over a 1 MHz to 100 GHz frequency range. The characteristic impedance of the differential line is 7.7 Ω, the effective line velocity, v_p is 0.0548 of c (the speed of light in vacuum), and the supply current is 200 mA. In the same technology, a conventional tree clock distribution architecture achieves 600 MHz clocking.

The energy in the wave does not need to be dissipated in terminating buffers as in a conventional clock distribution architecture. The buffers must still be driven, but now (in CMOS technology) reactive energy is used to charge the input capacitance of the buffers. Power is dissipated due to coupling effects, in transmission line loss, as well as by inverters dissipating a small amount of power at the input. It can be viewed that the energy required to charge and discharge capacitors is stored in, and recovered

[3] Actually, an AlCu interconnect of resistivity 4.05×10^{-8} Ω is used which has 50% higher resistance than pure Al, but this bimetallic combination reduces electromigration.

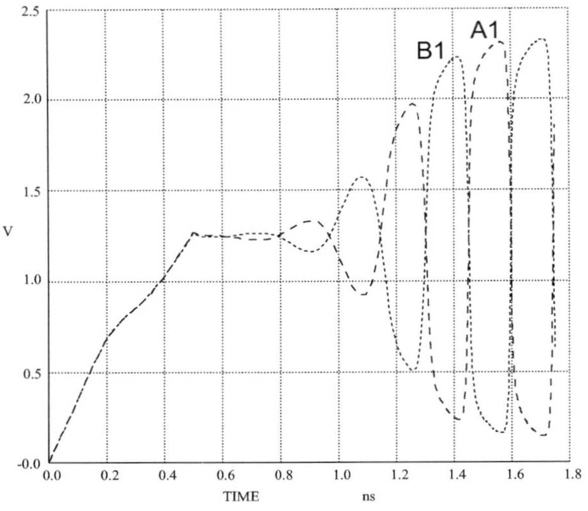

Figure 11.10 Start-up waveforms for the single ring rotary clock^{TM} oscillator shown in Figure 11.9. The waveforms are at two adjacent points on the differential line, and power supply ramp is 0.5 ns. (© 2000 Multigig, used with permission.)

by, the transmission line inductances, and so is not dissipated. This is known as an adiabatic process, but can be viewed as just a transmission line property, albeit here an artificial transmission line.

Figure 11.10 shows the start-up transient at two adjacent points, one on each conductor of the differential line. The curves are the waveforms at two nodes on the differential line, as indicated in Figure 11.9. After a short time steady-state conditions are reached with the waveform shown in Figure 11.11, and there is only a small differential balance error, see Figure 11.12[4]. Similar waveforms are obtained by plotting the voltages on the conductors of the differential line versus distance around the ring. The wavefronts rotate around the rings, and here rotation is clockwise enforced by the lateral displacement of the inverters of the anti-parallel inverter pair. Voltage waveforms at different locations around an individual ring are shown in Figure 11.13.

Thus the rotary oscillator clock establishes a travelling wave defined on the differential line that circulates around the ring as indicated in Figure 11.9. The ring has many inverter pairs, and the rotary clock^{TM} wavefront has a rising edge on one of conductors of the differential line conductors and a falling edge on the other conductor at the corresponding location. Initially, when power is applied or after reset, there is no signal on the differential line, and the travelling wave signal must establish itself. This will occur under any condition enforced by the anti-parallel inverter pair. Lateral displacement of the inverter pair along the line results in a preferred start-up direction. The differential wavefront travels around the ring with a defined skew. It takes two rotations of the ring to complete a full clock cycle, and the clock can be picked off at

[4] Again, this is for the four ring version, but identical results were obtained for the single ring clock.

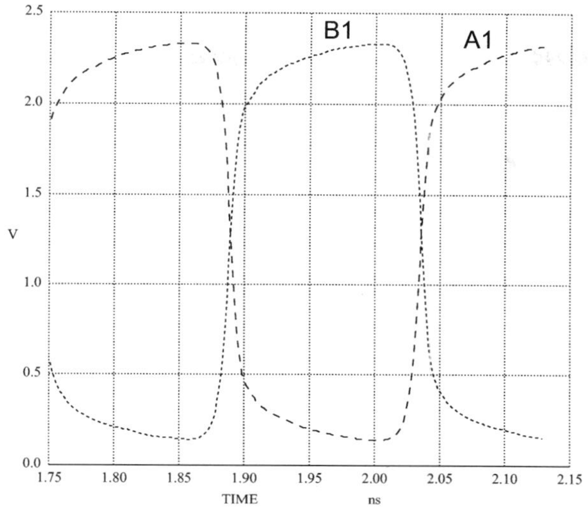

Figure 11.11 Steady-state waveforms for the rotary clock™ oscillator shown in Figure 11.9. The waveforms are at two adjacent points on the differential line. (© 2000 Multigig, used with permission.)

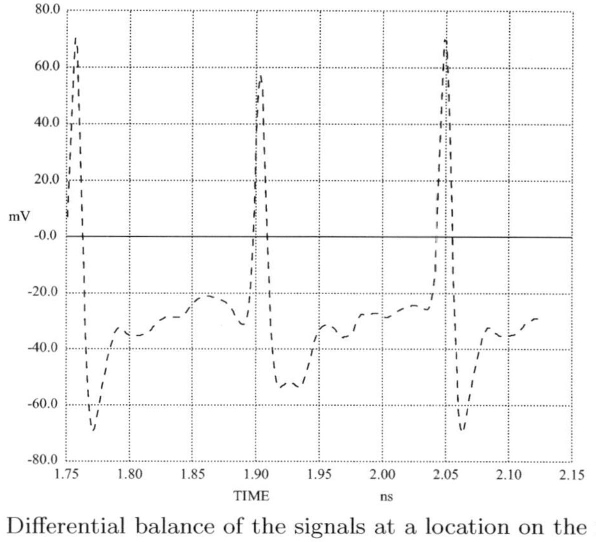

Figure 11.12 Differential balance of the signals at a location on the four ring rotary clock™ of Figure 11.9. The differential signal plotted is the sum of the A and B voltages offset by 2.5 V. (© 2000 Multigig, used with permission.)

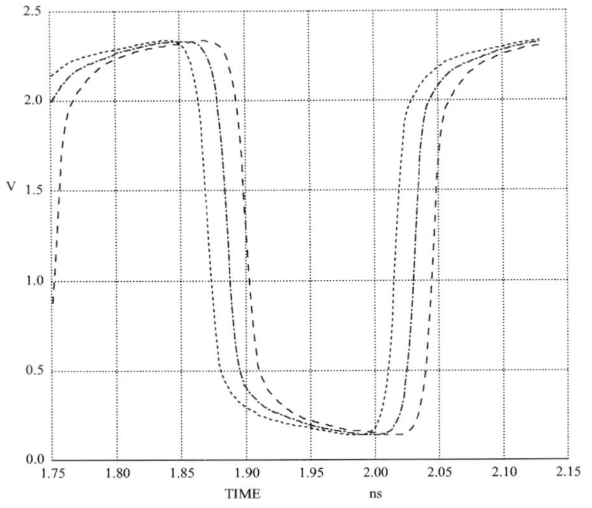

Figure 11.13 Voltages on the rotary clock^{TM} at different locations around the ring. (© 2000 Multigig, used with permission.)

any phase. The skew on the ring is precisely known and the clock reference can be extracted from any point on the ring. Thus, a multi-phase clock is supported. We will say more about this later.

The differential line currents corresponding to the voltage waveforms, in Figure 11.11, considered previously are shown in Figure 11.14. These current waveforms also travel around the ring and so the current drawn by the inverters, (for topping-up the rotating wave), is distributed in time so that current peaking is, at least due to the clock distribution, avoided. This is demonstrated by looking at the supply current versus time shown in Figure 11.15.

The locking behaviour of the four interconnected rings is well behaved, as shown in Figure 11.16. Even when the length of one of the rings is increased by a substantial amount, the lock is established, see Figure 11.17.

11.3.4 EFFECT OF COPPER INTERCONNECT

Implementing the above circuit using copper has the principal result that the resistance of the transmission lines decreases. The resistivity of copper as deposited is 2.0×10^{-8} Ω.m, and the lower resistance relative to the aluminium interconnect considered previously, leads to a change of frequency of less than 0.5%. The most significant change is that the total supply current is now 153 mA versus 203 mA with the aluminium interconnect. So the major effect is reduced power consumption. Figure 11.18 shows the steady-state waveform and Figure 11.19 are the line currents at the same locations. Here the total capacitance driven by the clock is 100 pF, due solely to the transistor input and output capacitances of the inverters.

With an additional 100 pF of load capacitance the clock frequency drops to 2.1 GHz and the voltage and current waveforms shown in Figures 11.20 and 11.21 were obtained. It is therefore important that buffering be used to redistribute the clock from the rotary distribution system. With the additional capacitive load the total

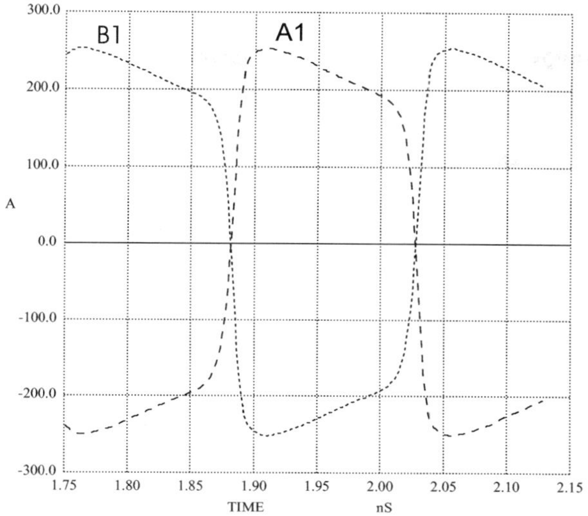

Figure 11.14 Currents on the differential line conductors of the rotary clock^{TM} shown in Figure 11.9. The voltage waveforms shown are at two adjacent points on the differential line. (© 2000 Multigig, used with permission.)

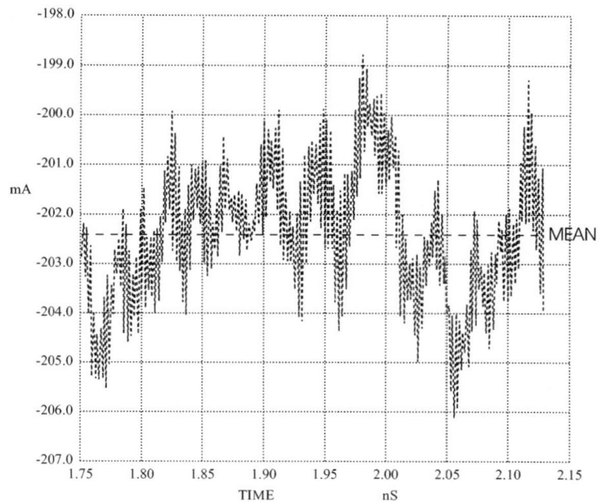

Figure 11.15 Power supply current on V_{DD} with the rotary clock^{TM} only and no driven circuitry. This is for the rotary clock^{TM} shown in Figure 11.9. (© 2000 Multigig, used with permission.)

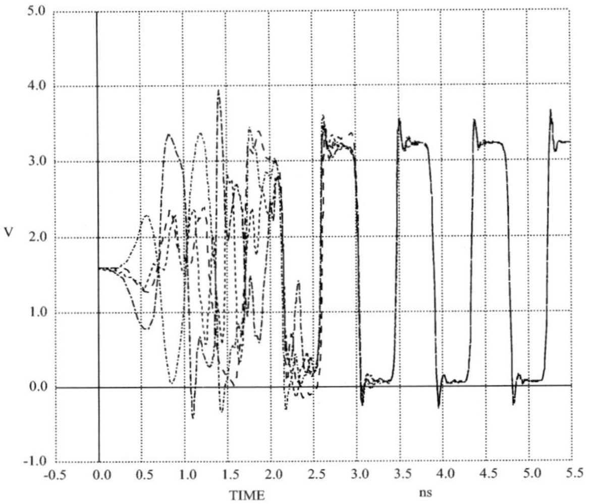

Figure 11.16 Start-up voltage waveforms at corresponding locations on each of the four rings of the rotary clock^{TM} shown in Figure 11.9. The power supply ramp is 0.5 ns. (© 2000 Multigig, used with permission.)

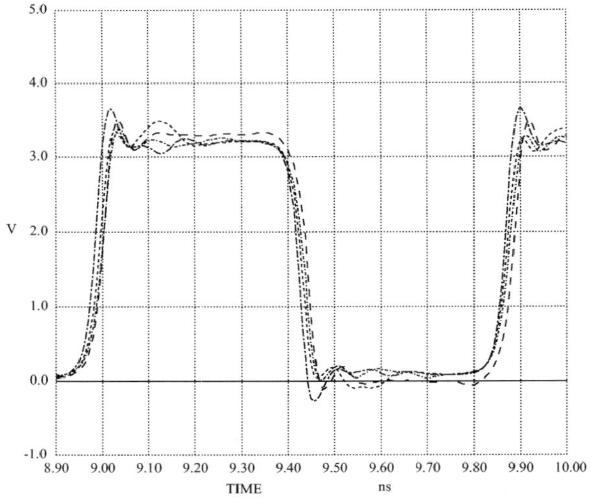

Figure 11.17 Voltage waveforms at corresponding locations on each of the four rings of the rotary clock^{TM} shown in Figure 11.9 when the length of the east wing is increased by 20%. (© 2000 Multigig, used with permission.)

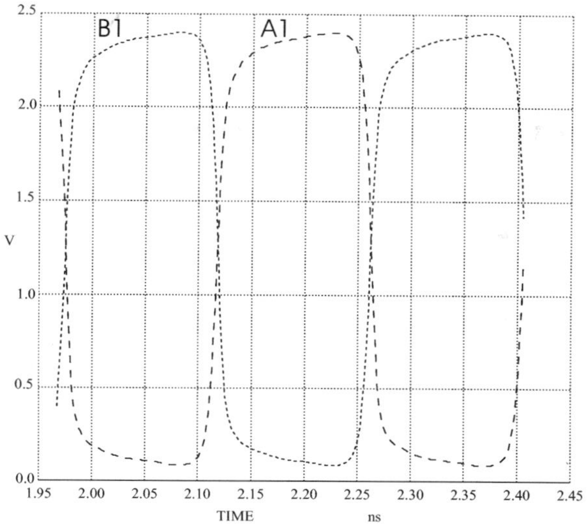

Figure 11.18 Steady-state waveform for a four ring rotary clock^{TM} in 0.25 μm Cu CMOS. The voltage waveforms shown are at two adjacent points on the differential line. (© 2000 Multigig, used with permission.)

current drawn from the supply is 235 mA. A conventional clock distribution system has a current draw proportional to the total capacitance to ground, C_{TOTAL}, the supply voltage, V, and the frequency, f_C. Thus the supply current

$$I_{\text{SUPPLY}} \approx C_{\text{TOTAL}} V f_C. \tag{11.3}$$

Under the same conditions as for the rotary clock^{TM}, i.e. at 2.1 GHz and now 200 pF total capacitance, the current drain of the conventional system would be 880 mA.

11.3.5 SUMMARY

The rotary clock^{TM} is an architecture that supports gigahertz speed clocks. It has the following attributes:

- No practical upper frequency limit. The upper frequency is limited by the f_T of the integrated circuit technology.
- It produces very good square waves, and with an f_T of approximately 30 GHz it produces square waves with rise and fall times of less than 20 ps.
- It is low power as it has an adiabatic characteristic in that the clock signal power is recirculated and not dissipated in an end termination.
- There is no limit to the size of the chip that can be clocked as the rotary rings can be linked indefinitely.
- The clock architecture inherently supports multiphase clocking. See, for example, the four phase rotary clock^{TM} system shown in Figure 11.22. It also supports non-overlapping differential clocking.
- The clock distribution uses the inductance of the interconnects to create a lumped element transmission line.

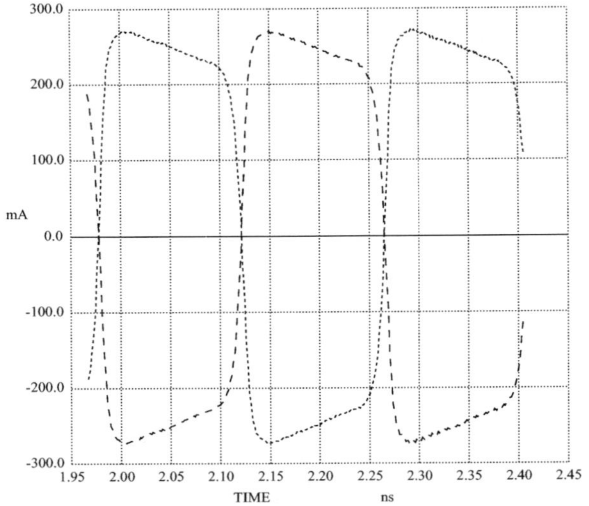

Figure 11.19 Steady-state current waveform for a four ring rotary clock^{TM} in 0.25 μm Cu CMOS. The voltage waveforms shown are at two adjacent points on the differential line. (© 2000 Multigig, used with permission.)

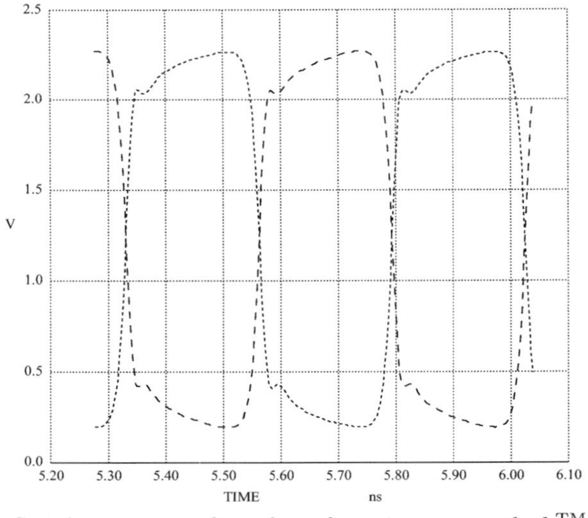

Figure 11.20 Steady-state waveform for a four ring rotary clock^{TM} in 0.25 μm Cu CMOS driving 100 pF of load capacitance. The voltage waveforms shown are at two adjacent points on the differential line. (© 2000 Multigig, used with permission.)

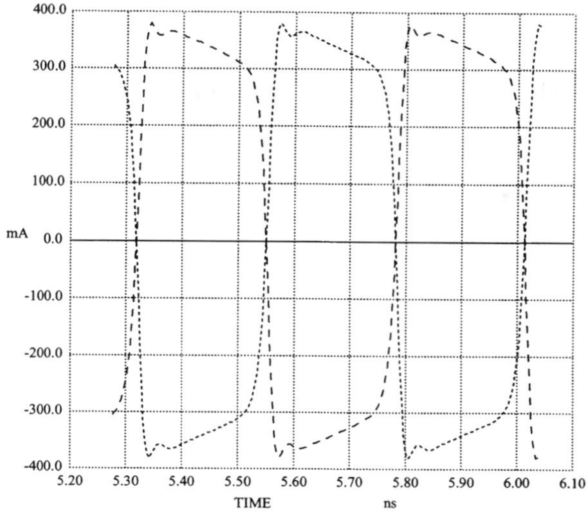

Figure 11.21 Steady-state current waveform for a four ring rotary clock^{TM} in 0.25 μm Cu CMOS driving 100 pF of load capacitance. The current waveforms shown are at two adjacent points on the differential line. (© 2000 Multigig, used with permission.)

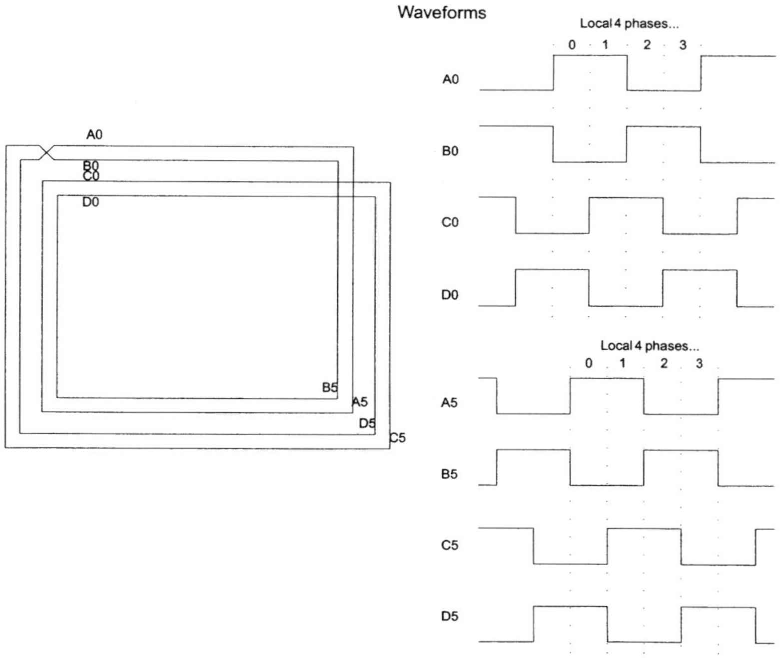

Figure 11.22 A four phase rotary clock^{TM} ring. (© 2000 Multigig, used with permission.)

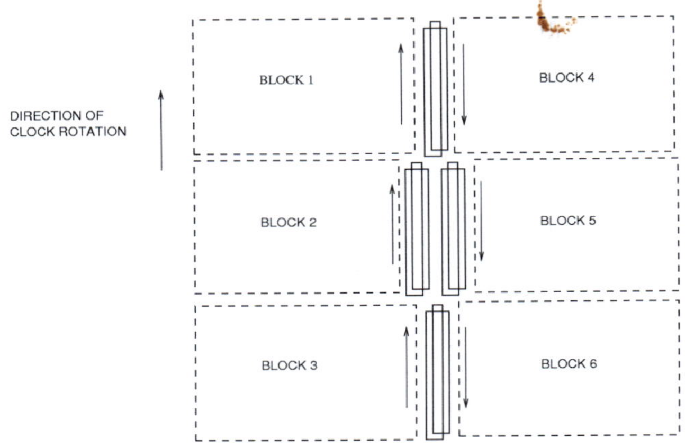

Figure 11.23 Elongated four ring rotary clock™. (© 2000 Multigig, used with permission.)

- The clock drives high total capacitance and still maintains squareness.
- It compensates fully for resistive and skin effect losses on the network through the regular squaring-up of the clock wave by the anti-parallel inverter pair.
- The rings can be compressed, as shown in Figure 11.23, with application to a broader range of circuits.

11.4 RF AND MICROWAVE ACTIVE DEVICES

Microstrip networks are particularly easily adapted to circuit design using transistors of various types. In the case of an amplifier, the circuit is appropriately optimized to provide low noise, moderate-to-high gain, or substantial output power. The specific types of transistor which can be implemented include:

- BJTs (Bipolar Junction Transistors)
- MESFETs (Metal Epitaxy Semiconductor Field Effect Transistors)
- HEMTs (High Electron Mobility Transistors)
- PHEMTs (Pseudomorphic HEMT)
- HBTs (Heterojunction Bipolar Transistors)
- LDMOS (Laterally Diffused Metal Oxide Semiconductor field effect transistor)
- CMOS (Complementary Metal Oxide Semiconductor field effect transistors).

There are a few other types of transistor devices, and it is likely that more will be developed during the lifetime of this text. Semiconductor materials broadly applicable to the above types of transistors include: silicon, silicon-on-insulator (SOI), silicon-germanium (SiGe), gallium arsenide (GaAs) and indium phosphide (InP). An extension of SiGe is silicon-germanium-carbon (SiGe:C), mainly for RF BiCOMS technology.

Prior to the mid-1980s, microwave amplifiers and oscillators were mainly implemented using various types of microwave active diodes, since transistors with sufficient available gain at microwave frequencies, or sufficiently low-noise or power capability, were simply not obtainable. Further, even when such transistors first became available, it was still several years before the requisite design tools were sufficiently refined for regular use. Now these obstacles have largely been overcome, and extensive CAD and computer simulation techniques can be used.

As a result, microwave transistor amplifiers, oscillators and many other types of circuits have been designed in both MIC and MMIC form. Active diode devices, such as Gunn (TED) and IMPATT, are still used in some instances, particularly at the high millimetre-wave frequencies (beyond approximately 100 GHz) where transistors with useful gain cannot yet be employed.

Amplifier circuits are generally designed for either narrowband or broadband applications, and microwave oscillators are also fully designable around the types of transistors listed above. Operating frequencies can range from below 1 GHz up to frequencies well beyond 90 GHz.

Hybrid microwave integrated circuits, generically termed MICs, are frequently used for many amplifier and oscillator configurations but Monolithic Microwave Integrated Circuits (MMICs), fabricated with any of the semiconductors mentioned above (choice depending upon several parameters), are becoming increasingly common for many applications. The present situation allows the co-existence of MICs and MMICs, and it is felt that, to some extent, this co-existence will probably continue for many years.

There are a number of existing texts on active MIC/MMIC design that provide excellent bases for the various design requirements [424, 439–441]. Reliability of GaAs MMICs has also been extensively reported on by Christou [442].

Computer-aided microwave design and engineering is now used extensively in this industry for both active MIC and MMIC design purposes, and some of the popular packages are available from the following companies:

- Agilent, http://www.agilent.com
- Ansoft, http://www.ansoft.com
- Applied Wave Research, http://www.appwave.com
- Cadence Design Systems (in particular Spectre), http://www.cadence.com
- Barnard Microsystems, http://www.barnardmicrosystems.com
- Jansen Microwave, http://www.linmic.com

During the lifetime of this book, other companies are sure to rise to prominence. In addition, many companies provide electromagnetic tools for the analysis of passive interconnect structures.

11.5 YIELD AND HYBRID MICs

Yield is one of the most important parameters in the integrated circuits industries, including the production of both MICs and MMICs. There are problems associated with this yield parameter for both types of technology and, in an attempt to improve the situation, Eda *et al.* [443] developed a novel thin-film technology to create miniature hybrid microwave integrated circuits (they term these MHMICs). The processes used in this technology allow for the full integration of thin-film resistors and capacitors together with conventional microstrip lines. This team reported their techniques for grounding the thin-film capacitors to the substrate, and demonstrated a 14 GHz power amplifier using this technology. The specific advantages of MHMIC technology include:

- size reduction (to two-thirds that of a conventional MIC)

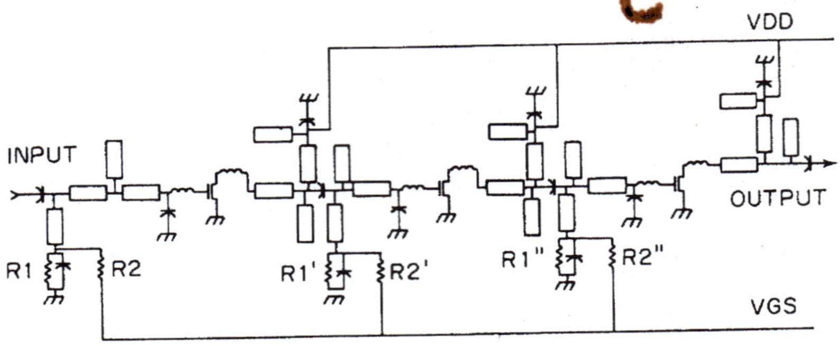

Figure 11.24 A Solid-State Power Amplifier (SSPA) module constructed with Miniature Hybrid Microwave ICs (MHMICs) fabricated by a novel thin-film technology. (© 1990 IEEE. Reprinted, with permission. from Eda *et al.* [443].)

- reduction in the number of discrete parts by a factor of three
- halving of the MIC mounting and bonding processes
- cost reduction due to the use of relatively inexpensive substrates.

There is also, and this is particularly important, a high yield obtained, and additionally an improved performance due to a significant decrease in the values of parasitic elements. The circuit diagram of the 14 GHz amplifier built on an MHMIC basis is shown in Figure 11.24.

In Section 11.6 of this chapter we consider, in detail the specific topic of amplifiers in both MIC and nominally MMIC technologies.

11.6 AMPLIFIERS

Amplifiers can be optimized for low-noise, moderate-to-high gain, or substantial power output. Using MESFETs, PHEMTs or HBTs, amplification can be obtained at frequencies extending well into the millimetre-wave regions (above 30 GHz). Oscillators can also be designed around the transistor devices.

There are many circuit configurations, and we shall only concentrate on a restricted number here — focusing on relatively simple single-ended designs and then describing distributed and balanced amplifiers. Another approach, namely feedback amplifiers, is also used quite often. However, we shall consider this topic outside the scope of the text and refer to several excellent references for this design approach.

Any microwave transistor amplifier requires the circuit arrangements indicated in the general block form of Figure 11.25. The DC biasing circuit is fairly standard, it does not involve any microwave constraints and it will not be discussed here. The lowpass filters ('bias circuits') can have one of several forms, some details of possibilities being given in Chapter 8. It is the realization of the input and output networks, and occasionally feedback networks, that is of principal significance to the designer of any amplifier. A slightly more detailed presentation of the device with its input and output networks is given in Figure 11.26. Here the intervening networks are labelled 'optimizing' rather than matching, since Γ_s in particular will be anything but the value for matching.

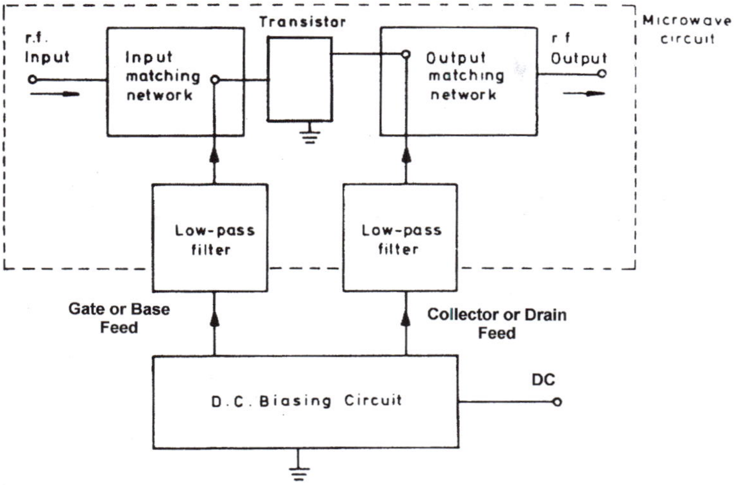

Figure 11.25 Block diagram indicating the microwave networks required for an amplifier.

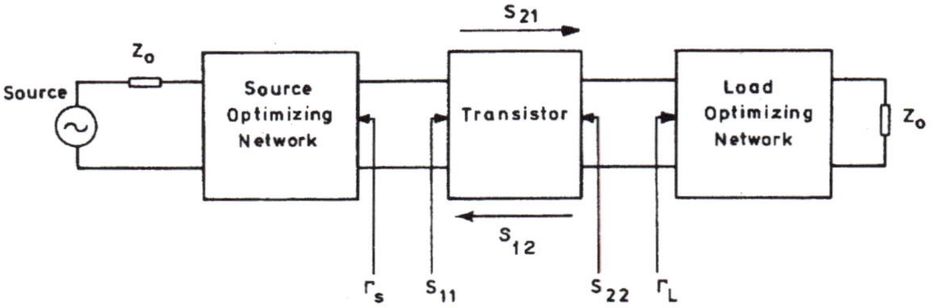

Figure 11.26 The scattering parameter (S_{nm}) and reflection coefficients (Γ) associated with a microwave transistor amplifier.

Frequently engineers use the generic term 'matching' to describe the requirement — regardless of whether power or noise optimization is actually to be designed for. It is, however, important to specify which form of optimization is meant by matching, as different input and output intervening networks will result for noise-matching and power-matching.

11.6.1 LOW-NOISE AMPLIFIER DESIGN STRATEGY

In this section we consider the general design strategy applicable to low-noise amplifiers. In general, the maximum noise figure (F) acceptable over the frequency band must be specified with the associated power gain and also the output VSWR (r).

The next requirement is to list the scattering parameters for the selected transistor device. These have to be measured at frequencies across the band of interest — and with the bias conditions (especially the collector or drain current) set for minimum noise figure. Each parameter must have both magnitude and phase specified.

Before any further design calculations are attempted, it is essential to check the stability and maximum available gain, and this can be accomplished using standard computer-based design software.

As with any amplifier, an optimum source impedance ($Z_{s,\mathrm{opt}}$) exists for a minimum noise figure. With $Z_{s,\mathrm{opt}}$, an associated optimum reflection coefficient Γ_o can be determined based upon the system characteristic impedance Z_0, which is usually 50 Ω. These quantities are best determined by measurement. It should also be carefully verified that the transistor will still yield some power gain under these conditions (a very low noise 'amplifier' which does not provide gain is not a very useful system element!). Computer routines yielding automatically plotted graphs can assist in this verification.

The problem is then to design an input network yielding the desired output reflection coefficient Γo.

The primary purpose of the input optimizing network is to realize the optimum source impedance for minimum noise. The main design steps are:

1. Convert Γ_o into an equivalent normalized admittance Y_{NF}.
2. Realize the susceptive component of Y_{NF}, i.e. B_{NF}.
3. Realize the real part of Y_{NF}, i.e. G_{NF}.

For an amplifier covering only a narrowband, an eighth-wavelength section of line yielding an inductive element can frequently be used to realize Step 2, and a quarter-wavelength impedance transformer may be selected to transform back to the 50 Ω source. The characteristic impedances of these microstrip or coplanar lines are determined, and the design approaches provided in this book enable the physical geometries of these lines to be computed for the chosen substrate. The denormalized noise impedance Z_F can be expressed in terms of the reflection coefficient Γ_o as

$$Z_F = \frac{(1-|\Gamma_o|^2)Z_0}{1+|\Gamma_o|^2 - 2|\Gamma_o|\cos\Phi_0} + j\frac{2Z_0|\Gamma_o|\sin\Phi_0}{1+|\Gamma_o|^2 - 2|\Gamma_o|\cos\Phi_0} \tag{11.4}$$

in which Φ_o is the phase angle of this reflection coefficient. Also, the usual impedance-to-admittance conversion is

$$Y_F = 1/Z_F \tag{11.5}$$

with the value of Z_F substituted after evaluation using Equation (11.4). The network topology is dictated within the computer program. This approach allows some flexibility in the design values and rapid indication of the sensitivities to various parameter changes. It is also important to check whether the characteristic impedances are capable of being realized without difficulty in microstrip or coplanar line. (For microstrip values below about 10 Ω lead to excessively wide lines, and when above 110 Ω they become difficult to manufacture.)

Discontinuity effects must be accounted for and, for example, right-angled bends should usually be mitred as described elsewhere (see Section 7.7 on page 237. Short-circuits, through the substrate or across to ground planes, can readily be realized using wire bonds or bridges.

Finally, the output matching network is designed. One of the design specification requirements would be a maximum allowable output VSWR (typically less than 1.5). Ideally, the VSWR would be that of a perfect match, i.e. unity, and that value is obtained when the load is conjugately matched — the condition for maximum power transfer. It will be assumed that true conjugate matching is desired, at least at the centre frequency. Now the output reflection coefficient is also dependent upon the input network reflection coefficient and the expression required, which is derived from S-parameter theory, is

$$S'_{22} = S_{22} + \frac{S_{12}S_{21}\Gamma_s}{1 - S_{11}\Gamma_s}. \tag{11.6}$$

It is important to note that, while $|S'_{22}|$ does not differ very much from $|S_{22}|$, the associated phase angles differ to a much greater extent and it is therefore of considerable importance to calculate the complex S'_{22} accurately so that the final computations are correct.

A practical approach is to start with the output impedance Z_{22} of the transistor, which it is desired to match. Often two-element networks will not provide sufficient bandwidth to meet the specification, and at least a five-element network is necessary. The steps are as follows:

1. Mainly for collector or drain bonding convenience, a short length of 50 Ω microstrip (or coplanar) line is extended from the collector location to the next element of the output matching network.
2. A short-circuited eighth-wavelength stub of characteristic impedance is connected to the network end of the line in Step 1. This is designed to tune out most, but not all, of the susceptance at this plane in the network. The remaining sections of the network will finally tune out all the susceptance.
3. Now, choose to employ just a length of microstrip or coplanar line between the load and the final shunt matching element. Thus the load is matched (real part of normalized admittance equal to 1.0) at all planes everywhere along this length of line, and it remains to properly connect a matching element somewhere along this line.
4. With the network designed as far as Step 2, the normalized admittance Y_N looking towards the transistor must be forced to equal

$$Y_N = 1 - jB_N. \tag{11.7}$$

In this expression the real part is unity, to force the matching condition, and the imaginary part is the susceptance that applies to the network and which must be tuned out by appropriate microstrip, lumped element or coplanar network design.

11.6.2 HIGH-GAIN NARROWBAND AMPLIFIER DESIGN

Many systems require relatively high-gain amplifiers in addition to low-noise or high-power stages, and in this section we indicate the design technique applicable to microstrip realization of such designs. By 'narrowband' we generally mean circuits having bandwidths less than 10% of their centre frequencies.

The overall configuration will be identical to that presented in Figures 11.25 and 11.26, and now the aim is to design the input and output microstrip networks so that high-gain is achieved. From transmission-line theory, the source reflection coefficient yielding maximum power gain is given by

$$\Gamma_{s,m} = \frac{Z_{s,m} - Z_0}{Z_{s,m} + Z_0} \tag{11.8}$$

where the quantities have been previously defined, except that now $\Gamma_{s,m}$ and $Z_{s,m}$ refer to maximum-yielding values of reflection coefficient and source impedance respectively. The source equivalent impedance for maximum power gain is therefore expressed as

$$Z_{s,\text{in}} = Z_0 \frac{1 + \Gamma_{s,m}}{1 - \Gamma_{s,m}} \tag{11.9}$$

where

$$\Gamma_{s,m} = |\Gamma_{s,m}| \cos \angle\Gamma_{s,m} + j \sin \angle\Gamma_{s,m} \tag{11.10}$$

is the source reflection coefficient yielding maximum available power gain.

When Equation (11.10) is substituted into Equation (11.9) and the resulting expression is rearranged, an expression of the form of the earlier Equation (11.4) is obtained, but of course with maximal quantities replacing the original low-noise quantities:

$$Z_{s,in} = Z_0 \frac{1 - |\Gamma_{s,m}|^2 + j2\,|\Gamma_{s,m}| \sin \angle\Gamma_{s,m}}{1 + \ |\Gamma_{s,m}|^2 - \ 2\,|\Gamma_{s,m}| \cos \angle\Gamma_{s,m}}. \tag{11.11}$$

The output matching network must also be designed for maximum power gain, and the load equivalent impedance for this network takes the following similar form:

$$Z_{lo} = Z_0 \frac{1 - |\Gamma_{l,m}|^2 + j2\,|\Gamma_{l,m}| \sin \angle\Gamma_{l,m}}{1 + \ |\Gamma_{l,m}|^2 - \ 2\,|\Gamma_{l,m}| \cos \angle\Gamma_{l,m}}. \tag{11.12}$$

We can now list the design procedure for this type of amplifier:

1. Convert the source and load reflection coefficients to the source and load equivalent impedances.
2. Find the source and load equivalent admittances from these impedances.
3. Realize the susceptance components with either a short or an open-circuited three-eighths-wavelength stub.
4. Realize the conductance components with quarter-wavelength transformers.

11.6.3 DESIGN EXAMPLE

As an example of high-gain narrowband amplifier design, in the next section we consider a 6 GHz amplifier originally given by Liao [439].

A GaAs MESFET amplifier is required to provide maximum power gain over a 400 MHz bandwidth centred on 6 GHz. The device's measured parameters (all at 6 GHz) are

$$\Gamma_{s,m} = 0.761\angle 177^\circ \qquad \Gamma_{l,m} = 0.719\angle 104^\circ \qquad G_{a,\max} = 12 \text{ dB}.$$

The design starts as usual with the input matching network. From Equation (11.10) the source equivalent impedance is:

$$\begin{aligned} Z_{s,in} &= \frac{50(1 - |0.761|^2) + j2 \times 50 \times \sin(177^\circ)}{1 + |0.761|^2 - 2|0.761|\cos(177^\circ)} \\ &= 6.77 + j1.28. \end{aligned} \tag{11.13}$$

Convert $Z_{s,\text{in}}$ to $Y_{s,\text{in}}$

$$Y_{s,\text{in}} = 1/Z_{s,\text{in}} = 1/6.77 + j1.28 \ \Omega = 0.15 - j0.02 \ \text{S}. \tag{11.14}$$

Use a three-eighths-wavelength open-circuited stub to represent this shunt admittance such that $Y = jY_0 \tan(\beta l)$. Such a stub behaves as a shunt inductor of susceptance $-jY_0$. We therefore have the characteristic impedance for this stub:

$$Z_{01} = 1/Y_0 = 1/\text{Im}[Y_{s,\text{in}}] = 1/0.02 = 50.0 \ \Omega. \tag{11.15}$$

Next, a quarter-wave transformer will transform the source resistance of 50 Ω to the source equivalent conductance of 0.15 S. This transformer has the characteristic impedance

$$Z_{02} = \sqrt{50/0.15} = 18.26 \ \Omega. \tag{11.16}$$

This completes the input matching network. We now require the output matching network. For this we first find the load equivalent impedance from Equation (11.12)

$$\begin{aligned} Z_{10} &= \frac{50(1 - |0.719|^2) + j2 \times 50 \times 0.719\sin(104^\circ)}{1 + |0.719|^2 - 2 \times 0.719\cos(104^\circ)} \\ &= 12.83 + j37.30 \ \Omega. \end{aligned} \tag{11.17}$$

Now convert Z_{10} to Y_{10}:

$$Y_{10} = \frac{1}{Z_{10}} = 1/12.83 + j37.30 = 0.008 - j0.024 \ S. \tag{11.18}$$

Next, we use precisely the same method of employing a three-eighths-wavelength stub that was used for the input network, to realize an equivalent shunt inductor of susceptance $-jY_0$. Its characteristic impedance must be

$$Z_{03} = 1/Y_0 = 1/\text{Im}[Y_{l0}] = 1/0.024 = 41.67 \ \Omega. \tag{11.19}$$

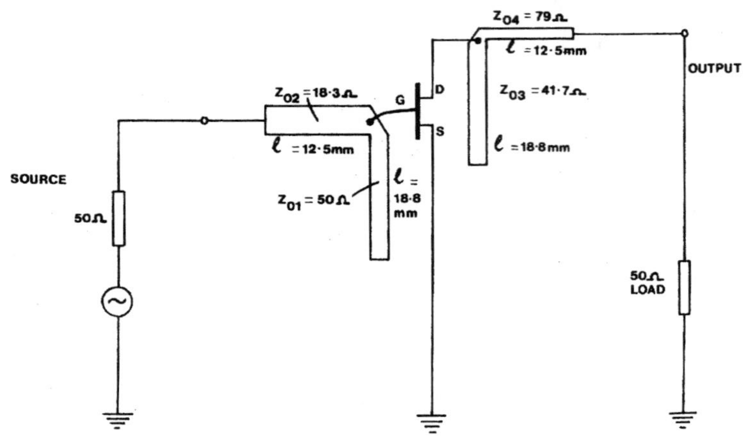

Figure 11.27 Circuit layout for the narrowband amplifier design example (after Liao and reprinted, with permission).

Finally, we use a quarter-wave transformer to transform between this network and the 50 Ω load impedance. The characteristic impedance of this transformer line section is

$$Z_{04} = \sqrt{\frac{50}{0.008}} = 79 \ \Omega. \tag{11.20}$$

The final circuit layout is shown in Figure 11.27, in which the source, output load and FET and their connections are shown symbolically. Actual physical designs of the microstrip sections for this circuit are left as an exercise for the reader.

High-Frequency Satellite Low-Noise Block

An important development that one must consider in amplifer design is the effect of significantly increasing the operating frequency. For example, the European satellite communications bands include 11.5 GHz and there is intense activity around bands including 22 GHz and even higher — into the millimetre-wave regions in fact. Millimetre-wave distribution systems (LMDS, MVDS) also demand frequencies of 28 GHz and beyond. Again, at this stage it is left to the reader to carefully enter into these considerations (issues such as choice of substrate and adjustments for discontinuities should be addressed on the basis of information provided earlier in this text).

We now have design steps for low-noise and high-gain amplifier stages. Such stages represent, for example, the first two cascaded requirements for 'low-noise block' (LNB) or 'down-converter' modules in satellite and other types of receivers and they are followed by a microwave mixer from which the Intermediate Frequency (IF) is extracted. After this comes the IF amplifier which drives the signal, now centred on about 1.35 GHz, via the down-lead cable to the TV set. In many designs of production LNBs, most of the circuitry is monolithically integrated within one or more MMICs.

The recent state-of-the-art is well represented by a 22 GHz LNB designed by Imai and Nakakita [444]. This circuit uses three cascaded HEMTs in the pre-amplifier, to achieve the required low-noise and gain at 22 GHz, and a local oscillator based upon a 21.4 GHz GaAs MESFET oscillator stabilized using a dielectric resonator. The overall noise figure is less than 2.8 dB over the 22.5–23.0 GHz frequency range, and the local oscillator achieves a frequency variation of less than 600 kHz over a −20°C to +60°C temperature excursion. The overall conversion gain achieved is 47–48 dB over the full frequency range. A simple E-field probe transition from the waveguide input to the microstrip circuit was used. This performed adequately, yielding a maximum insertion loss of only 0.2 dB over this narrow bandwidth.

The IF amplifier was realized with three cascaded packaged MMICs such that it provided a noise figure below 2.5 dB and a flat gain of 32 dB over the entire IF band. It is the pre-amplifier that is of primary interest here, and this is designed on a 0.38 mm thick alumina substrate. Input and output matching networks are designed around each HEMT, using the principles described above, and the resulting stages are then directly cascaded. Finally, the matching networks of the noise-critical first and second stages are trimmed to minimize the total noise figure.

11.7 CUSTOM HYBRID AMPLIFIERS

For a variety of technical and manufacturing reasons, custom hybrid (i.e. MIC) amplifiers remain an important concept and one which several companies continue to develop with added value. A further evolution of this approach comprises building amplifiers using 'standard modules' but a more important development is that of custom hybrid amplifiers.

Building amplifiers from standard modules has been the approach adopted by a number of manufacturers. This was because glass masks were expensive and difficult to justify, production quantities were usually small, and the approach allowed for the rapid evaluation of specific customer requirements. The advent of a low cost route for circuit mask production, coupled with an increase in manufacturing volumes, provided an opportunity for significant reductions in custom development cycle time and manufacturing costs.

11.7.1 STANDARD MIC AMPLIFIER MODULES

Standard MIC modules inherently provide a fixed range of gain, noise figure and output power options, covering many frequency ranges. With a selection of standard module types it is possible to hold fully built modules in stock, thereby reducing cycle time in development as the manufacturing time is reduced. Whilst satisfying these requirements this approach is not efficient in production, as each module has functionality built in that is not always required in every design. As a consequence, much productive effort is absorbed in building circuit components that can be largely redundant.

A 2–18 GHz 'gain stage' single 6 mm square standard module is illustrated in Figure 11.28. This module has typically 9 dB gain with a 4 dB noise figure and an output

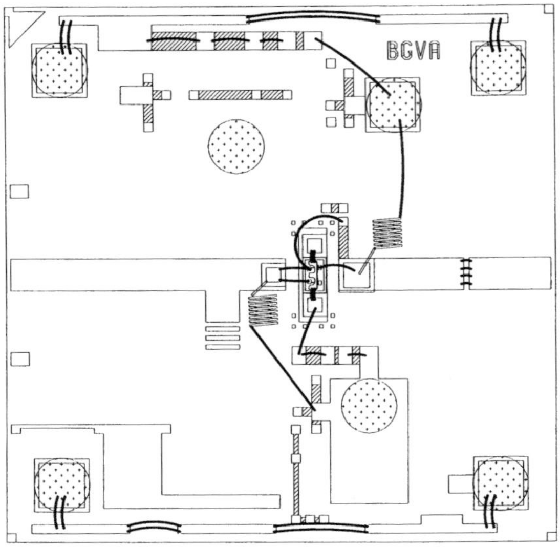

Figure 11.28 Single 6-mm standard module (courtesy: Filtronic Components Limited). This module is fabricated using a 6mm x 6mm x 0.38mm alumina substrate and the circuit is attached to the tungsten copper carrier using gold/tin solder.

power at 1 dB gain compression of 14 dBm [5].

This standard module has 33 wire bonds, five DC decoupling capacitors and eight through-circuit ground connections. It also requires eight wire bonds to interconnect it to modules placed at either the input or output. The capacitors in each corner are required to de-couple the power supply rail, which is formed by the interconnection of the separate sections on each module. It should be noted that the supply connections on the lower edge do not actually provide any supply for this module. The connections exist as a mandatory specification of the standard module format, as they may be required to link the supply through to another module. This necessary condition also requires the placing of a grounding connection at each corner of the module. The 50 Ω output RF line is broken and reconnected with four wire bonds. This is in case the module were to be used as the output stage of an amplifier in which case it would need a DC blocking capacitor fitted at this point. It is also possible to see on the circuit an unused ground connection — this was used in a now redundant temperature compensation network.

11.7.2 CUSTOM MIC AMPLIFIER MODULES

With this approach the customer-specific design route ensures that the amplifier contains only those parts that are required for functionality. Figure 11.29 shows a typical gain block that provides the same performance as that previously indicated in Figure 11.28. This is a section of a module built using a 24 mm × 4 mm × 0.38 mm

[5] For all information provided here on both the 'standard' and custom hybrid MIC modules, Filtronic Components (http://www.filtronic.com) are gratefully acknowledged [445].

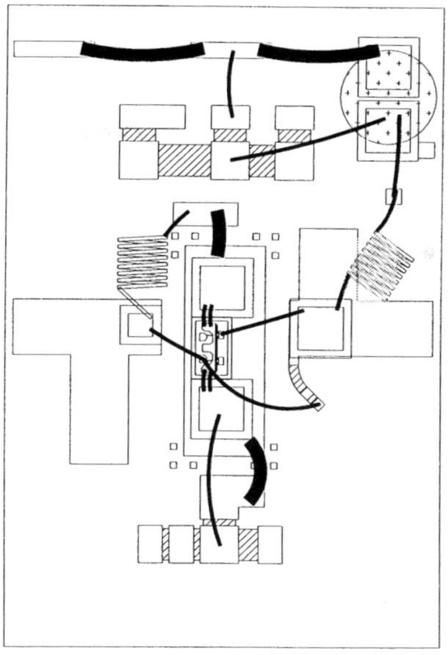

Figure 11.29 A portion of a custom-designed 24-mm × 4-mm module (courtesy: Filtronic Components Limited).

alumina substrate.

This gain block has only 15 wire bonds, one DC decoupling capacitor and two through circuit ground connections. The circuit layout occupies just 2.7 mm of the 24 mm long module. This functional block can now be cascaded with similar blocks to provide the required gain level. Other features such as temperature compensation for gain level and output power can also be realized on the 24 mm module. This module eliminates the requirement for multiple RF and DC module interconnections, with the consequent reduction in the number of DC decoupling capacitors and through circuit ground connections at the module interfaces.

The custom design route using a 24 mm long module also has benefits for the circuit designer. The major benefits result from designing the cascaded amplifier as a whole rather than as a collection of individual modules. This allows for the optimization of the layout to eliminate the excessive gain variation with frequency, that is a common feature with cascaded identical modules. This gain variation with frequency arises when the same gain block is cascaded with the same inter-device spacing. When a 1.3 mm length of 50 Ω line is inserted between the devices, the overall gain variation with frequency is 1.2 dB over the 1.5 to 19 GHz band — for a four-stage amplifier. This variation is less than 0.5 dB for a two-stage amplifier.

The selection of inter-device spacing also allows the amplifier interfaces to be better designed for manufacture. Each stage has a finite isolation, and therefore the spacing between the input device and the second device is critical in achieving the required input return loss. In this case, spacings of 1.4 mm and 0.4 mm, between the first two and last two stages, were found to be optima for reducing the input and output return losses respectively. Other spacings required tuning. The shift away from multiple 'standard' modules obviously removes the interconnections between each of

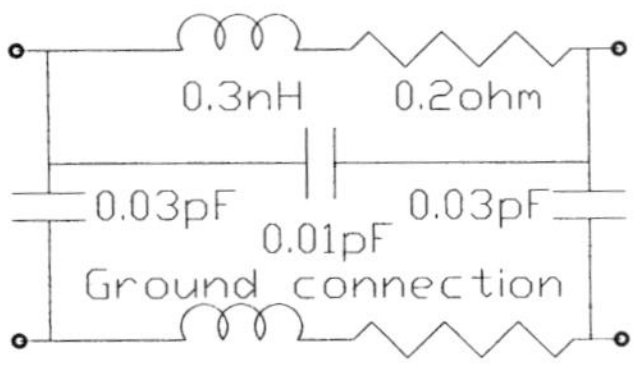

Figure 11.30 Equivalent circuit model for microstrip interconnects between standard modules (courtesy: Filtronic Components Limited [465]).

Table 11.2 Comparison of standard and custom MIC amplifier features.

Feature	Standard	Custom
Modules required	7	3
Ground posts	71	25
DC capacitors	30	11
Interconnects	56	7
Chip resistors	6	0
Unused rail bonds	60	3
DC rail bonds	101	31

the separate modules. A model used for such an interconnect is shown in Figure 11.30 [465]. The series inductor represents the worst case for two adjacent modules. The shunt capacitors are due to the end effects of the microstrip lines with the series capacitor formed between the two ends of the lines. The inductance in the ground plane is small compared with the bond wire inductance.

Taken alone, each interconnection may have a low insertion loss and small reflection coefficient. However, the insertion of this extra interface between two amplifiers can introduce significant extra gain variation. The interconnection itself has a worst case VSWR of 1.5:1 in the pass band.

Whilst in every instance the custom approach results in considerable reductions in the required numbers of features, the most dramatic instance is that of interconnects where a reduction by no less than a factor of eight applies. This is a highly significant aspect because of the substantially deleterious effects of these interconnects. The manufacturing costs are also approximately halved by adopting the custom design route.

11.8 BALANCED AMPLIFIERS

In 1964, several years before the advent of microwave transistors, Engelbrecht and Kurokawa invented the balanced transistor amplifier and during the following year the design theory of such amplifiers was developed [446]. Balanced amplifiers are available commercially from several vendors. This circuit configuration is now described since it possesses certain advantages over the single, cascaded or distributed stage approaches.

In this type of design, transistors are selected in pairs for operating in each amplifier stage and both the input and signal circuits are connected by means of 3 dB directional

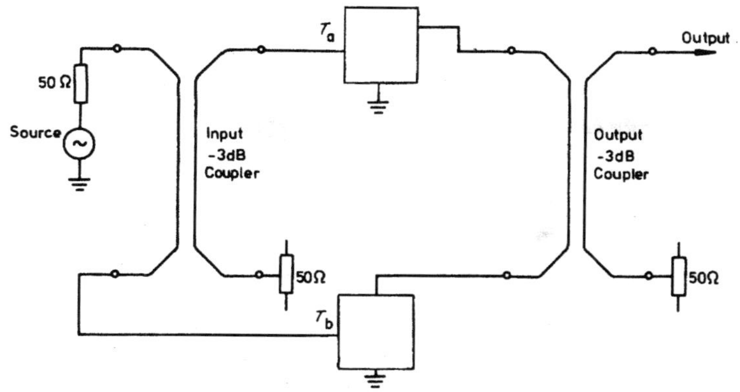

Figure 11.31 Topology for a balanced amplifier.

couplers. The general arrangement is shown in Figure 11.31 (the transistors, used in the active circuits T_a and T_b, may in principle be of any appropriate type). The complete practical circuit will of course require DC bias arrangements, and source and load optimization networks.

Balanced amplifiers have been manufactured which operate with only a few dB noise figure with substantial accompanying gain. These capabilities apply to operation over at least octave bandwidths, up to a maximum frequency of at least 50 GHz. Lange couplers are usually employed, and MESFETs, HBTs and PHEMTs form the favoured active devices.

The 50 Ω matched loads shown in Figure 11.31 are essential to preserve proper operation of the directional couplers, and it can also be shown that they assist in maintaining a low noise figure [446]. These loads may be formed in thin or thick-film resistive material and compensated if necessary to achieve good broadband performance.

It is useful to compare the balanced against the single-device amplifier stage; this is presented in the following lists:

Advantages for a single-device amplifier:

(a) Only one device is required, thus reducing component and assembly cost.
(b) Only about one-half of the DC supply power is consumed.
(c) Less circuitry and therefore less substrate area is required.
(d) There is no requirement for rather 'exotic' circuit elements such as the Lange coupler.

Advantages for the balanced amplifier:

(a) Input and output impedances are automatically matched in an amplifier optimized for noise figure or output power.
(b) Good phase linearity is obtained even when using 'conventional' 3 dB microstrip couplers. With Lange couplers the phase linearity is excellent and hence the group-delay distortion is very small (unless the load is significantly mismatched).

(c) Gain compression and intermodulation characteristics are good.
(d) Short-circuit and open-circuit stability is guaranteed.
(e) Overall sensitivity to variations in device parameters is low, provided that selected pairs of devices are used.
(f) The amplifier gain can be controlled over wide ranges by the DC bias with little effect on gain flatness or impedance matching.
(g) In some systems the reliability may be enhanced because of the two parallel circuits afforded by this arrangement. If one device fails a little finite gain is usually still available, although over a much reduced band, due to the second device.
(h) Power amplifiers can readily be designed in this topology.

For all these circuits, single-device as well as balanced, it is entirely likely that 'flip-chips' or gold-ball-bonded devices will increasingly be incorporated instead of the more familiar wire-bonded chips — for performance and production reasons. In such situations there is no facility for realizing inductances with bond wires, and it becomes essential to evolve circuits that avoid such inductances altogether (bearing in mind that microstrip characteristic impedances greater than about 110 Ω should also be avoided).

Computer-aided design, simulation (particularly important) and test are regular requirements in microwave laboratories and within the industry as a whole. Currently, in most situations, combinations of proprietary packages and in-house routines are run on sufficiently powerful desktop PC (or similar) computers or 'workstations'. With these facilities the designer can, for example, readily study the effects on overall performance of tolerances in the Lange couplers used in a balanced amplifier.

Millimetre-wave balanced amplifiers, using modified Lange couplers, are also feasible — typically covering the 26–40 GHz frequency range.

Ladbrooke [447] provides a scattering parameter analysis of the balanced amplifier, although this will not be repeated here. This analysis shows that, in the ideal case of perfect balance, both input and output reflection coefficients are zero, i.e. there are exact matches at both input and output.

Also, the magnitude of the forward parameter S_{21} is identical to that of each individual amplifying device, i.e. there is no increase in gain despite the use of two transistors instead of just one. In his text, Ladbrooke also points out that this type of circuit provides the capability of delivering almost twice the power available from each transistor as a consequence of additive amplification.

A design example of a balanced amplifier is given by Sweet [440, pp. 162–168]. The specification calls for a bandwidth of 4–10 GHz, which is particularly interesting because it is well over one octave. The minimum small-signal gain is specified as 5 dB, input and output return loss 10 dB minimum, saturated power output 1 W and DC bias 6 V with a maximum supply current of 450 mA. The final circuit is to be a hybrid MIC fabricated on an alumina substrate (relative permittivity 9.9).

A balanced reactively-matched amplifier is selected because it provides flexibility in terms of tailored output impedance for broadband power matching, and good input/output matches as described above. Each of the two elementary amplifying circuits, for insertion into the balanced topology, are single-ended GaAs MESFET circuits and the layout for each is shown in Figure 11.32. The L-type matching

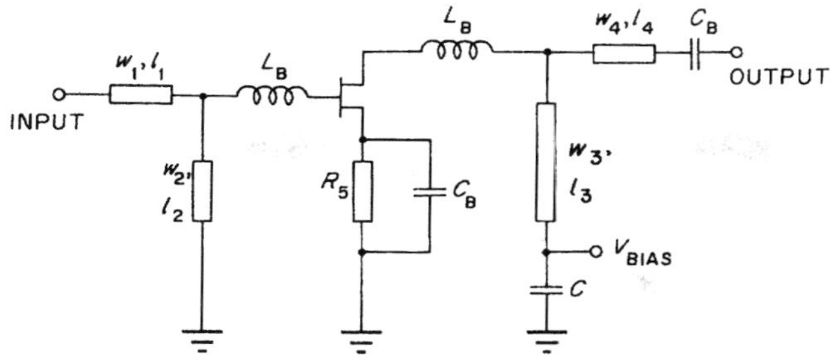

Figure 11.32 Single-ended GaAs MESFET amplifying circuit designed for later insertion into a balanced amplifier. (Two of these identical designs will be required.) (©1990 Artech House, from Sweet [440], reprinted with permission.)

networks in which the shorted stubs are connected close to the gate and drain are designed with this configuration to allow for matching the lower impedance levels here. The power FETs used in this design require relatively large gate widths and hence low input and output impedances. These circuit 'cells' (Figure 11.32) connect into the active circuit blocks, T_a and T_b, shown in Figure 11.31. The first step in the design is to consider mid-band, in this case 7 GHz, and to initially make the shorted stubs $\lambda_g/4$ long with the series lines $\lambda_g/8$ in length. For the alumina substrate, operated at 7 GHz, the effective microstrip permittivity is then 6.3 and the lengths are:

$$\lambda_g/4 = \frac{3 \times 10^{10}}{4 \times \sqrt{6.3} \times 7 \times 10^9} = 4.267 \text{ mm} \tag{11.21}$$

and $\lambda_g/8$ is of course half of this, i.e.

$$\lambda_g/8 \ = \ 2.134 \text{ mm}. \tag{11.22}$$

The stub widths are first chosen to be the minimum allowable by limitations of the thin-film on alumina MIC fabrication process, which is around 80 μm, and the series line widths are initially selected to be a midway compromise between the 50 Ω system impedance and the input or output impedance (as appropriate) of the transistor.

Detailed considerations for the active device are provided in Sweet's text and are essentially outside the scope of this book. Suffice it to say here that a Mitsubishi MGF2116 device was chosen (late-1980s technology), since it met the gain, power and frequency requirements. Each FET is self-biased in this circuit, so the source contacts are grounded by employing the usual parallel capacitor/resistor combination. The capacitor is effectively a short-circuit at 4 GHz, and therefore yields the RF ground whilst the 6 Ω resistor leaves the gate-source biased at 1.2 V. A 40 pF capacitor has a 1 Ω reactance at 4 GHz, and is therefore probably large enough in this design. In practice, the resistor and capacitor could be discrete chips, or they may comprise film layer and overlay structures, respectively.

The design of the Lange couplers follows the general method presented in Section 8.11.1. on page 291.

11.9 AMPLIFIERS USING MMIC TECHNOLOGY

Where relatively high volume manufacturing is required, as in many wireless applications, and also for several special design specifications — monolithic technology wins over hybrid. Where monolithic technology is applied to microwave applications the term MMIC is extensively used.

In this section we mainly concentrate on some examples of MMIC amplifier realizations.

11.9.1 DESIGN OF A DECADE-BANDWIDTH DISTRIBUTED AMPLIFIER

For certain applications, notably spread-spectrum, Electronic Warfare (EW) and instrumentation, solid-state microwave amplifiers capable of broadband operation are required. There are several circuit approaches leading to broadband MIC and MMIC design, dependent on the desired combinations of bandwidth, centre frequency, noise etc. One such design approach is the distributed amplifier and a specific example, originally due to Sweet [440, pp. 152–162] is outlined here.

Sweet initially calculates FET parameters leading to the individual device g_m, but we shall assume this has been done and state the outcome — which amounts to a circuit comprising four FET devices to be designed within a four-section distributed amplifier (for further details, Sweet's text is recommended). For distributed amplifiers the inductance per section of the 'artificial' lines is the square of the gate/drain characteristic impedance multiplied by the gate-source capacitance and, in this example, the result is 0.55 nH. The most economical method of realizing inductors having this value is with narrow microstrip transmission lines, and an expression for the length of a line for a specific value of inductance L was given as Equation 10.8 in Section 10.3.3 on page 361 with reference to Figure 10.15 on page 363. The expression is repeated here:

$$l = \frac{\lambda_g}{2\pi} \sin^{-1}\left(\frac{2\pi f L}{Z_0}\right). \tag{11.23}$$

Now it is fundamental for the distributed amplifier that the upper cut-off frequency determines the gate-line and drain-line inductances. So the upper cut-off frequency (20 GHz) must be used here and, assuming that 100 Ω microstrip lines with an effective microstrip permittivity of 8.5 (at 20 GHz) are employed, substitution into the above expression gives:

$$l = \frac{[3 \times 10^{10}/(\sqrt{8.5} \times 20 \times 10^9)]}{2\pi} \sin^{-1}\left[\frac{2\pi(20 \times 10^9)(0.55 \times 10^{-9})}{100}\right]. \tag{11.24}$$

So $l = 0.62$ mm.

The strip width is calculated from the narrow-strip synthesis expressions, given in Section 4.5.1 on page 93, and hence all the basic design criteria are known for the inductive microstrip lines.

The drain line loading capacitance value is the difference between the gate-source and drain-source capacitances and Sweet calculates this to be 0.15 pF. In practice, however, separate physical capacitors are not required because small drain line microstrip parasitic capacitances intrinsically serve the purpose to advantage.

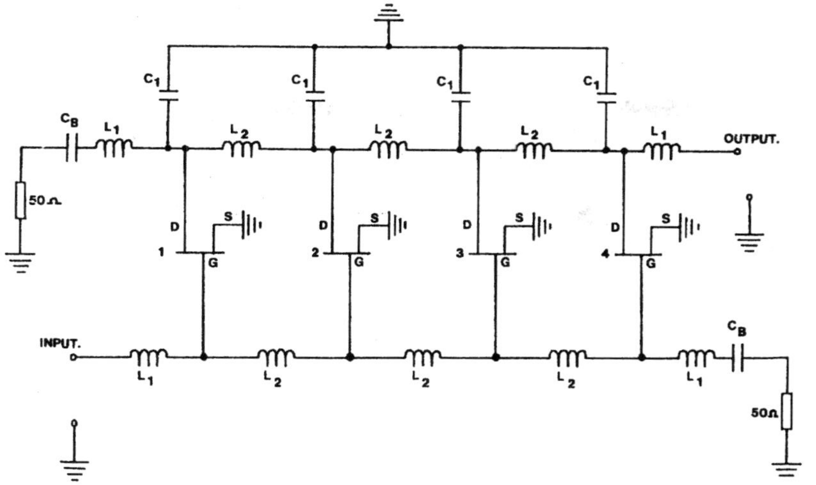

Figure 11.33 Overall topology of the 2–20 GHz MMIC distributed amplifier. (©1990 Artech House, from Sweet [440], reprinted with permission.)

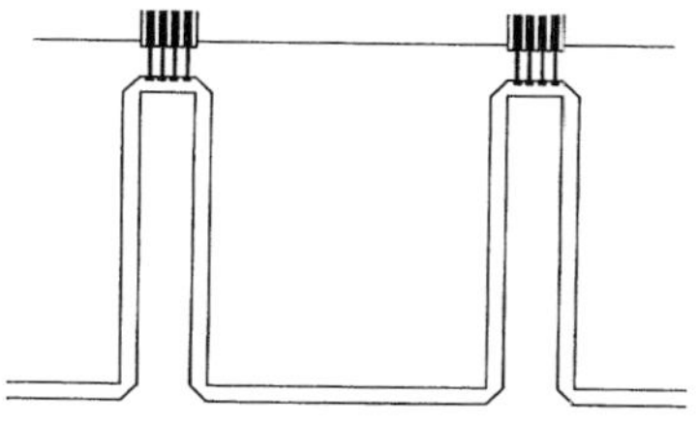

Figure 11.34 Detail of the gate-line sections for the amplifier of Figure 11.38. (©1990 Artech House, from Sweet [440], reprinted with permission.)

The overall topology of this amplifier is shown in Figure 11.33 and detail of the gate-line sections is indicated in Figure 11.34. The drain-line section layout is broadly similar, except that small strip fingers are provided for final connection to the drains instead of the four connections shown for each gate.

Having completed the preceding design procedures the next steps are: select an appropriate transistor (MESFET, PHEMT, HBT, etc), satisfy output power or noise figure specifications and choose a suitable bias circuit.

In general, the chosen device may have C_{gs} and C_{ds} values that differ from those used above, in which case iterations are necessary to finalize the design.

In his design example Sweet assumes that the actual device chosen has values for these capacitances that are identical to those calculated.

The next step is to check that the total DC current required by the amplifier is below the specified limit, and to determine the width of each gate finger of the chosen four-fingered FETs. From this data, combined with the equivalent circuit capacitances, resistances and g_m, a full model for the transistor is created using a suitable computer modelling and simulation tool.

Although the distributed amplifier described here is designed for relatively high gain (a 'gain block'), power considerations are always important and noise aspects may be significant. Therefore, appropriate power levels are calculated for the amplifier. The saturated drain current, maximum output power, DC supplied power and actual power output are all determined in the reference given [440].

The choice of bias circuit is application-specific to any amplifier design and, in this instance with such a broadband amplifier, Sweet shows that an inductor-capacitor bias circuit is ineffective because the inductor cannot be designed free of resonances over the full 2–20 GHz band. An active load bias technique is recommended, where an additional FET is used with its gate and source directly interconnected, and its gate width chosen such that I_{dss} is equal to the total current drawn by the entire amplifier. Such an active load provides an RF resistance in the region of several hundred ohms, and is much smaller than an 'equivalent' inductor.

11.9.2 W-BAND MMIC LNAs

With state-of-the-art technology InP-based HEMTs have demonstrated very high cut-off frequencies and low noise figures, making them very suitable InAlAs/InGaAs transistors for applications at least as high as W-band (75–110 GHz). The example presented here is taken directly from Hoel *et al.* [448] and it describes the fabrication and performance of a W-band monolithic 2-stage low noise amplifier based on lattice-matched HEMT devices. The gate process is specially developed for millimetre wave integrated circuits at W-band frequencies using coplanar waveguide technology.

The MMIC amplifier was fabricated on a lattice-matched InAlAs/InGaAs HEMT active layer grown by molecular beam epitaxy using a solid source MBE 2300 Riber system. The processing steps were optimized to provide high performance devices and a high yield. A SEM cross-section is shown in Figure 11.35.

To avoid reverse-side processing, coplanar waveguide technology is chosen for the passive circuitry. Metallic resistors consist of a 0.07 μm titanium (Ti) layer having a sheet resistance of 16 Ω/square. To define the CPW ground-to-ground spacing (d = 70 μm), a trade-off between losses and low dispersion up to W-band was considered. The line attenuation is about 0.4 dB/mm at 94 GHz for a 50 Ω transmission line. Electro-plated air-bridges were added at each discontinuity in order to suppress the undesired slot-line mode. These air-bridges were designed to introduce low parasitic effects over the entire W-band. The amplifier was designed by Thomson-Detexis (France) using accurate broadband models from a complete library of CPW passive elements developed at I.E.M.N. and validated up to 110 GHz [449]. This library includes matching networks (symmetric, asymmetric, bended stubs), metal resistors, MIM capacitors (series and shunt), DC-blocks and step impedance lines. These models do not use measured S-parameter fitting, but are based on Heinrich's CPW line model [450], which only needs geometrical data and physical process parameters. Each slice of transmission line, resistance or capacitance is modelled by a R-L-C-G equivalent circuit. For metal resistors and MIM capacitors, experimental DC resistance and low frequency capacitance measurements are also needed. The complete distributed model was implemented into Agilent's Advanced Design System® (ADS).

In order to design low noise circuits in the millimetre-wave range, reliable broadband small signal models of the active devices are required. The small signal equivalent circuit of the 0.1 μm LM-HEMT was determined from S-parameter measurements for

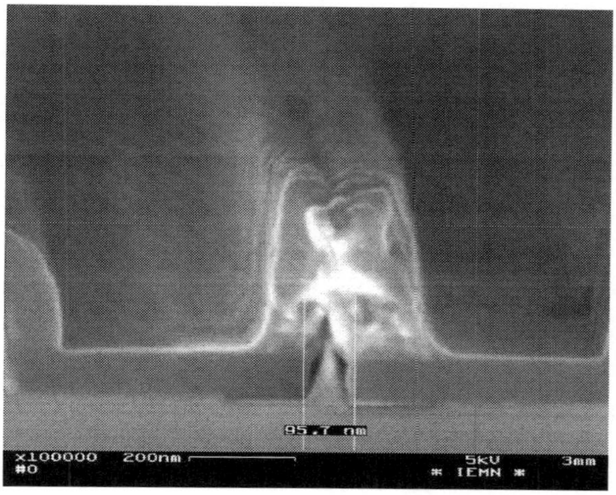

Figure 11.35 SEM image of a 0.1 μm LM-HEMT gate deposited on a SiN layer (courtesy of Hoel *et al.* [448])

a number of gate widths and DC drain current densities. A complete database was then developed enabling the optimization of the gate width and DC drain current of the LM-HEMT used in the two-stage amplifier.

Figure 11.36 shows a photograph of the complete realized W-band two stage CPW amplifier. The chip size is 1.7 mm $\times$ 1.5 mm. The S-parameters, noise figure and associated gain of the amplifiers were characterized using on-wafer techniques. Data were measured for S_{21} and S_{22} of the two-stage LNA. The transmission coefficient (S_{21}), which is 11.5 dB at 94 GHz, varies between 11 and 12.6 dB over the 75 to 95 GHz range and falls to 4.6 dB at 110 GHz. The output return loss value, better than 12 dB from 86 to 96 GHz, is 14 dB at 94 GHz. The isolation is better than 24 dB from 75 to 110 GHz. At 94 GHz, the best LNA manufactured demonstrated a minimum noise figure of 3.3 dB with an associated gain of 12 dB, which is excellent performance for a low noise amplifier using coplanar wave guide technology operating in W-band.

11.10 MICROWAVE OSCILLATORS

As with amplifiers, there also exists a wide range of possible types of oscillators. Various systems require swept oscillators, using varactors for a narrowband sweep or YIG for broadband. Voltage-Controlled Oscillators (VCOs) are required for tuners in cellular handsets, microwave FM, instrumentation and other purposes, and new developments are regularly reported on VCOs [451, 452]. For example, Dupois *et al.* [451] describe a hybrid MIC Ku-Band VCO, designed around an NEC NE71000 GaAs MESFET, with a tuning range extending from 12.6 to 15.5 GHz, an average power output of 12 dBm and a phase noise of -70 dBc/Hz at an offset of 10 kHz from

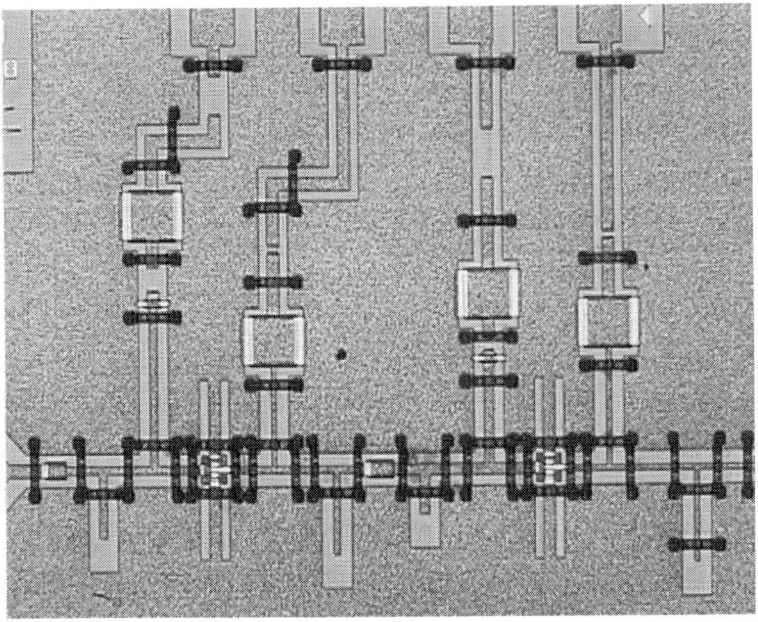

Figure 11.36 Photograph of the 1.7 mm × 1.5 mm 2-stage LM-HEMT W-band MMIC amplifier (courtesy of Hoel, V., *et al.* [448]).

the carrier. These researchers consider that the available relationships for designing microstrip (i.e. width and length synthesis) placed limitations on their design and that improvements should make their design approach viable for oscillators at frequencies up to at least 20 GHz.

It has long been appreciated that Surface-Acoustic-Wave (SAW) technology also provides a basis for stable oscillator design. Fleischmann *et al.* [452] have taken this technology a stage further by developing a low-noise 2.65 GHz VCO based on this type of stabilizing technology (they use the term acoustic Surface Transverse Wave, or STW). Fleishmann's team use STW delay lines operating at the fifth harmonic to achieve phase noise levels of −93 dBc/Hz at 1 kHz and −63 dBc/Hz at 100 Hz from the carrier. Their oscillator design, which has an output power of 3 dBm and a tuning range of 750 kHz, is centred on a low-noise silicon MMIC amplifier with a matched gain of 15.2 dB at 2.65 GHz.

Alternatively, fixed but highly frequency-stable oscillators are often required, for example, a circuit with low FM noise that is also stabilized using a dielectric resonator to form a 'Dielectric Resonator Oscillator' (DRO). Dielectric resonators, and filters using them, are described in Section 10.3.12. Most types of oscillator have been realized in MIC form, but we shall limit our considerations to the DRO and recent developments here.

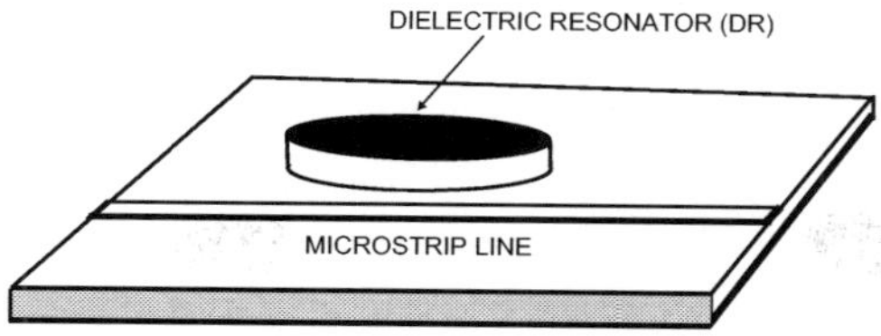

Figure 11.37 Structural diagram of dielectric resonator/microstrip coupling scenario.

11.10.1 EXAMPLE OF A DIELECTRIC RESONATOR OSCILLATOR

In many existing microwave and millimetre-wave systems, highly-stable oscillators use one of the following two techniques:

(a) Locking a relatively unstable microwave oscillator to a high stability but lower frequency quartz oscillator.
(b) Using a mechanically tuned microwave resonator embodying a low thermal expansion material such as Kovar. Again, a relatively unstable microwave oscillator is used.

The use of stable dielectric resonators represents a more recent technique — one that is much more compatible with MIC/MMIC realization. Dielectric resonators are available to operate at specific frequencies ranging from below 1 GHz to over 40 GHz. At the lower limit the resonator becomes rather large and tends to dominate the real estate, whilst at the higher frequencies the resonator 'puck' becomes somewhat 'uncontrollably small' and sensitive to production variations. These types of resonators have Q-factors in the 1000 to over 40 000 range, and thus oscillators controlled by such resonators are characterized by very low FM noise and an output frequency that is very temperature-stable.

It is important to couple the dielectric resonator to the appropriate microstrip line in some optimum way — either for maximum degree of coupling or for maximum output power from the oscillator. The general layout applicable to this situation is shown in Figure 11.37, where the dielectric resonator puck may have to slightly overlap the microstrip line for maximum coupling (in which case it would be necessary to include the layer of low-permittivity material).

Design Example

The example given here follows that originally provided by Sweet [440, pp. 211–216].

The specifications include: an operating frequency of 8.4 GHz, a minimum power output of +10 dBm, FM noise not to exceed −110 dBc at 1 MHz from the carrier in a 1 kHz bandwidth, and temperature stability to be +5 MHz over the temperature range 0 to +50°C.

This circuit is based on a GaAs MESFET. An effective method of providing the required negative resistance in a fixed-frequency oscillator is to incorporate an inductor between gate and ground, i.e. common-gate inductive feedback (as Sweet points out, this technique is not suitable for broadband tunable oscillators). It is not necessary for the negative resistance to be calculated and designed in detail, since it is only required that this resistance be significant for oscillation, which is easily obtained at these frequencies.

In the fundamental circuit topology for this DRO the resonator itself is modelled as a parallel tuned circuit, and its coupling to the GaAs MESFET's source is represented as an ideal transformer with an adjustable turns ratio. The drain-to-ground inductance and capacitance form a matching circuit for the output.

A specially selected dielectric resonator puck is required for the specified frequency of 8.4 GHz. In this case a Murata Erie type P/N DRD077SC034 is selected which has a diameter of 7.7 mm, a thickness of 3.4 mm and a relative permittivity of 28.6. The elements of this resonator's equivalent circuit are calculated from its Q-factor (5000), its resonant frequency (8.4 GHz), and an initially assumed value for either L or C. The choice of active device, calculation of noise and assessment of stability, and choice of bias circuit are beyond the scope of this text, but are all considered by Sweet. The gain/frequency requirements are well met by an NEC 71000 transistor, and the calculated noise-to-carrier ratio is 33 dB better than the specification demands. It is also shown in the same reference that the temperature drift with this type of oscillator is of the order of 4 ppm per degree Celsius.

The circuit is self-biased through the 50 Ω microstrip line to which the resonator is coupled, terminated in a resistor of approximately 50 Ω value which also acts as a source-bias resistor. A 5 V bias supply, fed through a choke inductor, is provided for the drain. In the practical layout of this oscillator two 0.025 mm thick alumina substrates are required, attached by a metal grounding rib on which the GaAs MESFET is die-attached. The resonator puck, coupled to the 50 Ω microstrip line, occupies one substrate, whereas the output capacitance/inductance matching circuit, together with the drain bias feed and RF output line, are located on the second substrate. The common-gate feedback, required to yield the negative resistance as described above, is provided simply by the bond wire from the gate to the rib ground.

This circuit can be optimized, including all parasitic elements associated with the practical layout, and simulated using a commercial Computer-Aided Engineering (CAE) package, available from organizations such as Agilent or Ansoft.

11.10.2 DRO OSCILLATOR DEVELOPMENTS

Seawright *et al.* [453], in their paper on the optimum synthesis of (in their example) a 13.8 GHz microstrip DRO, show that oscillator output power is a highly nonlinear function of DRO position relative to the microstrip. They also show that this position depends upon whether the circuit is unshielded, shielded (but untuned) or shielded and tuned. Assuming that such a circuit is likely to be both shielded and tuned in practice, then the results due to Seawright *et al.* indicate that the centre of the dielectric resonator should be less than 3.2 mm from the edge of the microstrip. At distances greater than 3.2 mm the output power decreases rapidly.

It should be noted that this is a different situation from the maximum coupling condition considered above.

One of the most significant attributes of a DRO is the temperature stability of its output frequency, and Massias *et al.* [454] have reported a novel analysis and design technique for highly temperature-stable DROs. They define a parameter involving the unloaded-Q of the dielectric resonator, the coupling and a circuit-related quantity and develop a fast method, based on measurements of the temperature stability of a given DRO, for analysing and designing new DROs.

It is sometimes necessary to synchronize several DROs simultaneously, as well as to further improve the oscillator phase noise (note that DROs inherently reduce FM noise but not necessarily phase noise, and this can become a serious problem). One method for achieving both objectives is to injection-lock a DRO and to synchronize it with a multitone signal, and such a technique has been described by Golubicic [455]. In this work an NEC NE71083 GaAs MESFET has its gate microstrip circuit (50 Ω) coupled to the dielectric resonator, and the multitone signal is injected to the source, matched by an L-topology microstrip circuit.

There are several alternative topologies and design approaches that have been described in the literature and significant new developments continue to be reported [456, 457]. A design method for DROs based upon parallel feedback is reported by Filicori *et al.* [456]. In this approach the dielectric resonator is coupled to two microstrip lines located in a parallel feedback path between source and drain of a GaAs MESFET. This oscillator runs at a stable frequency of 3 GHz and is simulated using a 'SPICE' circuit simulator.

Although DROs have achieved a strong position in this sector of solid-state microwave circuit realization, other approaches may be 'on the horizon'. For example, given the improvements in microstrip Q-factor achievable with superconductivity some oscillator (and indeed complete subsystem) designs using superconductive microstrip may emerge from current work. It would also be interesting to see whether SAW oscillators, perhaps combined with microstrip, could be improved using superconductivity.

The hybrid mode coupling of dielectric resonators to microstrip lines has been considered in detail by Liang and Zaki [457]. Their results apply to frequencies in the range 11.71–14.91 GHz and, in all cases, the external Q increases smoothly with the separation between the resonator and the microstrip line. However, the design implications reported by Seawright *et al.* [453] (described above) may be of more significance for oscillators.

11.10.3 MMIC OSCILLATOR EXAMPLE

By no means all millimetre-wave oscillators need to be manufactured using InP or GaAs semiconductors, or FET transistors for that matter, and advances continue to be made with silicon (or silicon-based) substrates.

As an example of this type of approach we describe here a 47 GHz SiGe-MMIC oscillator, developed at the DaimlerChrysler Research Center in Ulm (Germany) [458, 459]. This oscillator implements a six-finger SiGe HBT as the active element — integrated on a high-resistivity ($\rho > 4000$ Ω.cm) silicon substrate. The transistor has a transition frequency of 70 GHz and a maximum frequency of oscillation of 95 GHz.

Coplanar lines are used throughout the design and a continuous MBE process is used to fabricate the entire MMIC.

Regarding the transistor, in order to minimize the influence of parasitic elements (capacitances and inductances) a trench is etched around the intrinsic HBT and the base, collector and emitter are connected via air bridges. This device is operated in the common-base configuration.

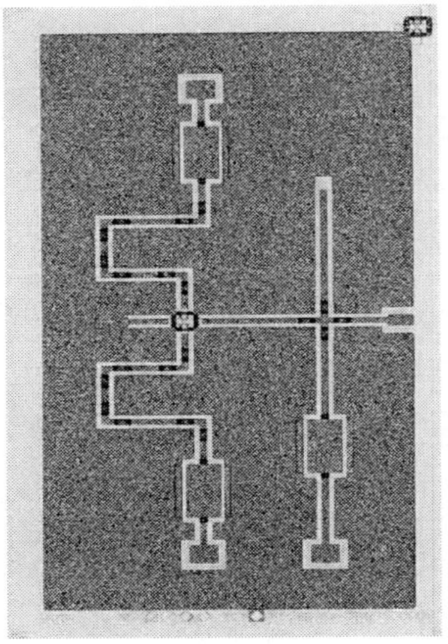

Figure 11.38 47 GHz SiGe-HBT MMIC oscillator chip (courtesy: DaimlerChrysler Research Center, Ulm, Germany).

The reflection-type circuit design is based upon coplanar lines and associated circuit elements. A key feature is that the rate of phase change with frequency, at the frequency of oscillation, is maximized for the output matching network so that the phase-noise performance is also maximized. The number of elements is kept as low as possible so as to reduce any additional uncertainties and the circuit layout is not constrained by chip size. Only 50 Ω CPW lines are used — with 50 μm ground-to-ground spacings. An in-house proprietary code was used to design the circuit, embodying finite-difference simulations in the frequency domain (FDFD). A photograph of the final MMIC chip is shown in Figure 11.38.

In this MMIC the SiGe-HBT is located to the left of centre — at the crossing between four sets of CPW lines. The base is taken out to the ground via the short CPW extending leftwards from the device, and the two independent emitter connections are symmetrically loaded with the identical circuit structures going upwards and downwards from the transistor — shown in the orientation of the chip as illustrated. These circuit structures form a part of the feedback required for oscillation. DC power is fed in via the DC pad which is evident towards the lower right-hand side, followed as usual by a lowpass filter incorporating a MIM capacitor between inductive CPW lines. Many air bridges, visible as more densely black structures around the MMIC, are evident. The output port is shown in the centre of the right-hand side of the chip.

The SiGe HBT was simulated under large-signal conditions using a modified Gummel-Poon description and the resulting model was implemented in a commercial CAE program. It was found that the simulated power differs by less than 1 dB from the measured value.

The output power spectrum generated by this oscillator is shown in Figure 11.39. With this circuit the SiGe-HBT is biased at a collector current of 32.59 mA and the collector-emitter voltage is 4.6 V. The measured maximum output power is just over 13 dBm (i.e. 20.4 mW) at the centre oscillation frequency of 47.02 GHz and the power levels lead to a DC-to-RF conversion efficiency of 13.6%.

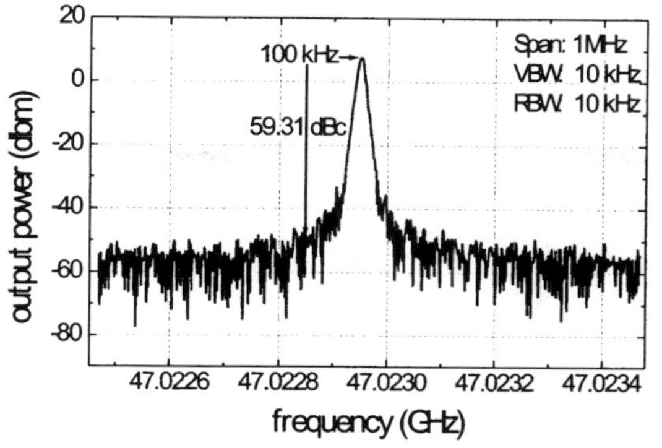

Figure 11.39 Measured power spectrum of the 47 GHz MMIC oscillator (courtesy: DaimlerChrysler Research Center, Ulm, Germany).

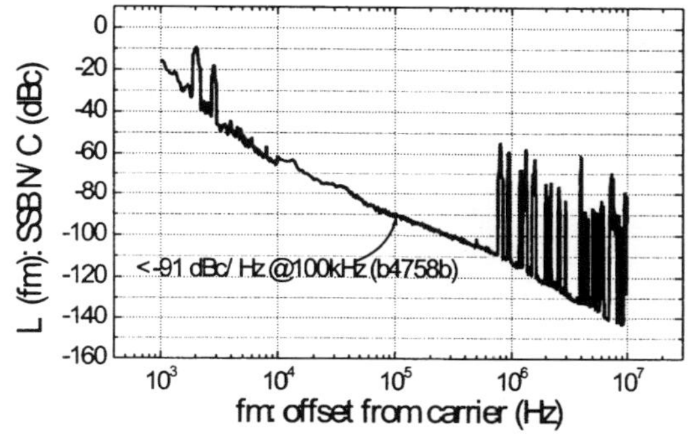

Figure 11.40 Single-sideband measured phase noise of the 47 GHz MMIC oscillator (courtesy: DaimlerChrysler Research Center, Ulm, Germany).

The measured phase noise, under single sideband conditions, is indicated in Figure 11.40. The phase-noise performance at 100 kHz off-carrier is estimated as −91 dBc/Hz.

The output power remains in the 10 to 13.1 dBm range for collector-base voltages varying from 2.8 to 4.5 V and for emitter-base voltages varying from −0.77 to −0.83 V. This indicates a sustained high output power over a wide range of bias voltages.

In contrast, the total output frequency variation over these voltage ranges (and indeed greater) is only a few MHz. This is very low when it is considered that a frequency deviation of (say) 2 MHz represents a variation of 0.004%.

This development was based upon an earlier reported research effort essentially by the same team [460].

11.11 ACTIVE MICROWAVE FILTERS

The filters described in Chapter 10 all embodied passive structures — microstrip MIC derived sections, coupled microstrip resonators, etc. By incorporating active elements in filters and related circuits inherently high Q-factors, voltage-tuning and internal gain may be obtained. Active filters free the designer from the restrictions of significant in-band insertion loss, corner rounding and poor skirt selectivity commonly encountered with the more familiar passive circuits. The active device is usually a MESFET or a HEMT, although it can in principle be any microwave amplifying device.

Two examples are now briefly presented describing active microwave filters. Chi-Yang and Itoh [461] describe a varactor-tuned, multipole active microwave bandpass filter based upon an NEC NE71083 GaAs MESFET. In order to obtain an improved aspect ratio for the coupled line section, the filters are designed with a characteristic impedance of 82 Ω. Transformation of this impedance to the 50 Ω external environment is achieved with tapered microstrips. The two-pole version of this filter exhibits a tuning range of 430 MHz with an 80 MHz 3 dB bandwidth centred on 10 GHz (i.e. a 0.8% bandwidth). In each case the passband insertion loss is typically 0±1 dB.

Jiao *et al.* report microwave frequency-agile active filters for MIC and MMIC applications [462]. Their designs use modified microstrip hairpin-shaped split-ring resonators, parallel-coupled to an active loop, and they have demonstrated both bandpass and bandstop filters. For the bandstop filters the entire split hairpin lines and their active loops are themselves parallel-coupled to a main through-connecting microstrip line. At the centre frequency of 3.1 GHz a rejection of 45.3 dB is obtained. The bandpass filters exhibit typically 0.03 dB insertion losses, together with a return loss of 35.9 dB in-band. The frequency tuning range is approximately 47 MHz and the 3 dB bandwidth is 3.17 MHz (i.e. 0.1%).

These types of developments offer the potential for effective high-Q resonators using entirely planar technologies (unlike dielectric resonators). Therefore, this direction of technology may hold a key for future all-MMIC filters and oscillators.

11.12 PHASE SHIFTERS

It has long been appreciated that these types of elements are of great significance in several systems — including phased array radars [463, 464]. Where such systems are operated up to lower tens of gigahertz frequencies the technology is reasonably straightforward. When, however, much higher frequencies are required (as in phased array seeker systems, for example), the technological constraints become considerable and new technologies have been explored.

In the following text we present, as an example, a coplanar 4-bit HEMT-based phase shifter approach applicable to W-band (94 GHz) phased array radar systems. This technology was originally announced by Zuefle *et al.* [463]. They developed MMIC technology coplanar 4-bit phase shifters for W-band operation, using PHEMTs as the switching elements. The actual phase shifting sections comprise stub-loaded CPW lines and branch-line couplers. To present equal power levels variable-gain amplifiers are interposed in both the transmit and receive paths — before the phase shifter and PA in the transmitter case, and after the LNA and phase-shifting section in the receive instance. The variable-gain elements comprise two-stage cascode amplifiers.

The overall combination of all these elements has an insertion gain between 2 and 8 dB and initially occupied chip dimensions of 1.5 × 4 mm. A redesigned version of the phase shifter reduced the chip dimensions to 1 × 2.5 mm.

Appendix A

TRANSMISSION LINE THEORY

This appendix presents an outline of microwave transmission line theory including the elements of impedance matching using transmission lines and also coupled transmission line theory that is used in Chapter 8. The background for the developments here is the TEM-Mode transmission line theory presented in Section 1.6 on page 14. Recall that a TEM-mode transmission line has its electric and magnetic field lines in the plane transverse to the direction of the line (direction of propagation). Thus TEM-mode stands for Transverse Electro-Magnetic mode. The coaxial line is the most common such line. Other lines, such as microstrip, have field distributions that are only approximately transverse,— and these are called quasi-TEM. Nevertheless, their characterization is referenced to a TEM-mode line.

A.1 HALF-, QUARTER-, AND EIGHTH-WAVELENGTH LINES

Transmission lines that are half of a wavelength long at some desired frequency of operation are often encountered in RF and microwave circuits. Various combinations of short-circuit or open-circuit terminations (both ends) may be used. Such lines then form 'half-wave resonators.' When both ends are short-circuit terminated, as in Figure A.1, coupling to the line would have to be provided by some separate probe. The actual length of such a resonator may be extended to $n\lambda_g/2$, where n is an integer. It is again left as an exercise to examine the behaviour of open-circuit terminated resonators. Lines which are a quarter of a wavelength long can provide useful properties. Clearly, for a short-circuit terminated quarter-wave line we obtain half, longitudinally, of the standing wave pattern shown in Figure A.1. At the input plane location P, the standing wave has a voltage maximum which corresponds to

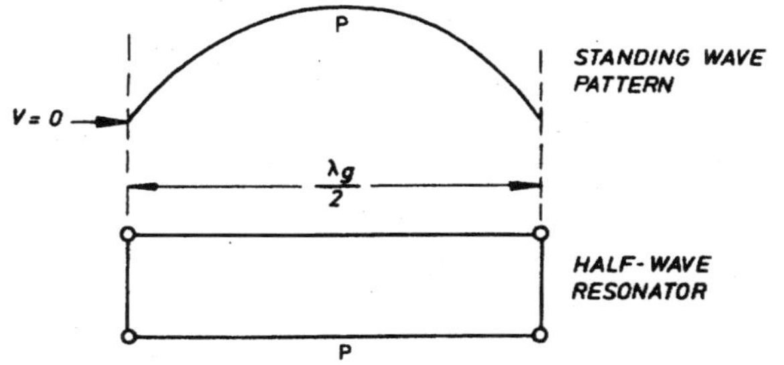

Figure A.1

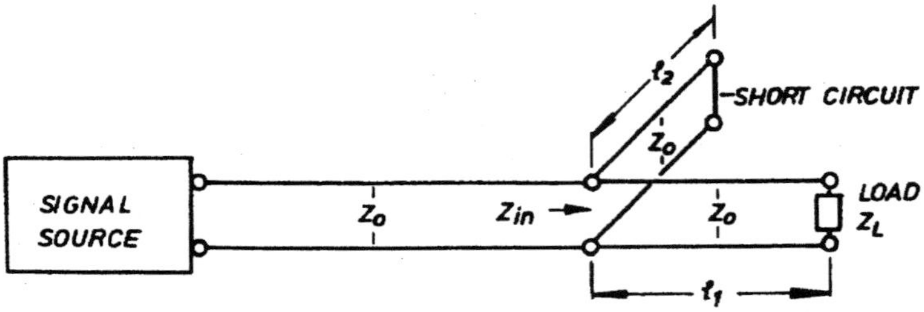

Figure A.2

an open-circuit condition. Thus we effectively transform from the short circuit to an open-circuit. This is consistent with an input impedance calculation using Equation (1.28), and substituting $\beta = 2\pi/\lambda_g$ yields

$$Z_{sc}|_{\lambda_g/4} = jZ_0 \tan(\pi/2) = \infty \tag{A.1}$$

which is certainly the impedance of an open-circuit. This feature can be used in many circuit designs to deliberately ensure that an open-circuit appears in a desired location — at one frequency.

A.2 SIMPLE (NARROWBAND) MATCHING

At RF and microwave frequencies the transfer of maximum power from one part of a circuit to another is critical. The aim is to achieve maximum power transfer, thus the load impedance is the complex conjugate of the effective input impedance. Matching is achieved by inserting a network called a matching network. When matching is implemented on-chip at a few GHz or less or on a circuit board at slightly lower frequencies, it is usual to implement lumped elements in the matching network. Lumped element matching network design is beyond the scope of this book. Above a few gigahertz, and possibly because of the lower losses that can be achieved even at megahertz and low gigahertz frequencies, connections of transmission lines are used. Transmission line based matching networks have higher power handling capability than their lumped element equivalents.

In most practical situations a transmission line ultimately feeds some load which has an impedance differing considerably from the characteristic impedance of the line, i.e. it is mismatched — possibly quite severely. Examples include feeds to solid-state amplifying devices such as transistors, and antenna feeders. By connecting a suitably designed network between the feed line and the load we can adjust impedances in order to obtain a good overall match. If this match is required to hold over a specified range of frequencies, the network design may become quite complicated.

This section briefly reviews two simple methods for narrowband matching: the single-shunt stub and the quarter-wave transformer.

A schematic representation of the single-shunt stub technique is shown in Figure A.2. Although a short-circuit termination is indicated here, an open-circuit termination could be used, provided that the termination quality is sufficiently good. In this arrangement all lines are shown with identical characteristic impedances. This may not always be so, and some extra flexibility is offered if the stub and main line

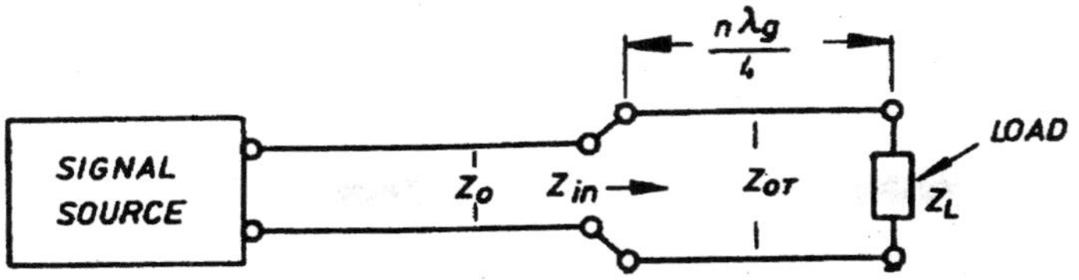

Figure A.3

impedances are of different values. The best method for 'hand calculating' the required lengths l_1 and l_2 to obtain the matched condition $Z_{in} = Z_0$ is to use a Smith chart. This is fully described in a number of standard texts and will not be outlined here.

For computer-aided design (CAD) suitable expressions are required, which are now stated for l_1 and l_2. The results are (for a resistive load):

$$l_1 = \frac{\lambda_g}{2\pi} \tan^{-1} \sqrt{\frac{Z_L}{Z_0}} \tag{A.2}$$

and

$$l_2 = \frac{\lambda_g}{2\pi} \tan^{-1} \left(\frac{\sqrt{Z_L Z_0}}{Z_L - Z_0} \right). \tag{A.3}$$

It is important to note that these results are satisfactory only when Z_L is real (resistive). If Z_L is complex the basic expressions must be re-examined in order to proceed with the analysis, which becomes quite lengthy. With the use of two, three, or sometimes more stubs the distance l_1 can be kept fixed and broader-band matching is achievable. Naturally, full analyses yielding lengths and separations become very unwieldy.

There is a simpler technique available for matching resistive loads: the quarter-wave transformer. In the single-section quarter-wavelength basic method this represents, again, a narrowband approach. The basic arrangement is shown in Figure A.3.

We set $Z_L = R_L$ (resistive) and $l = n\lambda_g/4$ in Equation (1.27) to obtain

$$Z_{in} = \frac{Z_{0T}^2}{R_L} \tag{A.4}$$

whence the required characteristic impedance of the quarter-wave line is

$$Z_{0T}^2 = Z_0 R_L. \tag{A.5}$$

For example, to match a 50 Ω line into a 10 Ω resistive load the impedance required is

$$Z_{0T}^2 = 500\ \Omega^2 \tag{A.6}$$

or

$$Z_{0T} = 22.4\ \Omega. \tag{A.7}$$

In stripline or microstrip circuits this would be relatively easy to implement.

A broader-band match is obtained if several quarter-wave sections are connected in cascade. The drawbacks are that space may be wasted in the layout, the match may remain poor at the band edges, and only resistive loads are accommodated.

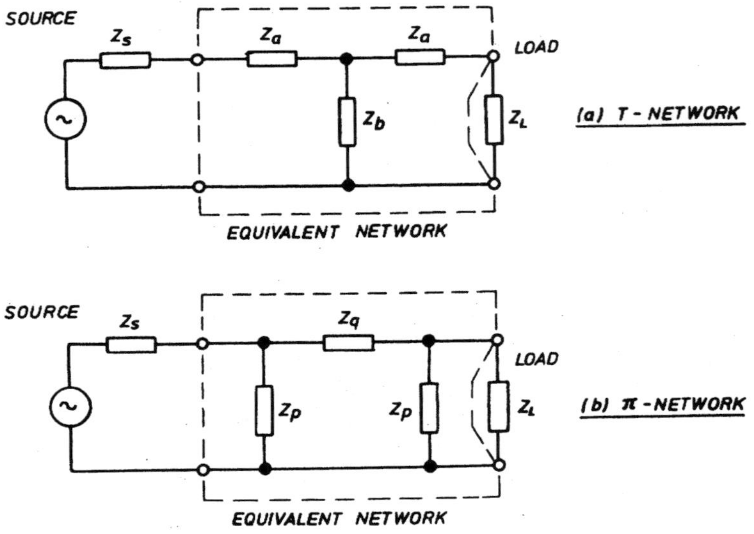

Figure A.4

A.3 EQUIVALENT TWO-PORT NETWORKS

There are applications where it is useful to employ a lumped-parameter, equivalent, two-port network for a particular length of line. One example is in the realization of filters consisting of cascaded lengths of transmission lines.

Any equivalent network must be constrained to have the same terminal characteristics as those of the length of transmission line it represents, and this will only hold at one single frequency.

Having said this, the behaviour may well be satisfactorily represented over at least a narrow band of frequencies. The propagation constant γ will usually be the most frequency sensitive parameter. The general equivalent T-network is given in Figure A.4(a), and the equivalent π-network is given in Figure A.4(b).

Expressions for the various impedances, given next, are readily obtained by equating transmission line terminal voltages and currents with network terminal voltages and currents:

$$Z_a = Z_0 \tanh \frac{\gamma l}{2} \tag{A.8}$$

$$Z_b = \frac{Z_0}{\sinh \gamma l} \tag{A.9}$$

and

$$Z_q = Z_0 \sinh \gamma l \tag{A.10}$$

$$Z_p = Z_0 \coth \frac{\gamma l}{2}. \tag{A.11}$$

For loss-free lines (so that $\gamma = j\beta$) these expressions become reactance terms as follows:

$$Z_a = jZ_0 \tan \frac{\beta l}{2} \tag{A.12}$$

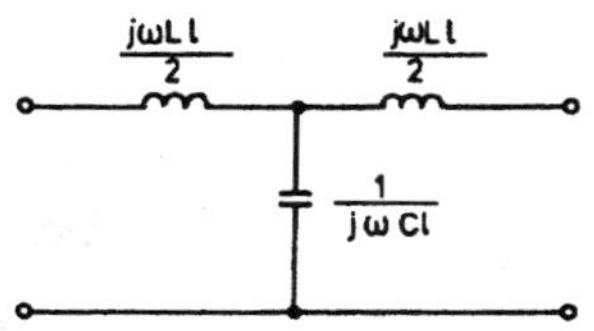

Figure A.5

$$Z_b = \frac{-jZ_0}{\sin \beta l} \tag{A.13}$$

and

$$Z_q = jZ_0 \sin \beta l \tag{A.14}$$

$$Z_p = -jZ_0 \cot \frac{\beta l}{2}. \tag{A.15}$$

This set of formulas is useful, for example, in some microstrip filter designs.

For very short lengths of loss-free transmission line, such that $l \ll \lambda_g/4$, a very simple approximate equivalent network, shown in Figure A.5, is adequate.

A.4 CHAIN (ABCD) PARAMETERS FOR A UNIFORM LENGTH OF LOSS-FREE TRANSMISSION LINE

The notation for the equivalent two-port network, to be characterized in terms of ABCD parameters, is shown in Figure A.6. With this choice of quantities, including directions, the independent variables are now assumed to be the output variables (V_2 and I_2). The ABCD parameters then directly provide the related input (dependent) variables, assuming the network is linear:

$$V_1 = AV_2 + BI_2 \tag{A.16}$$

$$I_1 = CV_2 + DI_2 \tag{A.17}$$

where A, B, C and D are constants that characterize the network.

In matrix form Equations (A.16) and (A.17) become

$$\begin{bmatrix} V_1 \\ I_1 \end{bmatrix} = \begin{bmatrix} A & B \\ C & D \end{bmatrix} \begin{bmatrix} V_2 \\ I_2 \end{bmatrix}. \tag{A.18}$$

This $ABCD$ matrix is often called the chain matrix because it is so useful in analysing cascaded (or 'chained') networks.

For a length l of uniform loss-free transmission line the voltages and currents are interrelated by

$$V_1 = V_2 \cos \beta l + jI_2 Z_0 \sin \beta l \tag{A.19}$$

and

$$I_1 = j\frac{V_2}{Z_0} \sin \beta l + I_2 \cos \beta l. \tag{A.20}$$

These expressions can be written in the following matrix form:

$$\begin{bmatrix} V_1 \\ I_1 \end{bmatrix} = \begin{bmatrix} \cos \beta l & jZ_0 \sin \beta l \\ jY_0 \sin \beta l & \cos \beta l \end{bmatrix} \begin{bmatrix} V_2 \\ I_2 \end{bmatrix}. \tag{A.21}$$

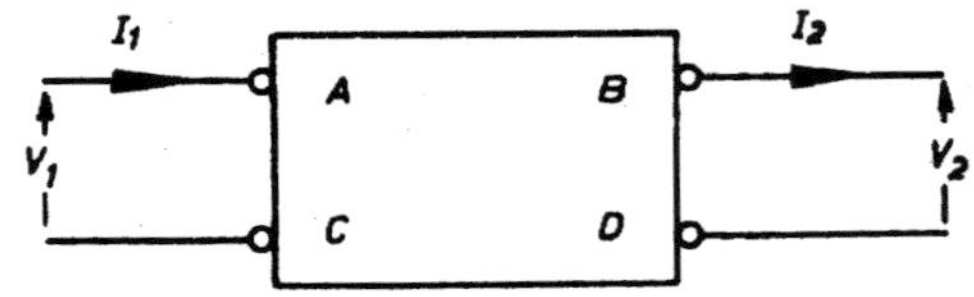

Figure A.6

Comparing Equation (A.18) with Equation (A.21) gives the final result for the transmission line chain matrix:

$$\begin{bmatrix} A & B \\ C & D \end{bmatrix} = \begin{bmatrix} \cos\beta l & jZ_0 \sin\beta l \\ jY_0 \sin\beta l & \cos\beta l \end{bmatrix}. \tag{A.22}$$

This result is useful in analysing many single and multiple transmission line structures. One area where it will be required in this work is in coupler considerations, leading to coupler design, as described in detail in Chapter 8.

A.5 PARALLEL COUPLED TRANSMISSION LINES

A.5.1 EVEN AND ODD MODES

Any parallel-coupled pair of transmission lines, regardless of their practical realization, may be described by the four-port configuration indicated schematically in Figure 1.14. The chain lines indicate mutual coupling; input and output connections and terminations have been omitted for simplicity.

To examine the coupling modes which are established, we now concentrate on a pure TEM-mode type of structure. The simplest structure to consider is the dual, coaxial, coupled-line system, in which electric and magnetic fields couple through a narrow slot cut into the adjacent outer conductor along a certain length. Such a structure is shown in Figure 1.15. Notice that the consistent field patterns correspond to waves travelling in opposite directions along the respective lines, and it may be stated that any 'parallel'-coupled pair of transmission lines yield contra-directional travelling waves.

The coupled coaxial arrangement constitutes a pure TEM-mode system, assuming either a uniform dielectric filling or an air-filled structure. At any instant the relative polarities of the voltages, taken at any specific plane along the structure, will either be alike or opposite. We refer to different field configurations set up by such polarities as the even mode and the odd mode, respectively. Figure 1.16 illustrates these effects. The complete behaviour of the coupled structure can be obtained by superposition of the effects due to these two modes.

Due to the assumed uniform dielectric filling, any signal travelling in such a pure TEM-mode system will always travel at the same velocity, $v = c/\sqrt{\varepsilon_r}$ (Equation (1.20)). Thus, the velocity associated with the even mode is identical to that associated with the odd mode here.

The equivalent primary constants for the coupled lines having even or odd modes, taken separately, must differ because of the different field distributions. As a result of this two distinct characteristic impedances may be defined, one for each mode, expressed in terms of the different primary constants by equations of the form of

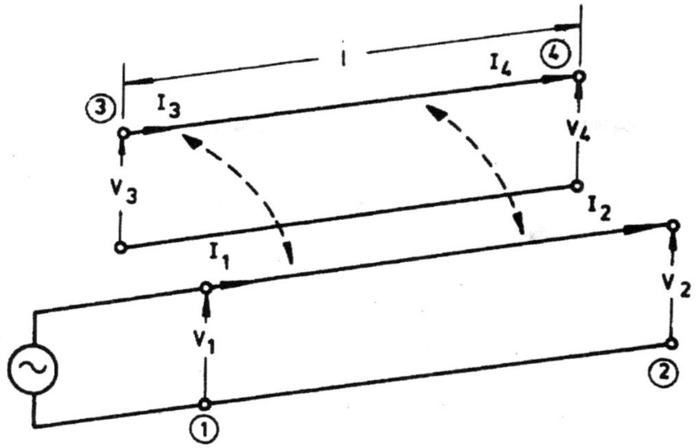

Figure A.7 Schematic representation of two parallel-coupled TEM transmission lines.

Equation (1.23). We term these impedances the even- and odd-mode characteristic impedances, denoted by Z_{0e} and Z_{0o}. This applies to any TEM-mode or quasi-TEM mode parallel-coupled structure, and the characteristic impedances are major parameters in design procedures.

A.5.2 OVERALL PARAMETERS FOR COUPLERS

A coupled-line circuit element may be specified in terms of the following overall parameters:

Coupling factor (or coupling parameter) =	C
Transmission factor =	T
Directivity factor =	D

The meaning of these parameters is clarified by Figure A.8. We can now define the quantities more closely as follows:

$C = V_3/V_1$	=	voltage fraction 'transferred' (coupled) across to the opposite arm
$T = V_2/V_1$	=	transmission directly through the 'primary' arm of the structure
$D = V_4/V_3$	=	measure of the undesirable coupling which can occur to port 4, which we shall call the directivity
$I = V_4/V_1$	=	degree of isolation between port 4 and port 1.

It is usual to quote all these quantities in decibels, so that we talk in terms of a '10 dB coupler,' for example. This is a somewhat loose description, since the voltage of port 3 is -10 dB relative to that on port 1. In ideal couplers $D = 0$, and the power is split evenly between ports 2 and 3. This applies quite accurately to coaxial and stripline couplers for which TEM modes exist, but it is not accurate for microstrip couplers of simple parallel-coupled construction, and the considerably reduced directivity can be a serious problem, which is described in detail in Chapter 8.

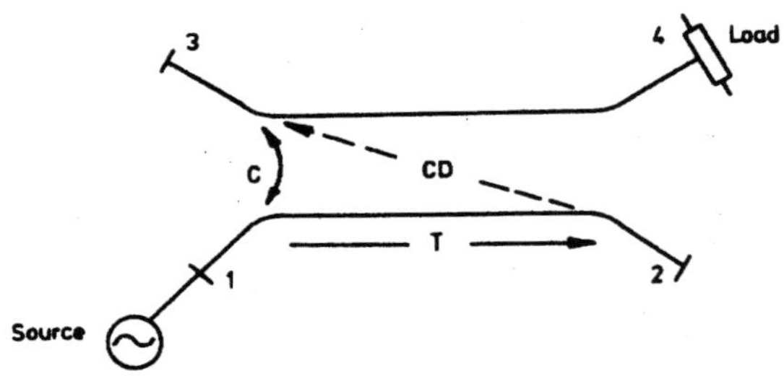

Figure A.8 Illustration of the defining parameters for coupled transmission lines: C, coupling factor; T, transmission factor; and D, directivity factor.

In practice, the specification for a coupled-line circuit element would include information on the following:

- Degree of coupling at the centre frequency (coupling C, in decibels).
- Type of transmission structure.
- Characteristic impedance of the 'external' environment (usually 50 Ω), bandwidth and centre frequency.
- Lowest acceptable 'directivity,' D (in decibels).

From data in this form, the circuit designer must eventually calculate and decide upon the structural dimensions of the coupler.

A.5.3 ANALYSIS OF PARALLEL-COUPLED TEM-MODE TRANSMISSION LINES

The material of Chapter 8 requires some concepts and expressions which are applicable to parallel-coupled TEM-mode transmission lines. Appropriate basic results are developed in this appendix.

The four-port configuration, Figure A.7, can readily be analysed by connecting matched terminations Z_0 at ports 2, 3 and 4, driving the circuit from a 2 V source having an impedance which is also equal to Z_0. The resulting 'correctly terminated' configuration is shown in Figure A.9. Later on in the analysis it is shown that Z_0 is explicitly related to the even- and odd-mode characteristic impedances associated with this coupled-line arrangement.

Note that for this analysis, which applies for the TEM mode only, we assume the lines to be totally embedded in a uniform dielectric. As a consequence of this, the odd- and even-mode characteristic impedances appear, but the phase velocity has a single value. This means that we can derive useful basic impedance relations and also the principal frequency dependence of voltage ratios which still apply quite accurately to cases where the propagation is not strictly TEM (especially microstrip). However, the even- and odd-mode phase velocities are of different values in the non-TEM situations and this leads to some considerable refinements of the theory. In particular, non-TEM couplers posses finite (i.e. imperfect) directivity. These features are discussed in more detail in Chapter 8.

We now proceed with a detailed analysis of the TEM-mode parallel coupled lines. The arrangement of Figure A.9 may be analysed by first considering excitation under

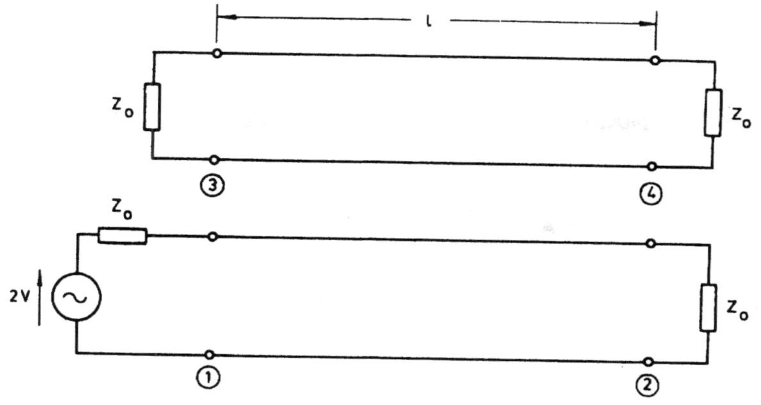

Figure A.9 Terminated coupled-line configuration.

even-mode conditions alone, then odd-mode conditions alone, and finally combining the results. The notation is given in Figure A.10. In his arrangement it is assumed that losses may be neglected so that the propagation coefficient, $\gamma = \alpha + j\beta$, reduces to the phase coefficient $j\beta$ only.

Since the structure is physically symmetrical, it can be seen (Figure A.10) that only two circuits actually have to be solved. We will take Figures A.10(a)(ii) and (b)(ii). Then, later, we can use the fact that the total voltages and currents on the original structure are a superposition of the even- and odd-mode solutions as follows:

$$\left.\begin{array}{ll} V_1 = V_{1e} + V_{1o} & I_1 = I_{1e} + I_{1o} \\ V_2 = V_{2e} + V_{2o} & I_2 = I_{2e} + I_{2o} \\ V_3 = V_{1e} - V_{1o} & I_3 = I_{1e} - I_{1o} \\ V_4 = V_{2e} - V_{2o} & I_4 = I_{2e} - I_{2o} \end{array}\right\}. \tag{A.23}$$

Now the ABCD matrix for any transmission line was given as Equation (A.22). This can be used to express the voltage-current relations for lines (a)(ii)— and (b)(i) in Figure A.10 as follows:

$$\begin{bmatrix} V_{1e} \\ I_{1e} \end{bmatrix} = \begin{bmatrix} \cos\theta & jZ_{0e}\sin\theta \\ jY_{0e}\sin\theta & \cos\theta \end{bmatrix} \begin{bmatrix} V_{2e} \\ I_{2e} \end{bmatrix} \tag{A.24}$$

and

$$\begin{bmatrix} V_{1o} \\ I_{1o} \end{bmatrix} = \begin{bmatrix} \cos\theta & jZ_{0o}\sin\theta \\ jY_{0o}\sin\theta & \cos\theta \end{bmatrix} \begin{bmatrix} V_{2o} \\ I_{2o} \end{bmatrix}. \tag{A.25}$$

Applying the terminal conditions $V_{2e} = Z_0 I_{2e}$, $V_{2o} = Z_0 I_{2o}$, $V_{1e} + I_{1e} Z_0 = 1$, and $V_{1o} + I_{1o} Z_0 = 1$ results in the voltage and current expressions:

$$V_{1e} = \frac{Z_{0e} Z_0 \cos\theta + jZ_{0e}^2 \sin\theta}{2Z_{0e} Z_0 \cos\theta + j\,(Z_{0e}^2 + Z_0^2)\sin\theta} \tag{A.26}$$

$$I_{1e} = \frac{Z_{0e}\cos\theta + jZ_0\sin\theta}{2Z_{0e} Z_0 \cos\theta + j\,(Z_{0e}^2 + Z_0^2)\sin\theta} \tag{A.27}$$

$$V_{2e} = \frac{Z_0 Z_{0e}}{2Z_{0e} Z_0 \cos\theta + j\,(Z_{0e}^2 + Z_0^2)\sin\theta} \tag{A.28}$$

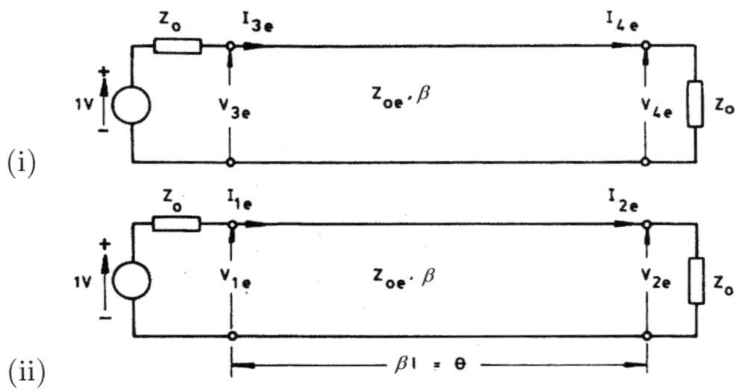

(a) Even mode

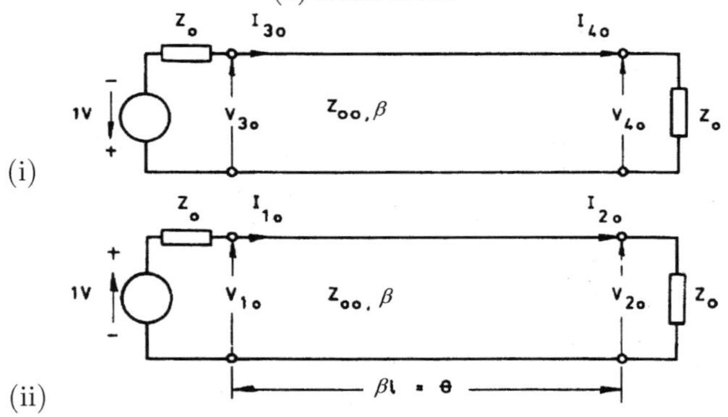

(b) Even mode

Figure A.10 Even and odd mode coupled-line representations.

$$I_{2e} = \frac{Z_{0e}}{2Z_{0e}Z_0\cos\theta + j\,(Z_{0e}^2 + Z_0^2)\sin\theta} \tag{A.29}$$

$$V_{1o} = \frac{Z_{0o}Z_0\cos\theta + jZ_0^2\sin\theta}{2Z_{0e}Z_0\cos\theta + j\,(Z_{0e}^2 + Z_0^2)\sin\theta} \tag{A.30}$$

$$I_{1o} = \frac{Z_{0o}\cos\theta + jZ_0\sin\theta}{2Z_{0o}\cos\theta + j\,(Z_{0o}^2 + Z_0^2)\sin\theta} \tag{A.31}$$

$$V_{2o} = \frac{Z_0 Z_{0o}}{2Z_{0o}Z_0\cos\theta + j\,(Z_{0o}^2 + Z_0^2)\sin\theta} \tag{A.32}$$

$$I_{2o} = \frac{Z_{0o}}{2Z_{0o}Z_0\cos\theta + j\,(Z_{0o}^2 + Z_0^2)\sin\theta}. \tag{A.33}$$

The next step consists of an important simplification and amounts to forcing the composite original circuit to become *matched* to the feed lines of characteristic impedance Z_0. Therefore, the total input impedance Z_{in} has to equal Z_0:

$$Z_{in} = Z_0 = \frac{V_1}{I_1} = \frac{V_{1e} + V_{1o}}{I_{1e} + I_{1o}}. \tag{A.34}$$

Substituting Equation (A.26), (A.27), (A.30) and (A.31) into Equation (A.34) gives

$$Z_0 = \frac{Z_0(Z_{0o}D_e + Z_{0e}D_o)\cos\theta + j(Z_{0o}^2 D_e + Z_{0e}^2 D_o)\sin\theta}{(Z_{0o}D_e + Z_{0e}D_o)\cos\theta + jZ_0\,(D_e + D_o)\sin\theta} \tag{A.35}$$

where

$$D_e = 2Z_{0e}Z_0\cos\theta + j(Z_{0e}^2 + Z_0^2)\sin\theta \tag{A.36}$$

and

$$D_o = 2Z_{0o}Z_0\cos\theta + j(Z_{0o}^2 + Z_0^2)\sin\theta. \tag{A.37}$$

The equality of Equation (A.35) will be established if

$$Z_0^2(D_e + D_o) = Z_{0o}^2 D_e + Z_{0e}^2 D_o \tag{A.38}$$

Substituting, using Equation (A.36) and (A.37) for D_e and D_o, results in

$$\begin{aligned}&2Z_0^3(Z_{0e} + Z_{0o})\cos\theta + j(Z_{0e}^2 Z_0^2 + Z_{0o}^2 Z_0^2 + 2Z_0^4)\sin\theta \\ &\quad = 2Z_{0e}Z_{0o}Z_0(Z_{0e} + Z_{0o})\cos\theta + j(Z_{0e}^2 Z_0^2 + Z_{0o}^2 Z_0^2 + 2Z_{0e}^2 Z_{0o}^2)\sin\theta.\end{aligned}$$

This equality will be satisfied if

$$Z_0^2 = Z_{0e}Z_{0o}. \tag{A.39}$$

Inserting this relationship into Equation (A.26) to (A.33) gives

$$V_{1e} = \frac{Z_0\cos\theta + jZ_{0e}\sin\theta}{2Z_0\cos\theta + j\,(Z_{0e} + Z_{0o})\sin\theta} \tag{A.40}$$

$$I_{1e} = \frac{\cos\theta + jZ_{0o}Y_0\sin\theta}{2Z_0\cos\theta + j\,(Z_{0e} + Z_{0o})\sin\theta} \tag{A.41}$$

$$V_{2e} = \frac{Z_0}{2Z_0\cos\theta + j\,(Z_{0e} + Z_{0o})\sin\theta} \tag{A.42}$$

$$I_{2e} = \frac{1}{2Z_0\cos\theta + j\left(Z_{0e} + Z_{0o}\right)\sin\theta} \tag{A.43}$$

$$V_{1o} = \frac{Z_0\cos\theta + jZ_{0o}\sin\theta}{2Z_0\cos\theta + j\left(Z_{0e} + Z_{0o}\right)\sin\theta} \tag{A.44}$$

$$I_{1o} = \frac{\cos\theta + jZ_{0e}Y_0\sin\theta}{2Z_0\cos\theta + j\left(Z_{0e} + Z_{0o}\right)\sin\theta} \tag{A.45}$$

$$V_{2o} = \frac{Z_0}{2Z_0\cos\theta + j\left(Z_{0e} + Z_{0o}\right)\sin\theta} \tag{A.46}$$

$$I_{2o} = \frac{1}{2Z_0\cos\theta + j\left(Z_{0e} + Z_{0o}\right)\sin\theta}. \tag{A.47}$$

Final expressions for the total (complete structure — Figure 1.14) voltages and currents can now be obtained by substituting Equation (A.40) to (A.47) into the group of Equation (A.23):

$$V_1 = V_{1e} + V_{1o} = \frac{2Z_0\cos\theta + j\left(Z_{0e} + Z_{0o}\right)\sin\theta}{2Z_0\cos\theta + j\left(Z_{0e} + Z_{0o}\right)\sin\theta} = 1 \tag{A.48}$$

$$I_1 = I_{1e} + I_{1o} = \frac{2\cos\theta + jY_0\left(Z_{0e} + Z_{0o}\right)\sin\theta}{2Z_0\cos\theta + j\left(Z_{0e} + Z_{0o}\right)\sin\theta} = \frac{1}{Z_0} \tag{A.49}$$

$$V_2 = V_{2e} + V_{2o} = \frac{2Z_0}{2Z_0\cos\theta + j\left(Z_{0e} + Z_{0o}\right)\sin\theta} \tag{A.50}$$

$$I_2 = I_{2e} + I_{2o} = \frac{2}{2Z_0\cos\theta + j\left(Z_{0e} + Z_{0o}\right)\sin\theta} \tag{A.51}$$

$$V_3 = V_{1e} - V_{1o} = \frac{j(Z_{0e} - Z_{0o})\sin\theta}{2Z_0\cos\theta + j\left(Z_{0e} + Z_{0o}\right)\sin\theta} \tag{A.52}$$

$$I_3 = I_{1e} - I_{1o} = \frac{jY_0(Z_{0o} - Z_{0e})}{2Z_0\cos\theta + j\left(Z_{0e} + Z_{0o}\right)\sin\theta} \tag{A.53}$$

$$V_4 = V_{2e} - V_{2o} = \frac{0}{2Z_0\cos\theta + j\left(Z_{0e} + Z_{0o}\right)\sin\theta} = 0 \tag{A.54}$$

$$I_4 = I_{2e} - I_{2o} = \frac{0}{2Z_0\cos\theta + j\left(Z_{0e} + Z_{0o}\right)\sin\theta} = 0. \tag{A.55}$$

$$\tag{A.56}$$

These voltages and currents can be further simplified by defining a coupling parameter C such that

$$C = \frac{Z_{0e} - Z_{0o}}{Z_{0e} + Z_{0o}}. \tag{A.57}$$

This definition plus the relationship $Z_0^2 = Z_{0e}Z_{0o}$, Equation (A.39), allows us to write, from Equation (A.48):

$$V_1 = 1 \text{ (just the input voltage, as defined)} \tag{A.58}$$

$$V_2 = \frac{\sqrt{1 - C^2}}{\sqrt{1 - C^2}\cos\theta + j\sin\theta} \tag{A.59}$$

$$V_3 = \frac{jC\sin\theta}{\sqrt{1 - C^2}\cos\theta + j\sin\theta} \tag{A.60}$$

$$V_4 = 0 \ . \tag{A.61}$$

The maximum degree of coupling occurs when the length of the coupled structure is $\lambda_g/4$, i.e. when $\theta = \pi/2$. This condition, set using Equation (A.58) to (A.61) yields

$$\begin{aligned} V_1 &= 1 \text{ just the input voltage, as in Equation A.48} && \text{(A.62)} \\ V_2 &= -j\sqrt{1-C^2} && \text{(A.63)} \\ V_3 &= C && \text{(A.64)} \\ V_4 &= 0 \text{ as in Equation (A.61).} && \text{(A.65)} \end{aligned}$$

Equation (A.64) gives the mid-band (maximum) extent of coupling C, expressed in terms of impedances by Equation (A.57). Equation (A.63) shows that the transmitted voltage V_2(not electromagnetically coupled) depends upon C and is 90° out-of-phase with V_1, and hence also with V_3. This feature gives rise to the names often used for couplers based upon this parallel-coupled line principle: 90°hybrid, or quadrature coupler (There is zero output on port 4, for a theoretically perfect structure.)

The principal frequency response characteristics, $|V_2(\theta)|$ and $|V_3(\theta)|$ may be obtained from Equation (A.59) and (A.60) respectively, where we note that the electrical length $\theta°$ is also given by

$$\theta = \frac{2\pi l}{\lambda_g} = \frac{2\pi f l}{v}. \tag{A.66}$$

Since the physical length l of the structure is fixed and we assume the velocity of propagation v to be constant, Equation (A.66) shows that θ is directly dependent on the frequency f.

In design procedures we shall usually be confronted with the problem of determining the physical dimensions of some proposed parallel-coupled structure. The cross-sectional dimensions, line separation and line diameters or widths, can be determined knowing where the precise relationships depend upon the proposed structure. Thus we require Z_{0e} and Z_{0o} as functions of Z_0 and C. These relationships may be obtained from Equation (A.57) and (A.39). The coupling factor C is often expressed in decibels. If we call this value $C'$and remember that we have a voltage ratio, then from Equation (A.57):

$$C' = 20 \log \left| \frac{Z_{0e} - Z_{0o}}{Z_{0e} + Z_{0o}} \right| \quad \text{dB}. \tag{A.67}$$

Combining this with Equation (A.39), $Z_0^2 = Z_{0e}Z_{0o}$, yields the initial design equations:

$$Z_{0e} = Z_0 \sqrt{\frac{1 + 10^{C'/20}}{1 - 10^{C'/20}}} \tag{A.68}$$

and

$$Z_{0o} = Z_0 \sqrt{\frac{1 - 10^{C'/20}}{1 + 10^{C'/20}}}. \tag{A.69}$$

The application of these equations is discussed in Chapter 8.

Appendix B

Q-Factor

B.1 DEFINITION

The Q, or Quality factor, of a component or circuit is defined as the ratio of 2π times the maximum energy stored to the energy lost per cycle. For a parallel resonant circuit with elements L, C and G (see Figure B.1(a)) the Q is

$$Q = \omega_r C/G = 1/(\omega_r LG). \tag{B.1}$$

where $\omega_r = 2\pi f$ is the resonant frequency and is the frequency at which the maximum amount of energy is stored. The conductance G describes the energy lost in a cycle. A series resonant circuit with L, C and R elements has a similar expression for Q:

$$Q = \omega_r L/R = 1/(\omega_r CR). \tag{B.2}$$

These resonant circuits have a bandpass transfer characteristic and the Q is a measure of how sharp the resonance is. The Q is then the inverse of the fractional bandwidth of the resonator. The fractional bandwidth Δf is measured at the half power points as shown in Figure B.2. (Δf is also referred to as the two-sided -3 dB bandwidth.) Then

$$Q = \frac{f_r}{\Delta f}. \tag{B.3}$$

Thus the Q is a measure of the sharpness of the bandpass filter frequency response. The pairs of definitions Equations (B.1), (B.3) and Equations (B.2), (B.3) are identical for parallel RLC and series RLC circuits respectively.

The determination of Q using the measurement of bandwidth together with Equation (B.3) is often not very sensitive so another definition that uses the much more sensitive phase change at resonance is preferred. With ϕ being the phase (in radians) of the transfer characteristic the definition of Q is now

$$Q = \frac{\omega_r}{2}\left|\frac{\mathrm{d}\phi}{\mathrm{d}\omega}\right|. \tag{B.4}$$

Figure B.1 Parallel (a) and series (b) resonant circuits.

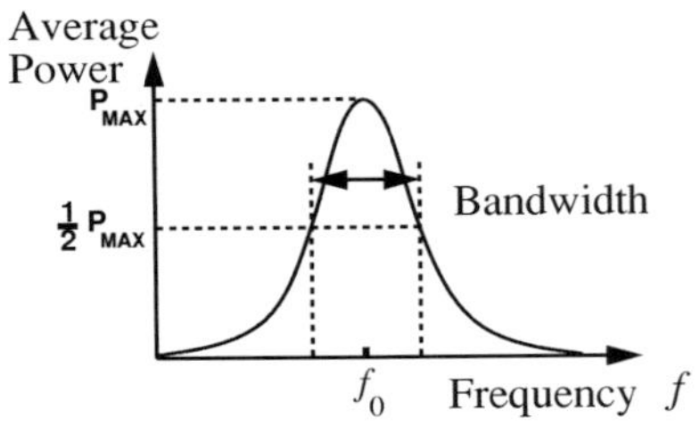

Figure B.2 Transfer characteristic of resonant circuit.

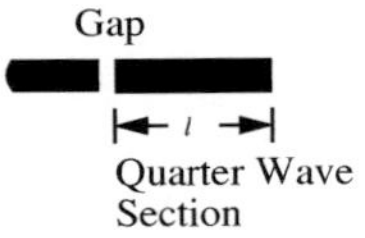

Figure B.3 Gap coupled quarter-wave section.

This is another equivalent definition for parallel RLC or series RLC resonant circuits. However, Q is used for more complicated circuits and then its meaning is always a ratio of the energy stored to the energy dissipated. Thus, it is meaningful to talk about the Q of circuits other than three element RLC circuits such as interconnects. The Q of these structures can no longer be determined by bandwidth, or by rate of phase change considerations.

B.2 LOADED Q-FACTOR

The Q of a component as defined in the previous section is called the unloaded Q, Q_u. However, if a component is to be measured or used in any way it is necessary to couple energy into and out of it. This leads to the use of the loaded Q, Q_l. So a parallel LCG circuit with elements L_r, C_r and G_r (at resonance) loaded by a shunt conductance G_l has

$$Q_u = \omega_r C_r / G_r = 1/(\omega_r L_r G_r) \tag{B.5}$$

and

$$Q_l = \omega_r C_r / (G_r + G_l). \tag{B.6}$$

So

$$\frac{1}{Q_l} = \frac{1}{Q_u} + \frac{1}{Q_x} \tag{B.7}$$

or

$$Q_x = \left(\frac{1}{Q_l} - \frac{1}{Q_u}\right). \tag{B.8}$$

Q_x is called the external Q, and it describes the effect of loading. Q_l is the Q that is actually measured. Q_u normally needs to be determined, but if the loading is kept very small $Q_l \approx Q_u$.

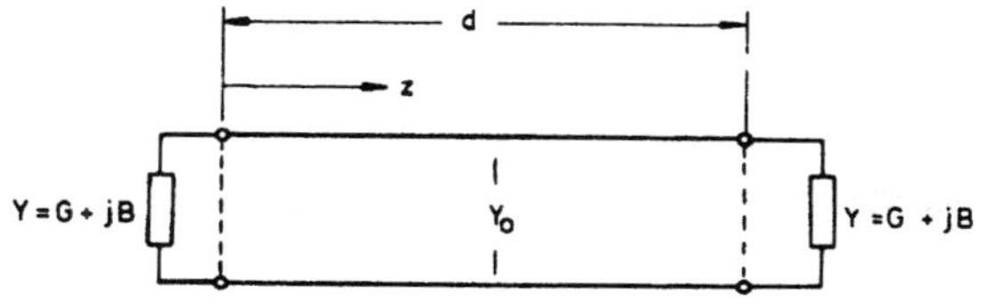

Figure B.4 Open-circuited microstrip resonator.

B.3 EXTERNAL Q-FACTOR OF A STRAIGHT-EDGED, OPEN-CIRCUITED MICROSTRIP RESONATOR

The open-circuited microstrip resonator is shown in Figure B.4. The reflection coefficients of the line terminations are equal and given by

$$\Gamma = \frac{Y_0 - Y}{Y_0 + Y} = \frac{(Y_0 - G) - jB}{(Y_0 + G) + jB}. \tag{B.9}$$

Rationalizing we obtain

$$\Gamma = \frac{(Y_0 - G)(Y_0 + G) - B^2 - j2BY_0}{(Y_0 + G)^2 + B^2} \tag{B.10}$$

or

$$\Gamma = |\Gamma| \exp(-j\theta) = \rho \exp(-j\theta) \tag{B.11}$$

where ρ and θ are the magnitude and phase angle of the reflection coefficient, respectively, given by

$$\rho = \frac{\left[(Y_0^2 - G^2 - B^2)^2 + 4B^2Y_0^2\right]^{1/2}}{(Y_0 + G)^2 + B^2}. \tag{B.12}$$

and

$$\theta = \tan^{-1}\left(\frac{2BY_0}{Y_0^2 - G^2 - B^2}\right). \tag{B.13}$$

These expressions will be required shortly, but first we must find the conditions for the maximum electric field $E_{\max}$ in the cavity, since this corresponds to the resonance condition. We now proceed to obtain the conditions for $E_{\max}$ in terms of the reflection coefficient Γ, the phase coefficient β, and the resonator length. It is initially necessary to consider the multiple reflections occurring within the resonator (see Figure B.5). The field amplitude E_0 is merely a reference field, so that the other relative field strengths may be developed. The phase coefficient $\beta = \omega_0\sqrt{\varepsilon_{\text{eff}}(f)}/c$. From the fields shown on the diagram (Figure B.5), we may add all the components at distance z to obtain the total field at that location:

$$E_x(z) = E_0 \exp(-j\beta z) + \Gamma E_0 \exp\{-j\beta(2d - z)\} + \cdots + \Gamma^2 E_0 \exp\{-j\beta(2d + z)\} + \cdots \tag{B.14}$$

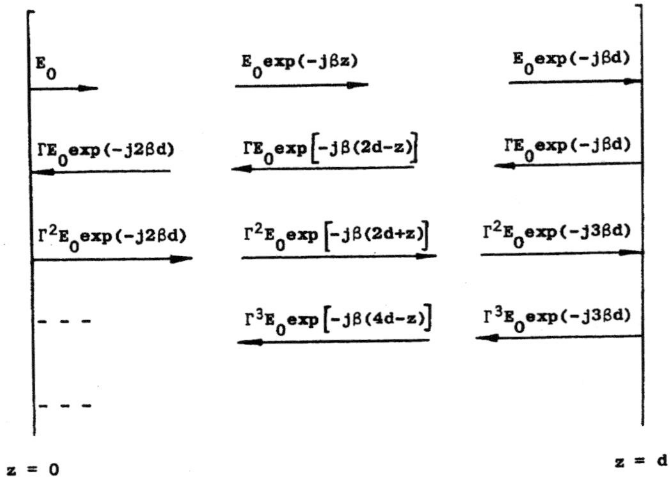

Figure B.5 Reflection diagram.

or

$$E_x = E_0 \exp(-j\beta z) \sum_{n=0}^{\infty} \Gamma^{2n} \exp(-j2\beta nd) + \cdots + \Gamma E_0 \exp\{-j\beta(2d - z)\} \times \sum_{n=0}^{\infty} \Gamma^{2n} \exp(-j2\beta nd) \tag{B.15}$$

whence

$$E_x = \frac{E_0\{\exp(-j\beta z) + \Gamma \exp[-j\beta(2d - z)]\}}{1 - \Gamma^2 \exp(-j2\beta d)}. \tag{B.16}$$

The resonance of the microstrip resonator, which corresponds to a maximum of this electric field E_x, occurs when the denominator (D, say) of Equation (B.16) is a minimum. Thus we require a minimum for

$$|D| = \left|1 - \Gamma^2 \exp[-j2(\beta d)]\right| \tag{B.17}$$

or, inserting the polar form of Γ given earlier,

$$|D| = \left|1 - \rho^2 \exp\{-j2(\beta_d + \theta)\}\right| \tag{B.18}$$

and this is a minimum when

$$2\,(\beta d + \theta) = 2\pi l \tag{B.19}$$

where l =1, 2, 3, ..., i.e. an integer. In the 'ideal' case where Y is either a true open-circuit ($G = 0$, $B = 0$) or a perfect short-circuit ($G = \infty$, $B = 0$, or $B \neq 0$), we obtain $\theta = 0$ and the almost trivial solution to Equation (B.19, that is $\beta d = \pi l$ — whence $d = l\lambda_o/2$. In our more practical case with finite Y, the full resonance condition of Equation (B.19) is required.

From Equations (B.17) and (B.18), the minimum value of $|D|$ is clearly

$$|D|_{\min} = 1 - \rho^2. \tag{B.20}$$

Now, as D increases, E_x will exhibit the usual resonant response curve as shown in Figure B.2. It is well known that the Q-factor is then given by

$$Q_e = \omega_0/(2\Delta\omega) \tag{B.21}$$

where $E_{\max}$, at ω_0, is reduced to $E_{\max}/\sqrt{2}$ at $(\omega \pm \Delta\omega)$. Equation (B.21) can also be written in terms of the phase coefficient

$$Q_e = \beta/(2\Delta\beta) \tag{B.22}$$

where β is the phase coefficient at frequency ω_0. We therefore now require an expression for $\Delta\beta$, in order to finally obtain a formula for Q_e as a function of ρ. The field $E_{\max}/\sqrt{2}$ occurs when $|D|$ is increased to the value $\sqrt{2}\left(1-\rho^2\right)$ and the corresponding $(\beta \pm \Delta\beta)$ and $(\theta \pm \Delta\theta)$ respectively. Therefore, we can write

$$\Delta(\beta d + \theta) = \Delta\beta d + \Delta\theta = \Delta\beta\left(d + \frac{\Delta\theta}{\Delta\beta}\right). \tag{B.23}$$

To find the required quantity $\Delta\beta$, we replace $(\beta + \theta)$ by

$$(\beta d + \theta) + \Delta\left(\beta d + \theta\right) \tag{B.24}$$

in the expression for $|D|^2$ and equate the result to

$$\left[\sqrt{2}\left(1-\rho^2\right)\right]^2 = 2\left(1-\rho^2\right)^2. \tag{B.25}$$

First, expand the expression in β, d, θ, using Equations (B.19) and (B.23):

$$(\beta d + \theta) + \Delta\left(\beta d + \theta\right) = \pi l + \Delta\beta\left(d + \Delta\theta/\Delta\beta\right). \tag{B.26}$$

Hence,

$$\begin{aligned}|D| &= \left|1 - \rho^2 \exp\{-2j\pi l + \Delta\beta(d + \Delta\theta/\Delta\beta)\}\right| \\ &= \left|1 - \rho^2 \exp\{-j2\Delta\beta(d + \Delta\theta/\Delta\beta)\}\right| \\ &= \left|1 - \rho^2 \cos 2\Delta\beta\left(d + \Delta\theta/\Delta\beta\right) + \rho^2 j \sin 2\Delta\beta\left(d + \Delta\theta/\Delta\beta\right)\right|. \end{aligned} \tag{B.27}$$

We may also expand $|D|^2$ as a complex product: $|D|^2 = DD^*$ or

$$\begin{aligned}|D|^2 &= \{1 - \rho^2 \cos 2\Delta\beta(d + \Delta\theta/\Delta\beta)\}^2 + \rho^4 j \sin^2 2\Delta\beta(d + \Delta\theta/\Delta\beta) \\ &= (1-\rho^2)^2 + 2\rho^2\{1 - \cos 2\Delta\beta(d + \Delta\theta/\Delta\beta)\}. \end{aligned} \tag{B.28}$$

This is now equated to $2\left(1-\rho^2\right)^2$ to give

$$(1-\rho^2)^2 = 2\rho^2\{1 - \cos 2\Delta\beta(d + \Delta\theta/\Delta\beta)\}. \tag{B.29}$$

Since, for moderate or high Q-factor resonators, $\Delta\beta$ must be small we may use the small-angle (δy) approximation:

$$1 - \cos\delta y \approx \frac{1}{2}(\delta y)^2 \tag{B.30}$$

giving

$$(1-\rho^2)^2 = 4\rho^2\Delta\beta^2(d+\Delta\theta/\Delta\beta)^2 \tag{B.31}$$

whence

$$\Delta\beta = \frac{1-\rho^2}{2\rho(d+\Delta\theta/\Delta\beta)}. \tag{B.32}$$

Substituting this into Equation (B.22) gives

$$Q_e = \frac{\beta\rho(d+\Delta\theta/\Delta\beta)}{1-\rho^2}. \tag{B.33}$$

The quantity $\Delta\theta/\Delta\beta$ is still required, and we obtain this by differentiating Equation (B.13):

$$\frac{\Delta\theta}{\Delta\beta} = \frac{\Delta}{\Delta\beta}\left(\left\{\tan^{-1}\left[\frac{2BY_0}{(Y_0^2-G^2-B^2)}\right]\right\}\right). \tag{B.34}$$

The result of this can be written down if we first note that

$$\frac{\mathrm{d}}{\mathrm{d}x}\left(\tan^{-1}\frac{x}{a}\right) = \frac{a}{x^2+a^2}\frac{\mathrm{d}}{\mathrm{d}x}\left(\frac{x}{a}\right). \tag{B.35}$$

Then

$$\frac{\Delta\theta}{\Delta\beta} = \frac{Y_0^2-G^2-B^2}{(Y_0^2-G^2-B^2)^2+4B^2Y_0^2}\frac{1}{\Delta\beta}\left[\Delta\left(\frac{2BY_0}{Y_0^2-G^2-B^2}\right)\right]. \tag{B.36}$$

In order to evaluate this we really need to know the functional dependence of B and G on β. However, because the susceptance is approximately of a capacitive nature ($B \approx \omega C$), B is approximately a linear function of β and then

$$\frac{\Delta B}{\Delta\beta} \approx \frac{\theta}{\beta}. \tag{B.37}$$

Taking this result and substituting, first in Equation (B.36) and finally in Equation (B.33), we find

$$Q_e = \frac{\rho(\beta d+\theta)}{1-\rho^2} \tag{B.38}$$

that is

$$Q_e = \frac{\pi l\rho}{1-\rho^2}. \tag{B.39}$$

This is the 'classic' result for the external Q-factor when a resonator is equally end-loaded, and it is easily evaluated. The calculation of B and G follow from ρ and θ as expressed by Equations (B.12) and (B.13). G is quite easily determined by combining Equations (B.12) and (B.13):

$$\rho = \frac{2BY_0\sqrt{1+\cot^2\theta}}{(Y_0+G)^2+B^2} \tag{B.40}$$

or

$$G = \left(\frac{2BY_0\sqrt{1+\cot^2\theta}}{\rho} - B^2\right)^{1/2}. \tag{B.41}$$

To obtain B we assume that $G \ll Y_0$ and determine the approximate $\tan\theta$ expression from Equation (B.13), yielding

$$\tan\theta \approx \frac{2BY_0}{Y_0^2 - B^2} \tag{B.42}$$

or

$$B = \frac{Y_0(-1 \pm \sqrt{1+\tan^2\theta})}{\tan\theta}. \tag{B.43}$$

We have to choose the correct sign on the root in this equation, and to decide this we study the effects of the extreme condition $B \ll Y_0$. This gives, from Equation (B.42),

$$\tan\theta \approx \frac{2B}{Y_0} \approx \theta \tag{B.44}$$

whence

$$B \approx \frac{Y_0}{2}\tan\theta \approx \frac{Y_0}{2}. \tag{B.45}$$

If we set the condition $\theta \ll 1$, then $\tan\theta \approx \theta$, and Equation (B.43) becomes

$$B \approx \frac{Y_0(-1 \pm \sqrt{1+\theta^2})}{\theta} \approx \frac{Y_0\{-1 \pm (1+\theta^2/2)\}}{\theta}. \tag{B.46}$$

This can only equal $Y_0\theta/2$, as in Equation (B.45), if the positive root is taken; hence the approximate expression for B is

$$B = \frac{Y_0(-1 + \sqrt{1+\tan^2\theta})}{\tan\theta}. \tag{B.47}$$

In a situation where the initial assumption, $G \ll Y_0$, is not valid, an iterative procedure must be carried out in order to find a corrected expression for $\tan\theta$. At each stage updated values of B and G are obtained, and the procedure is only halted when a satisfactory convergence to limiting values has been obtained.

In the above expressions Y_0, the characteristic admittance of the resonator, is equal to $1/Z_0$ ($Y_0 = 0.02$ S when $Z_0 = 50\ \Omega$), ($Y_0 = 0.1$ S when $Z_0 = 10\ \Omega$, etc.).

The phase angle θ of the reflection coefficient can be readily obtained from Equation (B.19):

$$\beta d + \theta = \pi l \tag{B.48}$$

where

$$\beta = c/\left(f\sqrt{\varepsilon_{\text{eff}}(f)}\right) \tag{B.49}$$

$$d = \text{known length (physically measured)} \tag{B.50}$$

$$l = \text{(integer) order of resonance.} \tag{B.51}$$

Appendix C

Outline of Scattering Parameter Theory

C.1 INTRODUCTION

Characterization and design at frequencies ranging from audio to some hundreds of megahertz proceeds on the basis of admittances, impedances and dimensionless factors. Transistors, for example, have input impedance, output impedance and transconductance. Above a few hundred megahertz it is neither practical nor meaningful to measure voltage, current or impedance — at least not in any direct manner. It is important to appreciate that the circuits are distributed, and so are the voltages and currents (even across a given plane in a transmission line). On the other hand, useful quantities such as *voltage reflection coefficient* and *microwave power* can be quite readily measured. A significant set of parameters known as 'scattering parameters' or more briefly S-parameters embody the effects of reflection or through transmission of power for any network. It matters not whether the network is passive or active, nor whether it comprises one port or one hundred ports; scattering parameters remain very convenient to use but, as will be seen, it is easy to convert these to more familiar network parameters.

In order to see why a new parameter set is highly desirable, we first review the more 'conventional' sets of parameters.

C.2 NETWORK PARAMETERS

It is well known that several parameter sets are available for the characterization of a two-port network (i.e. a 'four-terminal' network), as illustrated in Figure C.1. In an N-port network there are $2N$ terminals as each port has two terminals with one of the terminals used as a reference. For example, the voltage of a port is the voltage of the other terminal with respect to that of the reference terminal.

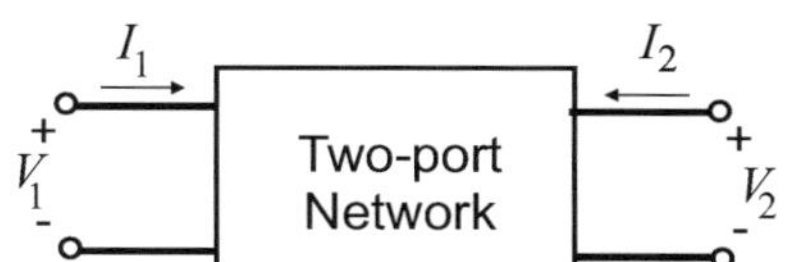

Figure C.1 A two-port network.

Although each parameter, in any set, should strictly be written as a function of complex frequency s (or at least as a function of the operator $j\omega$), this fact is generally left understood. For example, in admittance parameters, y_{11} really means $y_{11}(s)$, or $y_{11}(j\omega)$.

A few examples are now given of different network parameters and the way they describe the terminal characteristics of a two-port network.

Impedance parameters, or z-parameters:

$$\left.\begin{array}{rcl} V_1 & = & z_{11}I_1 + z_{12}I_2 \\ V_2 & = & z_{21}I_1 + z_{22}I_2 \end{array}\right\}. \tag{C.1}$$

Admittance parameters, or y-parameters:

$$\left.\begin{array}{rcl} I_1 & = & y_{11}V_1 + y_{12}V_2 \\ I_2 & = & y_{21}V_1 + y_{22}V_2 \end{array}\right\}. \tag{C.2}$$

Hybrid parameters, or h-parameters:

$$\left.\begin{array}{rcl} V_1 & = & h_{11}I_1 + h_{12}V_2 \\ I_2 & = & h_{21}I_1 + h_{22}V_2 \end{array}\right\}. \tag{C.3}$$

Chain parameters, or $ABCD$ parameters:

$$\left.\begin{array}{rcl} V_1 & = & AV_2 - BI_2 \\ I_1 & = & CV_2 - DI_2 \end{array}\right\}. \tag{C.4}$$

For active circuits, particularly discrete transistors, the h-parameters have been used quite extensively at low frequencies, and the y-parameters at frequencies ranging from low megahertz to hundreds of megahertz. The reasons for choosing h-parameters at lower frequencies and y-parameters at higher frequencies are largely to do with measurement convenience. It can be seen from Equation (C.3) that each h-parameter may be measured by alternately open-circuiting the input port and short-circuiting the output port, producing signal current zero and signal voltage zero, respectively. At lower frequencies this has proved a satisfactory procedure — especially for characterizing active devices in view of bias requirements and device resistances. Under radio-frequency conditions, some hundreds of kilohertz upwards, open-circuits become hard to produce and less reliable. The y-parameters have become attractive at such frequencies because, as can be seen from Equations (D.2), each y-parameter is obtainable by alternately short-circuiting output and input ports.

All the parameters are interrelated (any set of N-port parameters can be exactly related to any other set), and the relationships are given in matrix notation in Table C.1 where the quantity Δ is the determinant in each case. For example:

$$\Delta_y = y_{11}y_{22} - y_{12}y_{21}. \tag{C.5}$$

Each parameter, in every set of parameters, is sensitive to change in frequency, bias conditions, or temperature (especially for active devices), and this important fact must always be remembered. For design at higher frequencies, especially beyond a few hundred megahertz, y-parameters become less useful and a new concept is required. There are two principal reasons for this:

Table C.1 Two-port S-parameter conversion chart. The Z, Y and H parameters are normalized to Z_0. Z', Y' and H' are the actual parameters

	S	In terms of S
Z	$z'_{11} = z_{11}Z_0 \quad z'_{12} = z_{12}Z_0$	$z'_{21} = z_{21}Z_0 \quad z'_{22} = z_{22}Z_0$
	$S_{11} = [(z_{11}-1)(z_{22}+1) - z_{12}z_{21}]/\Delta_z$ $S_{12} = 2z_{12}/\Delta_z$ $S_{21} = 2z_{21}/\Delta_z$ $S_{22} = [(z_{11}+1)(z_{22}-1) - z_{12}z_{21}]/\Delta_z$	$z_{11} = [(1+S_{11})(1-S_{22}) + S_{12}S_{21}]/\Delta_S$ $z_{12} = 2S_{12}/\Delta_S$ $z_{21} = 2S_{21}/\Delta_S$ $z_{22} = [(1-S_{11})(1+S_{22}) + S_{12}S_{21}]/\Delta_S$
Y	$y'_{11} = y_{11}/Z_0 \quad y'_{12} = y_{12}/Z_0$	$y'_{21} = y_{21}/Z_0 \quad y'_{22} = y_{22}/Z_0$
	$S_{11} = [(1-y_{11})(1+y_{22}) + y_{12}y_{21}]/\Delta_y$ $S_{12} = -2y_{12}/\Delta_y$ $S_{21} = -2y_{21}/\Delta_y$ $S_{22} = [(1+y_{11})(1-y_{22}) + y_{12}y_{21}]/\Delta_y$	$y_{11} = [(1-S_{11})(1+S_{22}) + S_{12}S_{21}]/\Delta_S$ $y_{12} = -2S_{12}/\Delta_S$ $y_{21} = -2S_{21}/\Delta_S$ $y_{22} = [(1+S_{11})(1-S_{22}) + S_{12}S_{21}]/\Delta_S$
H	$h'_{11} = h_{11}Z_0 \quad h'_{12} = h_{12}$	$h'_{21} = h_{21} \quad h'_{22} = h_{22}/Z_0$
	$S_{11} = [(h_{11}-1)(h_{22}+1) - h_{12}h_{21}]/\Delta_h$ $S_{12} = 2h_{12}/\Delta_h$ $S_{21} = -2h_{21}/\Delta_h$ $S_{22} = [(1+h_{11})(1-h_{22}) + h_{12}h_{21}]/\Delta_h$	$h_{11} = [(1+S_{11})(1+S_{22}) - S_{12}S_{21}]/\Delta_S$ $h_{12} = 2S_{12}/\Delta_S$ $h_{21} = -2S_{21}/\Delta_S$ $h_{22} = [(1-S_{11})(1-S_{22}) - S_{12}S_{21}]/\Delta_S$

1. The difficulty in realizing accurate, well-defined, broadband short-circuits as required for y-parameters.
2. The considerable chance that the device or amplifier may become unstable and oscillate under short-circuit test.

As discussed in the introduction (Section C.1), the direct measurement of S-parameters is the preferred approach above a few hundred megahertz but one that can be used even down to a few hundred kilohertz. The only hurdle inhibiting most people adopting them for every situation is that they are defined in terms of travelling wave voltages and not total (conventional) voltages and currents, as are the other parameters considered above. As will be seen, S-parameters are particularly convenient to use in characterizing interconnects as they naturally relate to signals entering a line at one end and appearing at the other.

C.3 SCATTERING PARAMETERS

It will already be well known that the reflection coefficient of a 'one-port network' can quite easily be measured. This quantity (Γ) is simply related to the impedance (Z_L) of such a 'one-port':

$$\Gamma = \frac{Z_L - Z_0}{Z_L + Z_0} \tag{C.6}$$

where Z_0 is the characteristic impedance of the connecting transmission line. When a network is more complicated, that is one consisting of two or more ports, then quantities beyond reflection coefficients are required to fully characterize its behaviour. We shall be able to make considerable use of two-port networks and these may well include active (i.e. amplifying) elements. One basic way of identifying four parameters is suggested, rather fancifully, in Figure C.2. Here Γ_{in} and Γ_{out} are respectively the

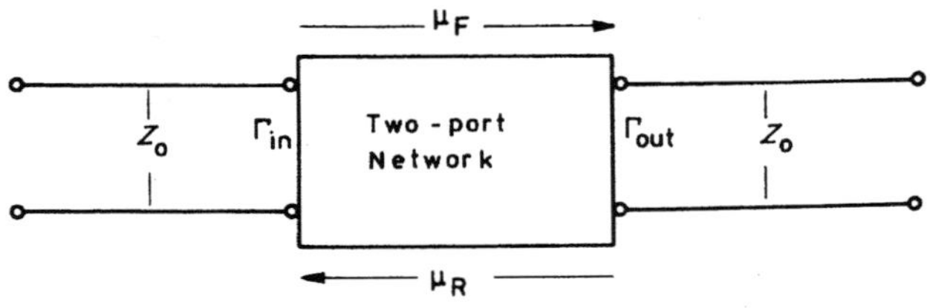

Figure C.2 A two-port network with input and output reference transmission lines.

input and output port reflection coefficients, μ_F is a forward transmission parameter and μ_R is a reverse transmission parameter. These four 'parameters' have only been mentioned in passing at this stage. They have not been properly defined and are not in the notation used for S-parameters.

It was previously mentioned that power is another quantity which can be directly measured at microwave frequencies. The S-parameters will lead explicitly to power relationships (such as interconnect power transmission loss) and, conceptually, these S-parameters are based upon the incident and reflected proportions of power around the network. More precisely, quantities having the dimensions 'square-root' of 'power' are considered and are generally referred to as 'waves'. A transmission parameter (or 'transfer' parameter) through a circuit or a junction is a relative measure of the amplitude and phase of the transmitted wave as compared with the amplitude and phase of the incident wave. In other words, the directly measurable quantities are the amplitudes and phase angles of the waves reflected, or scattered, from a junction or port relative to the incident wave amplitudes and phase angles. In view of the linearity of the field equations for interconnects and a variety of important microwave devices, the scattered-wave amplitudes are linearly related to the incident wave amplitudes. The matrix describing this linear relationship is called the scattering matrix.

Scattering matrices are therefore particularly significant and useful in a high speed transmission environment. Impedance as such cannot, however, be discarded, since the characteristic impedance Z_0, of the transmission lines will be vitally and inextricably involved with the elements of these scattering matrices. Other circuit impedances will be shown to be interrelated with certain elements of the matrices or 'scattering parameters' (transistor equivalent circuit impedances and mutual conductance are a good example).

A basic treatment, applicable to circuits where the characteristic impedance is real and positive, is given here.

C.3.1 SCATTERING PARAMETERS FOR A TWO-PORT NETWORK

Filters can be synthesized and interconnect networks can be characterized in terms of the scattering or S-parameters applicable to two-port networks. These restricted versions of S-parameters are therefore very important and worthy of detailed examination. Figure C.3 shows a two-port network connected to a load Z_L by means of a length of transmission line and fed from a source, with impedance Z_S. Z_S is also connected by means of a length of transmission line. The transmission line serves to

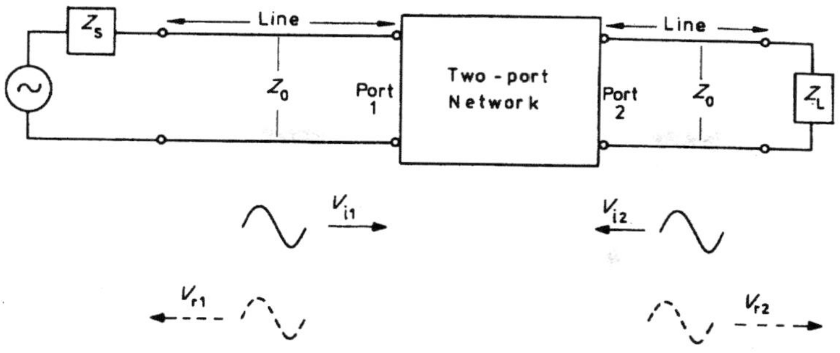

Figure C.3 A two-port network showing forward and backward travelling waves on the reference transmission lines.

separate forward and backward travelling waves. They can also be vanishingly short without affecting the following discussion.

The reference transmission lines will here be assumed to have identical characteristic impedances Z_0. (S-parameters are still defined if the reference impedances are different.) The equivalent voltages V_{in} are incident voltage waves travelling along the transmission line towards port n, and the equivalent voltages V_{rn} are reflected voltage waves travelling along the transmission line away from port n ($n = 1, 2$).

Since each reflected wave must be a linear combination of both a port 1 incident term and a port 2 incident term we have:

$$V_{r1} = x_{11}V_{i1} + x_{12}V_{i2} \tag{C.7}$$

$$V_{r2} = x_{21}V_{i1} + x_{22}V_{i2} \tag{C.8}$$

where x_{nm} depends upon the precise characteristics of the network. We shall not need to dwell upon the precise nature of these x_{nm} coefficients.

Now the incident power wave travelling along the transmission line towards port 1 has a value given by

$$P_{i1} = \frac{|V_{i1}^2|}{Z_0}. \tag{C.9}$$

The square root of this power, conventionally given the symbol a, is simply related to the incident voltage V_{i1} by

$$|a_1| = \sqrt{P_{i1}} \quad \text{as} \quad a_1 = V_{i1}/\sqrt{Z_0}. \tag{C.10}$$

Similar arguments apply to the power waves reflected from port 1, and incident upon and reflected from port 2 to yield:

$$b_1 = V_{r1}/\sqrt{Z_0} \qquad a_2 = V_{i2}/\sqrt{Z_0} \qquad b_2 = V_{r2}/\sqrt{Z_0}\ .$$

When both sides of both Equations C.7 and C.8 are divided through by $\sqrt{Z_0}$ the variables automatically become a_n and b_n and the coefficients are then called the scattering parameters.

The expressions can then be written:

$$b_1 = S_{11}a_1 + S_{12}a_2 \tag{C.11}$$

$$b_2 = S_{21}a_1 + S_{22}a_2. \tag{C.12}$$

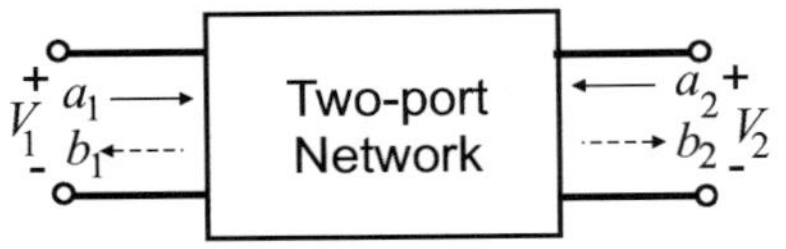

Figure C.4 Two-port network with root power variables.

In matrix form:

$$[\mathbf{b}] = [\mathbf{S}]\,[\mathbf{a}]\,. \tag{C.13}$$

As a reminder, it has to be appreciated that:

$$|a| = \sqrt{\text{Incident power}} \qquad \text{and} \qquad |b| = \sqrt{\text{Reflected power}}\,.$$

Note that a and b also have phase. The network diagram may now be re-sketched, with these 'root power' variables and scattering parameters included, and the result is Figure C.4. Using this diagram, and Equations (C.11) and (C.12), the following definitions can readily be established.

C.3.2 DEFINITIONS OF TWO-PORT S-PARAMETERS

S_{11}: If the output line is matched, $Z_L = Z_0$, the load cannot reflect power, $a_2 = 0$ and Equation (C.11) then yields

$$S_{11} = \left.\frac{b_1}{a_1}\right|_{a_2=0}. \tag{C.14}$$

By alternately matching the input and output the remaining three parameters are defined as follows:

S_{22}:

$$S_{22} = \left.\frac{b_2}{a_2}\right|_{a_1=0} \tag{C.15}$$

S_{21}:

$$S_{21} = \left.\frac{b_2}{a_1}\right|_{a_2=0} \tag{C.16}$$

S_{12}:

$$S_{12} = \left.\frac{b_1}{a_2}\right|_{a_1=0}. \tag{C.17}$$

Under these definitions, specific physical meanings can be ascribed to the S-parameters as follows:

S_{11} is the input reflection coefficient (i.e. Γ_{11}) when the output is matched.
S_{22} is the output reflection coefficient (i.e. Γ_{22}) when the input is matched.
S_{21} is the forward transfer (or 'transmission') coefficient with the output matched. (S_{21} strongly influences the characteristics of a filter, and indicates the loss of an interconnect and the gain of an amplifier.)
S_{12} is the reverse transfer (or 'transmission') coefficient with the input matched. (S_{12} becomes particularly important in amplifier stability considerations.) For an interconnect $S_{12} = S_{21}$ indicating that an the interconnect is reciprocal.

One of the more important things to note is that to measure the S-parameters it is necessary to use terminations that absorb all of the power delivered to them so that $a_1 = 0$ and $a_2 = 0$. At RF and microwave frequencies this is much easier to achieve than with either short or open-circuits.

Whereas S_{21} and S_{12} may be complicated functions of network characteristics, the reflection coefficient parameters S_{11} and S_{22} are readily expressed in terms of driving-point impedances. It is already known that the reflection coefficient Γ for an impedance Z loading a transmission line is

$$\Gamma = \frac{Z - Z_0}{Z + Z_0}. \tag{C.18}$$

Therefore

$$S_{11} = \frac{Z_{11} - Z_0}{Z_{11} + Z_0} \tag{C.19}$$

and

$$S_{22} = \frac{Z_{22} - Z_0}{Z_{22} + Z_0} \tag{C.20}$$

where Z_{11} and Z_{22} are, respectively, the input and output driving-point impedances for the two-port network when the appropriate opposite port is matched.

The scattering parameters are small-signal quantities and therefore to remain consistent with the other small-signal parameters, a lower-case s should strictly be used. However, we have retained the conventional upper-case S here.

Also, it is understood that each parameter is (like any small-signal parameter) a function of frequency, bias conditions and temperature. The strict frequency function $S(j\omega)$ is left understood and just S is used.

C.3.3 EVALUATION OF SCATTERING PARAMETERS

A simple example of a shunt susceptive element is used to illustrate this (see Figure C.5). To find S_{11} assume the output side to be matched so that $a_2 = 0$. The reflection coefficient on the input side is then:

$$S_{11} = \frac{Y_0 - Y_{\text{in}}}{Y_0 + Y_{\text{in}}}. \tag{C.21}$$

However, in this case $Y_{\text{in}} = Y_0 + jB$, because the (matched) output line shunts the susceptance jB, therefore:

$$S_{11} = \frac{Y_0 - Y_0 - jB}{2Y_0 + jB} = \frac{-jB}{2Y_0 + jB} \tag{C.22}$$

From symmetry considerations it is clear that:

$$S_{22} = S_{11}. \tag{C.23}$$

Next, S_{12} can be evaluated by finding the transmitted 'voltage' b_2 (i.e. $\sqrt{\text{power}}$) with the output line matched. This is the $a_2 = 0$ condition. For a pure shunt 1-port element

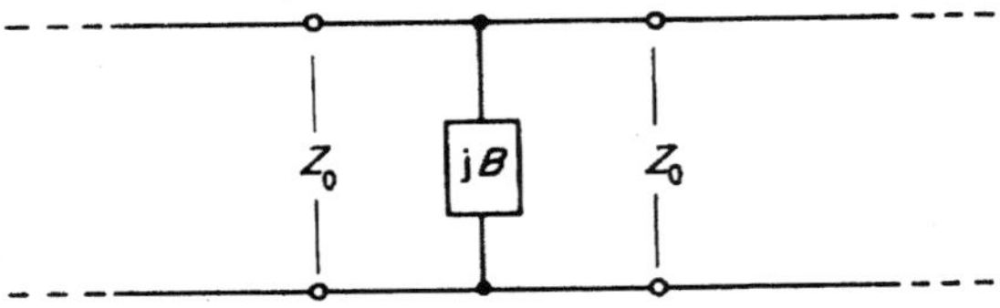

Figure C.5 A shunt susceptance element.

we must have: $b_2 = S_{21}a_1$ (see also Equation (C.16) here). We can combine expressions to obtain

$$\begin{aligned} b_2 &= a_1 + b_1 \\ &= a_1(1 + S_{11}). \end{aligned} \qquad \text{(C.24)}$$

Since, with this matched load situation $a_2 = 0$, we also have

$$S_{21} = 1 + S_{11} = S_{12} \qquad \text{(C.25)}$$

and we have already found S_{11} as Equation (C.22). Therefore, after substituting and clearing we obtain:

$$S_{12} = \frac{2Y_0}{2Y_0 + jB}. \qquad \text{(C.26)}$$

C.3.4 MEASUREMENT OF SCATTERING PARAMETERS

A network analyser system is usually employed for this purpose. This comprises principally:

1. A swept-frequency generator or frequency synthesizer.
2. A display plotting the S-parameters in various forms, principally on a Smith chart, a polar plot, or magnitude and phase on rectangular plots.
3. An S-parameter test set. This device has two measuring ports so that, when required, S_{11}, S_{12}, S_{21} and S_{22} can all be determined under program control.

An outline block diagram of a suitable system is shown in Figure C.6(a).

The initial calibration of the system is very important and it is necessary to calibrate the system using, for example:

(a) Precision short-circuits: to establish S_{11}= S_{22}= $1.0\angle 180°$, and
(b) 50 Ω through-line: to interconnect ports 1 and 2 and calibrate S_{12}= S_{21}= $1.0\angle 0°$.

Errors in the calibration standard are well characterized by the manufacturer, and used in the error correction software controlling the network analyser.

For interconnect characterization it is preferable to use calibration 'standards' that consist of combinations of transmission line lengths and repeatable reflections. These calibration procedures are known mostly by combinations of the letters T, R and L such as the standard TRL, for Through-Reflect-Line, procedure, [13, 355] or the TL, for Through-Line, procedure [14,15] that relies on symmetry to replace the third standard.

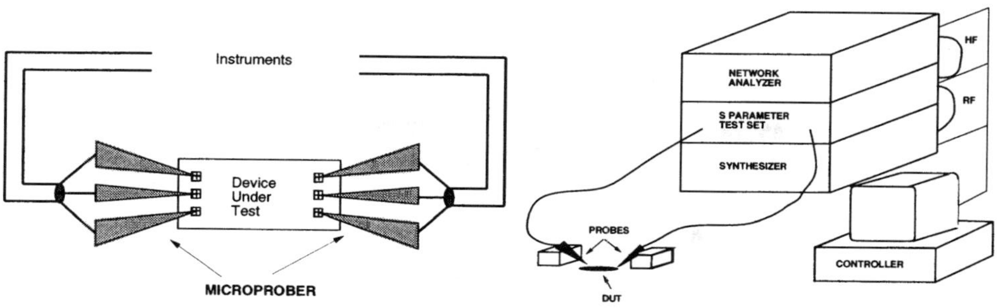

(a) (b)

Figure C.6 *S*-parameter measurement system: (a) shown in a configuration to do on-chip or MCM testing; and (b) details of coplanar probes with GSG (Ground-Signal-Ground configuration) contacting pads.

The calibration must be completed over the full bandwidth required and step (b) can be performed using a 50 Ω standard microstrip or coplanar line which is inserted into a special jig or 'fixture.' Several manufacturers supply such jig arrangements. The purpose of this special jig is to accept the package shape(s) of a transistor, or other device, whose *S*-parameters are desired.

For the measurement of complete circuits, whether active or passive, separate coaxial feed lines (and re-calibration) are often necessary. By using microprobes instead of coaxial connections, as indicated in Figure C.6(b) the need to recalibrate can be avoided by using a probe-compatible standard test substrate to establish calibration. The microprobes also enable low parasitic connections to be made to chips and MCMs. The microprobes are particularly convenient for characterizing interconnects. A typical microprobe is based on a micro-coaxial cable with the centre conductor (carrying the signal) extended a millimetre or less to form a needle-like contact. Two other needle-like contacts are made by attaching short extensions to the outer conductor of the coaxial line on either side of the signal connection. Such probes are called Ground-Signal-Ground, or GSG, probes. A similar probe can be made using a coplanar wavegude arrangement. The GSG probe connects to three pads on a substrate, each probe pad being as small as 50 μm square. The contact parasitic, which we want to keep small, is typically 100 fF or less [16]. GS and GS_1S_2G microprobes are available for measuring coplanar strip and differential lines.

C.3.5 SOME S-PARAMETER RELATIONSHIPS WHICH ARE PARTICULARLY USEFUL IN INTERPRETING INTERCONNECT MEASUREMENTS

(In this subsection only lossless, reciprocal junctions are treated. Therefore the following results are neither valid for transistors, nor for circuits with ferrite elements.)

Interconnects to be studied here are passive circuits with lossless and reciprocal properties. For any such networks:

$$S_{12} = S_{21}. \tag{C.27}$$

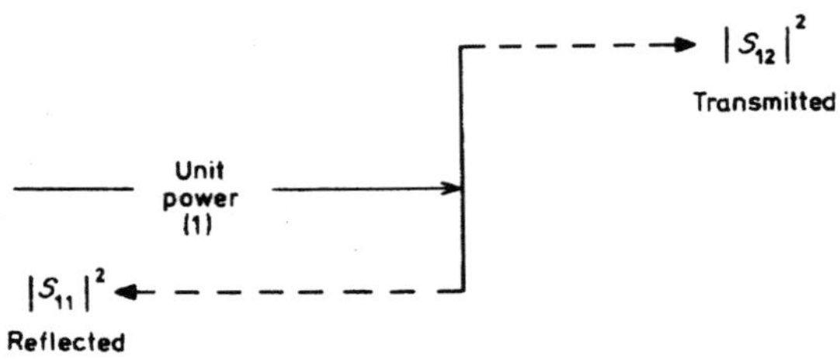

Figure C.7 Power flow breakdown.

(This is really a statement directly defining the 'reversibility' or reciprocity of the network.)

Now, if unit power flows into the network, a fraction $|S_{11}|^2$ is reflected and a further fraction $|S_{12}|^2$ is transmitted through the network. This is illustrated by the following power flow breakdown for this case as indicated in Figure C.7.

Thus, we may write:

$$|S_{11}|^2 + |S_{12}|^2 = 1. \tag{C.28}$$

It is customary to express each quantity in terms of complex conjugate products. Thus, Equation (C.28) becomes:

$$S_{11}S_{11}^* + S_{12}S_{12}^* = 1. \tag{C.29}$$

By similar arguments the following relationships also apply:

$$S_{22}S_{22}^* + S_{12}S_{12}^* = 1 \tag{C.30}$$

and

$$S_{11}S_{12}^* + S_{12}S_{22}^* = 0. \tag{C.31}$$

These equations can all be derived in a more rigorous mathematical manner known as the unitary condition:

$$[S][S]^* = 1. \tag{C.32}$$

From Equations (C.29) and (C.30): $S_{11} = S_{22}$ and substituting this in Equation (C.29), again:

$$[S_{12}]^2 = [S_{21}]^2 = 1 - [S_{11}]^2. \tag{C.33}$$

C.3.6 MULTIPORT S-PARAMETERS

The simple two-port scattering matrix extends naturally to cover multiport situations. An 'N-port' network is shown in Figure C.8. The scattering matrix for this becomes:

$$\begin{bmatrix} b_1 \\ b_2 \\ \cdots \\ b_N \end{bmatrix} = \begin{bmatrix} S_{11} & S_{12} & S_{13} & \cdots & S_{1N} \\ S_{21} & S_{22} & S_{23} & \cdots & S_{2N} \\ \vdots & \vdots & \vdots & \ddots & \vdots \\ S_{N1} & S_{N2} & S_{N3} & \cdots & S_{NN} \end{bmatrix} \begin{bmatrix} a_1 \\ a_2 \\ \vdots \\ a_N \end{bmatrix} \tag{C.34}$$

or, more briefly and concisely,

$$[\mathbf{b}] = [\mathbf{S}][\mathbf{a}]. \tag{C.35}$$

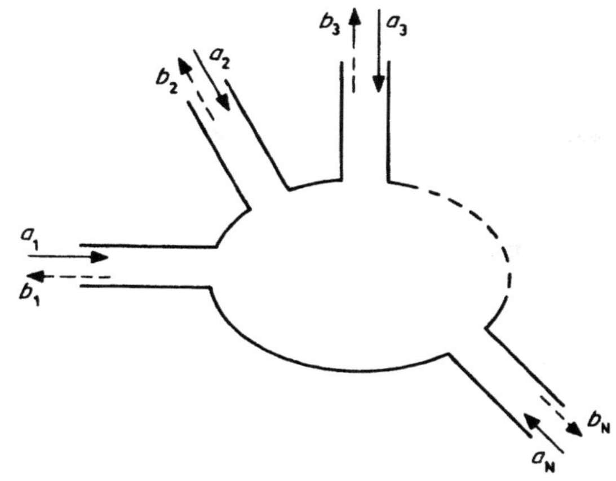

Figure C.8 An N-port network.

Three-port scattering matrices are required in microwave circuits involving circulators. Procedures for extracting the three-port parameters from two-port S-parameter measurements are described by Goldberg *et al.* [432, 433] The 3-port junction must be analysed here because the results are essential for reflection amplifier work. Specifically, we must determine the effect of attempting to simultaneously match all 3 ports of a lossless, reciprocal 3-port junction of transmission lines. We start by noting the complete 3-port scattering matrix:

$$[S] = \begin{bmatrix} S_{11} & S_{12} & S_{13} \\ S_{21} & S_{22} & S_{23} \\ S_{31} & S_{32} & S_{33} \end{bmatrix}. \tag{C.36}$$

If the junction can be matched then we have

$$S_{11} = S_{22} = S_{33} = 0. \tag{C.37}$$

Also, by reciprocity,

$$S_{21} = S_{12} \qquad S_{31} = S_{13} \qquad S_{32} = S_{23}. \tag{C.38}$$

Substituting the conditions described by Equations (C.37) and (C.38) into Equations (C.35) and (C.36) we obtain:

$$[S] = \begin{bmatrix} 0 & S_{12} & S_{13} \\ S_{12} & 0 & S_{23} \\ S_{13} & S_{23} & 0 \end{bmatrix}. \tag{C.39}$$

The conditions described under Section D.3.5 are also demanded for this 3-port junction. They may be expressed thus:

$$\sum_{m,n=1}^{3} |S_{mn}|^2 = 1 \tag{C.40}$$

and

$$\sum_{m,n=1}^{3} S_{1n}S_{mn}^* = 0 \tag{C.41}$$

where both Equations (C.40) and (C.41) can be conveniently combined into the unitary scattering matrix:

$$S\tilde{S}^* = \begin{bmatrix} 1 & 0 & 0 \\ 0 & 1 & 0 \\ 0 & 0 & 1 \end{bmatrix} \tag{C.42}$$

(another form of this is Equation (C.32) given earlier). We note from Equation (C.39) that only six terms are involved here, and expand the unitary matrix to yield:

$$S_{12}S_{12}^* + S_{13}S_{13}^* = 1 \tag{C.43}$$

$$S_{12}S_{12}^* + S_{23}S_{23}^* = 1 \tag{C.44}$$

$$S_{13}S_{13}^* + S_{23}S_{23}^* = 1. \tag{C.45}$$

We also have

$$S_{13}S_{23}^* = 0 \tag{C.46}$$

$$S_{12}S_{13}^* = 0 \tag{C.47}$$

$$S_{12}S_{23}^* = 0. \tag{C.48}$$

To satisfy these Equations (C.46)–(C.48) we must have:

$$S_{13} = 0 \quad \text{or } S_{23} = 0 \quad \text{or } S_{12} = 0 \ .$$

If we initially suppose that $S_{12} = 0$, then Equation (C.43) gives:

$$S_{13}S_{13}^* = 1 \tag{C.49}$$

and Equation (C.44) gives

$$S_{23}S_{23}^* = 1 \tag{C.50}$$

hence

$$S_{13}S_{13}^* + S_{23}S_{23}^* = 2. \tag{C.51}$$

This directly contradicts the necessary result expressed as Equation (C.45), being double the required quantity, and thus an intolerable inconsistency exists. We therefore conclude that a lossless, reciprocal 3-port junction cannot be simultaneously matched at all 3-ports. In fact, a special nonreciprocal arrangement is required, and this realized in practice by employing a microwave ferrite component called a circulator.

C.3.7 SIGNAL-FLOW GRAPH TECHNIQUES AND S-PARAMETERS

So-called signal-flow graphs form a logical and convenient means for analysing microwave circuits of virtually any form and complexity, provided only that they can be treated as linear and S-parameters are applicable. Typical problems requiring this kind of analysis include measurement error assessment and amplifier gain calculations.

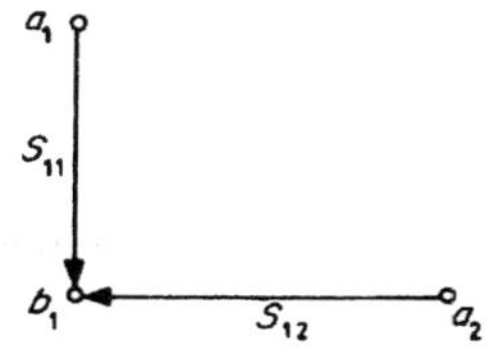

Figure C.9 Signal flow graph showing distribution of input reflection and output transfer to the response leaving port 1 of a two-port network.

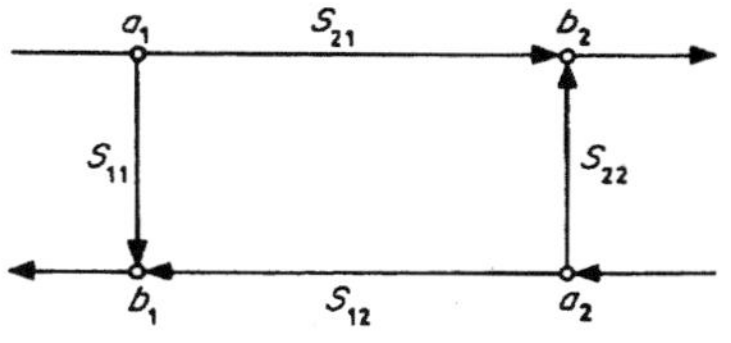

Figure C.10 Complete signal flow graph of two-port network.

To set up equivalent S-parameter signal-flow graphs, each port of the network is given two 'nodes' — one for the incident wave (a_k) and the other for the reflected wave (b_k). Branches, represented as lines between these nodes, carry the appropriate scattering parameters for each particular section of the network. Thus, the input reflection and reverse transfer sections appear as shown in Figure C.9. When the forward transfer and output reflection parameters are included, the full flow-graph for the two-port network appears as shown in Figure C.10.

C.4 SCATTERING TRANSFER (OR T) PARAMETERS

Because the basic S-parameters relate reflected waves to incident waves (both at the output and the input) they are not immediately useful when dealing, as often in communication circuits and systems, with cascaded networks. Instead, there is a related set of parameters based upon the output waves as the independent variables and the input waves as the dependent variables. These are called the scattering transfer (or 'T') parameters, and their matrix is, for a two-port network:

$$\begin{bmatrix} b_1 \\ a_1 \end{bmatrix} = \begin{bmatrix} T_{11} & T_{12} \\ T_{21} & T_{22} \end{bmatrix} \begin{bmatrix} a_2 \\ b_2 \end{bmatrix} \tag{C.52}$$

or

$$[\mathbf{i}] = [\mathbf{T}][\mathbf{o}] \tag{C.53}$$

where i means 'input' and o means 'output'. Since the same network, with the same conditions and the same 'root power' waves is involved, these T-parameters are exactly and simply related to the original S-parameters. This relationship is:

$$\begin{bmatrix} T_{11} & T_{12} \\ T_{21} & T_{22} \end{bmatrix} = \begin{bmatrix} (-S_{11}S_{22} - S_{12}S_{21}/S_{21}) & (S_{11}/S_{21}) \\ -(S_{22}/S_{21}) & (1/S_{21}) \end{bmatrix}. \tag{C.54}$$

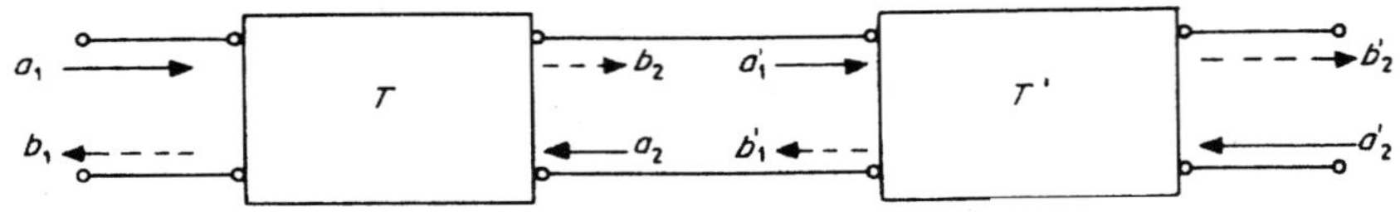

Figure C.11 Cascaded two-port networks.

C.4.1 CASCADED TWO-PORT NETWORKS: THE UTILITY OF T-PARAMETERS

Amplifier units are invariably multi-stage, with the individual stages being connected in cascade. Also, each stage will necessarily feature input and output matching networks. If each stage and matching network can be characterized in T-parameters, then the overall response can be evaluated readily. This is indicated by considering two networks in cascade as shown in Figure C.11.

Here, the individual T-matrices are:

$$\begin{bmatrix} b_1 \\ a_1 \end{bmatrix} = \begin{bmatrix} T_{11} & T_{12} \\ T_{21} & T_{22} \end{bmatrix} \begin{bmatrix} a_2 \\ b_2 \end{bmatrix} \tag{C.55}$$

and

$$\begin{bmatrix} b'_1 \\ a'_2 \end{bmatrix} = \begin{bmatrix} T'_{11} & T'_{12} \\ T'_{21} & T'_{22} \end{bmatrix} \begin{bmatrix} a'_2 \\ b'_2 \end{bmatrix}. \tag{C.56}$$

However, because of the uninterrupted flow between the networks,

$$\begin{bmatrix} a_2 \\ b_2 \end{bmatrix} = \begin{bmatrix} b'_1 \\ a'_1 \end{bmatrix}. \tag{C.57}$$

Therefore, Equations (C.55) and (C.56) can be combined to yield the overall result:

$$\begin{bmatrix} b_1 \\ a_1 \end{bmatrix} = \begin{bmatrix} T_{11} & T_{21} \\ T_{21} & T_{22} \end{bmatrix} \begin{bmatrix} T'_{11} & T'_{12} \\ T'_{21} & T'_{22} \end{bmatrix} \begin{bmatrix} a'_2 \\ b'_2 \end{bmatrix}. \tag{C.58}$$

Two aspects are important here:

(a) The matrices must be multiplied in the proper order (matrix multiplication is not commutative).
(b) The proper order to use is that in which the networks are actually connected.

Appendix D

Capacitance Matrix Extraction

This appendix describes how the capacitance matrix of a number of coupled conductors can be extracted from measurements or measurement-like simulations. Measurements and many programs that extract capacitance of an interconnect network do so by evaluating one capacitance at a time for different connections of interconnects to each other. What is presented here was developed by Biswas and Steer and described in Reference [7].

Without loss of generality consider the three line structure shown in Figure D.1 where there are four conductors including the ground. While three parallel lines are shown here this could be a much more complicated structure. A 3×3 capacitance matrix is required to described the capacitive coupling of this structure, but measurements and most electromagnetic simulators can only determine one capacitance at a time. The solution is to connect the invidual interconnects into two connected structures and determine the capacitance between the pairs of connected conductors. This procedure is repeated until every combination is considered. In this case there are seven possible, convenient, combinations as shown on the left hand side of Figure D.2. These combinations can be realized experimentally by connecting together the groups of conductors to that the capacitance, here, is measured between the black conductor group and the white group. The measurement of two similar structures, but with some known difference, such as lines of different length, can be used to calibrate out the error introduced by the electrical connections, or they can be simply ignored which is usually reasonable. In an electromagnetic simulation the connections are conveniently realized by holding the black group at one voltage, say 1 V, and the white group at 0 V. The corresponding capacitance connections are shown on the right of this figure, and the seven individual capacitance measurements

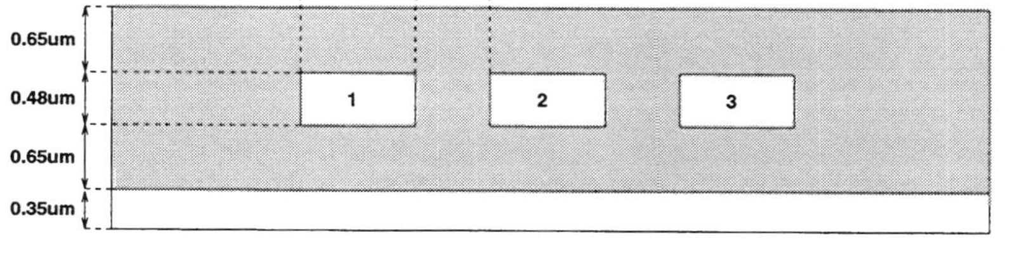

Figure D.1 Three lines structure with four conductors.

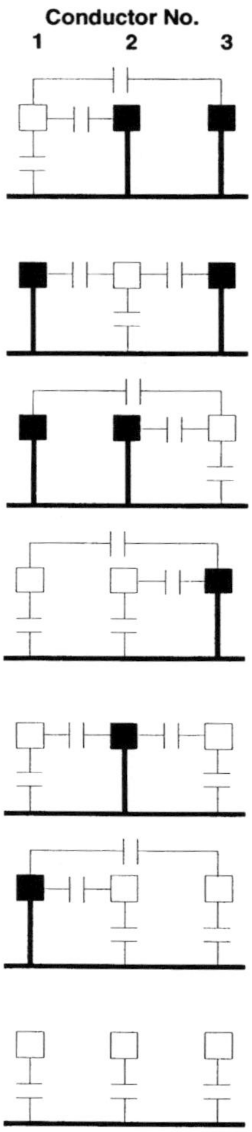

$$C_A = C_{11} + C_{12} + C_{13}$$

$$C_B = C_{22} + C_{12} + C_{23}$$

$$C_C = C_{33} + C_{13} + C_{23}$$

$$C_D = C_{11} + C_{13} + C_{22} + C_{23}$$

$$C_E = C_{11} + C_{12} + C_{33} + C_{23}$$

$$C_F = C_{22} + C_{12} + C_{33} + C_{13}$$

$$C_G = C_{11} + C_{22} + C_{33}$$

Figure D.2 Combinations of conductors leading to various capacitance measurements.

are C_A, C_B, C_C, C_D, C_E, C_F, C_G. Note that because of reciprocity, $c_{ij} = c_{ji}$. Thus the capacitance matrix

$$\mathbf{C} = \begin{bmatrix} C_{11} & C_{12} & C_{13} \\ C_{21} & C_{22} & C_{23} \\ C_{31} & C_{32} & C_{33} \end{bmatrix} = \begin{bmatrix} C_{11} & C_{12} & C_{13} \\ C_{12} & C_{22} & C_{23} \\ C_{13} & C_{23} & C_{33} \end{bmatrix} \tag{D.1}$$

can be derived. Note that there are seven measurements that can be conveniently made, but there are only six unknown quantities in the capacitance matrix. Thus, only six measurements are required to determine the capacitance matrix and the redundancy (the extra measurement) can be used as a check on the results. Using the first six measurements:

$$\begin{bmatrix} C_A \\ C_B \\ C_C \\ C_D \\ C_E \\ C_F \end{bmatrix} = \begin{bmatrix} 1 & 1 & 1 & 0 & 0 & 0 \\ 0 & 1 & 0 & 1 & 1 & 0 \\ 0 & 0 & 1 & 0 & 1 & 1 \\ 1 & 0 & 1 & 1 & 1 & 0 \\ 1 & 1 & 0 & 0 & 1 & 1 \\ 0 & 1 & 1 & 1 & 0 & 1 \end{bmatrix} \begin{bmatrix} C_{11} \\ C_{12} \\ C_{13} \\ C_{22} \\ C_{23} \\ C_{33} \end{bmatrix}. \tag{D.2}$$

Thus

$$\begin{bmatrix} C_{11} \\ C_{12} \\ C_{13} \\ C_{22} \\ C_{23} \\ C_{33} \end{bmatrix} = \begin{bmatrix} 1 & 1 & 1 & 0 & 0 & 0 \\ 0 & 1 & 0 & 1 & 1 & 0 \\ 0 & 0 & 1 & 0 & 1 & 1 \\ 1 & 0 & 1 & 1 & 1 & 0 \\ 1 & 1 & 0 & 0 & 1 & 1 \\ 0 & 1 & 1 & 1 & 0 & 1 \end{bmatrix}^{-1} \begin{bmatrix} C_A \\ C_B \\ C_C \\ C_D \\ C_E \\ C_F \end{bmatrix}. \tag{D.3}$$

It is important to note that the driving point capacitance[1] of an interconnect is affected by the signals or voltages on surrounding structures. For example, if each conductor (other than ground) is driven by a driver and, referring to Figure D.2 the output of the driver for the two left conductors is held at 0 V, then this corresponds to the situation for C_C. Then C_C is the effective capacitance seen by the driver of the rightmost interconnect (and is the capacitance that must be charged in changing voltage levels). Thus, C_C is dependent on the data on the neighbouring nets.

A similar procedure can be used to extract the elements of the inductance matrix.

[1] The driving point impedance, or here capacitance, is what would be measured without knowledge of the behaviour of surrounding structures. It is the Thevenin equivalent input impedance (capacitance) of the structure.

References

[1] Bakoglu, H. B., *Circuits, Interconnections and Packaging for VLSI*, Addison-Wesley Publishing Co., New York, 1990.

[2] Doane, D. A. and Franzon, P. D., *Multichip Module Technologies and Alternatives: The Basics*, Van Nostrand Reinhold, New York, 1993.

[3] Harper, C. A., *Electronic Packaging & Interconnection Handbook,* Second Edition, McGraw-Hill Inc., New York, 1997.

[4] Heaviside, O., 'Electromagnetic induction and its propagation,' *Electrician*, 3 June 1887.

[5] Gutmann, R. J., 'Advanced silicon IC interconnect technology and design: present trends and RF wireless implications,' *IEEE Trans. Microwave Theory and Techn.*, Vol. 47, June 1998, pp. 667–674.

[6] Biswas, B., Glasser, A., Steer, M., Franzon, P. , Griffis, D. and Russel, P., 'Experimental electrical characterization of on-chip interconnects,' *Proc. IEEE 6th Topical Meeting on Electrical Performance of Electronic Packaging*, Oct. 1997, pp. 57–59.

[7] Biswas, B., *Modeling and Simulation of High Speed Interconnects*, M.S. Thesis, North Carolina State University, 1998.

[8] Deutsch, A., Kopscay, G., Ranieri, V.A., Cataldo, J.K., Galligan, E. A., Graham, W. S., McGouey, R. P., Nunes S. L., Paraszczak, Ritsko, J. J., Serino, R. J., Shih, D. Y. and Wilczynski, J. S., 'High speed signal propagation on lossy transmission lines,' *IBM J. Res. Devel.*, Vol. 34, 1990, pp. 601–615.

[9] Deutsch, A., Kopscay, G., Surovic, C. W., Rubin, B. J., Terman, L. M., Dunne, Jr., R. P., Gallo, T. A. and Dennard R. H., 'Modeling and characterization of long on-chip interconnections for high-performance microprocessors,' *IBM J. Res. Devel.*, Vol. 39, 1995, pp. 547–567.

[10] Jong, J.-M., Janko, B. and Tripathi, V., 'Equivalent circuit modeling of interconnects from time-domain measurements,' *IEEE Trans. Components, Hybrids and Manufacturing Techn.* Vol. 16, Feb. 1993, pp. 119–126.

[11] Restle, P. J., 'Measurement and modeling of on-chip transmission line effects in a 400 MHz microprocessor,' *IEEE J. of Solid-State Circuits*, Vol. 33, Apr. 1998, pp. 662–665.

[12] Deutsch, A., 'Electrical characteristics of interconnectios for high-performance systems,' *Proc. IEEE*, Vol. 86, Feb. 1998, pp. 315–355.

[13] Engen, G. F. and Hoer, C. A., "Thru-reflect-line: an improved technique for calibrating the dual six-port automatic network analyzer," *IEEE Trans. on Microwave Theory and Techn.*, Vol. 27, Dec. 1979, pp. 987-993.

[14] Steer, M. B., Goldberg, S., Rinne, G., Franzon, P. D., Turlik, I. and Kasten, J. S., "Introducing the through-line deembedding procedure," *1992 IEEE MTT-S Int. Microwave Symp. Dig.,*, June 1992, pp. 1455–1458.

[15] Goldberg, S. B., Steer, M. B., Franzon, P. D. and Kasten, J. S., 'Experimental electrical characterization of interconnects and discontinuities in high speed digital systems,' *IEEE Trans. on Components Hybrids and Manufacturing Techn.*, Vol. 14, Dec. 1991, pp. 761–765.

[16] Lipa, S., Steer, M. B., Morris, A. S. and Franzon, P. D., 'Comparison of methods for determining the capacitance of planar transmission lines with application to multichip module characterization,' *IEEE Trans. on Components Hybrids and Manufacturing Techn., Part B: Advanced Packaging*, Vol. 16, May 1993, pp. 247–252.

[17] Lipa, S, Steer, M. B., Cangellaris, A. C. and Franzon P. D. , 'Experimental characterization of transmission lines in thin-film multichip modules,' *IEEE Trans. Components, Packaging and Manufacturing Techn.*, Vol. 19, Mar. 1996, pp. 122–126.

[18] Elmore, W. C., 'The transient response of a damped linear network with particular regard to wideband amplifier,' *J. Applied Physics*, Vol. 19, No. 1, 1948, pp. 55–63.

[19] Restle, P., Ruehli, A. and Walker, S.G., 'Dealing with inductance in high-speed chip design,' *Proc. 36th Design Automation Conf.*, 1999, pp. 904–909.

[20] Deutsch, A. *et al.*, 'Design guidelines for short, medium and long on-chip interconnections,' *Proc. 5th IEEE Topical Meeting Electrical Performance of Electronic Packaging*, 1996, pp. 30–32.

[21] Weber, E., *Linear Transient Analysis*, Vol. 2, John Wiley & Sons, New York, 1956, pp. 381–384.

[22] Ho, C. W., Chance, D. A., Bajonek, C. H. and Acosta, R. E., 'The thin film module as a high performance semiconductor package,' *IBM J. Res. Devel.*, Vol. 26, 1982, pp. 286–296.

[23] Wang, J., Tausch, J. and White, J., 'A wide frequency range surface integral formulation for 3-D RLC extraction,' *1999 IEEE/ACM Int. Conf. on Computer-Aided Design Dig. of Technical Papers*, 1999, pp. 453–457.

[24] Kamon, M., Marques, N., Massoud, Y., Silveira, L., and White, J., 'Interconnect analysis: from 3-D structures to circuit models,' *Proc. 36th Design Automation Conf.*, 1999, pp. 910–914.

[25] Ying L., Pileggi, L.T. and Strojwas, A.J., 'A sparse macromodeling method for RC interconnect multiports,' *Proc. of the IEEE 1997 Custom Integrated Circuits Conf.*, 1997, pp. 163–166.

[26] Ling, D. D. and Ruehli A. E., 'Interconnect modeling,' in *Advances in CAD for VLSI, Vol. 3 — Part II: Circuit Analysis, Simulation and Design,* Edited by A. E. Ruehli, Amsterdam, Holland: North Holland, 1987, pp. 211–291.

[27] Kamon, M., Wang, F. and White, J., 'Recent improvements for fast inductance extraction and simulation,' *Proc. 7th IEEE Topical Meeting Electrical Performance of Electronic Packaging*, Oct. 1998, pp. 281–284.

[28] Ruehli, A., 'Inductance calculation in a complex integrated circuit,' *IBM J. Res. Devel.*, Vol. 16, Sept. 1972, pp. 470–481.

[29] Bernstein, K., Carrig, K. M. *et al.*, *High Speed CMOS Design Styles*, Kluwer Academic Publishers, 1998.

[30] Adler, V. and Freidman, E. G., 'Repeater design to reduce delay and power in resistive interconnect,' *IEEE Trans. on Circuits and Systems—II: Analog and Digital Signal Processing*, Vol. 45, May 1998, pp. 607–616.

[31] Hannemann, R. J., Kraus, A. D. and Pecht, M., *Physical Architecture of VLSI Systems*, Wiley Interscience, New York, 1994.

[32] Gu R. X., Sharaf K. M. and Elmasry M. I., *High-Performance Digital VLSI Circuit Design*, Kluwer Academic Publishers, 1996.

[33] Chandrakasan A. and Brodersen R. (Eds), *Low-Power CMOS Design* IEEE Press, 1998.

[34] Yeap G. K. and Najm F. N. (Eds), *Low Power VLSI Design and Techn.,* World Scientific, 1996.

[35] Barnoski, M. K. (Ed.), *Introduction to Fibre Optics*, Plenum Publishing Corporation, New York, 1974.

[36] Knox, R. M., 'Dielectric waveguide microwave integrated circuits — an overview,' *IEEE Trans. Microwave Theory and Techn.*, Vol. 24, Nov. 1976, pp. 806–814.

[37] Caulton, M. and Sobol, H., 'Microwave integrated-circuit technology — a survey,' *IEEE*

J. Solid-State Circuits, Vol. 5, Dec. 1970, pp. 292–303.

[38] Hinken, J. H., 'Waveguides for microwave integrated circuits,' *Proceedings of Ninth European Microwave Conference*, Sept. 1979, pp. 656–660.

[39] Chi C.-Y. and Rebeiz, G.M., 'Design of Lange-couplers and single-sideband mixers using micromachining techniques,' *IEEE Trans. on Microwave Theory and Techn.*, Vol. 45, Feb. 1997, pp. 291–294.

[40] Weller, T.M., Katehi, L.P.B. and Rebeiz, G.M. , 'High performance microshield line components ,' *IEEE Trans. on Microwave Theory and Techn.*, Vol. 43, Mar. 1995, pp. 534–543 .

[41] Chi, C.-Y. and Rebeiz, G.M. , 'Planar microwave and millimeter-wave lumped elements and coupled-line filters using micro-machining techniques,' *IEEE Trans. on Microwave Theory and Techn.*, Vol. 43, April 1995, pp. 730–738.

[42] Stotz, M., Gottwald, G., Haspeklo, H. and Wenger, J., 'Planar millimeter-wave antennas using SiN/sub x/-membranes on GaAs,' *IEEE Trans. on Microwave Theory and Techn.*, Vol. 44, Sept. 1996, pp. 1593–1595.

[43] Kiang, J.-F., 'Characteristic impedance of microshield lines with arbitrary shield cross section,' *IEEE Trans. on Microwave Theory and Techn.*, Vol. 46, Sept. 1998, pp. 1328–1331.

[44] Iordanescu, S., Bartolucci, G., Simion, S. and Dragoman, M., 'Coplanar waveguide stub/filters on thin membranes and standard substrates,' *Proc. 1997 Int. Semiconductor Conference,* Vol. 2, 1997, pp. 357–360.

[45] Blondy, P., Brown, A.R., Crost, D. and Rebeiz, G.M. , 'Low loss micromachined filters for millimeter-wave telecommunication systems ,' *1998 IEEE MTT-S Int. Microwave Symp. Digest,* Vol. 3, June 1998, pp. 1181–1184.

[46] Sajin, G., Matei, E. and Dragoman, M., 'Microwave straight edge resonator (SER) on silicon membrane,' *Proc. 1999 Int. Semiconductor Conference,* Vol. 1, 1999, pp. 283–286.

[47] Katehi, L.P.B. and Rebeiz, G.M., 'Novel micromachined approaches to MMICs using low-parasitic high-performance transmission media and environments,' *IEEE MTT-S Int. Microwave Symp. Digest*, Vol. 2, June 1996, pp. 1145–1148.

[48] Herrick, K.J., Schwarz, T.A. and Katehi, L.P.B., 'Si-micromachined coplanar waveguides for use in high-frequency circuits,' *IEEE Trans. on Microwave Theory and Techn.,* Vol. 46, June 1998 , pp. 762–768

[49] Herrick, K.J., Yook, J.G. and Katehi, L.P.B., 'Microtechnology in the development of three-dimensional circuits,' *IEEE Trans. on Microwave Theory and Techn.,* Vol. 46, Nov. 1998, pp. 1832–1844.

[50] Drayton, R.F., Henderson, R.M. and Katehi, L.P.B., 'Monolithic packaging concepts for high isolation in circuits and antennas,' *IEEE Trans. on Microwave Theory and Techn.,* Vol. 46, July 1998, pp. 900–906.

[51] Henderson, R.M. and Katehi, L.P.B., 'Silicon-based micromachined packages for high-frequency applications,' *IEEE Trans. on Microwave Theory and Techn.,* Vol. 47, Aug. 1999, pp. 1563–1569.

[52] Meier, P. J., 'Integrated finline millimeter components,' *IEEE Trans. Microwave Theory and Techn.*, Vol. 22, Dec. 1974, pp. 1209–1216.

[53] Meier, P. J., 'Millimeter integrated circuits suspended in the E-plane of rectangular waveguide,' *IEEE Trans. Microwave Theory and Techn.*, Vol. 26, Oct. 1978, pp. 726–733.

[54] Cohen, L. D. and Meier, P. J., 'E-plane mm-wave circuits,' *Microwave Journal*, Aug. 1978, pp. 63–66.

[55] Bates, R. N. and Coleman, M. D., 'Millimetre wave finline balanced mixers,' *Proc. of 9th European Microwave Conf.*, Sept. 1979, pp. 721–725.

[56] El Hennawy, H. and Schunemann, K., 'Analysis of fin-line discontinuities', *Proc. 9th European Microwave Conf.*, Sept. 1979, pp. 448–453.

[57] Gupta, K. C., Garg, R. and Bahl, I. J., *Microstrip Lines and Slotlines*, second edition, Artech House Inc., Dedham, MA, 1996.

[58] Schneider, M. V., 'Microstrip lines for microwave integrated circuits,' *The Bell System Technical Journal*, May–June 1969, pp. 1421–1444.

[59] Gunston, M. A. R., *Microwave Transmission-line Impedance Data*, Van Nostrand Reinhold Company, London, 1973.

[60] Villotte, J. P., Aubourg, M. and Garault, Y., 'Modified suspended striplines for microwave integrated circuits,' *Electron. Lett.*, Vol. 14, No. 18, Aug. 1978, pp. 602–603.

[61] Simons, N., Dib, N. I. and Katehi, L. P. B., 'Modeling of coplanar stripline discontinuities,' *IEEE Trans. on Microwave Theory and Techn.*, Vol. 44, May 1996, pp. 711–716.

[62] Prieto, D. *et al.*, 'CPS structure potentialities for MMICs: a CPS/CPW transition and a bias network,' *1998 IEEE MTT-S Int. Microwave Symp. Dig.*, June 1998, pp. 111–114.

[63] Goverdhanam, K., Simons, R. N. and Katehi, L. P. B., 'Micro-Coplanar Striplines — a new transmission Media for Microwave Applications,' *1998 IEEE MTT-S Int. Microwave Symp. Dig.*, June 1998 pp. 1035–1038.

[64] Goverdhanam, K., Simons, R. N. and Katehi, L. P. B., 'Coplanar Stripline Components for High-Frequency Applications,' *IEEE Trans. on Microwave Theory and Techn.*, Vol. 45, Oct. 1997, pp. 1725–1729.

[65] Knorr, J. B. and Kuchler, K.-D., 'Analysis of Coupled Slots and Coplanar Strips on Dielectric Substrate,' *IEEE Trans. on Microwave Theory and Techn.*, Vol. 23, July 1975, pp. 541–548.

[66] Pintzos, S. G., 'Full-Wave Spectral-Domain Analysis of Coplanar Strips,' *IEEE Trans. on Microwave Theory and Techn.*, Vol. 39, Feb. 1991, pp. 239–246.

[67] Holder, P. A. R., 'X-band microwave integrated circuits using slotline and coplanar waveguide,' *The Radio and Electronic Engineer*, Vol. 48, No. 1/2, Jan./Feb. 1978, pp. 38–43.

[68] Spielman, B.E. 'Computer-aided analysis of dissipation losses in isolated and coupled transmission lines for microwave and millimetre-wave applications,' *NRL Formal Report*, No. 8009, Washington, D.C., 1976.

[69] Ganguly, A. K. and Spielman, B. E., 'Dispersion characteristics for arbitrarily configured transmission media,' *IEEE Trans. Microwave Theory and Techn.*, Vol. 25, Dec. 1977, pp. 1138–1141.

[70] Ladbrooke, P. H., England, E. H. and Potok, M. H. N., 'A specific thin-film technology for microwave integrated circuits,' *IEEE Trans. on Manufacturing Techn.*, Vol. 2, Dec. 1973, pp. 53–59.

[71] Rickard, D. C., 'Thick film MIC components in the range 10–20 GHz,' *Proc. 6th European Microwave Conf.*, Rome, Italy, 1976, pp. 687–691.

[72] Jensen, R. J., 'Thin film multilayer interconnection technologies for multichip modules,' *Multichip Module Technologies and Alternatives: The Basics*, Doane, D. A. and Franzon, P. D. (Eds.), Van Nostrand Reinhold, New York, 1993, pp. 255–309.

[73] Sergent, J. E., 'The hybrid microelectronics technology,' *Electronic Packaging & Interconnection Handbook,* second edition Harper, C. A. (Ed.), Second Edition, McGraw-Hill Inc., New York, 1997.

[74] Harper, C. A. (Ed.), *Handbook of Thick Film Hybrid Microelectronics*, McGraw-Hill Book Co., New York, 1974.

[75] Ramy, J.-P *et al.*, 'Optimization of the thick- and thin-film technologies for microwave circuits on alumina and fused silica substrates,' *IEEE Trans. Microwave Theory and Techn.*, Vol. 26, Oct. 1978, pp. 814–820.

[76] Goloubkoff, M. and Thebault, C., 'M.I.C. technology of fused silica substrate: Ku band transponder application,' *Proc. 9th European Microwave Conf.*, Sept., 1979, pp. 651–

655.

[77] Rasshofer, R. H. *et al.*, 'Long-Term Stability of Passive Millimeterwave Circuits on High-Resistivity Silicon Substrates,' *1999 IEEE MTT-S Int. Microwave Symp. Dig.*, June 1999, pp. 585–588.

[78] Strohm, K. M., Nuechter, P., Rheinfelder, C. N. and Guehl, R., 'Via hole technology for microstrip transmission lines and passive elements on high resistivity silicon,' *IEEE MTT-S Int. Microwave Symp. Dig.*, June 1999, pp. 581–584.

[79] Zarkesh-Ha, P., Bendix, P., Loh, W., Lee J.-J. and Meindl, J. D., 'The impact of Cu/low κ on chip performance,' *Proc. Twelth Annual IEEE Int. ASIC/SOC Conference*, Sept. 1999, pp. 257–261.

[80] Sweet, A., *MIC and MMIC Amplifier and Oscillator Circuit Design*, (Particularly Chapter 9), Artech House, Inc., 1990.

[81] Ponchak, G. E., Matloubian, M. and Katehi, L., 'A measurement-based design equation for the attenuation of MMIC-compatible coplanar waveguides,' *IEEE Trans. on Microwave Theory and Techn.*, Vol. 47, Feb. 1999, pp. 241–243.

[82] Brzozowski, V. J., 'Rigid and flexble printed wiring boards,' *Electronic Packaging & Interconnection Handbook,* second edition, Harper, C. A. (Ed.), Second Edition, McGraw-Hill Inc., New York, 1997.

[83] Van Deventer, T. E., Katehi, P. B. and Cangellaris, A. C., 'An integral equation method for the evaluation of conductor and dielectric losses in high-frequency interconnects,' *IEEE Trans. Microwave Theory Techn.*, Vol. 37, Dec. 1989, pp. 1964–1972.

[84] Oliner, A.A., 'Leakage from higher order modes on microstrip line with application to antennas,' *Radio Science*, Vol. 22, No. 6, Nov. 1987, pp. 907–912.

[85] Harokopus, W. P. and Katehi, P. B., 'Radiation properties of microstrip discontinuities,' *IEEE Antennas and Propagation Symp. Dig.*, June 1989, pp. 26–30.

[86] Horng, T. S., Wu, S. C., Yang, H. Y., Alexopoulos, N. G. and Katehi, P. B., 'A generalized method for the distinction of radiation and surface wave losses in microstrip discontinuities,' *1990 IEEE MTT-S Int. Microwave Symp. Dig.*, May 1990, pp. 1055–1058.

[87] Harokopus, W. P., Katehi, P. B., Ali-Ahmad, W. Y. and Rebwiz, G. M., 'Surface wave excitation from open microstrip discontinuities,' *IEEE Trans. Microwave Theory Techn.*, Vol. 39, July 1991, pp. 1098–1107.

[88] Gupta, K. C., Garg, R., Bahl, I. J. and Bhartia, P., *Microstrip Lines and Slotlines*, second edition, Artech House Inc., Norwood, MA, 1996.

[89] Greig, D. D. and Engelmann, H. F., 'Microstrip — a new transmission technique for the kilomegacycle range,' *Proc. Institute of Radio Engineers,* Vol. 40, No. 12, Dec. 1952, pp. 1644–1650.

[90] Fromm, W. E. and Fubini, E. G., 'Characteristics and some applications of stripline components,' *Proc. National Electronics Conf.,* Oct. 1954, pp. 58–59.

[91] Dukes, J. M. C., 'The application of printed circuit techniques to the design of microwave components,' *Proc. Institution of Electrical Engineers*, Vol. 105, Part B, Mar. 1958, pp. 155–172.

[92] Yamashita, E. and Atsuki, K., 'Design of transmission-line dimensions for a given characteristic impedance,' *IEEE Trans. Microwave Theory and Techn.*, Vol. 17, Aug. 1969, pp. 638–639.

[93] Wheeler, H. A., 'Transmission-line properties of parallel wide strips by a conformal mapping approximation,' *IEEE Trans. Microwave Theory and Techn.*, Vol. 12, May 1964, pp. 280–289.

[94] Wheeler, H. A., 'Transmission-line properties of parallel strips separated by a dielectric sheet,' *IEEE Trans. Microwave Theory and Techn.*, Vol.13, , Mar. 1965, pp. 172–185.

[95] Owens, R. P., 'Accurate analytical determination of quasi-static microstrip line parameters,' *The Radio and Electronic Engineer*, 46, No. 7, July 1976, pp. 360–364.

[96] Presser, A., 'RF properties of microstrip lines,' *MicroWaves*, Vol. 7, Mar. 1968, pp.

53–55.

[97] Hammerstad, E. O. and Bekkadal, F., 'A microstrip handbook,' ELAB Report, STF44 A74169, University of Trondheim, Norway, Feb. 1975.

[98] Cahill, L. W., 'Approximate formulas for microstrip transmission lines,' *Proc. Institute of Radio Engineers*, Vol. 3, Oct. 1974, pp. 317–321.

[99] Bryant, T. G. and Weiss, J. A., 'Parameters of microstrip transmission lines and of coupled pairs of microstrip lines,' *IEEE Trans. Microwave Theory and Techn.*, Vol. 16, Dec. 1968, pp. 1021–1027.

[100] Owens, R. P., Aitken, J. E. and Edwards, T. C., 'Quasi-static characteristics of microstrip on an anisotropic sapphire substrate,' *IEEE Trans. Microwave Theory and Techn.*, Vol. 24, Aug. 1976, pp. 499–505.

[101] Hayashi, Y. and Kitazawa, T., 'Analysis of microstrip transmission lines on a sapphire substrate,' *Trans. Inst. Electron & Common. Eng. Japan*, Sect. E (Japan), Vol. E62. No. 6, June 1979, pp. 426–427.

[102] Mariki, G. E. and Yeh, C., 'Dynamic Three-Dimensional TLM Analysis of Microstriplines on Anisotropic Substrate,' *IEEE Trans. Microwave Theory and Techn.*, Vol. 33, Sept., 1985, pp. 789–799.

[103] Lewin, A., 'A resonance absorption isolator in microstrip at 4Gc/s, *Proc. IEE, Part B* Vol. 104, Suppl. 6, 1957, pp. 364–365.

[104] Brodwin, M. E., 'Propagation in ferrite-filled microstrip,' *IRE Trans. Microwave Theory Tech.*, Vol. 6, 1958, pp. 153–155.

[105] Pucel, R. A. and Masse, D. J. 'Microstrip propagation on magnetic substrates, Part 1. Design theory. Part 11. Experiment,' *IEEE Trans. Microwave Theory and Techn.*, Vol. 20, 1972, pp. 304–314.

[106] Bahl, I. J. and Garg, R., 'Simple and accurate formulas for microstrip with finite strip thickness,' *Proc. IEEE*, Vol. 65, 1977, pp. 1611–1612.

[107] Schwan, K. P., 'Take the guesswork out of thick microstrip K,' *MicroWaves*, Dec. 1977, pp. 184–185.

[108] Bahl, I. J., 'Use exact methods for microstrip design,' *MicroWaves*, Dec. 1978, pp. 61–62.

[109] Kowalski, G. and Pregla, R., 'Dispersion characteristics of shielded microstrips with finite thickness,' *Arch. Elek. Ubertragung.*, Vol. 25. 1971, pp. 193–196.

[110] Daly, P., 'Hybrid-mode analysis of microstrip by finite-element methods,' *IEEE Trans. on Microwave Theory and Techn.*, Vol. 19, Jan. 1971, pp. 19–25.

[111] Gelder, D., 'Numerical determination of microstrip properties using the transverse field components,' *Proc. IEE*, Vol. 117, Apr. 1970, pp. 699–703.

[112] Corr, D. G. and Davies, J. B., 'Computer analysis of the fundamental and higher order modes in single and coupled microstrip,' *IEEE Trans. on Microwave Theory and Techn.*, Vol. 20, Oct. 1972, pp. 669–677.

[113] Corr, D G., *Finite Difference Analysis of Hybrid Modes in Microstrip Structures*, PhD Dissertation, Univ. of London, Aug. 1970.

[114] Grunberger, G. K., Keine, V. and Meinke, H. H., 'Longitudinal field components and frequency-dependent phase velocity in the microstrip transmission line,' *Electron. Lett.*, No. 21, 15 Oct. 1970, pp. 683–685.

[115] Grunberger, G. K. and Meinke, H. H., 'A theory of the microstrip line including longitudinal components,' *Proc. 1971 European Microwave Conf.*, Aug. 1971, pp. C4/2:1–C4/2:4.

[116] Laloux, A. and Vander Vorst, A., 'The dispersive character of microstrip lines,' *Proc. 9th European Microwave Conf.*, Aug. 1971, pp. C3/6:1 to C3/6:3.

[117] Van de Capelle, A. R. and Luypaert, P. J., 'An investigation of the higher order modes in open microstrip lines,' *Proc. 5th Colloquium on Microwave Communication*, 24–30 June 1974, Vol. 3, Budapest, pp. ET-31–ET-32.

[118] Schmitt, J. H. and Sarges, K. H., 'Wave propagation in microstrip,' *Nachrichtentech*

NTZ, 5, May 1971, pp. 260–264.

[119] Hornsby, J. S. and Gopinath, A., 'Numerical solutions of inhomogeneously filled guides with symmetrical microstrip line,' *Proc. 1969 European Microwave Conf.*, IEE Conf. Publ. 58, Sept. 1969, pp. 114–117.

[120] Hornsby, J. S. and Gopinath, A., 'Fourier analysis of a dielectric-loaded waveguide with a microstrip line,' *Electron. Lett.*, Vol. 5, 12 June 1969, pp. 265–267.

[121] Hornsby, J. S. and Gopinath, A., 'Numerical analysis of a dielectric-loaded waveguide with a microstrip line — finite-difference methods,' *IEEE Trans. on Microwave Theory and Techn.*, Vol. 17, Sept. 1969, pp. 684–690.

[122] Denlinger, E. J., 'A frequency-dependent solution for microstrip transmission lines,' *IEEE Trans. on Microwave Theory and Techn.*, Vol. 19, Jan. 1971, pp. 30–39.

[123] Chudobiak, W. J., Jain, O. P. and Makios, V., 'Dispersion in microstrip,' *IEEE Trans. on Microwave Theory and Techn.*, Vol. 19, Sept. 1971, pp. 783–784.

[124] Jain, O. P., Makios, V. and Chudobiak, W. J., 'The dispersive behaviour of microstrip transmission lines,' *Proc. European Microwave Conf.*, Aug. 1971, pp. C3/5:1 to C3/5:4.

[125] Jain, O. P., Makios, V. and Chudobiak, W. J., 'Coupled-mode model of dispersion in microstrip,' *Electron. Lett.*, Vol. 7, 15 July 1971, pp. 405–407.

[126] Schneider, M. B., 'Microstrip dispersion,' *Proc. IEEE*, Jan. 1972, pp. 144–146.

[127] Carlin, H. L., 'A simplified circuit model for microstrip,' *IEEE Trans. on Microwave Theory and Techn.*, Vol. 21, No. 9, Sept. 1973, pp. 589–591.

[128] Getsinger, W. J., 'Microstrip dispersion model,' *IEEE Trans. on Microwave Theory and Techn.*, Vol. 21, Jan. 1973, pp. 34–39.

[129] Getsinger, W. J., 'Microstrip dispersion model,' COMSAT Laboratories, Technical Memorandum CL-9-72, 10 Mar. 1972.

[130] Edwards, T. C. and Owens, R. P., '2–18 GHz dispersion measurements on 10-100 ohm microstrip lines on sapphire,' *IEEE Trans. on Microwave Theory and Techn.*, Vol. 24, Aug. 1976, pp. 505–513.

[131] Yamashita, E., Atsuki, K. and Ueda, T., 'An approximate dispersion formula of microstrip lines for computer-aided design of microwave intergrated circuits,' *IEEE Trans. on Microwave Theory and Techn.*, Vol. 27, Dec. 1979, pp. 1036–1038.

[132] Kirschning, M. and Jansen, R. H., 'Accurate model for effective dielectric constant of microstrip with validity up to millimetre-wave frequencies,' *Electron. Lett.*, 18 Mar. 1982, pp. 272–273.

[133] Kobayashi, M., 'A dispersion formula satisfying recent requirements in microstrip CAD,' *IEEE Trans. on Microwave Theory and Techn.*, Vol. 36, Aug. 1988, pp. 1246–1250.

[134] York, R. A. and Compton, R. C., 'Experimental evaluation of existing CAD models for microstrip dispersion,' *IEEE Trans. on Microwave Theory and Techn.*, Vol. 38, Mar. 1990, pp. 327–328.

[135] Atwater, H. A., 'Tests of microstrip dispersion formulas,' *IEEE Trans. on Microwave Theory and Techn.*, Vol. 36, Mar. 1988, pp. 619–621.

[136] Kurushin, E. P. *et al.*, 'Dispersion in microstripline on a ferrite substrate,' *Radio Physics and Quantum Electronics,* Mar. 1982, pp. 251–254.

[137] Hsia, I. Y., Yang, H.-Y. and Alexopoulos, N. G., 'Basic Properties of Microstrip Circuit Elements on Nonreciprocal Substrate-Superstrate Structures,' *1990 IEEE MTT-S Int. Microwave Symp. Dig.*, June 1990, pp. 665–668.

[138] Kobayashi, M., 'Frequency dependent characteristics of microstrips on anisotropic substrates,' *IEEE Trans. on Microwave Theory and Techn.*, Vol. 31, Nov. 1982, pp. 2054–2059.

[139] Mariki, G. E., "Dynamic three-dimensional TLM analysis of microstriplines on anisotropic substrate,' *IEEE Trans. on Microwave Theory and Techn.*, Vol. 33, Sept. 1984, pp. 789–798.

[140] Kowalski, G. and Pregla, R., 'Dispersion characteristics of single and coupled

microstrips,' *Arch. Elek. Ubertragung.*, Vol. 26, June 1972, pp. 276–280.

[141] Krage, M. K. and Haddad, G. I., 'Frequency-dependent characteristics of microstrip transmission lines,' *IEEE Trans. on Microwave Theory and Techn.*, Vol. 20, Oct. 1972, pp. 678–688.

[142] Itoh, T. and Mittra, R., 'Spectral-domain approach for calculating the dispersion characteristics of microstrip lines,' *IEEE Trans. on Microwave Theory and Techn.*, Vol. 21, No. 7, July 1973, pp. 496–499.

[143] Itoh, T. and Mittra, R., 'A technique for computing dispersion characteristics of shielded microstrip lines,' *IEEE Trans. on Microwave Theory and Techn.*, Vol. 22, Oct. 1974, pp. 896–898.

[144] Jansen, R. H., 'High-speed computation of single and coupled microstrip parameters including dispersion. high-order modes, loss and finite strip thickness,' *IEEE Trans. on Microwave Theory and Techn.*, Vol. 26, No. 2, Feb. 1978, pp. 75–82.

[145] Wu, Ke. and Vahideick, R., 'A novel space-spectral domain approach and its application to three-dimensional MIC/MMIC circuits,' *Proc. 19th European Microwave Conf.*, London, Sept. 1989, pp. 751–756.

[146] Dumbell, K. D., Pennock, S. R. and Rozzi, T. E., 'Effects of asymmetry in shielded microstrip transmission lines,' *Proc. 19th European Microwave Conf.,* London, Sept. 1989, pp. 761–766.

[147] Qian, Y. and Yamashita, E., 'Compression and reshaping of picosecond electrical pulses using dispersive microwave transmission lines,' *1990 IEEE MTT-S Int. Microwave Symp. Dig.*, May 1990, pp. 951–954.

[148] Rittweger, M. and Wolff, I., 'Analysis of complex passive (M)MIC-components using the finite difference time-domain approach,' *1990 IEEE MTT-S Int. Microwave Symp. Dig.*, My 1990, pp. 1147–1150.

[149] Bianco, B., *et al.*, Some consideration about the frequency dependence of the characteristic impedance of uniform microstrips,' *IEEE Trans. on Microwave Theory and Techn.*, Vol. 26, Mar. 1978, pp. 182–185.

[150] Getsinger, W. J., 'Microstrip characteristic impedance,' *IEEE Trans. on Microwave Theory and Techn.*, Vol. 27, Apr. 1979, p. 293.

[151] Owens, R. P., 'Predicted frequency dependence of microstrip characteristic impedance using the planar waveguide model,' *Electron. Lett.*, Vol. 12, 1976, pp. 269–270.

[152] Owens, R. P., *The Electrical Characteristics of Transferred-Electron Device Package and Mounts and Their Effects on Oscillator Performance,* PhD Dissertation, University of London, 1977.

[153] Kompa, G. and Mehran, R., 'Planar waveguide model for calculating microstrip components,' *Electron. Lett.,* Vol. 11, 1975, pp. 459–460.

[154] Bianco, B. et al., 'Frequency dependence of microstrip parameters,' *Alta Frequenza,* Vol. 43, 1974, pp. 413–416.

[155] Howe, Jr, H., *Stripline Circuit Design,* Artech House Inc., Dedham, MA, 1974.

[156] Bahl, I. J. and Garg, R., 'A designer's guide to stripline circuits,' *MicroWaves*, Jan. 1978, pp. 90–96.

[157] Knorr, J. B. and Tufekcioglu, A., 'Spectral-domain calculation of microstrip characteristics impedance,' *IEEE Trans. on Microwave Theory and Techn.*, Vol. 23, 1975, pp. 725–728.

[158] Mirshekar-Syahkal, D. and Davies, J. B., 'Accurate solution of microstrip and coplanar structures for dispersion and for dielectric and conductor losses,' *IEEE Trans. on Microwave Theory and Techn.*, Vol. 27, July 1979, pp. 694–699.

[159] Brews, J., 'Characteristic impedance of microstrip lines,' *IEEE Trans. on Microwave Theory and Techn.*, Vol. 35, Jan. 1987, pp. 30–34.

[160] Vendelin, G. D., 'Limitations on stripline Q,' *Microwave Journal,* May 1970, pp. 63–69.

[161] Pucel, R. A., Masse, D. J. and Hartwig, C. P., 'Losses in microstrip,' *IEEE Trans. on Microwave Theory and Techn.*, Vol. 16, 1968, pp. 342–350. Also see correction in *IEEE*

Trans. on Microwave Theory and Techn., Vol. 16, 1968, p. 1064.

[162] Wheeler, H. A., 'Formulas for the skin effect,' *Proc. IRE,* Vol. 30, 1942, pp. 412–426.

[163] Welch, J. D. and Pratt, H. J., 'Losses in microstrip transmission systems for integrated microwave circuits,' *NEREM Rec.*, Vol. 8, 1966, pp. 100–101.

[164] Horton, R., Easter, B. and Gopinath, A., 'Variation of microstrip losses with thickness of strip,' *Electron. Lett.*, Vol. 7, 1971, p. 490.

[165] Hammerstad. E. O. and Bekkadal, F., 'A microstrip handbook,' *ELAB Report*, STF 44 A74169, University of Trondheim, Norway, 1975, pp. 98–110.

[166] Gupta, K. C., Garg, R., Bahl, I. J. and Bhartia, P., *Microstrip Lines and Slotlines*, Second Edition, Artech House Inc., Norwood, MA, 1996, pp. 91–92.

[167] James, J. R. and Henderson, A., 'High-frequency behaviour of microstrip open-circuit terminations,' *IEE J. on Microwaves, Optics and Acoustics,* Vol. 3, No 5, Sept. 1979, pp. 205–218.

[168] Armstrong, A. F. and Cooper. P. D., 'Technique for investigating spurious propagation in enclosed microstrip,' *Radio & Electron. Eng,* Vol. 48, 1978, pp. 64–72.

[169] James, J. R. and Ladbrooke, P. H., 'Surface-wave phenomena associated with open-circuited stripline terminations,' *Electron. Lett.,* Vol. 9, No. 24, 29 Nov. 1973, pp. 570–571.

[170] Horng, T. S., Yang, H. Y. and Alexopoulos, N. G., 'A full-wave analysis of shielded microstrip line-to-line transitions' *1990 IEEE MTT-S Int. Microwave Symp. Dig.*, May 1990, pp. 251–254.

[171] Goldfarb, M. E. and Aryeh Platzker, 'Losses in GaAs Microstrip,' *1990 IEEE MTT-S Int. Microwave Symp. Dig.*, May 1990, pp. 563–565.

[172] Lee, H.-Y., Kong, K.-S. and Itoh, T., 'Conductor loss calculation of superconducting microstrip line using a phenomenological loss equivalence method,' *Proc. 19th European Microwave Conf.,* Sept. 1989, pp. 757–760.

[173] Bhasin, K. B. *et al.,* Performance and modelling of superconducting ring resonators at millimeter-wave frequencies,' *1990 IEEE MTT-S Int. Microwave Symp. Dig.*, May 1990, pp. 269–276.

[174] van Deventer, T. E., *et al.*, 'High frequency characterization of high-temperature superconducting thin film lines,' *1990 IEEE MTT-S Int. Microwave Symp. Dig.*, May 1990, pp. 285–288.

[175] Hammond, R. B., *et al.*, 'Superconducting Tl-Ca-Ba-Cu-O thin film microstrip resonator and its power handling performance at 77K,' *1990 IEEE MTT-S Int. Microwave Symp. Dig.*, May 1990, pp. 867–870.

[176] Cohn, S. B., 'Slot line on a dielectric substrate,' *IEEE Trans. on Microwave Theory and Techn.*, Vol. 17, Oct. 1969, pp. 768–778.

[177] Sheen, J.-W. and Lin, Y.-D., 'Propagation characteristics of the slotline first higher order mode,' *IEEE Trans. on Microwave Theory and Techn.*, Vol. 46, Nov. 1998, pp. 1774–1781.

[178] Wen, C. P., 'Coplanar waveguide: a surface strip transmission line suitable for nonreciprocal gyromagnetic device applications,' *IEEE Trans. on Microwave Theory and Techn.*, Vol. 17, Dec. 1969, pp. 1087–1090.

[179] Houdart, M., 'Coplanar lines: application to broadband microwave integrated circuits,' *Proc. 6th European Microwave Conf.*, Sept. 1976, pp. 49–53.

[180] Gupta, K. C., Garg, R., Bahl, I. J. and Bhartia, P., *Microstrip Lines and Slotlines*, second edition, Artech House, Inc., 1996.

[181] Veyres, C. and Fouad-Hanna, V., 'Extension of the application of conformal mapping techniques to coplanar lines with finite dimensions,' *Int. J. of Electronics*, Vol. 48, Jan. 1980, pp. 47–56.

[182] Ghione, G. and Naldi, C., 'Parameters of coplanar waveguides with lower ground plane,' *Electron. Lett.*, Vol. 19, Sept. 1983, pp. 734–735.

[183] Ghione, G. and Naldi, C., 'Analytical formulas for coplanar lines in hybrid and

monolithic MIC's,' *Electron. Lett.*, Vol. 20, 16 Feb. 1984, pp. 179–181.

[184] Fang, S.-J. and Wang, B-S., 'Analysis of asymmetric coplanar waveguide with conductor backing,' *IEEE Trans. on Microwave Theory and Techn.*, Vol. 47, Feb. 1999, pp. 238–240.

[185] Rayit, A. K., *Characteristics and Applications of Coplanar Waveguide and its Discontinuities,* PhD Dissertation, University of Bradford (UK), 1997.

[186] Hilberg, W., 'From approximations to exact relations for characteristic impedances,' *IEEE Trans. on Microwave Theory and Techn.*, Vol. 17, May 1969, pp. 259–265.

[187] Koshiji, K., Shu, E. and Miki, S., 'Dielectric and conductor losses in coplanar waveguides,' *Electron. and Communications in Japan*, Vol. 65-b, 1982.

[188] Jackson, R. W., 'Coplanar waveguide vs. microstrip for millimeter wave integrated circuits,' *1986 IEEE MTT-S Int. Microwave Symp. Dig.*, June 1986, pp. 699–702.

[189] Zhang, M., Wu, C., Wu, K. and Litva, J., 'Losses in GaAs microstrip and coplanar waveguide,' *1992 IEEE MTT-S Int. Microwave Symp. Dig.*, June 1992, pp. 971–974.

[190] Vietzorreck, L. and Pasher, W., 'Modeling of conductor loss in coplanar circuit elements by the method of lines,' *IEEE Trans. on Microwave Theory and Techn.*, Vol. 45, Dec. 1997, pp. 2474–2478.

[191] Ghione, G., Goano, M. and Pirola, M., 'Exact, conformal-mapping models for the high-frequency losses of coplanar waveguides with thick electrodes of rectangular or trapezoidal cross section,' *1999 IEEE MTT-S Int. Microwave Symp. Dig.*, June 1999, pp. 1311–1314.

[192] Wu, Y., Gamble, H. S., Armstrong, B. M., Fusco, V. F. and Stewart, J. A. C., 'SiO_2 interface layer effects on microwave loss of high-resistivity CPW line,' *IEEE Microwave and Guided Wave Letters*, Vol. 9, Jan. 1999, pp. 10–12.

[193] Papapolymerou, J., East, J. and Katehi, L. P. B., 'GaAs versus quartz FGC lines for MMIC applications,' *IEEE Trans. on Microwave Theory and Techn.*, Vol. 46, Nov. 1998, pp. 1790–1793.

[194] Simons, R. N. and Ponchak, G. E., 'Modelling of some coplanar waveguide discontinuities,' *1988 IEEE MTT-S Int. Microwave Symp. Dig.*, June 1988, pp. 297–300.

[195] Kuo, C. W. and Itoh, T., 'Characterization of the coplanar step discontinuity using the transverse resonance method,' *Proc. 19th European Microwave Conf.*, Sept. 1989, pp. 662–665.

[196] Chen, Z.-Q. and Gao, B.-X., 'Full-wave analysis of step discontinuities of planar transmission lines by the method of lines,' *2nd Symp. on Recent Advances in Microwave Techn.*, ISRAMT, Bejing, China, 1989, pp. 24–27.

[197] Kuo, C. W., Kitazawa, T. and Itoh, T., 'Analysis of shielded coplanar waveguide step discontinuity considering the finite metalization thickness effect,' *1991 IEEE MTT-S Int. Microwave Symp. Dig.*, June 1991, pp. 473–475.

[198] Naghead, M. *et al.*, 'A new method for the calculation of the equivalent inductances of coplanar waveguide discontinuities,' *1991 IEEE MTT-S Int. Microwave Symp. Dig.*, June 1991, pp. 747–750.

[199] Mirshekar-Syahkal, D., 'Computation of equivalent circuits of cpw discontinuities using quasi-static spectral domain method,' *IEEE Trans. on Microwave Theory and Techn.*, Vol. 44, June 1996, pp. 979–984.

[200] Sewell, P. and Rozzi, T., 'Characterization of air-bridges in mm-wave coplanar waveguide using the complete mode spectrum of CPW,' *IEEE Trans. on Microwave Theory and Techn.*, Vol. 42, Nov. 1994, pp. 2078–2086.

[201] Lee, C.-Y., Liu, Y. and Itoh, T., 'The effects of the coupled slotline mode and air-bridges on CPW and NLC waveguide discontinuities,' *IEEE Trans. on Microwave Theory and Techn.*, Vol. 43, Dec. 1995, pp. 2759–2765.

[202] Hettak, K., Dib, N. and Omar, A. 'A new class of miniature radiationless CPW Shunt stubs printed on the center conductor,' *1999 IEEE MTT-S Int. Microwave Symp. Dig.*,

June 1999, pp. 1335–1338.

[203] Ponchak, G. E. and Tentzeris, E. 'Development of finite ground coplanar (FGC) waveguide 90 degree crossover junctions with low coupling conductor,' *2000 IEEE MTT-S Int. Microwave Symp. Dig.*, June 2000.

[204] Malherbe, J. A. G., *Microwave Transmission Line Filters*, Artech House, Dedham, MA, 1979, pp. 142–148.

[205] Omar, A. A. and Chow, Y. L., 'Coplanar waveguide with top and bottom shields in place of air bridges,' *IEEE Trans. on Microwave Theory and Techn.*, Vol. 41, Sept. 1993, pp. 1559–1563.

[206] Mernyei, F., Aoki, I. and Matsuura, H., 'MMIC bandpass filter using parallel-coupled CPW lines,' *Electron. Lett.*, 27th Oct. 1994, pp. 1862–1863.

[207] Ghione, G. and Naldo, M., 'A closed-form CAD-oriented model for the high-frequency conductor attenuation of symmetrical coupled coplanar waveguides,' *IEEE Trans. on Microwave Theory and Techn.*, Vol. 45, July 1997, pp. 1065–1070.

[208] Ho, C.-H., Fan, L. and Chang, K., 'Broad-band uniplanar hybrid-ring and branch-line couplers,' *IEEE Trans. on Microwave Theory and Techn.*, Vol. 41, No. 12, Dec. 1993, pp. 2116–2125.

[209] Fan, L., Ho, C.-H., Kanamaluru, S. and Chang, K., 'Wide-band reduced-size uniplanar magic-T, hybrid-ring and de Ronde's CPW-slot couplers,' *IEEE Trans. on Microwave Theory and Techn.*, Vol. 43, No. 12, Dec. 1995, pp. 2749–2758.

[210] Ho, C.-H., Fan, L. and Chang, K., 'New uniplanar coplanar waveguide hybrid-ring couplers and magic-T's,' *IEEE Trans. on Microwave Theory and Techn.*, Vol. 42, Dec. 1994, pp. 2440–2448.

[211] Gu, H. and Wu, K., 'Broadband design consideration of uniplanar double-y baluns for hybrid and monolithic integrated circuits,' *1999 IEEE MTT-S Int. Microwave Symp. Dig.*, June 1999, pp. 863–866.

[212] Fan, L. and Chang, K., 'Uniplanar power dividers using coupled cpw and asymmetric CPS for MIC's and MMIC's,' *IEEE Trans. on Microwave Theory and Techn.*, Vol. 44, Dec. 1996, pp. 2411–2420.

[213] Budimir, D., Wang, Q. H., Rezazadeh, A. A. and Robertson, I. D., 'V-shaped CPW transmission lines for multilayer MMICs,' *Electron. Lett.*, 26th Oct. 1995, pp. 1928–1930.

[214] Herrick, K. J., Henderson, R. M. and Katehi, L. P. B., 'Wave effects in Si-micromachined multilayer guiding structures operating at W-band,' *1999 IEEE MTT-S Int. Microwave Symp. Dig.*, June 1999, pp. 61–64.

[215] Yang, S., Hu, Z., Buchanan, N. B., Fusco, V. F., Stewart, J. A. C., Wu, Y., Armstrong, M., Armstrong, G. A. and Gamble, H. S., 'Characteristics of trenched coplanar waveguide for high-resistivity Si MMIC applications,' *IEEE Trans. on Microwave Theory and Techn.*, Vol. 46, May 1998, pp. 623–631.

[216] Jin, H. and Vahldieck, R., 'Full-wave analysis of coplanar waveguide discontinuities using the frequency domain TLM method,' *IEEE Trans. on Microwave Theory and Techn.*, Vol. 41, Sept. 1993, pp. 1538–1542.

[217] Ellis, T. J. *et al.*, 'A wideband CPW-to-microstrip transition for millimeter-wave packaging,' *1999 IEEE MTT-S Int. Microwave Symp. Dig.*, June 1999, pp. 629–632.

[218] Lin, Y.-S. and Chen, C. H., 'Novel lumped-element coplanar waveguide-to-slotline transitions,' *1999 IEEE MTT-S Int. Microwave Symp. Dig.*, June 1999, pp. 96–102.

[219] Ma, K-P., Qian, Y. and Itoh, T., 'Analysis and applications of a new CPW-slotline transition,' *IEEE Trans. on Microwave Theory and Techn.*, Vol. 47, Apr. 1999, pp. 426–432.

[220] Kaneda, N., Qian, Y. and Itoh, T., 'A broadband microstrip-to-waveguide transition using quasi-Yagi antenna,' *1999 IEEE MTT-S Int. Microwave Symp. Dig.*, June 1999, pp. 1431–1434.

[221] Simons, R. N., Lee, R. Q. and Perl, T. D., 'New techniques for exciting linearly tapered

slot antennas with coplanar waveguide,' *Electron. Lett.*, 26 Mar. 1992, pp. 620–621.

[222] Simons, R. N. and Lee, R. Q., 'Coplanar waveguide aperture coupled patch antennas with ground plane/substrate of finite extent,' *Electron. Lett.*, 2 Jan. 1992, pp. 75–76.

[223] Zhu, L. and Wu, K., 'Unified CAD-oriented circuit model of finite-ground coplanar waveguide gap structure for uniplanar M(H)MICs,' *1999 IEEE MTT-S Int. Microwave Symp. Dig.*, June 1999, pp. 39–42.

[224] Zhu, L. and Wu, K., 'Hybrid FGCPW/CPS scheme in the building block design of low-cost uniplanar and multilayer circuit and antenna,' *1999 IEEE MTT-S Int. Microwave Symp. Dig.*, June 1999, pp. 867–870.

[225] Nadan, T. *et al.*, 'Optimization and miniaturization of a filter/antenna multi-function module using a composite ceramic-foam substrate,' *1999 IEEE MTT-S Int. Microwave Symp. Dig.*, June 1999, pp. 219–222.

[226] Vaupel, T. and Hansen, V., 'Electrodynamic coplanar/slotline structure analysis with 3-D components based on a surface/volume integral equation approach,' *1999 IEEE MTT-S Int. Microwave Symp. Dig.*, June 1999, pp. 1681–1684.

[227] Lee, G.-A. and Lee, H.-Y., 'Suppression of the CPW leakage in common millimeter-wave flip-chip structures,' *IEEE Microwave and Guided Wave Lett.*, Vol. 8, Nov. 1998, pp. 366–368.

[228] Heinrich, W., Jentzsch, A. and Baumann, G., 'Millimeter-wave characteristics of flip-chip interconnects for multichip modules,' *IEEE Trans. on Microwave Theory and Techn.*, Vol. 46, Dec. 1998, pp. 2264–2268.

[229] Hirose, T. *et al.*, 'A flip-chip mmic design with coplanar waveguide transmission line in the W-band,' *IEEE Trans. on Microwave Theory and Techn.*, Vol. 46, Dec. 1998, pp. 2276–2282.

[230] Jentzsch, A. and Heinrich, W., 'Optimization of flip-chip interconnects for millimeter-wave frequencies,' *1999 IEEE MTT-S Int. Microwave Symp. Dig.*, June 1999, pp. 637–640.

[231] Feng, Z. *et al.*, 'RF and mechanical characterization of flip-chip interconnects in CPW circuits with underfill,' *IEEE Trans. on Microwave Theory and Techn.*, Vol. 46, No. 12, Dec. 1998, pp. 2269–2275.

[232] Goverdhanam, K., Simons, R. N. and Katehi, L. P. B., 'Novel three-dimensional vertical interconnect technology for microwave and RF applications,' *1999 IEEE MTT-S Int. Microwave Symp. Dig.*, June 1999, pp. 641–644.

[233] Soliman, E. A. *et al.*, 'Suppression of the parasitic modes in CPW discontinuities using MCM-D technology – application to a novel 3-dB power splitter,' *IEEE Trans. on Microwave Theory and Techn.*, Vol. 46, Dec. 1998, pp. 2426–2430.

[234] Kerssenbrock, T. V. and Heide, P., 'Novel 77 GHz sensor modules for automotive radar applications,' *1999 IEEE MTT-S Int. Microwave Symp. Dig.*, June 1999, pp. 289–292.

[235] Cahana, D., 'A new coplanar waveguide/slotline double-balanced mixer,' *1989 IEEE MTT-S Int. Microwave Symp. Dig.*, June 1989, pp. 967–968.

[236] Hsu, P.-C., Nguyen, C. and Kintis, M., 'A new uniplanar broad-band singly balanced diode mixer,' *IEEE Trans. on Microwave Theory and Techn.*, Vol. 46, Nov. 1998, pp. 1782–1784.

[237] Thiel, W. and Menzel, W., 'Full-wave design and optimization of mm-wave diode-based circuits in finline technique,' *1999 IEEE MTT-S Int. Microwave Symp. Dig.*, June 1999, pp. 787–790.

[238] Wasige, E. *et al.*, 'GaAs FET characterization in a quasi-monolithic Si environment,' *1999 IEEE MTT-S Int. Microwave Symp. Dig.*, June 1999, pp. 1889–1892.

[239] Yun, T.-Y. and Chang, K., 'One-dimensional photonic bandgap resonators and varactor tuned resonators,' *1999 IEEE MTT-S Int. Microwave Symp. Dig.*, June 1999, pp. 1629–1632.

[240] Milanovic, V., Ozgur, M, DeGroot, D. C., Jargon, J. A., Gaitan, M. and Zaghoul, M. E., 'Characterization of broad-band transmission for coplanar waveguides on CMOS

silicon substrates,' *IEEE Trans. on Microwave Theory and Techn.*, Vol. 46, May 1998, pp. 632–640.

[241] Milanovic, V. *et al.*, 'Micromachined coplanar waveguides in CMOS technology,' *IEEE Microwave and Guided Wave Lett.*, Vol. 6, Oct. 1996, pp. 380–382.

[242] Guillon, B. *et al.*, 'Design and realization of high Q millimeter-wave structures through micromachining techniques,' *1999 IEEE MTT-S Int. Microwave Symp. Dig.*, June 1999, pp. 1519–1522.

[243] Hesselbarth, J. and Vahldieck, R., 'Leakage suppression in coplanar waveguide circuits by patterned backside metalization,' *1999 IEEE MTT-S Int. Microwave Symp. Dig.*, June 1999, pp. 871–874.

[244] Kleveland, B., Lee, T. H. and Wong, S., '50 GHz interconnect design in standard silicon technology,' *1998 IEEE MTT-S Int. Microwave Symp. Dig.*, June 1998, pp. 1913–1916.

[245] Yakovlev, A. B., 'On the nature of critical points in leakage regimes of a conductor-backed coplanar strip line,' *IEEE Trans. Microwave Theory and Techn.*, Vol. 45, Jan. 1997, pp. 87–94.

[246] Nghiem, D., Williams. J. T., Jackson, D. R. and Oliner, A. A., 'Existence of a leaky dominant mode on microstrip line with an isotropic substrate: theory and measurement,' *1993 IEEE MTT-S Int. Microwave Symp. Dig.*, June 1993, pp. 1291–1294.

[247] Nghiem, D., Williams. J. T., Jackson, D. R. and Oliner, A. A., 'Leakage of the dominant mode on stripline with a small air gap,' *IEEE Trans. Microwave Theory and Techn.*, Vol. 43, Nov. 1995, pp. 2549–2556.

[248] Hotta, M., Qian, Y. and Itoh, T., 'Efficient FDTD analysis of conductor-backed CPW's with reduced leakage loss,' *IEEE Trans. Microwave Theory and Techn.*, Vol. 47, Aug. 1999, pp. 1585–1587.

[249] Shigesawa, H., Tsuji, M. and Oliner, A. A., 'Conductor-backed slot line and coplanar waveguide: danger and full-waves analyses,' *1989 IEEE MTT-S Int. Microwave Symp. Dig.*, June 1989, pp. 783–786.

[250] Tsuji, M., Shigesawa, H. and Oliner, A. A., 'Printed-circuit waveguide with anisotropic substrates: a new leakage effect,' *1989 IEEE MTT-S Int. Microwave Symp. Dig.*, June 1989, pp. 450–457.

[251] Dib, N.I., 'Effects of ground plane equalization on the electrical performance of asymmetric CPS shunt stubs,' *1993 IEEE MTT-S Int. Microwave Symp. Dig.*, June 1993, pp. 701–704.

[252] Freire, M. J., Mesa, F. and Horno M., 'Excitation of complex and backward mode on shielded lossless printed lines,' *IEEE Trans. Microwave Theory and Techn.*, Vol. 47, July 1999, pp. 1098–1105.

[253] Carrin, L., Slade, G. W. and Webb, K. J., 'Mode coupling and leakage effects in finite-size printed interconnects,' *IEEE Trans. Microwave Theory and Techn.*, Vol. 47, July 1999, pp. 450–457.

[254] Ponchak, G. E. and Katehi, L. P. B., 'Measured attenuation of coplanar waveguide on CMOS grade silicon substrates with polyimideinterface layer,' *Electron. Lett.*, Vol. 34, No. 13, 25 June 1998, pp. 1327–1329.

[255] Spiegel, S. J. and Madjar, A., 'Light dependence of silicon FGCPW transmission lines,' *IEEE MTT-S Int. Microwave Symp. Dig.*, June 1999, pp. 1801–1804.

[256] Crampagne, R. and Khoo, G., 'Synthesis of certain transmission lines employed in microwave integrated circuits,' *IEEE Trans. on Microwave Theory and Techn.*, Vol. 25, May 1977, pp. 440–442.

[257] Tsuji, M. and Shigesawa, H., 'Mode-coupling effect between different two types of the surface-wave leaky mode on conductor-backed coplanar strips (CBCPS),' *2000 IEEE MTT-S Int. Microwave Symp. Dig.*, June 2000.

[258] Akello R. J., Easter, B. and Stephenson, I. M., 'Effects of microstrip discontinuities on GaAs MESFET amplifier gain performance,' *Electron. Lett.*, Vol. 13, No. 6, 17 Mar.

1977, pp. 160–162.

[259] Slaymaker, N. A., Soares, R. A. and Turner, J. A., 'GaAs MESFET small-signal X-band amplifiers,' *IEEE Trans. Microwave Theory Techn*, Vol. 24, June 1976, pp. 329–337.

[260] Silvester, P. and Benedek, P., 'Equivalent capacitances of microstrip open circuits,' *IEEE Trans. Microwave Theory Techn.*, Vol. 20, Aug. 1972, pp. 511–576.

[261] Thomson, A. F. and Gopinath, A., 'Calculation of microstrip discontinuity inductances,' *IEEE Trans. Microwave Theory Techn.*, Vol. 23, Aug. 1975, pp. 648–657.

[262] Cohn, S. B., 'Problems in strip transmission lines,' *IRE Trans. PG Microwave Theory Techn.*, Vol. 3, No. 2, Mar. 1955, pp. 119–126.

[263] Garg, R. and Bahl, I. J., 'Microstrip discontinuities,' *Int. J. Electronics*, Vol. 45, No. 1, 1978, pp. 81–87.

[264] Costamagna, E., 'Some notes on a method of calculating gap capacitances in microstrip structures,' *Alta Frequenza*, Vol. XLIII, No. 6, June 1974, pp. 362–364.

[265] Benedek, P. and Silvester, P., 'Equivalent capacitances for microstrip gaps and steps,' *IEEE Trans. Microwave Theory Techn.*, Vol. 20, Nov. 1972, pp. 729–733.

[266] Silvester, P. and Benedek, P., 'Microstrip discontinuity capacitances for right-angle bends, T-junction and crossings,' *IEEE Trans. Microwave Theory and Techn.*, Vol. 21, No. 5, May 1973, pp. 341–346.

[267] Anders, P. and Arndt, F., 'Moment method of designing matched microstrip bends,' *Proc. 9th European Microwave Conf.*, 1979, pp. 430–434.

[268] Easter, B. *et al.*, 'Theoretical and experimental methods for evaluating discontinuities in microstrip,' *Radio and Electronic Engineer*, Vol. 48, No. 1/2, Jan./Feb. 1978, pp. 73–84.

[269] Hoefer, W. J. R., 'Equivalent series inductivity of a narrow transverse slit in microstrip,' *IEEE Trans. Microwave Theory Techn.*, Vol. 25, Oct. 1977, pp. 822–824.

[270] Wheeler, H. A., 'Coupling holes between resonant cavities or waveguides evaluated in terms of volume ratios,' *IEEE Trans. Microwave Theory Techn.*, Vol. 12. Mar. 1964, pp. 231–244.

[271] Hall, P. S. and James, J. R., Private communication, Dec. 1978.

[272] Kompa, G., 'Reduced coupling aperture of microstrip stubs provides new aspects in stub filter design,' *Proc. 6th European Microwave Conf.*, 1976, pp. 39–43.

[273] Dydyk, M., 'Master the T-junction and sharpen your MIC designs,' *MicroWaves*, May 1977, pp. 184–186.

[274] Akello, R. J., Easter, B. and Stephenson, I. M., 'Equivalent circuit of the asymmetric crossover junction,' *Electron. Lett.*, Vol. 13, No. 4, 17 Feb. 1977, pp. 117–118.

[275] Itoh, T., 'Analysis of microstrip resonators,' *IEEE Trans. Microwave Theory Techn.*, Vol. 22, Nov. 1974, pp. 946–952.

[276] Katehi, P. B. and Alexopoulos, N. G., 'Frequency-dependent characteristics of microstrip discontinuities in millimeter-wave integrated circuits,' *IEEE Trans. Microwave Theory Techn.*, Vol. 33, Oct. 1985, pp. 1029–1035.

[277] Ozmehmet, K., 'New frequency dependent equivalent circuit for gap discontinuities of microstriplines,' *IEEE Proc.*, June 1987, pp. 333–335.

[278] McLean, J., Ling, H. and Itoh, T., 'Full wave modeling of electrically wide microstrip open end discontinuities via a deterministic spectral domain method,' *1990 IEEE MTT-S Int. Microwave Symp. Dig.*, May 1990, pp. 1155–1158.

[279] Achkar, J. *et al.*, 'Frequency-dependent characteristics of an asymmetric suspended stripline and its open end discontinuity for millimeter wave applications,' *Proc. 19th European Microwave Conf.*, Sept. 1989, pp. 767–772.

[280] Menzel, W. and Wolff, I., 'A method for calculating the frequency-dependent properties of microstrip discontinuities,' *IEEE Trans. Microwave Theory Techn.*, Vol. 25, Feb. 1977, pp. 107–113.

[281] Kompa, G., 'Design of Stepped microstrip components,' *Radio and Electronic Engineer*, 48, Jan./Feb. 1978, pp. 53–63.

[282] Mehran, R., 'Computer-aided design of microstrip filters considering dispersion, loss and discontinuity effects,' *IEEE Trans. Microwave Theory Techn.*, Vol. 27, Mar. 1979, pp. 239–245.

[283] Harokopus, W. P. Jr. and Katehi, P. B., 'Analysis of multilayer irregular microstrip discontinuities,' *Proc. 19th European Microwave Conf.*, Sept. 1989, pp. 745–750.

[284] Zheng, J. X. and Chang, D. C., 'Numerical modelling of chamfered bends and other microstrip junctions of general shape in MMICs,' *1990 IEEE MTT-S Int. Microwave Symp.*, May 1990, pp. 709–712.

[285] Wu, S.-C., Yang. H.-Y. and Alexopoulos, N. G., 'A rigorous dispersive characterization of microstrip cross and tee junctions,' *1990 IEEE MTT-S Int. Microwave Symp.*, May 1990, pp. 1151–1154.

[286] Chang, S. S., Huang, W. X. and Itoh, T., 'Computer aided design for planar microstrip circuit analysis by boundary-integral method,' *Proc. 19th European Microwave Conf.*, Sept. 1989, pp. 975–978.

[287] Giannini, F., Bartolucci, G. and Ruggieri, M., 'An improved equivalent model for microstrip cross-junction,' *Proc. 19th European Microwave Conf.*, Sept. 1989, pp. 1226–1231.

[288] Weisshaar, A. *et al.*, 'Modelling of radial microstrip bends,' *1990 IEEE MTT-S Int. Microwave Symp.*, May 1990, pp. 1051–1054.

[289] Finch, K. L. and Alexopoulos, N. G., 'Shunt posts in microstrip transmission lines,' *IEEE Microwave Theory Techn.*, Vol. 38, Nov. 1990, pp. 1585–1594.

[290] Garg, R. and Bahl, I. J., 'Characteristics of coupled microstriplines,' *IEEE Trans. Microwave Theory and Techn.*, Vol. 27, July 1979, pp. 700–705.

[291] Garg, R., Private communication concerning coupled-line formulas, 1980.

[292] Akhtarzad, S., Rowbotham, T. R. and Jones, P. B., 'The design of coupled microstrip lines,' *IEEE Trans. Microwave Theory and Techn.*, Vol. 23, June 1975, pp. 486–492.

[293] Tripathi, Vijay, K. and Hill, Achim, 'Equivalent Circuit Modeling of Losses and Dispersion in Single and Coupled Lines for Microwave and Millimeter-Wave Integrated Circuits,' *IEEE Trans. Microwave Theory and Techn.*, Vol. 36, Feb. 1988, pp. 256–262.

[294] Getsinger, W. J., 'Dispersion of parallel coupled microstrip,' *IEEE Trans. Microwave Theory and Techn.*, Vol. 21, 1973, pp. 145–146.

[295] Easter, B. and Gupta, K C., 'More accurate model of the coupled microstripline section,' *Proc. IEE, Part H, Microwaves, Optics and Acoustics*, Vol. 3, May 1973, pp. 99–103.

[296] Richings, J. C. and Easter, B., 'Measured odd and even mode dispersion of coupled microstrip lines,' *IEEE Trans. Microwave Theory and Techn.*, Vol. 23, 1975, pp. 826–828.

[297] Kirschning, M. and Jansen, R. H., 'Accurate wide-range design equations for the frequency-dependent characteristic of parallel coupled microstrip lines,' *IEEE Trans. Microwave Theory and Techn.*, Vol. 32, Jan. 1984, pp. 83–89.

[298] Napoli, L. S. and Hughes, J. J., 'Characteristics of coupled microstrip lines,' *RCA Rev.*, Vol. 31, Sept. 1970, pp. 479–498.

[299] Lange, J., 'Interdigitated strip-line quadrature hybrid,' *1969 G-MTT Symp., Dig. of Papers*, 1969, pp. 10–13. (See also the paper by Lange, J., in *IEEE Trans. Microwave Theory and Techn.*, Vol. 20, Nov. 1972, pp. 777–779.)

[300] Presser, A., 'Interdigitated microstrip coupler design,' *IEEE Trans. Microwave Theory and Techn.*, Vol. 26, Oct. 1978, pp. 801–805.

[301] Ou, W. P., 'Design equations for interdigitated directional coupler,' *IEEE Trans. Microwave Theory and Techn.*, Vol. 19, Feb. 1975, pp. 253–255.

[302] Osmani, R. M., 'Synthesis of Lange couplers,' *IEEE Trans. Microwave Theory and Techn.*, Vol. 29, Feb. 1981, pp. 168–170.

[303] Fusco, V. F., *Microwave Circuits*, Prentice-Hall, Reading, MA, 1987.

[304] Waugh, R. W., 'Sensitivity analysis of the Lange coupler,' *Microwave J.*, Nov. 1989, pp. 121–129.

[305] Tajima, Y. and Kamihashi, S., 'Multiconductor couplers,' *IEEE Trans. Microwave Theory and Techn.*, Vol. 26, Oct. 1978, pp. 795–801.

[306] Waugh, R. and Lacombe, D., 'Unfolding the Lange coupler,' *IEEE Trans. Microwave Theory and Techn.*, Vol. 20, Nov. 1972, pp. 777–779.

[307] Weiss, J. A., 'Microwave propagation on coupled pairs of microstrip transmission lines,' in *Advances in Microwaves*, L. Young and H. Sobol (Eds.), Vol. 8, Academic Press Inc., New York, 1974, pp. 295–320.

[308] Kouki, A. B., Khebir, A. and Mittra, R., 'Finite element modeling of two-dimensional transmission line structures using a new asymptotic boundary condition,' *1990 IEEE MTT-S Int. Microwave Symp. Dig.*, June 1990, pp. 717–720.

[309] Shelad B. and Spielman, B. E., 'Broadband (7–18 GHz) 10 dB overlay coupler for MIC application,' *Electron. Lett.*, Vol. 11, No. 8, 17 Apr. 1975, pp. 175–176.

[310] Dydyk, M., 'Accurate design of microstrip directional couplers with capacitive compensation,' *1990 IEEE MTT-S Int. Microwave Symp. Dig.*, May 1990, pp. 581–584.

[311] Szentkuti, B. T., 'Simple analysis of anisotropic microstrip lines by a transform method,' *Electron. Lett.*, Vol. 12, 1976, pp. 672–673.

[312] Uysal, S., Aghvami, A. H. and Mohamed, S. A., 'Microstrip quadruplexer using -3dB hybrids,' *Proc. 19th European Microwave Conf.*, Sept. 1989, pp. 905–910.

[313] Nakajima, M., Yamashita, E. and Asa, M., 'New broad-band 5-section microstrip-line directional coupler,' *1990 IEEE MTT-S Int. Microwave Symp. Dig.*, May 1990, pp. 383–386.

[314] Pavio, A. M. and Sutton, S. K., 'A microstrip re-entrant mode quadrature coupler for hybrid and monolithic circuit applications,' *1990 IEEE MTT-S Int. Microwave Symp. Dig.*, May 1990, pp. 573–576.

[315] Fusco, V. F. and Merugu, L. N., 'Full wave analysis of 94 GHz patch coupler,' *1990 IEEE MTT-S Int. Microwave Symp. Dig.*, May 1990, pp. 649–652.

[316] Rizzoli, V., 'Highly efficient calculation of shielded microstrip structures in the presence of undercutting,' *IEEE Trans. Microwave Theory and Techn.*, Vol. 17, Feb. 1979, pp. 150–157.

[317] Shamasundra, S. D. and Gupta, K. C., 'Sensitivity analysis of coupled microstrip directional couplers,' *IEEE Trans. Microwave Theory and Techn.*, Vol. 26, Oct. 1978, pp. 788–794.

[318] Gunton, D. l. and Paige, E. G. S., 'Directional coupler for gigahertz frequencies, based on the coupling properties of two planar comb transmission lines,' *Electron. Lett.*, Vol. 11, No. 17, 21 Aug. 1975, pp. 406–409.

[319] Rigg, P. R. and Carroll, J. E., 'The canonical coupler: a new approach to microwave coupler design,' *Proc. 9th European Microwave Conf.*, 1979, pp. 490–494.

[320] Islam, S., 'An analysis and a design technique for microstrip comblines,' *Microwave J.*, Nov. 1989, pp. 79–91.

[321] Catt, T., 'Crosstalk (noise) in digital systems,' *IEEE Trans. on Dig. Comp.*, EC-16, No. 6, Dec. 1967, pp. 743–763.

[322] Okugawa, S. and Hagiwara, H., 'Analysis and computation of crosstalk noise between microstrip transmission lines,' *Electron. and Communications in Japan*, Vol. 53-C, No. 7, 1970, pp. 128–138.

[323] Coekin, J. A., *High-Speed Pulse Techniques*, Pergamon Press, Oxford, 1975, pp. 203–205.

[324] Gilb, J. P. and Balanis, C. A., 'Transient analysis of distortion and coupling in lossy coupled microstrips,' *1990 IEEE MTT-S Int. Microwave Symp. Dig.*, May 1990, pp. 641–644.

[325] Orhanovic, N., Tripathi, V. K. and Wang, P., 'Time domain simulation of uniform and nonuniform multiconductor lossy lines by the method of characteristics,' *1990 IEEE MTT-S Int. Microwave Symp. Dig.*, May 1990, pp. 1191–1194.

[326] Sabban, A. and Gupta, K. C., 'A planar-lumped model for coupled microstrip line discontinuities,' *1990 IEEE MTT-S Int. Microwave Symp. Dig.*, May 1990, pp. 247–254.

[327] Wertgen, W. and Jansen, R. H., 'Novel Green's function database technique for the efficient full-wave analysis of complex irregular (M)MIC-structures,' *Proc. 19th European Microwave Conf.*, Sept. 1989, pp. 199–204.

[328] Kaajfez, D. and Govind, S., 'Effects of difference in odd- and even-mode wavelengths on a parallel-coupled bandpass filter,' *Electron. Lett.*, Vol. 11, No. 5, 6 Mar. 1975, pp. 117–118.

[329] Lange, L. 'Interdigitated strip-line quadrature hybrid,' *IEEE Trans. Microwave Theory Techn.*, Vol. 17, Dec. 1969, pp. 1150–1151.

[330] Gunton, D. J., *Electron. Lett.*, Vol. 11, No. 24, 1975, pp. 607–608.

[331] Operating manual for the Hewlett-Packard 11608A microwave transistor measurement fixture (now Agilent).

[332] England, E. H., 'A coaxial to microstrip transition,' *IEEE Trans. Microwave Theory Techn.*, Vol. 26, Jan. 1976, pp. 47–48.

[333] Schneider, M. V., Glance, B. and Bodtmann, W. F., 'Microwave and millimetre wave hybrid integrated circuits for radio systems,' *Bell System Tech. J.*, July/Aug. 1969, pp. 1703–1726.

[334] Cohn, S. B., 'Optimum design of stepped transmission-line transformers,' *IRE Trans. Microwave Theory Techn.* Vol. 3, Apr. 1955, pp. 16–21.

[335] Hopfer, S., 'The design of ridged waveguides,' *IRE Trans. Microwave Theory Techn.*, Vol. 3, Oct. 1955, pp. 20–29

[336] Menzel, W. and Klaassen, A., 'On the transition from ridged waveguide to microstrip,' *Proc. 19th European Microwave Conf.*, Sept. 1989, 1265–1269.

[337] Van Heuven, J. H. C., 'A new integrated waveguide-microstrip transition,' *IEEE Trans. Microwave Theory Techn.*, Vol. 24, Mar. 1976, pp. 144–147.

[338] Wu, Y.-S., Schneider, M. V. and Trambarulo, R., 'Waveguide-to-microstrip power splitter,' *1990 IEEE MTT-S Int. Microwave Symp. Dig.*, May 1990. pp. 475–478.

[339] Grabherr, W., Huder, B. and Menzel, W., 'Microstrip to waveguide transition compatible with MM-wave integrated circuits,' *IEEE Trans. Microwave Theory Techn.*, Vol. 42, Sept. 1994, pp. 1842–1843.

[340] Hall, P. S. and James, J. R., 'Analysis and design of triplate corporate feeds at high frequencies,' *Proc. 9th European Microwave Conf.*, 1979, pp. 106–110.

[341] Robinson, G. H. and Allen J. L., 'Slotline application to miniature ferrite devices,' *IEEE Trans. Microwave Theory Techn.*, Vol. 17, 1969, pp. 1097–1101.

[342] Cullen, A. L., 'The six-port and the microprocessor in microwave measurements,' *Proc. 9th European Microwave Conf.*, 1979, pp. 74–82.

[343] Qin, T.-K. and Wang, L.-F., 'Design of dielectric loaded slot-line-microstrip transitions,' *Proc. 2nd Int. Symp. Recent Advances in Microwave Techn.*, Sept. 4–8 1989, pp. 332–335.

[344] Climer, B. J. 'A broadband single-sideband mixer with low spurious levels,' *Proc. 19th European Microwave Conf.*, London, Sept. 1989, pp. 737–742.

[345] Viitanen, A. J. and Kuismann, P. A., 'Analysing the asymmetric coupled microstrip line and the microstrip-coplanar guide transition,' *Proc. of 2nd Int. Symp. on Recent Advances in Microwave Techn.*, Sept 4–8, 1989, pp. 887–892.

[346] Howell, J. Q., 'A quick accurate method to measure the dielectric constant of microwave integrated-circuit substrates,' *IEEE Trans. Microwave Theory Techn.*, Vol. 21, Mar. 1973, pp. 142–143.

[347] Ladbrooke, P. H., Potok, M. H. N., England, E. H., 'Coupling errors in cavity-resonance measurements on MIC dielectrics,' *IEEE Trans. Microwave Theory Techn.*, Vol. 21, Aug. 1973, pp. 560–562.

[348] Owens, R. P., Aitken, J. E. and Edwards, T. C., 'Quasi-static characteristics of

microstrip on an anisotropic sapphire substrate,' *IEEE Trans. Microwave Theory Techn.*, Vol. 24, Aug. 1976, pp. 499–505.

[349] Troughton, P., 'High Q-factor resonators in microstrip,' *Electron. Lett.*, Vol. 4, No. 24, 29 Nov. 1968, pp. 520–522.

[350] Troughton, P. 'Measurement techniques in microstrip, *Electron. Lett.*, 1969, pp. 25–26.

[351] Owens, R. P., 'Curvature effect in microstrip ring resonators,' *Electron. Lett.*, Vol. 12, No.14, 8 July 1976, pp. 356–357.

[352] Richings, J. G., *The Realization of Filter Networks in Microstrip*, PhD Dissertation, University College of North Wales, Caerns, UK. June 1973.

[353] Deutsch, J. and Jung, H. J., 'Measurement of the effective dielectric constant of microstrip lines in the frequency range from 2 GHz to 12 GHz,' *Nachrichtentech. Z.*, Vol. 12, 1970, pp. 620–624.

[354] Edwards, T. C., *Microwave design parameters for microstrip transmission lines on sapphire substrates*, M. Phil. Thesis, Royal Military College of Science, Shrivenham, Mar. 1976.

[355] Woodin, C. and Goff, M., 'Verification of MMIC on-wafer microstrip TRL calibration,' *1990 IEEE MTT-S Int. Microwave Symp. Dig.*, May 1990, pp. 1029–1032.

[356] Easter, B., 'The equivalent circuit of some microstrip discontinuities,' *IEEE Trans. Microwave Theory Techn.*, Vol. 23, Aug. 1975, pp. 655–660.

[357] Aitken, J. E., 'Swept-frequency microwave Q-factor measurement,' *Proc. IEE*, Vol. 123, Sept. 1976, pp. 855–862.

[358] Owens, R. P., Private communication, 1976.

[359] Rizzoli, V., 'Resonance measurement of even and odd mode propagation constants in coupled microstrips,' *1975 IEEE MTT-S Int. Microwave Symp.*, May 1975.

[360] Babu, G. R. and Banmali, 'Measurement of coupled microstrip line parameters,' *Int. J. Electron.*, Vol. 45, No. 5, 1978, pp. 551–560.

[361] Ladbrooke, P. H., 'A novel standing-wave indicator in microstrip,' *Radio and Electronic Engineer*, Vol. 44, May 1974, pp. 273–280.

[362] Piller, U., 'Time-domain immittance measurements,' *Proc. 4th European Microwave Conf.*, 1974, pp. 61–65.

[363] Lee, T. H., *The Design of CMOS Radio-Frequency Integrated Circuits*, Cambridge University Press, Cambridge, 1998.

[364] McCreary, J. L., 'Matching properties and voltage and temperature dependence of MOS capacitors,' *IEEE J. Solid State Circuits*, Vol. 16, No. 6, Dec. 1981, pp. 608–616.

[365] Yuan, C. P. and Trick, T. N., 'A simple formula for the estimation of the capacitance of two-dimensional interconnects in VLSI circuits,' *IEEE Electron Device Lett.*, Vol. 3, 1982, pp. 391–393.

[366] Barke, E., 'Line-to-ground capacitance calculation for VLSI: a comparison,' *IEEE Trans. Computer Aided Design,* Vol. 7, No. 2, Feb. 1988, pp 295–298.

[367] Sakurai, T. and Tamaru, K., 'Simple formulas for two- and three-dimensional capacitances,' *IEEE Trans. Electron Devices*, Vol. 30, Feb. 1983, pp. 183–185.

[368] Meijs, N. v.d., and Fokkema, J. T., 'VLSI circuit reconstruction from mask topology,' *Integration,* Vol. 2, No. 2, 1984, pp. 85–119.

[369] Wheeler, H A., 'Simple inductance formulas for radio coils,' *IRE Proc.,* 1928, p. 1398.

[370] You, C. P. and Wong, S. S., 'On-chip spiral inductors with patterned ground shields fo Si-based RF IC's,' *Dig. Tech. Papers Sym. VLSI Circuits*, 1997, pp. 85–86.

[371] Mernyei, F., Darrer, F., Pardoen, M. and Sibrai, A., 'Reducing the substrate losses of RF integrated inductors,' *IEEE Microwave Guided Wave Lett.*, Vol. 8, No. 9, Sept. 1998, pp. 300–301.

[372] Bahl, I. J., 'High current handing capacity multilayer inductors for RF and microwave circuits,' *Int. J. RF and Microwave Computer-Aided Engineering*, Vol. 10, No. 2, March 2000, pp. 139–146.

[373] Burghartz, J. N., 'Spiral inductors on silicon — status and trends,' *Int. J. of RF and*

Microwave Computer-Aided Engineering, Vol. 8, No. 6, Nov. 1998, pp. 422–432.

[374] Burghartz, J. N., Soyuer, M., Jenkins, K. A. and Hulvey, D., 'High-Q inductors in standard silicon interconnect technologies and its application to an integrated RF power amplifier,' *Tech. Dig. Int. Electron Devices Meeting*, 1995, pp. 1015–1017.

[375] Burghartz, J. N., Jenkins, K. A. and Soyuer, M., 'Multilevel-spiral inductors using VLSI interconnect technology,' *IEEE Electron Device Lett.*, Vol. 17, No. 9, 1996, pp. 428–430.

[376] Groves, R., Stein, K., Harame, D. and Jadus, D., 'Temperature dependence of Q in spiral inductors fabricated in a silicon-germanium/BiCMOS technology,' *Proc. Bipolar/BiCMOS Circuits and Techn. Meeting*, 1996, pp. 153–156.

[377] Kim, K. and O, K., 'Characteristics of an integrated spiral inductor with an underlying N-well,' *IEEE Trans. Electron Devices*, Vol. 44, July 1997, pp. 1565–1567.

[378] Merrill, R. B., Lee, T. W., You, H., Rasmussen, R. and Moberly, L., 'Optimization of high-Q integrated inductors for multi-level metal CMOS,' *Tech. Dig. Int. Electron Devices Meeting*, 1995, pp. 983–986.

[379] Nguyen, N. M. and Meyer R. G., 'Si IC-compatible inductors and LC passive filters,' *IEEE J. Solid State Circuits*, Vol. 25, No. 4, 1990, pp. 1028–1031.

[380] Soyuer, M., Burghartz, J. N., Jenkins, K. A., Ponnapalli, S., Ewen, J. F. and Pence W. E., 'Multilevel monolithicinductors in silicon technology,' *Electron Lett.*, Vol. 31, No. 5, 1995, pp. 359–360.

[381] Ashby, K. B., Finley, W. C., Bastek, J. J. and Moinian, S., 'High Q inductors for wireless applications in a complementary silicon bipolar process,' *Proc. BCTM*, 1994, pp. 179–182.

[382] Burghartz, J. N., Edelstein, D. C., Jenkins, K. A., Jahnes, C., Uzoh, C., O'Sullivan E. J., Chan, K. K., Soyuer, M., Roper, P. and Cordes, S., 'Monolithic spiral inductors using a VLSI Cu-damascene interconnect technology and low-loss substrates,' *Tech. Dig. Int. Electron Devices Meeting*, 1996, pp. 99–102.

[383] Burghartz, J. N., Edelstein, D. C., Soyuer, M., Ainspan, H. and Jenkins, K. A., 'RF circuit design aspects of spiral inductors on silicon,' *Dig. Tech. Papers 1998 Int. Solid-State Circuits Conf.*, 1998, pp. 246–247.

[384] Chang, J. Y.-C., Abidi, A. A. and Gaitan, M., 'Large suspended inductors on silicon and their use in a 2-μm CMOS RF amplifier,' *IEEE Electron Device Lett.*, Vol. 14, No. 5, 1993, pp. 246–248.

[385] Dekker, R., Baltus, P., van Deurzen, M., v.d. Einden, W., Maas, H. and Wagemans, A., 'An ultra low-power RF bipolar technology on glass,' *Tech. Dig. Int. Electron Devices Meeting*, 1997, pp. 921–923.

[386] Johnson, R. A., Chang, C. E., Asbeck, P. M., Wood, M. E., Garcia, G. A. and Lagnado, I., 'Comparison of microwave inductors fabricated on silicon-on-sapphire and bulk silicon,' *IEEE Microwave Guided Wave Lett.*, Vol. 6, No. 9, 1996, pp. 323–325.

[387] Park, M., Kim, C. S., Park, J. M., Yu, H. K. and Nam, K. S., 'High Q microwave inductors in CMOS double-metal technology and its substrate bias effects for 2 GHz RF ICS application,' *Tech. Dig. Int. Electron Devices Meeting*, 1997, pp. 59–62.

[388] Sun, Y., van Zeijl, H., Tauritz, J. L. and Baets, R. G. F., 'Suspended membrane inductors and capacitors for application in silicon MMIC's,' *1996 IEEE MTT-S Int. Microwave Symp. Dig.,* June 1996, pp. 99–102.

[389] Jiang, H., Wang, Y., J.-L. A. and Tien, N.C., 'Fabrication of high performance on-chip suspended spiral inductors by micromachining and electroless copper plating,' *2000 IEEE MTT-S Int. Microwave Symp. Dig.,* June 2000, pp. 279–282.

[390] Volant, R., Malinowski, J., Subbanna, S. and Begle, E., 'Fabrication of high frequency passives on BiCMOS silicon substrates,' *2000 IEEE MTT-S Int. Microwave Symp. Dig.,* June 2000, pp. 209–212.

[391] Yamaguchi, M., Baba, M., Suezawa, K., Moizumi, T., Arai, K. I., Shimada, Y., Haga, A., Tanabe, S. and Ito, K., 'Magnetic RF integrated thin-film inductors,' *2000 IEEE*

MTT-S Int. Microwave Symp. Dig., June 2000, pp. 205–208.

[392] Gouker, M., Konistis, K., Knecht, J., Kushner, L. and Travis, L., 'Multi-layer spiral inductors in a high precision, fully-planar MCM-D process,' *2000 IEEE MTT-S Int. Microwave Symp. Dig.,* June 2000, pp. 1055–1058.

[393] Kuhn, W. B., 'Approximate analytical modeling of currrent crowding effects in multi-turn spiral inductors,' *2000 IEEE MTT-S Int. Microwave Symp. Dig.,* June 2000, pp. 405–408.

[394] Pun, A. L. L., Tau, J., Clement, J. R. and Su, D. K., 'Substrate noise coupling through planar spiral inductor,' *IEEE J. of Solid-State Circuits,* Vol. 33, June 1998, pp.877–884.

[395] Kim, C. S., Park, M., Kim, C.-H., Park, M.-Y., Kim S.-D., Youn, Y.-S., Park, J.-W., Han, S.-H., Yu, H. K. and Cho, H., 'Design guide of coupling between inductors and its effect on reverse isolation of a CMOS LNA,' *2000 IEEE MTT-S Int. Microwave Symp. Dig.,* June 2000, pp. 225–228.

[396] Lacombe, D., 'A multi-octave microstrip 50 Ω termination,' *IEEE Trans. Microwave Theory and Techn.,* Vol. 20, Apr. 1972, pp. 290–292.

[397] Finlay, H. J., Hopkins, L. G. T. and Ozamiz, J. M., 'Design and applications of precision microstrip multi-octave attenuators and loads', *Proc 6th European Microwave Conf.*, 1976, pp. 692–696.

[398] Gupta, K. C. and Singh, A., *Microwave Integrated Circuits*, Wiley Eastern Limited, New Delhi.

[399] Ho, Chen Y., 'Transform impedance with a branch-line coupler,' *MicroWaves*, May 1976, pp. 47–52.

[400] Howe, H., *Stripline Circuit Design*, Artech House, Inc., Norwood, MA, 1973.

[401] Wright, A. S. and Judah, S. K., 'Very broadband flat coupling hybrid ring,' *Electron. Lett.*, Vol. 23, 1987, pp. 47–49.

[402] Ashforth, J.V., 'Design equations to realise a broadband hybrid ring or a two-branch guide coupler of any coupling coefficient', *Electron. Lett.*, Vol. 24, 1988, pp. 1276–1277.

[403] Mayer, B. and Knochel, R., Branchline-couplers with improved design flexibility and broad bandwidth,' *1990 IEEE MTT-S Int. Microwave Symp. Dig.*, May 1990, pp. 91–398.

[404] Knochel, R. and Mayer, B., 'Broadband printed circuit 0°/180° couplers and high power in-phase power dividers,' *1990 IEEE MTT-S Int. Microwave Symp. Dig.*, May 1990, pp. 471–478.

[405] Meaney, P., 'A novel branch-line coupler design for millimeter-wave applications,' *1990 IEEE MTT-S Int. Microwave Symp.*, May 1990, pp. 585–588.

[406] Roy, J. S. *et al.*, 'Analysis of curved microstrip transmission lines and its applications,' *Proc. 2nd Int. Symp. on Recent Advances in Microwave Techn. (ISRAMT '89)*, Sept 4–8, 1989.

[407] Tokumitsu, T. *et al.*, 'Multilayer MMIC using a 3 μm×3-layer dielectric film structure,' *1990 IEEE MTT-S Int. Microwave Symp. Dig.*, May 1990, pp. 831–834.

[408] Nakamoto, H., Tokumitsu, T. and Aikawa, M., 'A monolithic, port-interchanged rat-race hybrid using a thin film microstrip line crossover,' *Proc. 19th European Microwave Conf.*, Sept. 1989, pp. 311–316.

[409] Pavio, M. and Kikel, A., 'A Monolithic or Hybrid broadband compensated balun,' *1990 IEEE MTT-S Int. Microwave Symp. Dig.*, May 1990, pp. 483–486.

[410] Barber, R. G., 'Enhanced coupled, even mode terminated baluns and mixers constructed therefrom,' *1990 IEEE MTT-S Int. Microwave Symp. Dig.*, May 1990, pp.495–498.

[411] Cunningham, G. J., Blenkinsop, P. A. and Palmer, J. H., 'Microstrip end-coupled filter design at MM-Wave frequencies,' *Proc. 19th European Microwave Conf.*, Sept. 1989, pp. 1210–1213.

[412] Matthaei, G. L., Young, L. and Jones, E. M. T., *Microwave Filters, Impedance-Matching Networks and Coupling Structures*, McGraw-Hill, New York, 1965. (Back in print from

Oct. 1980: Artech House Inc., Norwood, MA.)

[413] Bahl, I. J., 'Broadbanding microstrip filters using capacitive compensation,' *Applied Microwave*, 1989, pp. 72–76.

[414] Kompa, G. and Mehran, R., 'Microstrip filter analysis using a microstrip waveguide model,' *Radio and Electronic Engineer*, Vol. 50, No. 1/2, Jan./Feb. 1980, pp. 54–58.

[415] Tzuang, C.-K. C. *et al.*, 'Design of a quasi-planar broadside end-coupled bandpass filter,' *1990 IEEE MTT-S Int. Microwave Symp. Dig.*, May 1990, pp. 407–410.

[416] Tzuang, C.-K. C. and Lo, W.-T., 'Printed-circuit realization of tapped combline bandpass filter,' *1990 IEEE MTT-S Int. Microwave Symp. Dig.*, May 1990, pp. 131–134.

[417] Pang, K. K., 'Design of microwave filters by sine-plane approach,' *IEEE Trans. Microwave Theory and Techn.*, Vol. 21, Oct. 1973, pp. 607–611. (Also private communications.)

[418] Mara, J. F. and Schappacher, J. B., 'Broadband microstrip parallel-coupled filters using multi-line sections,' *Microwave J.*, Apr. 1979, pp. 97–99.

[419] Nakamura, A. H. *et al.*, 'Expert system for microwave filter design,' *1990 IEEE MTT-S Int. Microwave Symp.*, May 1990, pp. 1183–1186.

[420] Bonetti, R. R. and Williams, A. E., 'Preliminary design steps for thin-film superconducting filters,' *1990 IEEE MTT-S Int. Microwave Symp. Dig.*, May 1990, pp. 273–276.

[421] Giannini, F., Sorrentino, R. and Vrba, J., 'Planar circuit analysis of microstrip radial stub,' *IEEE Trans. Microwave Theory and Techn.*, Vol. 32, Dec. 1984, pp. 1652–1655.

[422] Giannini, F., Paoloni, C. and Ruggieri, M., 'CAD-oriented lossy models for radial stubs,' *IEEE Trans. Microwave Theory and Techn.*, Vol. 36, Feb. 1988, pp. 305–313.

[423] Gardiol, F., *Introduction to Microwaves*, Artech House, Inc., Norwood, MA, 1984.

[424] Bahl, I. and Bhartia, P., *Microwave Solid State Circuit Design*, Wiley-Interscience, John Wiley & Sons, 1988.

[425] Bates, R. N., 'Design of microstrip spur-line band-stop filters,' *IEE Journal on Microwaves, Optics and Acoustics*, Vol. 1, No. 6, Nov. 1977, pp. 209–214.

[426] Bates, R. N. and Pearson, R. E., 'Designing bandstop filters for micro-wave frequencies,' *Electronic Engineering*, Apr. 1978, pp. 39–41.

[427] Maagt, P. De., *et al.*, 'PBG crystals: periodic dielectric materials that control EM wave propagation,' *Microwave Engineering Europe*, Oct. 1999, 35–43

[428] Powell, S. *et al.*, 'Broad band monolithic cross point switch matrices,' *1990 IEEE MTT-S Int. Microwave Symp. Dig.*, June 1990, pp. 461–464.

[429] Kane, R. C. and Wong, T., 'An edge-guide mode microstrip isolator with transverse slot discontinuity,' *1990 IEEE MTT-S Int. Microwave Symp.*, May 1990, pp. 1007–1010.

[430] Pan, J. J., Shih, M. and Riley, L., 'High performance millimeter-wave microstrip circulator for deep space communications,' *1990 IEEE MTT-S Int. Microwave Symp.*, May 1990, pp. 1015–1017.

[431] How, H. *et al.*, 'Theory and experiment of thin-film junction circulator,' *IEEE Trans. Microwave Theory and Techn.*, Vol. 46, Nov. 1998, pp. 1645–1653.

[432] Goldberg, S. B., Steer, M. B. and Franzon, P. D., 'Accurate experimental characterization of three-ports,' *1991 IEEE MTT-S Int. Microwave Symp. Dig.*, June 1991, pp. 241–244.

[433] Goldberg, S. B., Steer, M. B., Franzon, P. D. and Kasten J. S., 'Experimental electrical characterization of interconnects and discontinuities in high speed digital systems,' *IEEE Trans. on Components Hybrids and Manufacturing Techn.*, Vol. 14, Dec., 1991, pp. 761-765.

[434] *The International Semiconductor Roadmap for Semiconductors*, 1999.

[435] Multigig, Patent Pending, John Wood PCT/GB00/00175 1999/2000, assigned, http://multigig.com

[436] Chi, V. L., 'Salphasic distribution of clock signals for synchronous systems,' *IEEE*

Trans. Computers, Vol. 43, May, 1994, pp. 597–602.

[437] 'Salphasic distibution of timing signals for the synchronization of physically separated entities,' U.S. Patent No. 5,387,885, Feb. 7, 1995. Inventor: Chi, V. L.

[438] Kourtev, I. S. and Friedman, E. G., *Timing Optimization Through Clock Skew Scheduling*, Kluwer Academic Publishers, Boston, 2000

[439] Liao, S. Y., *Microwave Circuit Analysis and Amplifier Design,* Prentice-Hall, Inc., Reading, MA, 1987.

[440] Sweet, A., *MIC and MMIC Amplifier and Oscillator Circuit Design,* Artech House, Inc., 1990.

[441] Goyal, R., *High-Frequency Analog Integrated Circuit Design,* John Wiley & Sons, Inc. 1995.

[442] Christou, A., *Reliability of Gallium Arsenide MMICs,* John Wiley & Sons, Inc. 1992.

[443] Eda, K. *et al.*, 'Miniature hybrid microwave ICs using a novel thin-film technology,' *1990 IEEE MTT-S Int. Microwave Symp. Digest of Papers,* June 1990, pp. 419–422.

[444] Imai, K. and Nakakita, H., 'A 22GHz-band low noise down converter for satellite broadcast receivers,' *Proc. 19th European Microwave Conf.,* London, Sept. 1989, pp. 549–554.

[445] Filtronic Components: private communication (2000).

[446] Kurokawa, K., 'Design theory of balanced transistor amplifiers,' *Bell System Tech. J.,* Oct. 1975, pp. 1675–1698.

[447] Ladbrooke, P. H., *MMIC design: GaAsFETs and HEMTs,* Artech House, Inc., 1989.

[448] Hoel, V. *et al.*, '94 GHz low noise amplifier on InP in coplanar technology,' *Microwave Engineering Europe,* Nov. 1999, pp. 51–54 (also private communication, 2000).

[449] Boret, S., 'Modeling of passive coplanar elements for W-band ICs, experimental verification,' *Proc. European Microwave Conf.*, 1998, pp. 190–195.

[450] Heinrich, W., 'Quasi-TEM description of MMIC coplanar lines including conductor-loss effects,' *IEEE Trans. Microwave Theory Techn*, Vol. 41, Jan. 1993, pp. 45–52.

[451] Dupois, A., Hausner, J. and Russer, P., 'Hybrid integrated Ku-Band VCO,' *Proc. 19th European Microwave Conference,* Sept. 1989, pp. 1009–1014.

[452] Fleischmann, B. *et al.*, 'A high frequency low noise acoustic STW VCO,' *Proc. 19th European Microwave Conf.*, Sept. 1989, pp. 1015–1020.

[453] Seawright, S. W. J. *et al.*, 'Optimum DRO Synthesis,' *Proc. 19th European Microwave Conf.,* Sept. 1989, pp. 406–411.

[454] Massias, M., Howes, M. and Postoyalko, V., 'A novel analysis and design technique for temperature stable DROs,' *Proc. 19th European Microwave Conf.*, Sept. 1989, pp. 418–423.

[455] Golubicic, Z., 'Injection locked DRO synchronized by multitone signal,' *Proc. 19th European Microwave Conference,* Sept. 1989, pp. 395–400.

[456] Filicori, F., Monaco, V. A. and Vannini, G., 'A design method for parallel feedback dielectric resonator oscillators,' *Proc. 19th European Microwave Conf.*, Sept. 1989, pp. 412–417.

[457] Liang, X-P. and Zaki, K. A., 'Hybrid mode coupling of dielectric resonators to microstrip lines,' *1990 IEEE MTT-S Int. Microwave Symp. Dig.,* May 1990, pp. 395–398.

[458] Rheinfelder, C. N. *et al.*, '47 GHz SiGe-MMIC oscillator,' *1990 IEEE MTT-S Int. Microwave Symp. Dig.,* May 1990.

[459] Rheinfelder, C., private communication (2000).

[460] Rheinfelder, C. N. *et al.*, 'A coplanar 38 GHz SiGe-MMIC oscillator,' *IEEE Microwave and Guided Wave Lett.*, Vol. 6, Nov. 1996, pp. 398–400.

[461] Chi-Yang, C. and Itoh, T., 'A varactor-tuned, active microwave band-pass filter,' *1990 IEEE MTT-S Int. Microwave Symp. Digest of Papers,* pp. 499–502.

[462] Jiao, X. H. *et al.*, 'Microwave frequency agile active filters for MIC and MMIC applications,' *1990 IEEE MTT-S Int. Microwave Symp. Digest of Papers,* June 1990,

pp. 503–506.

[463] Zuefle, K., *et al.*, 'Coplanar 4-bit HEMT phase shifters for 94 GHz phased array radar systems,' *1990 IEEE MTT-S Int. Microwave Symp. Digest of Papers,* June 1990.

[464] Edward, B. J., Webb, R. S. and Weinreb, S., 'A W-band active phased array antenna,' *Microwave J.*, May 1996, pp. 254–262.

[465] Nelson, S., Youngblood, M., Pavio, J., Larson, B. and Kottman, R., 'Optimum Microstrip Interconnects,' *1991 IEEE MTT-S Int. Microwave Symp. Dig.*, June 1991, pp. 1071–1074.

Index